W9-AEZ-913

THE
ABC-CLIO
COMPANION TO

The Environmental Movement

Uncle Sam is depicted shorn of his forests in this 1908 cartoon.

THE
ABC-CLIO
COMPANION TO

The
Environmental
Movement

Mark Grossman

ABC-CLIO

Copyright © 1994 by Mark Grossman

All rights reserved. No part of this publication may be reproduced, stored in a retrieval system, or transmitted, in any form or by any means, electronic, mechanical, photocopying, recording, or otherwise,
except for the inclusion of brief quotations in a review, without prior permission in writing from the
publishers.

Library of Congress Cataloging-in-Publication Data

Grossman, Mark.
    The ABC-CLIO companion to the environmental movement / Mark
Grossman.
    p. cm. — (ABC-CLIO companions to key issues in American
history and life)
    Includes bibliographical references and index.
    1. Environmentalism—United States. 2. Environmental policy-
making—United States. I. Title. II. Series.
    GE197.G76   1994   363.7'00525'0973—dc20   94-36186

ISBN 0-87436-732-8

00 99 98 97 96 95 94   10 9 8 7 6 5 4 3 2 1

ABC-CLIO, Inc.
130 Cremona Drive, P.O. Box 1911
Santa Barbara, California 93116-1911

This book is printed on acid-free paper ∞.
Manufactured in the United States of America

Ref.
GE
197
.G76
1994

This book is dedicated to my parents, Larry and Lois Grossman, without whose love and un-demanding support I would not have completed this work; to Sherwood "Woody" Wilkes, cura-tor of the Museum of Discovery and Science in Fort Lauderdale, Florida, whose friendship and indomitable support of the environment and wildlife is a constant inspiration; and to Mrs. Paula Herbst, my mentor and, in many ways, my savior, who knew years ago that I would be a writer.

DEC 0 1 1995

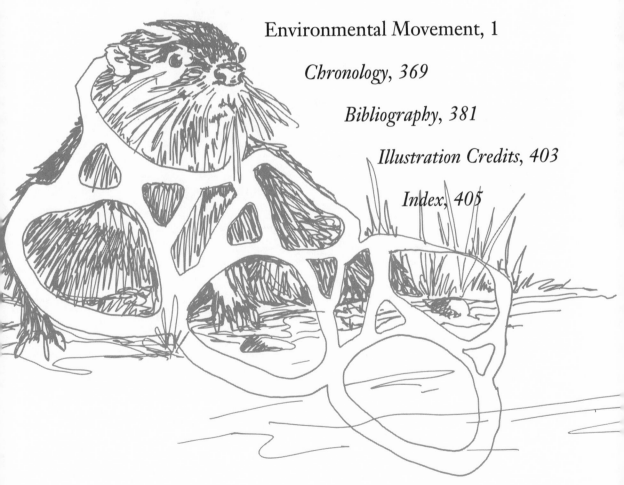

Contents

Preface

The *ABC-CLIO Companion to the Environmental Movement* is a guide to the conservation movement and environmentalism from the nation's early naturalists to the present. It assembles in one volume information on conservationists, environmental activists, key government officials, governmental and private institutions and organizations, landmark legislation and court decisions, concepts, and watershed events.

The ABC-CLIO Companions series is designed to provide the nonspecialist with concise, encylopedic guides to key movements, major issues, and revolutions in American history. The encyclopedia entries are arranged in alphabetical order. Cross-references connect related terms and entries. A chronology of key events provides a handy overview, and a bibliography is provided to facilitate further research.

I would like to thank the following people for their generous assistance in the preparation of this work: the staffs of the Manuscript Division and Law library at the Library of Congress, Washington, D.C.; the Civil Reference Branch and Legislative Branch staffs of the National Archives, Washington, D.C., particularly Richard Fusick and Ann Cummings; Cindy Brown and the staff of the Wyoming State Archives, Cheyenne, Wyoming; T. H. Watkins of the Wilderness Society, Washington, D.C., who allowed me access to valuable material in the Wilderness Society archives and whose hospitality on my two visits to their offices was of great comfort to a weary traveler; the staff of the Regional Oral History Research Office at the University of California at Berkeley; Lisa Backman and the staff of the Western History Department of the Denver Public Library, Denver, Colorado; Colleen Fiest, public affairs specialist at the Forest Products Laboratory in Madison, Wisconsin; the staff of the Alderman Library at the University of Virginia, Charlottesville; Paula Smith and the staff of the State Historical Society of Iowa in Iowa City; Brian Williams, archivist at the Oberlin College Archives, Oberlin, Ohio; Wayne J. Waller and the staff of the Sacramento Public Library, Sacramento, California; Eric M. Moody, curator of manuscripts, and the staff of the Nevada Historical Society, Reno, Nevada, for their help in gaining access to the papers of Francis G. Newlands and William Morris Stewart; Cliff Nelson and Joanne Taylor at the United States Geological Survey in Reston, Virginia; Rob Kenneson and the staff of the Lawrence Public Library, Lawrence, Kansas; Joan Reinertson, media assistant for the Izaak Walton League, who researched valuable material from the league's archives on Will Dilg; Janice Flug and the reference staff at the Bender Library at American University in

Washington, D.C.; Lois Archuleta, reference librarian at the Salt Lake City, Utah, Main Library, for her efforts in finding newspaper records of the Western Governors' Conference of 1913; Clifford G. Reed, chief, Branch of Forest Resources Planning at the U.S. Department of the Interior, Bureau of Indian Affairs, Division of Forestry Central Office in Portland, Oregon; Bob Miller of Polson, Montana; Kim Awbrey, conservation biology coordinator at Conservation International, Washington, D.C.; the staff of the Oral History Research Office at Columbia University in New York City; Terry West, chief historian for the U.S. Forest Service, who was an incredible morale booster; John White and the staff of the Louis Round Wilson Library at the University of North Carolina at Chapel Hill for all their assistance in helping me navigate through the manuscripts in the North Carolina Collection; Sara Swenson of the Reference Division of the Minnesota Historical Society, St. Paul; Harriet Rusin at the Natural Resources Library of the Department of the Interior in Washington, D.C., whose assistance enabled me to persevere through the library's stacks of material; Debbie Zills at the Tennessee Valley Authority; Elaine Engst, curator of manuscripts at the Carl A. Kroch Library at Cornell University in Ithaca, New York, for the alumni records of Jay P. Kinney; Rosa Ortiz de Gentry, a member of the Missouri Botanical Garden in St. Louis, who kindly supplied information on her late husband, Alwyn Gentry; Dr. J. V. Remsen, curator of birds and adjunct professor of zoology at Louisiana State University, for his papers on the late Theodore Parker III; the staffs of the Arizona State Archives and State Library at the State Capitol Building in Phoenix, Arizona; Wendy Helen Petry, of Rochester, New York, for the use of her thesis on solid waste disposal; the staffs of the Scottsdale Public Library and the Mustang Library, both in Scottsdale, Arizona; the staff of the reference division of the Maricopa County Library, Main Branch; Eloise Konzak at the Carnegie Library in Devils Lake, North Dakota; Denise Haberstroh, reference specialist at the State Historical Society of North Dakota in Bismarck; Carrie Yourd, staff member of the U.S. House Committee on Natural Resources; Gregory LePore, reference archivist at the Division of Archives and Records Services of the Utah State Department of Administrative Services; Joseph R. Dugan, assistant press secretary, the Committee on Agriculture, U.S. House of Representatives; Pam Wasmer, manuscript librarian, Indiana State Library, Indianapolis; Cheryl Oakes and Pete Steen at the Forest History Society in Durham, North Carolina, who arranged a Bell Fellowship grant that allowed me to take advantage of their wonderful resources; Ron Arnold, head of the Center for the Defense of Free Enterprise in Bellevue, Washington, who provided "another perspective" on environmental matters; Daisy I. Hermanson of Webster, North Dakota, whose enthusiasm about, and research into, the life of Henry Clay Hansbrough was both encouraging and exciting; and finally, John Mattson, forest archaeologist at the Chugach National Forest in Alaska, and Dave Wanderaas, forester at the Lewis & Clark National Forest in Montana, for their help in obtaining information on the Afognak Wildlife Refuge, as well as Irene Voit at the Forest Service Intermountain Research Station in Ogden, Utah, and Lori Erbs of the Forest Service Forestry Sciences Laboratory in Juneau, Alaska.

Introduction

People have long been concerned with the state of their environment. As early as 1661, a work entitled "Fumifugium: or the Inconvenience of the Aer and Smoake of London Dissipated" appeared. English economists Thomas Robert Malthus, David Ricardo, and John Stuart Mill in their earliest writings discussed the impact of industrial advancement on the environment. These works include Malthus's *An Essay on the Principle of Population As It Affects the Future Improvement of Society* (1798), Ricardo's *Principles of Political Economy and Taxation* (1817), and Mill's *Principles of Political Economy* (1848), all of which discuss how the agrarian society of Europe was giving way to the Industrial Revolution—and higher death rates from the resulting pollution. Malthus's theory was that industrial construction and population growth would strip the earth of land and vegetation and lead to periods of mass starvation.

The founding fathers knew the importance of the public lands to the nation's future development. Congress was given explicit powers over the public lands in the property clause of the U.S. Constitution, Article IV, section 3, clause 2, which reads, "The Congress shall have the power to dispose of and to make all needful rules and Regulations respecting Territory and other Property belonging to the United States." Thus, the environ-

ment has always been important to American history. In the last three hundred years, men and women inside and outside the political realm have fought to first preserve and protect, and then later to undo the damage to, the environment. This is the dividing line between the conservation movement (and in some cases the preservation movement) and the environmental movement. Those who came before stressed that to conserve meant to save what was there; today's activists seek to undo the damage done to lakes, streams, and the wilderness.

As early as 1626, settlers of the Plymouth Colony were so concerned about the lumber resources under their control that they enacted an ordinance forbidding the cutting of timber on public lands without the colony's consent. The earliest explorations of the nation by the father-and-son naturalist team of John and William Bartram served to uncover just a fraction of the nation's land and wildlife resources. Thus, the movement to conserve the environment stretches back even before the founding of the Republic.

Scientists and naturalists like the Bartrams, John Bradbury, and Thomas Nuttall were doing work that embraced environmental studies in the seventeenth and eighteenth centuries. Naturalist and wildlife illustrator John James Audubon made the biggest impact with his "elephant folios" of drawings of birds that to this day are pioneer works. Just before the turn of the eighteenth century, in 1799, Congress was passing laws to protect the nation's timber reserves for use by the military. One of the first acts of the Continental Congress was the Land Ordinance Act of 1785, which established the national interest in opening up mineral deposits in the undiscovered and uncharted western reaches of the country. This consideration of what natural resources lay out west served to pave the way for the great surveys (the Wheeler, Hayden, King, and Powell Surveys are among the best known) of the late nineteenth century.

Other concerns entered into the picture. The drive to colonize the West led to the wholesale destruction of native species and their habitats. In 1808, the ornithologist Alexander Wilson viewed a flock of passenger pigeons in Kentucky that took several hours to pass. He estimated there were 2 million birds in the one flock alone. A little over a century later later the last passenger pigeon died in captivity. This latter story is but one in the history of human abuse of the environment. In 1828, one Henry M. Brackenridge was planting acorns to grow trees to be used by the navy to build ships. The Timber Reservation Act of 1827 solidified this attitude into law. The establishment of the General Land Office in 1812 and the creation of the Department of the Interior 37 years later provided a forum for national policymaking and management of the country's natural resources and wildlife, although in many cases Congress was slow in legislating what could and could not be protected. The idea that later evolved into the national park system was expressed in 1832 by artist George Catlin, who wanted to set aside a preserve in the Missouri River country. In 1844 the first known wildlife conservation group, the New York Association for the Protection of Game, was founded. The following year, naturalist and civil protester Henry David Thoreau began an investigation into the sedimentary life at Walden Pond in Massachusetts, with the subsequent work on his studies resulting in *Walden, or Life in the Woods*, a pioneer work in a love of nature. The landmark Homestead Act of 1862 opened the vast plains and arid regions of the West to settlement by pioneers who answered the government's call to head west. The passage of the Lode Mining Law in 1866 and the General Mining Law six years later established rules and guidelines for mining in the west. The passage of the Timber Culture Act, the Desert Land Act, and the Free Timber Act, among other legislative actions, opened these areas for abuse and led to the subsequent enactment of the Forest Reserve Act, the General Revision Act, and the Creative Act of 1891, which authorized the president to set aside forest reserves.

The fight to save wildlife populations has been ongoing since the nineteenth century. The major early pieces of legislation in this area include the Lacey Act of 1900, the Alaska Game Act of 1902, the Act for the Protection of Game in Alaska (1908), and the Migratory Bird Act (Weeks-McLean Act) of 1913. Irrigation of the West was also a major concern. Government agencies such as the Powell Irrigation Survey and the Bureau of Reclamation, and laws such as the Carey Act of 1894 and the Newlands Act of 1902, enticed private individuals, farmers, and corporations to reclaim and irrigate portions of the arid western lands. Governmental entities, from the Department of the Interior to the Bureau of Land Management, created in the 1930s from the ashes of the General Land Office, have instituted far-ranging policies that affect all dimensions of the conservation question. The 1930s saw the formation of such groups as the Wilderness Society and the National Wildlife Federation. Secretary of the Interior Harold L. Ickes, whose 13 years in office was the longest tenure of any interior secretary, had a major impact on national conservation policies. Prior to the

1960s, the government's hand in private land dealings was controversial. Constitutional scholar Mark Tushnet writes, "Until the 1960s, control of private property in the United States was based on centuries old common-law doctrines in each state and by a few local and state statutory restrictions." After the beginning of what is known as the "conservation movement," this principle fundamentally changed.

In 1962, writer Rachel Carson's landmark work *Silent Spring* called attention to the excessive use of pesticides and their effect on the environment. The publication of her work is the line that separates the conservation movement from the environmental movement. Hereafter, activists sought to undo the damage people had done to their environment. Streams, lakes, and rivers that were polluted were targeted for cleanup; recycling programs to eliminate landfill pollution were instituted; polluted lands became areas of concern under such acts as the Superfund. Groups that originally formed part of the conservation movement, such as the Sierra Club and the Wilderness Society, have been joined by newer organizations such as Earth First!, Greenpeace, and the Earth Island Institute, which have injected a new radicalism into the dialogue on the environment and what steps society should take as a whole to "save the Earth." Such ideas as mitigation banking, inholders' rights, and conservation easements, drawing against the age-old theory of land ownership, are recent developments and have not as yet been challenged in court.

Environmental racism, "ecotourism," recycling, compost heaps, ozone depletion, endangered species, acid rain, DDT, "ecotage," CFCs—these are just a few of the terms used in the new area of environmental awareness that the American public has become accustomed to in the last two or three decades. Today's children are reminded daily—by their peers, their parents and other adults, and television shows (including the cartoon *Captain Planet and the Planeteers*)—that the Earth and their environment are immensely important and deserve their attention.

As opposed to other movements, such as the civil rights, union, and women's rights movements, the conservation and environmental movements have never had a leader per se, although men like John Burroughs and Theodore Roosevelt, and women like Rachel Carson and Anna Botsford Comstock, have risen to become effective spokespeople for the cause. This lack of cohesion, however, has been the environmental movement's Achilles' heel. In the 1980s, Norwegian Prime Minister Gro Harlem Brundtland rose to become a leader in the world environmental movement, but her name has been tarnished recently with the resumption by Norway of illegal whaling. Still, the names of the people who have utilized various forms of activism to speak on environmental matters is vast. From William John McGee to Louis Bromfield, Horace Marden Albright to John James Audubon, the history of America is replete with those of the green persuasion. Hundreds of federal laws have been passed since the founding of the nation to stem the tide of reckless environmental damage. Government agencies have been at the forefront of federal activities in this area, while Congress' many committees with jurisdiction over lands and the environment have come and gone.

This work highlights not just the major organizations and people that made up the conservation and environmental movements, but includes pertinent acts of Congress, important members of Congress who helped draft (and, in some cases, circumvent) these laws, governmental agencies, and key Supreme Court decisions.

Acid Rain

A matter of grave concern since the 1970s, the issue of acid rain has taken its place among a constellation of environmental hazards associated with air pollution. Generally, acid rain is any precipitation with a pH factor lower than that of normal rain, which is about 5.6. A pH of 7 is considered neutral. Nitric acid, sulfuric acid, and hydrochloric acid are formed in the atmosphere from the interaction of water vapor, sunlight, oxygen, fossil fuel emissions, and hydrogen chloride gas. These acids return to the earth's surface in the form of rain, snow, or other precipitation. Fish, aquatic insects, and other organisms in lakes, particularly where the regional geology already produces a low-pH environment, may be extremely susceptible to further acidification of the water, and acid rain has been implicated in the decline of thousands of lakes in the northeastern United States, Canada, and northern Europe. Streams, forests, and historic buildings also suffer damage.

Acid rain was first found in some areas downwind from the industrial city of Manchester, England, in 1852. It became a serious issue in Scandinavia in the 1950s, when traces began to be measured. By the 1980s, scientists monitoring water quality in northern New York State and southeastern Canada were finding that some lakes and streams were highly acidic. They linked their observations to emissions from eastern and midwestern smokestacks. Some recent studies, however, have shown that acid precipitation can occur naturally, and fluctuations in the pattern cannot be definitely traced to any one industry.

Adams, Ansel Easton (1902–1984)

Ansel Adams captured much of the beauty of the American West in photographs. He was born on 20 February 1902 in San Francisco and grew up as an only child in the lush valleys of California. It was there that Adams discovered his love of nature. Visits to an art show at the 1915 World's Fair and to Yosemite National Park the following year led him to meld his devotion to nature and his love of fine art into one pursuit. In 1919, he joined the Sierra Club. His fervor for conservation and his growing knowledge of photographic techniques led to a visit to Taos, New Mexico, in 1927, where he met such artists as Georgia O'Keefe. That same year, Adams published his first book, a series of photographs called *Parmelian Prints in the High Sierras*. In 1930, he published a second volume in his photographic portfolio series, *Taos Pueblo*. He wrote a total of 15 books, including *The Camera, Illustrated Guide to Yosemite Valley*, and *My Camera in the National Parks*, leaving an unmatched legacy of breathtaking western photography. Ansel Adams died at his home in Carmel, California, on 22 April 1984.

Afognak Forest and Fish Culture Reserve

This fishery and forest refuge was established by President Benjamin Harrison on 24 December 1892.

According to historian Hubert Howe Bancroft in his *History of Alaska*, Afognak Island, a small spit in the southern part of Alaska, was noted for its knotty timber, which was used by Eskimos and other native Alaskans for shipbuilding. Following the passage of legislation on 2 March 1889 requiring salmon reserves in Alaska to be examined by the U.S. Commission of Fish and Fisheries, Commissioner W. J. McDonald sent a team of specialists in fish examination and

Photographer and conservationist Ansel Adams at work in California's Owens Valley in the late 1940s

conservation to Afognak Island. Tarleton H. Bean, the commission's ichthyologist, headed the assemblage, which also included Dr. Livingston Stone, a fish hatcheries expert from California, and Franklin Booth, a cartographer. They discovered that the salmon fisheries on the island were being destroyed by nets that covered the entire salmon course in the area. Writes historian Dave Wanderaas, "The Karluk River fishery on nearby Kodiak Island had been reduced from one of the largest salmon runs of the entire coast of Alaska to an almost nonexistent stock of red salmon." At a New York meeting of the American Fisheries Society in May 1892, Stone recommended that Afognak be set aside as a wildlife refuge. After consultation with Secretary of the Interior John Willock Noble and Commissioner McDonald, President Harrison signed a presidential order establishing Afognak as a refuge.

Benjamin Harrison, grandson of President William Henry Harrison, was an odd man to take an interest in the conservation of wildlife. According to historian John Reiger, the president "had a penchant for duck hunting that knew no bounds, particularly when the canvasbacks and redheads were flying." Yet, on 24 December 1892, he signed Proclamation No. 39. Although Afognak was not officially called a "wildlife refuge," it had virtually the same designation—an area set aside as a refuge for particular forms of wildlife. "[T]he public lands in the Territory of Alaska, known as Afognak Island, are in part covered with timber, and are required for public purposes, in order that salmon fisheries in the waters of the Island, and salmon and other fish and sea animals, and other animals and birds, and the timber, undergrowth, grass, moss and other growth in, on, and about said Island may be protected and preserved unimpaired, and it appears that the public good would be promoted by setting apart and reserving said lands as a public reservation," wrote Harrison in his order. George Bird Grinnell, editor of *Forest and Stream* magazine, editorialized:

> The salmon canneries on the Afognak bay will be evacuated, but the natives will retain whatever fishing rights they had during the Russian occupation. It is hoped that several important marine species including the sea otter may be preserved on this reservation, and there is no doubt that the half dozen or more kinds of salmon and trout that now abound in Afognak River will be saved from extermination, and will soon increase the supplies of the surrounding region.... From the accounts of Messrs. Stone and Booth, we learn that Afognak (or Litnik) River, is admirably adapted to salmon hatching, being near a safe harbor in a region where skilled labor is cheap, abounding with fairly good timber, and visited by salmon and trout in large numbers. Moreover, the river is not subject to great changes of level and will furnish [an] ample supply of water by gravity.... The President, on the suggestion of the Interior Department and the Fish Commission, has by a stroke of his pen effected an object, the importance of which cannot be easily overestimated.

On 2 June 1908, an executive order by President Theodore Roosevelt merged Afognak with the nearby Chugach National Forest, which had been created by Roosevelt on 23 July 1907. Under the Alaska National Interest Lands Conservation Act of 2 December 1980, portions of what remained of the refuge at Afognak were incorporated into the Kodiak National Wildlife Refuge and the Alaska Maritime Wildlife Refuge.
See also Executive Order 1014 of 14 March 1903.

Agricultural Adjustment Act of 1933, First (48 Stat. 31)

This federal legislation was enacted on 12 May 1933 as a means of establishing a federal agency to handle crop management for farmers. The agency, the Agricultural Adjustment Administration (AAA), was empowered to "limit acreage on specified crops at the farmers' option and to pay benefits to farmers." The funds it provided to the farmers were to be raised by a process tax. The entire act was declared unconstitutional by the Supreme Court on 16 January 1936 in the cases of *United States v. Butler* (297 U.S. 1) and *Rickert Rice Mills, Inc. v. Fontenot* (297 U.S. 110). Through the enactment of the Soil Conservation Act of 1935 and the Second Agricultural Adjustment Act of 1938, however, the act's provisions and the AAA were revived.
See also Agricultural Adjustment Act of 1938; Soil Conservation Act.

Agricultural Adjustment Act of 1938 (7 U.S.C. 1282)

This act, the second of its kind, was passed by Congress to supplement the Soil Conservation Act of 1935. Enacted on 16 February 1938, this legislation reinforced the national policy "of conserving national resources, preventing the wasteful use of soil fertility, and of preserving, maintaining, and rebuilding the farm and ranch land resources in the national interest, to accomplish these through the encouragement of soil-building and soil-conserving crops and practices." The act recreated the Agricultural Adjustment Administration, which was eventually consolidated with the Soil Conservation Service and other conservation agencies into the Agricultural Conservation and Adjustment Administration under Franklin D. Roosevelt's Executive Order 9069.
See also Soil Conservation Act.

Agricultural Experiment Stations
See Hatch Act.

Agricultural Settlement Act
See Forest Homestead Act.

Air Quality Act (Public Law 90-148)
This federal legislation, which amended the Clean Air Act of 1965, was enacted on 21 November 1967. The Air Quality Act (AQA) added enforcement requirements to earlier air pollution legislation. The aims of this act were (1) to protect and enhance the quality of the nation's air resources so as to promote the public health and welfare and the productive capacity of its population; (2) to initiate and accelerate a national research and development program to achieve the prevention and control of air pollution; and (3) to provide technical and financial assistance to state and local governments in connection with the development and execution of their air pollution prevention and control programs. One section of the AQA established the National Air Pollution Control Administration.

See also National Air Pollution Control Administration.

Alabama v. Texas (347 U.S. 272 [1954])
In this case, the Supreme Court decided that Congress had the right to apportion lands to certain states and not to others under the Submerged Lands Act of 1953. The states of Alabama and Rhode Island (a coplaintiff) sued the state of Texas because their borders had not been extended under the Submerged Lands Act, while Texas claimed upwards of 150 miles into the Gulf of Mexico. The Supreme Court held 8–1 (Justice Hugo Black dissenting) that under the Constitution, Congress had the right to apportion any and all "submerged lands" any way it saw fit. The per curiam (no known author) opinion cited Article 4, section 3, clause 2 of the U.S. Constitution, which says that "Congress shall have Power to dispose of and make all needful Rules and Regulations respecting the Territory and other Property belonging to the United States." In recalling *United States v. California* (332 U.S. 19), the Court held that "the power over the public land thus entrusted to Congress is without limitations, and it is not for the courts to say how that trust shall be administered."

Alaska Game Act (32 Stat. 327)
This act was the congressional response to the poaching of wild animals in the Alaska Territory. Passed by Congress on 7 June 1902, it prohibited the "wanton destruction of wild game animals or wild birds, the destruction of nests and eggs, of such birds, or the killing of any wild birds other than a game bird, or wild game animal, for the purposes of shipment from Alaska."

Alaska National Interest Lands Conservation Act (94 Stat. 2371)
Known as the Alaska Lands Act, ANILCA, or the D-2 Act, this federal legislation was enacted on 2 December 1980 to expand the acreage in Alaska's wildlife refuge and park systems to some 97 million acres. Further, 25 free-flowing rivers were put under federal protection, and 56 million additional acres were categorized as wilderness. The passage of this act led to the establishment of the Alaska Independence Party, an autonomous political entity that has called for the act's retraction.

See also Hickel, Walter Joseph.

Albright, Horace Marden (1890–1987)
Horace Albright was, with Stephen Tyng Mather, one of the founders of the National Park Service. He was born on 6 January 1890 in Bishop, California. Both his grandfathers had arrived in California in 1851 to mine gold; his father, George Albright, was a miner as well. Horace Albright grew up among the spectacular scenery of California. He graduated from

Mount McKinley, centerpiece of what is now Denali National Park and Preserve, towers above the debris-covered Muldrow Glacier in Alaska. The Alaska National Interest Lands Conservation Act of 1980 changed the name of Mount McKinley National Park, created in 1917, to Denali and increased the park's size to 6 million acres. The act more than doubled the National Park System by adding some 43.6 million acres.

the University of California in 1912. A year later, Secretary of the Interior Franklin K. Lane enlisted Adolph C. Miller, an economics professor from the University of California at Berkeley, to come to Washington and establish a national park service. In turn, Miller brought a young student of his, Horace Albright, to work as a clerk in the Interior Department. In 1915, Albright was named special assistant to Stephen Tyng Mather, a prosperous Illinois businessman who had been put in charge of straightening out the woeful condition of the U.S. park system. Mather and Albright formed a two-man lobbying team that helped Congress pass the National Park Service Act of 1916, which created the national park system. The two men then set out to create park rules, laws, and administration.

Albright served as assistant director of the National Park Service until 1919. That year, he became superintendent of Yellowstone National Park. In 1929, he succeeded Mather as director of the National Park Service, holding the post until 1933, when he left government for the private sector. In his final days at the Park Service, he counseled President Franklin D. Roosevelt on his ideas for a civilian workforce to be used for environmental work. From this idea came the Civilian Conservation Corps.

In 1933, Albright became president of the United States Potash Company, a mining concern in New Mexico. He remained in private business until his

Horace Albright, National Park Service director 1929–1933, works in his Washington, D.C., office in 1933.

retirement in 1956, but he was always a leading voice in the conservation movement. In 1980, President Jimmy Carter awarded him the Medal of Freedom, the highest award given to civilians. In his final years, besides coauthoring a book, *The Birth of the National Park Service: The Founding Years, 1913–1933* (1985), Albright was on the boards of many environmental groups, including the Nature Conservancy and the Sierra Club. He died on 28 March 1987 at the age of 97.

See also Mather, Stephen Tyng; National Park Service.

Allen, Arthur Augustus (1885–1964)

A leading conservationist and ornithologist, Arthur Allen was an early advocate for the preservation and protection of vanishing bird species. He was born on 28 December 1885 in Buffalo, New York, the son of a lawyer and naturalist. He received bachelor's and master's degrees in zoology from Cornell University and was awarded a Ph.D. from Cornell in 1911.

Allen began to teach zoology at Cornell in 1910 and founded its ornithology laboratory in 1911. He was one of the first professors of ornithology, and in 1926 he became a full professor in the field. During his lengthy career, he studied the behavior and mannerisms of birds, collecting pictures of most North American bird species as well as recording their calls. One of the founding members of the Wildlife Society, Allen was a member of Cornell's faculty for 46 years and became not only the nation's leading expert on birds but also one of the leaders in the call for their protection and preservation. He died of a heart attack at his home in Ithaca, New York, on 17 January 1964 at the age of 78.

Allen, Robert Porter (1905–1963)

Robert Allen, whose monumental work *Birds of North America* was cut short by his untimely death, was a noted ornithologist who worked to save threatened bird species. He was born on 24 April 1905 in South Williamsport, Pennsylvania. He attended Lafayette College in Easton, Pennsylvania, and Cornell University.

In 1930, Allen began work as a staff assistant for the Audubon Society. Four years later, he was elevated to sanctuary director, then to research associate in 1939 and to research director in 1955. His interest in ornithology began in 1939, when he began to study the endangered whooping crane. By 1946, there were only 33 cranes alive, and he commenced an intensive program to find their nesting areas and habitats and try to nurture the few birds that were left. In 1955, he found the crane's summer nesting grounds at Great Slave Lake near the Arctic Circle. With the backing of the Audubon Society, he was able to get governmental protection for both the summer grounds and the winter grounds in Texas sanctuaries. In 1957, Allen published *On the Trail of Vanishing Birds*, in which he discussed the crane as well as the roseate spoonbill and the flamingo. The work earned him the John Burroughs

Medal for nature book of the year. He was also awarded the William Brewster Memorial Award by the American Ornithologists' Union. His work on saving the whooping crane was featured in a November 1959 *National Geographic* article, "Whooping Cranes Fight for Survival."

Near the end of his life, Allen began the projected 16-volume work *Birds of North America*, but the series was cut short by his death on 18 June 1963 from a heart attack. He was 58 and had suffered a similar attack in 1962.

American Forest Congress

This gathering of forestry experts was held in Washington, D.C., from 2 to 6 January 1905. Called by the federal Bureau of Forestry, the meeting was actually a symposium for the views of Gifford Pinchot, chief forester of the United States. As announced officially, the purpose of the conference was to "establish a broader understanding of the forest in its relation to the great industries depending upon it; to advance the conservative use of forest resources for both the present and the future needs of these industries, and to stimulate and unite all efforts to perpetuate the forest as a permanent resource of the nation." In addition to Pinchot, attendees included James Wilson, secretary of agriculture; Frederick Haynes Newell, chief engineer of the U.S. Reclamation Service; Joseph Trimble Rothrock, member of the Pennsylvania Reservation Commission; Charles Doolittle Walcott, head of the U.S. Geological Survey; Howard Elliott, president of the Northern Pacific Railroad; George K. Smith of the National Lumberman's Association; J. E. Defebaugh, editor of the *American Lumberman* journal; William A. Richards, General Land Office commissioner; Francis E. Warren of the National Woolgrowers' Association; Rep. John Lacey; and Overton W. Price, Pinchot's right-hand man. The conference called for the enactment of the Transfer Act (which would transfer control of the administration of the nation's forestry reserves from the Department of the Interior to the Forestry Division of the Department of Agriculture), and the repeal of the Timber and Stone Act. The Transfer Act, which was sponsored in the House by Lacey, was passed by both houses of Congress and was signed into law by President Theodore Roosevelt on 1 February 1905.

See also United States Forest Service.

American Forestry Association

The American Forestry Association (AFA) is the largest U.S. organization dedicated to the conservation of forests

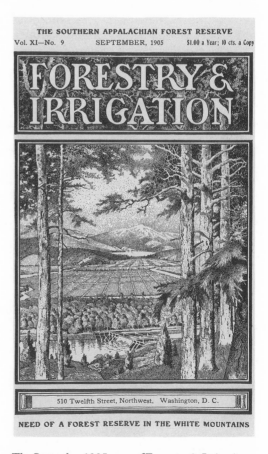

THE SOUTHERN APPALACHIAN FOREST RESERVE
Vol. XI—No. 9   SEPTEMBER, 1905 $1.00 a Year; 10 cts. a Copy

FORESTRY & IRRIGATION

510 Twelfth Street, Northwest, Washington, D. C.

NEED OF A FOREST RESERVE IN THE WHITE MOUNTAINS

The September 1905 cover of Forestry & Irrigation, *published by the American Forestry Association, represented the organization's dual interest in woodlands and farmland.*

and forestry resources. It was founded in 1875 by John Aston Warder, an Ohio physician and naturalist. The *Chicago Tribune* of 10 September 1875 contained a small announcement: "A number of gentlemen interested in forest-culture assembled in the ladies' ordinary at the Grand Pacific yesterday at 8 o'clock, for the purpose of organizing a Forestry Association." Under the direction of Warder and James Douglas, an English naturalist, the AFA was founded for the "protection of the existing forests of the country from unneccesary waste" and to encourage "the propagation and planting of useful trees." Writes historian William G. Robbins, "Although the AFA was not very visible at first, it became more active when it merged with another Warder organization, the American Forestry Congress, in the 1880s." In fact, the AFA absorbed the American Forestry Council in 1876, the Southern Forestry Congress in 1888, and the National Conservation Association in 1923. Today's organization is an amalgamation of all these groups.

In its 120-year history, the American Forestry Association has been instrumental in the formation of the National Forest Commission (1896–1897); the passage of the 1897 Pettigrew Amendment, which provided for the consolidation and management of national forest reserves under one central authority in 1905 (which it called for at the 1905 American Forest Congress); and the advocacy of sustained-yield status for the Oregon and California forestlands. Although it is one of the oldest and most influential conservation organizations, the AFA is also one of the least well known. The organization is now called American Forests.

See also American Forest Congress; Warder, John Aston.

American Game Protective and Propagation Association

Formed on 25 September 1911 with author John Bird Burnham as its president, the American Game Protective and Propagation Association (AGPPA) tried to establish itself as a conservation organization. One of its founding members was Frederic Collin Walcott (1869–1949), who was later elected to the U.S. Senate from Connecticut. According to author James Trefethen, "the avowed purpose of the new organization was to promote wildlife restoration on a national and international scale. One of its first announced goals was the enactment of legislation based on the principles of the Weeks Bill, which was then dormant in Congress."

The AGPPA was soon hobbled by its fight with such men as William Temple Hornaday, who accused the organization (correctly, it turned out) of being financed by the gun companies. Its influence ruined, the group eventually became known as the American Wildlife Institute, a short-lived federation of sportsmen, in August 1935.

See also Burnham, John Bird.

American Wild Fowlers

According to author James Trefethen, this short-lived conservation group's goals were "to interest the public in waterfowl conservation, to help preserve the authority of the Bureau of Biological Survey over migratory birds, to cooperate with the state agencies and the Bureau in making waterfowl censuses and enforcing the laws, and to assist in scientific research on waterfowl biology and movements." The group was headed by John C. Phillips, author of *A Natural History of the Ducks* and *American Waterfowl*; noted writer and naturalist George Bird Grinnell; and Nash Buckingham, one of the nation's best-known outdoor writers at the time. Founded in 1927, the group opposed the McNary-Haugen bird limit bill (which would have limited the number of ducks and geese hunters could bag). In 1931, soon after helping to defeat that legislation, the organization became the More Game Birds in America Foundation, which is now called

Ducks Unlimited and works in Canada to preserve waterfowl habitats there.

Amoco Production Company et al. v. Village of Gambell, Alaska, et al.
See Hodel, Secretary of the Interior, et al. v. Village of Gambell et al.

Anadromous and Great Lakes Fish Act (79 Stat. 1125)
This federal legislation, enacted on 30 October 1965, was sponsored in the House by Rep. John D. Dingell of Michigan. Anadromous fish are fish that are born in fresh water, spend all their lives in salt water, then return to fresh water to spawn. By the early 1960s, stocks of such fish had been depleted by overfishing. This act established a five-year schedule (until 1970) for state and federal monies (no more than $50 million) to be used to establish programs that would conserve the dwindling stocks of these fish.

Anderson, Clinton Presba (1895–1975)
Secretary of agriculture in the Truman administration and later a U.S. senator from New Mexico, Clinton P. Anderson may be best known for his work in getting the Wilderness Act of 1964 enacted. He was born in Centerville, South Dakota, on 28 October 1895, the son of Andrew Jay Anderson, a Swedish immigrant and farmer, and Hattie Belle (nee Presba) Anderson. After public schooling, Clinton Anderson attended Dakota Wesleyan University in Mitchell, South Dakota (1913–1915), and studied law at the University of Michigan for a year. When the United States entered World War I, Anderson volunteered, but a lung condition kept him out of the service and forced him to a warmer climate. He eventually settled in Albuquerque, New Mexico, where he worked for a number of years as a reporter and editor for the

Albuquerque Journal. He later entered the insurance business and eventually owned his own insurance firm. Anderson was present at the historic 1922 Santa Fe meeting, attended by then Secretary of Commerce Herbert Hoover, which resulted in the Colorado River Compact. Another attendee was Wayne N. Aspinall, who later became a congressman and fought Anderson on conservation issues.

Anderson entered the political field when, in 1933, the governor of New Mexico appointed him state treasurer. This work led to his being named head of the New Mexican Relief Organization and the state field representative of the Federal Emergency Relief Administration, two agencies established to aid victims of the Great Depression. In 1940, he was elected as a representative at large to the 77th Congress. He served on the powerful Ways and Means Committee and had a generally liberal record. On 23 May 1945, President Harry S Truman named Anderson secretary of agriculture. During his three-year tenure, he carried out Truman's Nine-Point Famine Relief Program by conserving wheat stocks. Anderson resigned from the cabinet to enter the 1948 Democratic primary for the U.S. Senate. Anderson's campaign was successful, and he was ultimately reelected in 1954, 1960, and 1966. His key accomplishments came during the 88th Congress (1963–1964), labeled by writer Richard A. Baker as the "Conservation Congress." Passed during this session were the Land and Water Conservation Fund Act, the Outdoor Recreation Act, and the Public Land Law Review Commission Act. However, the landmark of that Congress was the enactment of the Wilderness Act. Anderson, chairman of the Senate Committee on the Interior and Insular Affairs from 1961 to 1963, butted heads with his counterpart in the House, Wayne N. Aspinall of Colorado. Aspinall was one of the leading spokesmen for the development of the West and did not favor passage of the Wilderness

Act. During the debate on the legislation, Sen. Robert S. Kerr of Oklahoma died, opening up the chairmanship of the Senate Aeronautical and Sciences Committee, which Anderson could take only by relinquishing his other chairmanship. This factor played heavily in the Wilderness Act debate, as Anderson did not have the same clout he had had before the move. Working effectively with Presidents John F. Kennedy and Lyndon Johnson, Anderson and Aspinall crafted a bill that met their own separate demands. The provision declaring that a Public Land Law Review Commission would be established to clear the way for the selling of federal lands in the West removed Aspinall's opposition to the final bill. The death of advocate Howard Zahniser clinched the legislation's passage. On 3 September 1964, President Johnson signed the Wilderness Act into law.

Anderson retired from the Senate in 1972. He died on 11 November 1975, two weeks after his eightieth birthday.

See also Aspinall, Wayne Norviel; Wilderness Act; Zahniser, Howard Clinton.

Anderson, Harold Cushman (1879–1967)

Harold Anderson was one of the eight organizers of the Wilderness Society. Born at West Burlington, New York, on 20 July 1879, he was educated in the public schools and in 1910 was granted a certified public accounting license by the University of the State of New York. Although he was a staunch environmentalist, at his death he was recognized more for his work in accounting.

Although the exact date of his famed wilderness travels cannot be ascertained, his diary, *On Some Adirondack Trails*, was written in 1922. Five years later, Anderson helped organize the Potomac Appalachian Trail Club to preserve that section of wilderness, which runs from Harrisburg, Pennsylvania, south to the southern end of Shenandoah National Park. Anderson also served for several years as secretary of the club. Later, he was one of the founding members of the Wilderness Society. He served as president of the D.C. Institute of Certified Public Accountants and was a tax specialist in an accounting firm in Scarsdale, New York. Harold Anderson died in Washington, D.C., on 15 February 1967.

See also Wilderness Society.

Andrus, Cecil Dale (1931–)

Twice governor of Idaho, Cecil D. Andrus served during the Carter administration as secretary of the interior. He was born on 25 August 1931 in Hood River, Oregon, the son of Hal Stephen Andrus, a sawmill operator, and Dorothy (nee John) Andrus. Growing up in the Pacific Northwest, Cecil Andrus learned the conservation ethic that he would later use on a national scale. He went to local schools before attending Oregon State University for two years (1948–1949), although he did not graduate.

In 1951, Andrus joined the navy, where he served in the Pacific theater during the Korean War but saw no combat. He rose to the rank of aviation electronic technician second class before being discharged in 1955. That year he returned to the United States and began working for the Tru-Cut Lumber Company in Orofino, Idaho. As Jack Shepherd reported in the *New York Times*, "He started as a lumberjack, operated haulers to bring out the large trees, then moved up to millwright and production manager. Later, he ran his own small wood-products firm, and went into the insurance business." In 1961, inspired by John F. Kennedy, Andrus ran for and was elected to the Idaho state senate. At the time, he was the youngest man ever elected to that body. He was eventually elected to a second term, serving until 1966. That year, he ran for governor of Idaho but was defeated by Republican Don Samuelson. In 1968, he won back his old state senate seat and served until

1970. In the meantime, he rose to become general manager of the Idaho office of Paul Revere Insurance, based in Massachusetts.

In 1970, Andrus opposed Samuelson again. The main issue of the campaign was American Smelting and Refining Company's plans to dig an open-pit molybdenum mine in the White Cloud Mountains and the Challis National Forest. (Molybdenum is a hard, gray, metallic element used to toughen steel alloys and to soften tungsten alloy. It is also used in fertilizers, enamels, and other industrial applications.) Andrus opposed the company's plans and was backed by environmentalists. He won the election by some 10,000 votes and became the first Democratic governor of Idaho in 24 years. Reelected in 1974, he served until early 1977. His administration was marked by a concern for the environment, although some disagreed with Andrus's total program. Wrote journalist Bruce Hamilton:

Andrus is for preserving wilderness and wild rivers, but often not as large an area as some environmentalists would like. He is for controlling predators, but against the indiscriminate use of 1080 (a poison used by ranchers to kill predators). He sees some strip mining as inevitable, but insists on adequate reclamation. He believes clearcutting is a sound management practice for some forests, but opposes abuses, which have led to accelerated soil erosion, loss of wildlife habitat, polluted streams, and denuded hillsides.

On 18 December 1976, President-elect Jimmy Carter named Andrus as his secretary of the interior. Carter was impressed with Andrus's environmental record and, as one source speculated, was looking for western support because he had failed to carry a single western state in the 1976 election. Andrus was sworn in on 23 January 1977. His support for halting many federal construction projects, such as the Tennessee-Tombigbee Waterway and Central Arizona Project, earned him the enmity of westerners, and Carter had to reverse Andrus's decision to cut the projects. One source noted, "On 6 November 1978, Andrus withdrew 110 million acres of public land in Alaska 'to preserve values that would otherwise be lost' under section 204(e) of the Federal Land Policy and Management Act of 1976." In 1981, when the Carter administration left office, Andrus departed from the Interior Department. From then until 1987, he served as director of the Albertson's food chain and chairman of the board of trustees of the College of Idaho. In 1987, he was elected governor of Idaho for the second time. This term in the governor's house was marked by his battle with antiabortion activists, who were disappointed when he failed to sign a ban on abortion that the state legislature had passed. He was reelected in 1991 but announced his intention not to run for a third term in 1994.

See also Department of the Interior.

Andrus, Secretary of the Interior v. Sierra Club (442 U.S. 347 [1979])

This Supreme Court case dealt with the question of whether the National Environmental Policy Act of 1969 (NEPA) mandated the preparation of an environmental impact statement (EIS) by federal agencies requesting appropriations for projects. Under section 102(c) of NEPA, all federal agencies must prepare EISs when proposed legislation and major federal performances will "significantly [affect] the quality of the human environment." Secretary of the Interior Cecil Andrus requested appropriations for the Interior Department. The Sierra Club sued Andrus because he had not issued an impact statement with the request for appropriations. An appeals court held that Andrus must submit an EIS with his appropriations request, so Andrus sued to the U.S. Supreme Court. The Council on Environmental Quality, which had issued guidelines for the preparation of EISs, reversed itself and declared that an EIS was not necessary for appropriations requests. With such a change in guidelines, the Supreme Court unanimously ruled that under NEPA, an EIS is necessary only for "proposals for legislation" and not for appropriations requests. Writing for the Court, Justice William Brennan differentiated between requests for legislative action, which require proposals, and appropriations requests, which merely fund legislation already enacted.

See also Department of the Interior; Environmental Impact Statements.

Antiquities Act (34 Stat. 225)

This congressional action was sponsored by Rep. John Fletcher Lacey, known as the "father of American conservation" and the "father of federal game legislation." Twice before he had attempted to get Congress to pass a law establishing federal protection for the Petrified Forest in Arizona and the Cliff Dwellers region near Santa Fe, New Mexico. Lacey visited New Mexico in the summer of 1902 to see firsthand the site of early American Indian habitation. Following his visit, he introduced the Act for the Preservation of American Antiquities, known as the Antiquities Act or the Lacey Antiquities Act. Although his earlier proposals had received only lukewarm support, this time Lacey was backed by such notables as geologist W J McGee, who had accompanied him on his New Mexico trip, as well as scientists at numerous U.S. museums, universities, and other scientific institutions, including the American Archaeological Institute. With the endorsement of these groups, the bill passed both houses of Congress and was enacted into law on 8 June 1906. It stated:

[A]ny person who shall appropriate, excavate, injure, or destroy any historic or prehistoric ruin or monument, or any object of antiquity, situated on lands owned or controlled by the Government of the United States, without the permission of the Secretary of the Department of the Government having jurisdiction over the lands on which said antiquities are situated, shall, upon conviction, be fined a sum of not more than five hundred dollars or be imprisoned for a period of not more than ninety days, or shall suffer both the fine and imprisonment, in the discretion of the court.

Under this act, the Petrified Forest in Arizona, the Cliff Dwellers Monument in New Mexico, a section of the Olympic Mountains in Washington State, and more than 200 other areas have been preserved and protected. In 1935, Congress enacted the Historic Sites and Buildings Act, amending the Antiquities Act to transfer the administration of national historic sites to the National Park Service, permit the protection and in some cases the purchase of new sites, and set up an advisory board on national parks, historic sites, buildings, and monuments, which today is part of the Interior Department and serves as a screening body for proposed additions to the National Park System.

See also Lacey, John Fletcher.

Arizona v. California (373 U.S. 546 [1963])

This Supreme Court case was the fifth of its kind involving these two states and rights to the waters of the Colorado River. Other cases had been tried in 1931, 1934, 1935 (with the United States as a plaintiff), and 1936. In this particular case, the Court dealt with the matter by assigning a "special master," Simon Rifkind, to "find facts, state conclusions of law, and recommend a decree, all subject to consideration, revision, or approval by the Court." The Court's opinion states that "after argument, the Court agreed with the master's conclusions 1–3, 5, and 7–10, and disagreed with conclusions 4 and 6." These conclusions were: (1) the issues in the case were not controlled by the Colorado River Compact of 1922, the law of prior appropriation, or the doctrine of equitable apportionment; (2) the only waters apportioned by the Boulder Canyon Project Act were the mainstream waters of the Colorado River, so that diversions within Arizona and Nevada of tributary waters flowing in those states would not be regarded as part of either state's allocation of Colorado River water; (3) although the states of the lower Colorado River basin had failed to allocate among themselves the first 7,500,000 acre-feet of water apportioned to them from the Colorado River mainstream, the contracts of the secretary of the interior, together with the statutory limitation on California's share, validly apportioned the water 4,400,000 acre-feet to California, 2,800,000 acre-feet to Arizona, and 300,000 acre-feet to Nevada; (4) in the case of a shortage, the secretary of the interior must make a pro rata reduction in each state's share; (5) the secretary was without power to charge Arizona and Nevada for diversions made to them from Colorado River tributaries above Lake Mead; (6) the secretary was without power to charge Arizona and Nevada for diversions from the Colorado River mainstream in the lower basin above Lake Mead; (7) mainstream water could not be delivered to Nevada users unless contracts with such users were made with the secretary, notwithstanding Nevada's contract with the secretary; (8) in creating the Chemehuevi, Cocopah, Fort Yuma (or Quechan), Colorado River, and Fort Mojave Indian Reservations, the United States had reserved enough water from the Colorado River to irrigate the irrigable parts of the reserved lands, for future as well as present needs, and such water rights were

present perfected rights entitled to priority; (9) the United States also intended to reserve water sufficient for the future requirements of the Lake Mead National Recreation Area, the Havasu Lake National Wildlife Refuge, the Imperial National Wildlife Refuge, and the Gila National Forest; and (10) all uses of mainstream Colorado River water within a state, including the uses of the United States, were chargeable to that state's apportionment. Justice William O. Douglas dissented in part, claiming that diversions of tributary water should be charged to a state's share. Justice John Marshall was joined by Justices Douglas and Potter Stewart in dissenting that the secretary of the interior had the power to apportion the Colorado River's waters among the states and did not have the power to reduce shares during times of shortage.

See also Arizona v. California (1983).

Arizona v. California
(460 U.S. 605 [1983])

This case, an extension of the 1963 litigation over the apportionment of the waters of the Colorado River among several western states, dealt exclusively with the water rights of the five Indian tribes located in these states—the Chemehuevi, Cocopah, Fort Yuma (or Quechan), Colorado River, and Fort Mojave Indian tribes. In 1979, in a decree signed by all the states and the federal government, the states agreed to apportion waters on the Indian reservations according to their acreage, with the apportionment to be increased as states worked out land disputes with the Indians. The tribes themselves did not sign the decree; their interests were represented by the U.S. government. The Indians then sued the government to establish their own conditions and terms in the decree. The Supreme Court held (5–3) on 30 March 1983 that the decree was "not relitigable in the interest of finality." Justices Byron White, Lewis Powell, William Rehnquist,

Sandra Day O'Connor, and Chief Justice Warren Burger voted in the majority; Justices William Brennan, Harry Blackmun, and John Paul Stevens dissented. Justice Thurgood Marshall did not participate.

Aspinall, Wayne Norviel (1896–1983)

Colorado Democrat Wayne Aspinall, dubbed "Mr. West" by the *Denver Post*, was for a number of years one of the most influential politicians in Washington in the area of the development of the West's natural resources. He was born in Middleburg, Ohio, on 3 April 1896, the eldest child of Mack Aspinall, a farmer, and Jessie Edna (nee Norviel) Aspinall. Because of Jessie Aspinall's lung disease, the family moved to Palisade, Colorado, in 1904, where Mack Aspinall ran a peach farm. Wayne Aspinall was educated in the public schools of Palisade and then at the University of Denver. When the United States entered World War I in 1917, Aspinall enlisted in the U.S. Army Signal Corps and served in a training squadron. After the armistice, he continued his education at the university, receiving a bachelor of arts degree in 1919 and later a law degree. He taught school in Palisade and eventually returned there to practice law.

In 1930, Aspinall won a seat in the Colorado state assembly. During his eight years there, he served as Democratic whip and as Speaker. In the state senate (1939–1948), he was the Democratic whip, majority floor leader, and minority floor leader. In 1948, he won a seat in the U.S. House of Representatives from the Fourth Colorado District, the first of 12 terms (1949–1973). His most important committee assignment was on the House Public Lands Committee (now the Committee on Interior and Insular Affairs), which he would later chair (1959–1973). His tenure was marked by a zeal to develop the West through water projects. Wrote William Ritz in the *Denver Post*, "Aspinall introduced and pushed through Congress more than a dozen water projects, including the Vega Reservoir, the Silt Project, Smith Forks and the Frying Pan–Arkansas. He was the architect and chief sponsor of the Colorado River Storage Project Act, which authorized the federal government to build and operate reservoirs at Glen Canyon, Navajo, Flaming Gore and Curecanti, and he is called the father of the Upper Colorado River Basin Compact." Author Philip Fradkin noted, "His colleagues in Congress called Aspinall 'Mr. Reclamation.' He was certainly 'Mr. Water' as far as Colorado was concerned. ... If Aspinall had a basic weakness, it was his unrelenting attitude toward the conquest of the [Colorado] river." His opposition to the Wilderness Act of 1964 was lessened only when his idea of a Public Land Law Review Commission was brought to fruition. On 18 February 1963, Aspinall introduced H.R. 3846, the Land and Water Conservation Fund bill. Later enacted into law, this legislation funded the acquisition of lands and the management of programs at public recreation areas.

Aspinall's pro-development agenda angered environmentalists, and they opposed him for reelection in 1972. He was defeated in the Democratic primary by a liberal attorney, Alan Merson, who lost in the general election. Aspinall returned to his home in Palisade, where he spent the last ten years of his life in retirement. He died of cancer on 9 October 1983 at the age of 87.

See also Land and Water Conservation Fund Act; Wilderness Act.

Audubon, John James (1785–1851)

John James Audubon's name remains the best known among nineteenth-century naturalists and ornithologists. He was born an illegitimate child in Santo Domingo (now Haiti) on 26 April 1785 with the name Jean Rabin Audubon. His father was Jean Audubon, a Frenchman from Les Sables d'Olonne on the Bay of Biscay who was a seaman and traveler. The elder

Nineteenth-century wildlife artist John James Audubon, depicted here in a portrait by John W. and Victor G. Audubon (circa 1841), painted the birds and mammals of North America in their natural settings.

Audubon attained some wealth by shipping slaves to the New World. Audubon's mother, known only as Mademoiselle Rabin, was apparently a Creole. Audubon left Santo Domingo with his father and half-sister soon after the death of his mother. They returned to France, where Audubon was brought up by his father's wife, Anne Moynet Audubon, who legally adopted Jean when he was nine. His name was legally changed to Jean Jacques, which was later anglicized to John James.

Audubon's education included mathematics and Latin, but he became expert in nature studies. He came to the United States in 1803, when his father purchased a farm, Mill Grove, near what is now Norristown, Pennsylvania. Audubon spent much of his time studying and drawing the numerous birds that inhabited the area. Although he engaged in private business for a time and even returned to

France for a short period, by 1807 he was back in the United States and devoting all his time to nature studies. Bankruptcy resulting from the failure of his brother-in-law's grist and lumber mill drove him to travel, but he began earning a decent living by selling his bird drawings. Unable to publish his works in the United States, Audubon went to England, where *Birds of America* was published in 1827 in huge volumes known as elephant folios. In 1830, he and his wife bought a home in Edinburgh, Scotland, where he began to write the text portion of *Birds*, which he later published separately as *Ornithological Biography* (1838). *Synopsis of the Birds of America* (1839), a review of the five volumes of his momentous work, appeared the following year. When he returned to the United States, Audubon collaborated with naturalist John Bachman on the three-volume *Viviparous*

Quadrupeds of North America (1845–1849). Two of Audubon's sons later married two of Bachman's daughters.

In his final years, Audubon worked on color plates of his *Quadrupeds* series. He purchased a large estate, Minnies' Land, on the Hudson River in New York, which today is known as Audubon Park. He also began to tutor several up-and-coming ornithologists, among them Spencer F. Baird. Audubon suffered a crippling stroke in early January 1851 and succumbed on 27 January. He was 65.

See also Bachman, John; Baird, Spencer Fullerton.

Audubon Movement

Encouraged by various reformers during the latter part of the nineteenth century, the conservation movement—under the direction of George Bird Grinnell, writer and editor of the sporting magazine *Forest and Stream*—attempted to start an organization that could lobby for the protection of wild birds and their eggs. In 1886, Grinnell formed the Audubon Society, named after the man whose artwork had introduced the images of many unknown species of birds to the public. Within two years, however, Grinnell became swamped by more than 50,000 applications for membership, and he disbanded the organization. In 1896, a movement formed to protest the mass destruction of heron territories in the South. This protest movement evolved into the Massachusetts Audubon Society, with ornithologist William Brewster as the organization's first president. The Lacey Act of 1900, which set up federal protection for birds and their habitats, was a key piece of legislation that was lobbied for by a number of state Audubon societies. Onto the scene stepped insurance salesman William E. Dutcher, who had been encouraging the growth of state Audubon societies nationwide. In 1905, with more than 40 clubs in existence, Dutcher founded the Association of Audubon Societies, which led to the creation of the modern-day Audubon Society, an important information and lobbying group. In 1940, the group's name was changed to the National Audubon Society.

See also Brewster, William; Dutcher, William E.; Grinnell, George Bird; National Audubon Society.

Babbitt, Bruce Edward (1938–)

Bruce Babbitt served as Arizona's attorney general and governor, was a presidential candidate in 1988, and was appointed secretary of the interior in the Clinton administration. Born in Los Angeles, California, on 27 June 1938, Babbitt is descended from a prestigious Arizona pioneer family. Edward Bobet arrived in the Massachusetts Colony in 1639, and descendants known as the five Babbitt brothers, David, Billy, Edward, Charles, and George, came to the Arizona Territory in the 1880s and opened the Babbitt Brothers Trading Company in Flagstaff. Babbitt's father, Paul Babbitt, Jr., was the mayor of Flagstaff.

Bruce Babbitt attended Notre Dame University, where he received a bachelor's degree in geology. He then enrolled at the University of Newcastle in England as a Marshall scholar and received a master's degree in geophysics. Babbitt also was awarded a law degree from Harvard Law School in 1965. From 1965 to 1967, he worked in the Office of Economic Opportunity (OEO), a Great Society program, as a special assistant in civil rights to the director of VISTA (Volunteers in Service to America). He then joined a private law firm in Phoenix, Arizona, where he was counsel for the Arizona Wildlife Federation and the Arizona Newspaper Association. He also represented Navajo Indian claims of political gerrymandering that prevented Native Americans from serving in the state legislature.

In 1974, Babbitt was elected attorney general of Arizona. In this post, he was most noted for extending the powers of the attorney general to bring criminal charges against citizens (a right previously held only by county attorneys) and executing state antitrust laws that had been loosely enforced. On 4 March 1978, the death of Gov. Wesley Bolin led to Babbitt's succession as governor. His tenure lasted for nine years (1978–1987), during which he supported Ronald Reagan's New Federalism (the theory that the federal government should give more power to the states). "Western governors have been playing this theme for years," Babbitt said of Reagan's policies in a 1982 interview. "Now the President has shoved a huge pile of chips into the game and said, 'O.K., let's play for keeps.'" Dan Goodgame wrote in *Newsweek* that, as governor, Babbitt "proved himself a popular and shrewd executive in a deeply conservative state. He balanced his budgets, refused to throw money at problems and avoided fights he couldn't win. He pressed the legislature to improve health care for the poor, while holding taxes down and deregulating business. Says House Majority Whip Jane Hull, a conservative Republican and frequent Babbitt critic, 'I guess he did drag us kicking and screaming into the twentieth century.'"

In 1988, after leaving the governor's mansion, Babbitt ran a short and unsuccessful campaign for the Democratic presidential nomination. Following the campaign, he served from 1988 to 1993 as president of the League of Conservation Voters. Babbitt was chosen in December 1992 as Clinton's secretary of the interior. As of this writing, his tenure has been marked by battles with westerners over mining and ranching grazing fees. Michael Murphy of the *Phoenix Gazette* wrote, "Although damaged by losing a major fight in Congress over grazing fees, Babbitt is making a strong push for fundamental changes in the management

Bruce Edward Babbitt, forty-seventh secretary of the interior, serves in the administration of President William Clinton.

of millions of acres of public lands across the West.... Canvassing Western states with a zeal that marked his 1988 presidential bid, Babbitt is trying to stem a groundswell of opposition by ranchers and to finesse disgruntled environmentalists with Olympic-quality political maneuvers." Wrote George Gobbe, "Babbitt is a quick study whose first few months on the job included significant victories that were somewhat obscured by the media coverage of the public hearings in Oregon on the fates of the spotted owl and old-growth forests." Babbitt's initiatives in his first two years on the job include the establishment of the National Biological Survey. In 1993 and again in 1994, he was considered for a seat on the U.S. Supreme Court.

See also Department of the Interior; National Biological Survey.

Bachman, John (1790–1874)

Collaborator with John James Audubon on some of the greatest ornithological histories of the nineteenth century, naturalist John Bachman was also a prominent American botanist and zoologist. The youngest son of farmer Jacob Bachman, he was born in the village of Rhinebeck in upstate New York on 4 February 1790. "The Bachman family

is traditionally supposed to have come to Pennsylvania with William Penn seven generations earlier, though this is not a matter of precise record. The family took its origin in the canton of Berne, Switzerland, and one of its members, one Lieutenant Bachman, was killed with the Swiss guards who died in the defense of the Tuileries" during the French Revolution, reports Bachman biographer Donald C. Peattie. John Bachman reportedly attended Williams College in Williamstown, Massachusetts, but a check of the records there did not show his name. Other sources report that he was educated in theology through private tutors and later was licensed to preach, becoming a Lutheran clergyman as well as a naturalist.

Bachman related that "from my earliest childhood I had an irrepressible desire for the study of Natural History." Prior to 1813, he met and was befriended by ornithologist Alexander Wilson, who introduced him to the German naturalist Alexander von Humboldt. Bachman taught in a school in Ellwood, Pennsylvania, and entered the Lutheran ministry in 1813. For the next two decades he devoted his life to the church. Because of a bout with tuberculosis, he was forced to leave Pennsylvania for the milder climate of Charleston, South Carolina. In his pastorate there, writes Peattie, "he was tireless in building up his congregation, giving especial attention and benevolence to the negroes."

In 1831, Bachman met John James Audubon. Describing the meeting, Audubon's biographer Alice Ford writes, "Audubon could have imagined nothing better than this chance meeting with a great-hearted man of science, one not seeking personal glory, but only eager to be of assistance." Over the next twenty years, Bachman worked closely with Audubon in collecting and preparing specimens for Audubon's paintings of birds. In fact, Bachman is credited with 134 references in Audubon's 1838 work *Ornithological Biography*. After a short time in Europe, Audubon returned to the United States in 1836 and began work on his massive enterprise, the three-volume *Viviparous Quadrupeds of North America* (1845–1849); Bachman provided assistance in composing the scientific text. The two became such close friends that the Bachman and Audubon families intertwined, and two of Audubon's sons married two of Bachman's daughters.

Bachman also worked on behalf of black Americans at a time when slavery was a burning issue. A staunch Unionist, he nonetheless served in the South during the Civil War, giving aid to the sick and dying. The war left him propertyless and penniless. He worked to rebuild his place in society, but several strokes left him paralyzed. He died on 24 February 1874.

See also Audubon, John James.

Baird, Spencer Fullerton (1823–1887)

Spencer F. Baird, naturalist and zoologist, was secretary of the Smithsonian Institution as well as the first commissioner of the U.S. Fish Commission (now the U.S. Fish and Wildlife Service). Born on 3 February 1823 in Reading, Pennsylvania, Baird earned bachelor's and master's degrees in medicine from Dickinson College in 1840 and 1843, respectively. He later gave up medicine for the study of nature. Having met and befriended naturalist and illustrator John James Audubon in 1840, Baird became immersed in the writings of earlier naturalists. In 1846, he was made a professor of natural history at Dickinson College. There, he took a leading role in establishing zoology and naturalism as important subjects for study. In 1848, after he was also made chairman of Dickinson's chemistry department, Baird received the Smithsonian Institution's first scientific exploration grant to study bone caves in Pennsylvania. A year later, he was working as an editor of *The Iconographic Encyclopedia*.

In 1850, Baird was named assistant secretary to the head of the Smithsonian, noted physicist Joseph Henry. He served with Henry until the latter's death in 1878. Baird was then named secretary of the Smithsonian. His time there was spent pursuing government-funded explorations of the western United States and thus developing the Smithsonian into a true national museum. In 1871, President Ulysses S. Grant appointed Baird head of the newly created U.S. Fish Commission. Under his leadership, fish species in the Atlantic and Pacific Oceans were examined and studied; the nation's first marine laboratory at Woods Hole, Massachusetts, was established; and fish and fish habitat protection was championed. Baird appointed scientist James W. Milner to examine the issue of fish stock depletion in the Great Lakes and what could be done to solve the problem. Milner's summary, *Report on the Fisheries of the Great Lakes: The Result of Inquiries Prosecuted in 1871 and 1872* (1873), eventually led to the formation of the United States–Canada Joint Fish Commission in 1892.

In his lifetime, Baird was an organizer of the National Academy of Sciences, as well as the author of several hundred books and articles, including the classic works *A History of North American Birds* (1874), *North American Reptiles* (1853), and *Catalogue of North American Mammals* (1857). He was also a mentor to several other conservationists, scientists, and environmentalists, including Clinton Hart Merriam of the Bureau of Biological Survey. Baird died on 19 August 1887 at the age of 64.

See also Merriam, Clinton Hart; United States Fish and Wildlife Service; United States–Canada Joint Fish Commission of 1892.

Bald and Golden Eagle Protection Act (16 U.S.C. 668)

This federal legislation was enacted on 8 June 1940 in response to the growing threat to the survival of the bald and golden eagle populations in the United

The Bald and Golden Eagle Protection Act of 1940 preserved eagles from being hunted or captured.

States. The act declared that "by tradition and custom during the life of this Nation, the bald eagle is no longer a mere bird of biological interest but a symbol of the American ideals of freedom" and thus was deserving of federal protection. The act made it illegal to take; possess; sell; purchase; barter; offer to sell, purchase, or barter; transport; export; or import any bald or golden eagles, any bald or golden eagle eggs, or any bald or golden eagle nests.

Ballinger, Richard Achilles (1858–1922)

Richard Ballinger was secretary of the interior during the Taft administration. He was born on 9 July 1858 in Boonesboro, Iowa. His father was a lawyer who read law in the offices of Abraham Lincoln and later commanded an infantry of soldiers during the Civil War. Ballinger attended local schools and graduated from Williams College in Williamstown, Massachusetts, in 1884. He read the law in Kankakee, Illinois, and was admitted to the bar in 1886. At that time he moved to Port Townsend in the Washington Territory, which became Washington State in 1890. Later, Ballinger moved to Seattle, where he began his upward climb through U.S. politics. In 1894, he served as a superior court judge, overseeing Jefferson County. He was the mayor of Seattle from 1904 to 1906. A self-taught expert in mining law, Ballinger authored *A Treatise on the Property Rights of Husband and Wife Under the Community or Ganancial System* (1895).

In 1907, Ballinger's schoolmate at Williams, James R. Garfield, then secretary of the interior, asked his old friend to become commissioner of the General Land Office. Ballinger accepted but served only a year; he spent much of the time dealing with land claims in Alaska. In 1908, he returned to his law practice in Seattle. In 1909, President William Howard Taft appointed Ballinger secretary of the interior. There had been wide speculation that Taft would retain Garfield, a friend of Teddy Roosevelt's who had worked hand in hand on environmental matters with Gifford Pinchot.

Almost immediately, Ballinger ran head-on into the matter of land-fraud claims from Alaska. When Louis R. Glavis, a field-worker for the General Land Office, claimed that Ballinger was impeding the investigations into these claims, President Taft ordered Glavis fired. Glavis's dismissal was the first shot fired in the internal war between the conservative and progressive forces in the Republican Party. When Congress conducted an inquiry into the Glavis firing, Gifford Pinchot worked behind the scenes on Glavis's behalf. Pinchot eventually lost his job as a result of the controversy. Although the final congressional report found that Glavis's firing was proper, the repercussions of the affair damaged Taft within his own party. In order to appease the Progressives, he asked for and received Ballinger's resignation in March 1911. Ballinger returned home to Seattle, where he picked up his law practice. He died there on 6 June 1922. In 1940, Secretary of the Interior Harold L. Ickes looked into the Ballinger-Pinchot affair and found that Ballinger had been treated like "an American Dreyfus." Ickes, in an article in the *Saturday Evening Post*, called the inquiry and forced resignation of Ballinger "the most cruel persecution I am familiar with in modern times."

See also Ballinger-Pinchot Affair; Department of the Interior.

Ballinger-Pinchot Affair

This 1910–1911 incident during the Taft administration over environmental policy and the direction of the Republican Party was based on ideology. It led to a split in the party that cost it the White House in 1912. The fight was between Richard Achilles Ballinger, secretary of the interior under William Howard Taft, who led the conservative faction of the Republican Party, and Gifford Pinchot, chief of the U.S. Forestry Service. Ballinger sided with developers who had laid claims to land in Alaska. Pinchot, who was a close friend of Teddy Roosevelt, the conservationist president who led the progressive wing of the party, sided with Louis R. Glavis, a field-worker from the General Land Office. Glavis believed that the wealthy investors Ballinger was siding with—men known as the Cunningham group—were exploiting Alaskan lands. Glavis accused Ballinger of impeding investigations into these claims, and President Taft dismissed Glavis. Congress looked into the matter, and Glavis published his charges in an article in *Collier's* magazine. When it was discovered that Pinchot, who was supposed to be backing the administration, had in fact helped Glavis write his article, Taft fired the chief forester and his assistant, Overton Price. Although the congressional inquiry cleared Ballinger of any improper conduct, Taft asked for and received the interior secretary's resignation in March 1911. The affair led to a split in the Republican Party and gave rise to Teddy Roosevelt's Bull Moose presidential run in 1912.

See also Ballinger, Richard Achilles; Pinchot, Gifford; Price, Overton Westfeldt.

Barton, Benjamin Smith (1766–1815)

Benjamin Smith Barton and his nephew, William Paul Crillon Barton (1786–1856), were responsible for early American works on the botany of North America. Benjamin Smith Barton was born in Lancaster, Pennsylvania, on 10 February 1766. His father, the Reverend Thomas Barton, had been rector of St. James's Episcopal Church in Lancaster and was an enthusiastic botanist, mineralogist, and scientist in his own right. Both his parents died before Barton reached the age of 15, leaving him with a large estate. He attended York Academy in York, Pennsylvania, for a time, where he was immersed in the field of botany and became interested in nature. After his parents' deaths, he removed to the College of Philadelphia, where he studied medicine and literature. In 1785, he assisted his uncle, David Rittenhouse, in a survey of the western boundary of Pennsylvania. The following year, he began three years of study in the field of medicine at the University of Edinburgh, Scotland.

While in Europe, Barton came under the influence of naturalist Sir Joseph Banks. To finish his education in medicine, Barton traveled to Göttingen, Germany, where he was awarded a medical degree in 1789. Although he returned to the United States and began a medical practice, within a year Barton was named professor of natural history and botany at the College of Philadelphia (renamed in 1791 the University of Pennsylvania), a post he held until 1795. In 1795, he was advanced to professor of materia medica (pharmacology), and, finally, with the death of Dr. Benjamin Rush in 1813, professor of theory and the practice of medicine.

During the years he taught, Benjamin Barton traveled extensively, collecting specimens for such noted naturalists as Thomas Nuttall and Frederick Pursh (1774–1820). Barton was a prolific author in his own right, penning such works as *Memoir Concerning the Fascinating Faculty Which Has Been Ascribed the*

Rattlesnake and Other North American Serpents (1796; supplemented in 1800 and revised in 1814), *Collections Towards a Materia Medica of the United States* (two volumes, 1798–1804), a treatise on the use of natural medicinal plants in pharmacology, *Fragments of the Natural History of Pennsylvania* (1799), and *Elements of Botany* (1803), the first such work on American flora, which included plates drawn by naturalist William Bartram.

Benjamin Barton suffered from ill health all his life, and, as biographer George Blumer relates, "he was never robust and his naturally delicate constitution was impaired by his habits of work. Barton realized this himself, for he wrote of 'the pernicious consequences of his midnight and injurious toils.'" He developed tuberculosis and died on 19 December 1815 at the age of 49. His work was carried on by his nephew, William Paul Crillon Barton, who edited *A Biographical Sketch of Professor Barton* (1816).

Born in Philadelphia on 17 November 1786, William Paul Crillon Barton was the son of Benjamin Barton's brother, Judge William Barton, and Elizabeth (nee Rhea) Barton. He received his bachelor's degree from Princeton University in 1805 and his medical degree from the University of Pennsylvania three years later, while studying under his famed uncle.

After leaving medical school, Barton began his own practice, but in 1809 he was appointed a U.S. Navy surgeon upon the recommendation of the famed physician Benjamin Rush. He served on the frigates *United States*, *Essex*, and *Brandywine* and in marine hospitals (now naval hospitals) in Philadelphia, Norfolk, and Pensacola. He was later responsible for organizing the United States Naval Bureau of Medicine and served as the bureau's first chief clerk. Barton eventually achieved the rank of senior surgeon of the navy. In 1815, when his uncle died, Barton was named to replace him as professor of botany at the University of Pennsylvania. He later acted as professor

of materia medica at Philadelphia's Jefferson Medical School.

Because of his close contact with his uncle at the University of Pennsylvania, William Barton had a love of nature and botany that blended with his interest in medicine. Early studies in pharmacology allowed the medical field to find new drugs and cures in the flora and fauna that covered North America. Barton wrote such works as *Flora Philadelphicæ Prodrumus* (1815), *Vegetable Materia Medica of the United States* (two volumes, 1817–1818), *Syllabus of the Lectures Delivered on Vegetable Materia Medica and Botany in the University of Pennsylvania* (1819), *Compendium Floræ Philadelphicæ* (1824), and *Flora of North America* (three volumes, 1821–1823). Of the latter, biographer Donald C. Peattie writes, "[it was] magnificently illustrated by his wife," and, "if not complete as its title might imply, is at least a fine piece of popularization of the work of earlier systematists." Barton was also president of the Linnæan Society and a member of the American Philosophical Society. A later work, *Hints to Naval Officers Cruising in the West Indies* (1830), was directed to the military portion of his career. Barton died in Philadelphia on 29 February 1856 at the age of 69.

Barton, William Paul Crillon
See Barton, Benjamin Smith.

Bartram, John (1699–1777)
John Bartram and his son William (1739–1823) are regarded as America's first naturalists. John Bartram was born on his family's farm near Darby in Delaware County, Pennsylvania, on 23 March 1699, to devout Quaker parents, William and Elizabeth (nee Hunt) Bartram. John Bartram's ancestors had fought in the Norman invasion of England; his grandfather, also named John, had left England in 1682 and settled near Darby. John Bartram received very little education; it was

while farming the land that he discovered his love of nature in general and botany in particular. He soon began to absorb the works of such naturalists as James Logan and Carolus Linnæus. He built a botanic garden in Kingsessing, near Philadelphia, where he collected the seeds of various plants and hybridized them to create new strains. His correspondence with British merchant Peter Collinson included his sending cuttings of these new plants to Europe. One such species that he studied is today known as the *John Bartram Franklinia Alatamaha*, found by Bartram on a spot along Georgia's Altamaha River and named after Bartram's friend Benjamin Franklin. The species is now extinct in the wild and is grown only in arboretums. John Bartram died on 22 September 1777. His name is remembered through *Bartramia*, a genus of wild mosses.

John's son William was born on 9 February 1739. William, who was also a botanist and naturalist, toured southern North America from 1773 until 1778, adding specimens to his father's garden. His *Travels Through North & South Carolina, Georgia, East and West Florida*, known as *The Travels of William Bartram* (1791), became one of the most widely read works in the field of botany. After William Bartram's death on 22 July 1823 in Philadelphia, the botanic garden he and his father had dedicated their lives to creating became part of the Philadelphia park system. It is one of the oldest of its kind in the United States.

Bartram, William (1739–1823)
See Bartram, John.

Bell, John Calhoun (1851–1933)
John C. Bell of Colorado led the 1891 charge in the House of Representatives against the establishment of forestry reserves. Born in the village of Sewannee (or Suwannee) in Franklin County, Tennessee, on 11 December 1851, he was the son of Harrison Bell, a prominent busi-

nessman, and Rachel (nee Laxson) Bell. He attended the schools of Franklin County and studied law in Winchester, Tennessee, before being admitted to the bar in 1874. In June of that year, he moved west to Colorado, where he opened a law practice in the town of Saguache. He served as county attorney of Saguache County from 1874 until 1876, when he moved to Lake City, Colorado, which one source calls "then the most thriving town in the great San Juan mining region." He was twice mayor of Lake City but resigned in 1885 when he opened a law practice in Montrose, Colorado. From 1888 to 1892, he was the judge for the seventh Colorado judicial district. In 1892, he was elected to the U.S. House of Representatives and served for five terms (1893–1903). His most famous actions include helping to open the southern Ute Indian reservation for settlers, getting appropriations for the Uncompaghre Reclamation Project (he was a member of the powerful Appropriations Committee), and cosponsoring the Newlands Reclamation Act in 1902.

In his 1893 message to Congress, President Grover Cleveland asked for legislation protecting the existing forestry reserves in the nation. The major bill was introduced by Rep. Thomas C. McRae of Arkansas. Because such reserves would deny entry to settlers and proprietors, many western senators and representatives, particularly those from Colorado, fought the measure bitterly. Among them was Bell. Historian G. Michael McCarthy said of him, "A transplanted Tennessean who had practiced law in some of Colorado's most important mining districts, Bell was practically a replica of [Sen. Henry Moore] Teller—an irascible, determined enemy of conservation." Once debate on the McRae bill began, Bell stepped forward and condemned it as an infringement on the rights of farmers and ranchers. He angrily said of the legislation: "In the state of Colorado these reservations . . .

take in farms, take in settlers, and in one county in particular they do not leave a single stick of timber outside the reservation line, so that from the day that reservation was declared every settler in the county has had to steal every stick that has gone into his fireplace.... Not only that, but there have been declared in a state reservation over a quarter million acres of land that has not a stick of timber or brush on it.... These reservations ought to be knocked out of existence.... In Colorado, they have done us no good whatsoever."

Although Bell and his colleagues were successful in killing the bill, it was reintroduced in the House in 1894 and passed with several amendments. In the Senate, however, Bell's fellow Coloradan Henry Moore Teller helped kill the bill. The McRae bill was never resurrected. When President Cleveland circumvented Congress and created 13 new forest reserves, totaling more than 21 million acres of land (an action known as "The Midnight Reserves"), western politicians in the Congress arose in anger to pass legislation later known as the Pettigrew Act. Although supportive of this action, Bell used the debate to again attack forest reserves, particularly the Battlement Mesa Reserve: "In Rouett County [Colorado] there was not left any valuable timber in that county outside the line of reservation.... [T]he reserve took in agricultural lands, pastoral lands, and mining lands that had no timber.... That is why these reservations are so unpopular; that is why this cry has come from the West. It is of this injustice and this indiscriminate setting aside of reservations that the people of my district complain." The Pettigrew Act eventually passed, effectively ending debate on forest reserves in the Congress for several years.

Denied reelection to a sixth term, Bell returned to Montrose and picked up his law practice. He was considered the state's leading Populist, and his work to endorse reforms in the body politic was hailed by the *Rocky Mountain News* when

he said, "Before things are right at Washington, you must get them right in your precincts at home." Back in Montrose, his last duties were to serve on the Court of Appeals of Colorado (1913–1915) and as a member of the state Board of Agriculture (1931–1933), work he finished shortly before his death on 12 August 1933 at the age of 81. The *Rocky Mountain News* hailed him as a "State Pioneer."

See also McRae, Thomas Chipman; Pettigrew Act; Shafroth, John Franklin; Teller, Henry Moore.

Bennett, Hugh Hammond (1881–1960)

Hugh Bennett was a leading scientist and advocate in the field of soil conservation. Born near Wadesboro, North Carolina, on 15 April 1881, he was the son of William Osborne Bennett and Rosa May (nee Hammond) Bennett, both farmers. Hugh Bennett grew up on his father's expansive farm, and the experience educated him in the techniques of farming and soil conservation. Bennett attended small rural schools, then chopped wood to procure the necessary money to enroll at the University of North Carolina in 1897. Although a lack of funds forced him to leave the school for a time, he returned to earn a bachelor's degree in science in 1903.

That same year, Bennett was employed by the Bureau of Soils in the Department of Agriculture, using his talent for soil examination and application. Six years later, he was named the supervisor for soil surveys for the eastern and southern divisions of the Bureau of Soils, a position he held until 1928. During his investigations, he concluded that soil erosion was "the most serious agricultural problem" facing the nation, author Henry Clepper writes. Bennett's various inquiries into soil-saving techniques earned him high praise in the government and private sector.

In 1928, Bennett wrote the book *The Soils of Cuba* with Robert V. Allison and coauthored with William R. Chapline the important Agriculture Department paper *Soil Erosion, A National Menace*. Bennett's testimony before Congress on the problem led to legislative action. On 16 February 1929, Rep. James P. Buchanan (1867–1937) of Texas attached an amendment, later called the Buchanan Amendment, to the Department of Agriculture's appropriations for 1930. The amendment called for soil-erosion studies to be done by Bennett with an appropriation of $160,000. In 1933, when the New Deal's "alphabet soup" of programs set up the Soil Erosion Service in the Department of Agriculture, Bennett was named its chief. On 27 April 1935, following the release of the Buchanan Amendment's soil-erosion studies, Congress enacted the Soil Conservation Act of 1935, which authorized the creation of the Soil Conservation Service (SCS). Bennett headed the SCS from its inception until his retirement in 1951. As chief, Bennett oversaw the establishment of some 1,700 soil conservation districts around the United States. With the help of Civilian Conservation Corps (CCC) workers, the SCS was able to establish programs at a minimal cost. The SCS soon became one of the most successful New Deal agencies.

Bennett was also a key member of the Great Plains Drought Area Committee, which was set up during the Depression to study arid parts of the nation. Awarded various honors for his work, he was nominated in 1948 for the Nobel Prize for his services to humanity. He was the author of *The Agricultural Possibilities of the Canal Zone* (1914), *The Soils and Agriculture of the Southern States* (1921), *The Cost of Soil Erosion* (1934), and *Soil Conservation* (1939). Hugh Hammond Bennett died on 7 July 1960 at the age of 79.

See also Great Plains Drought Area Committee.

Biodiversity

The term *biodiversity* has come to mean many things and symbolize many actions within the environmental movement. One source defines it as "the sum total of all the plants, animals, fungi, and microorganisms in the world, or in a particular area; all of their individual variations(s), and all of the interactions between them." Another describes it as "the richness of species and the range of their genetic makeup." Overall, it is agreed that the biodiversity of a designated area is the totality of the natural animals, plants, fungi, and their subspecies and subgroups situated in a fixed setting, and how they survive in this environment. The sheer number of any one species does not denote a successful or failing environment. Many nonnative species can throw into shock the environment of areas into which they have been introduced. Laws such as the one that created the Endangered Species List have been used to preserve the biodiversity of certain areas.

Bland, Thomas (1809–1885)

English naturalist and conchologist Thomas Bland became a leading authority on American mollusks. Born in Newark, Nottinghamshire, England, on 4 October 1809, he was the son of Dr. Thomas Bland. He was educated at the Charterhouse School in London, where he was a classmate of William Makepeace Thackeray. Bland later studied the law and for a time ran his own practice.

In 1842, Bland traveled to Barbados and, later, Jamaica. After meeting Professor Charles B. Adams of Amherst College, he set out on what became a ten-year odyssey across the Caribbean in an attempt to educate himself in the area of conchology, the study of mollusks and their shells. The journey ended when Bland came to New York in 1852 and began to work with William G. Binney, who was in the beginning stages of preparing the later volumes of his father Amos Binney's work, *Terrestrial Air-Breathing Mollusks*, published posthumously (1851–1878). Bland and William Binney subsequently worked together on a number of scientific papers dealing with conchology, including part 1 of *Land and Freshwater Shells of North America* (in *Smithsonian Institution Miscellaneous Collections* no. 8, 1869), which one biographer called an "important contribution to the growth and development in the field and for many years a recognized authority" in the area of conchology. Bland also produced *On the Geographical Distribution of the Genera and Species of Land Shells of the West India Islands, and a Catalogue of the Species of Each Island* (in *Annals of the New York Lyceum of Natural History* 7, 1862). Many more of his papers were published by the *Lyceum* and the *American Journal of Conchology*.

Bland has not received the recognition he deserves for his early work in the study and categorization of North American mollusks. Referring to *Land and Freshwater Shells of North America*, biographer Frank E. Ross reports in the *Dictionary of American Biography* that Bland's work "systemized, expanded and put in manual form the knowledge of the land shells of this continent and placed this information within the reach of students everywhere. For many years it was the chief authority in its particular field, and even at this late date it must still be consulted by all who study this fauna." Thomas Bland died in Brooklyn, New York, on 20 August 1885, at the age of 75.

Bliss, Cornelius Newton (1833–1911)

Secretary of the Interior Cornelius Bliss served in the McKinley administration from 1897 to 1899. He was born in Fall River, Massachusetts, on 26 January 1833. When his father died, Cornelius was left in the care of his grandparents. Until he was 14, Bliss worked at odd jobs while attending Fish Academy. In 1847, he joined his mother and stepfather in New Orleans, taking up work in the counting room of his stepfather's dry-goods store.

Bliss soon tired of the South, and he returned to Massachusetts, where he was employed by James M. Beebe & Co., the largest dry-goods store in the United States at that time. In 1866, he went to New York, where he joined the Boston textile firm of J. S. & E. Wright. When Bliss was made a partner, the firm became Wright, Bliss and Fabyan, then Bliss, Fabyan & Company. Interested in politics, Bliss served for a time as New York State Republican Chairman, which brought him great notice within the party. From 1892 until 1904, he was the treasurer of the Republican National Committee.

Having supported William McKinley for president in 1896, Bliss was targeted by the new administration for a cabinet post. He refused to be secretary of the treasury but accepted the interior post, which he held for two years. Bliss returned to private business in New York in 1899 and turned down several opportunities to be drafted for mayor and other elective posts. He served as a member of several charitable societies, including a term as president of the Society of the New York Hospital. Bliss suffered from a weakening of the heart before his death on 9 October 1911.

Boone and Crockett Club

This conservation group was founded in 1888 by Theodore Roosevelt and George Bird Grinnell, among others. In 1884, Roosevelt journeyed to the western United States and was made aware of the mass slaughter of wildlife in the area. This butchery caused Roosevelt to call for the establishment of a sportsmen's organization to lobby for the enactment of laws to preserve much of the animal life in the West.

At the end of 1887, Roosevelt returned to New York, where he met with *Forest and Stream* editor George Bird Grinnell and other sportsmen and called for the formation of an association that would "work for the preservation of the large game of this country, further legislation for that purpose, and assist in enforcing existing laws." Roosevelt proposed that the group should honor frontiersmen Daniel Boone and Davy Crockett. The following month, January 1888, the Boone and Crockett Club was established. Writes forest historian Terry West, "Although never large (about 100 core members), the club's membership was influential: powerful opinion molders such as Senator Henry Cabot Lodge, novelist Owen Wister, General William Tecumseh Sherman, and renowned landscape painter Albert Bierstadt, as well as many who became active in the conservation movement, such as Gifford Pinchot, Madison Grant, founder of the Save the Redwoods League, Arnold Hague of the U.S. Geological Survey, and Clinton Hart Merriam, head of the Bureau of Biological Survey."

Never a large organization, the Boone and Crockett Club, now headquartered in Alexandria, Virginia, has worked with many environmental groups to lobby for the enactment of conservation laws and to purchase land to be used as wildlife refuges.

See also Grinnell, George Bird; Roosevelt, Theodore.

Bowers, Edward Augustus (1857–1924)

Writes environmental author Henry Clepper, "One of America's pioneer forest conservationists, Edward Bowers was a director and twice secretary of the American Forestry Association." Gifford Pinchot called him "one of the most devoted and effective friends of the forest in America." Born in Hartford, Connecticut, on 2 August 1857, Bowers attended Yale University, where he was awarded an A.B. degree in 1879 and a law degree two years later. Traveling west, he served as an attorney and judge in the Dakota Territory from 1882 to 1886. In 1886, he was appointed by Secretary of the Interior Lucius Q. C. Lamar as a special inspector in the public land service. He held that post until 1893, when he was appointed assistant commissioner in

the General Land Office, a post Bowers held until 1895. After 1898, he had a law practice in New Haven.

At a joint meeting of the American Forestry Association and the American Economic Association held in Washington, D.C., in December 1890, Bowers read a landmark report called *The Present Condition of the Forests on the Public Lands.* Wrote Gifford Pinchot, "It was the paper of a man who knew his business.... Bowers declared that although 38,000,000 acres had been entered under the Timber Culture Act (taken up by settlers on condition that they plant trees), not over 50,000 acres had been successfully covered with young tree plantations. His paper was a complete demonstration that the laws relating to public timber, as administered, were utterly inadequate." Bowers drew up proposed federal legislation that was ultimately enacted in 1891 as the General Revision Act, or the Payson Act. Bowers was also responsible for the idea that led to the passage in 1897 of the landmark Pettigrew Act, which established rules for the administration of the nation's forests.

In his final years, Bowers remained a member of the American Forestry Association, as well as a lecturer at Yale University on forest administration and forestry law. He died on 8 December 1924 at the age of 67.

See also Payson Act; Pettigrew Act.

Brewer, Thomas Mayo (1814–1880)

Thomas Brewer was a noted ornithologist and oologist (one who specializes in the study of birds' eggs). He was born in Boston, Massachusetts, on 21 November 1814, the son of Col. Thomas Brewer, a patriot who participated in the Boston Tea Party. Thomas Brewer's early education was in the Boston schools, and he graduated from Harvard University in 1835 and Harvard Medical School in 1838. Although a practicing physician for several years, he soon found nature writing to be more to his taste. He also became the editor of the Whig organ the *Boston Atlas.*

It was his study of birds and their eggs that made Brewer famous. A friend of John James Audubon, Brewer is credited with supplying some of the information that was accurately reflected in Audubon's watercolors. Brewer was also an acquaintance of Spencer Fullerton Baird, the naturalist and zoologist. A member of the Boston Society of Natural History for much of his life, Brewer wrote several articles and books on ornithology, including *North American Oology* (1857) and *North American Birds* (1875); the latter was written in conjunction with Baird. Brewer retired from the publishing business in 1875 to devote all his time to bird-watching, but he suffered from ill health soon after and died on 23 January 1880.

Brewer, William Henry (1828–1910)

William Brewer was a forestry expert concerned with the destruction of the nation's timberlands. He was born in Poughkeepsie, New York, on 14 September 1828 and spent much of his early life on a farm near Ithaca, New York. In 1848, he enrolled at the scientific school at Yale (now known as the Sheffield Scientific School), from which he received a Ph.D. in 1852. Brewer then went abroad for several years, studying nature in Heidelberg and Munich. He returned to the United States and taught chemistry and geology at Washington and Jefferson College in Pennsylvania from 1858 until 1860.

From 1860 to 1864, Brewer traveled across the western United States, mapping and assisting in the geologic survey of the area. His work on a California plant species, *Polypetalae,* was published in volume 1 of the *Geological Survey of California.* In 1864, he returned to the Sheffield School as a professor of agriculture and served there until his retirement in 1903.

Brewer was at the forefront of a movement that called for a halt to the destruc-

tion of the nation's timberlands. For the ninth census in 1870, he wrote *The Woodlands and Forest Systems of the United States,* which was a complete survey of those areas. Over the next three decades, through speeches, writings, and lobbying, he helped change governmental attitudes toward forest conservation. An intimate of geologist and surveyor Clarence King, Brewer lobbied President Rutherford B. Hayes in 1879 to establish the U.S. Geological Survey and recommended that King serve as the agency's first director. In 1900, Brewer was one of the founders of the Yale University Forest School. He died on 2 November 1910 at the age of 82.

See also National Forest Commission.

Brewster, William (1851–1919)

A noted ornithologist, William Brewster was born in Wakefield, Massachusetts, on 5 July 1851. He was educated in the public schools of nearby Cambridge but, because of ill health, never attended college. The only degrees he ever received were honorary ones from Amherst College in 1880 and Harvard University in 1899. In 1870, Brewster followed his father's wishes and became a banker, but he soon discovered that his love for nature was overwhelming and resigned his job to study birds full time. He joined the Boston Society of Natural History and helped develop a collection of birds and mammals. In 1873, he was a founding member of the Nuttall Ornithological Club; ten years later, he was one of three founding members of the American Ornithologists' Union. In 1885, he was named head of the Department of Mammals and Birds at the Museum of Comparative Zoology in Cambridge. In 1896, he was the first president of the Massachusetts Audubon Society, one of the clubs that was a forerunner of the modern National Audubon Society.

In his lifetime, Brewster authored more than 300 works on birds, including the multivolume *Descriptions of the First*

Plumage in Various North American Birds (1878–1887), his classic work *Bird Migration* (1886), *Birds of the Cape Region of Lower California* (1902), and the unfinished and posthumously published *Birds of the Lake Umbagog Region, Maine* (1924). Brewster died on 11 July 1919, just six days after his sixty-eighth birthday. To honor his work, the American Ornithologists' Union named its annual award for nature book authorship the Brewster Medal.

Bromfield, Louis (1896–1956)

Louis Bromfield was the Pulitzer Prize–winning author whose experimental farm in Ohio was and continues to be a model for conservation and nature studies. Bromfield was born in Mansfield, Ohio, on 27 December 1896, the son of a banker and the grandson of an abolitionist agitator. Mansfield later became the setting for many of Bromfield's novels. After being educated in the public schools of Mansfield, Bromfield attended Cornell, Ohio Northern, and Columbia Universities (studying at both the agriculture and journalism schools of the latter) before joining the French army during the early years of World War I as an ambulance driver. He later served with the American Expeditionary Force when it arrived in Europe.

Following the war and his return to the United States, Bromfield worked as a journalist for the Associated Press; a foreign editor for the journal *Musical America*; theater, art, and music critic for the *Bookman*; and music critic for *Time* magazine. It was at this time that he began to write full-length novels. His first and second works were rejected by various publishers, and he destroyed both of them. His next two were also rejected, but his fifth effort, *The Green Bay Tree*, a slice of life in a fictional American steel town, was published in 1924. Then followed *Possession* (1925), the story of a girl who leaves her Ohio hometown for the big city and a career in music. The following year saw *Early Autumn*, which earned him the Pulitzer Prize for fiction in 1927.

After the publication of *Possession*, Bromfield and his family moved to France, where they lived until 1938. They returned to the United States as the rumblings of World War II were beginning to sound. Bromfield purchased and settled on a farm near his boyhood home in Ohio. This is where he created Malabar Farm. Malabar—the name is from a coastal region in India that Bromfield had visited—became a one-man experiment in the self-sufficiency of the average American farmer (although Bromfield was quite wealthy and did the work only as a hobby). He documented his efforts in building the farm into a model of conservation in *Pleasant Valley* (1945), *Malabar Farm* (1948), *Out of the Earth* (1950), and *From My Experience* (1955). From Malabar came ideas relating to soil-saving techniques, conservation, and cheaper farming methods. In addition, Bromfield transmitted radio lectures from the farm to spread his ideas. A member of various conservation groups, including the Audubon Society, and a founder of Friends of the Land, Bromfield was awarded the Audubon Medal in 1952 for his work. Upon his death on 18 March 1956, the state of Ohio turned his experiment into the Louis Bromfield Ecological Center, an arm of the state. In 1988, author Charles E. Little edited Bromfield's works in *Louis Bromfield at Malabar.*

Brower, David Ross (1912–)

Known as the "archdruid" of the American environmental movement, David Brower was a leading member of the Sierra Club and founder of the Earth Island Institute. Born in Berkeley, California, on 1 July 1912, Brower was the third of four children of Ross Brower, a professor of mechanical drawing at the University of California at Berkeley, and Mary Grace (nee Barlow) Brower. David

David Ross Brower

Brower was exposed to the natural surroundings of his home in Berkeley, which awakened his love of the environment. On one of his excursions, he identified a previously unknown butterfly that has been designated *Anthocaris sara reakirtii broweri*.

After attending local schools, Brower attended the University of California at Berkeley from 1929 to 1931, before dropping out for financial reasons. For the next two years, he worked in a candy store to make ends meet. In 1933, his association with photographer Ansel Adams led to his entry into the Sierra Club. In a history of the Sierra Club, Brower is called "one of the Club's leading climbers.... His first ascent of New Mexico's Shiprock in 1939 helped introduce the use of expansion bolts on otherwise unclimbable faces." Brower was also the first person to scale the Vasquez Pinnacle in the Pinnacles National Monument. His work on the 1939 film *Ski-Land Trails of the Kings* led, ac-

cording to some, to the establishment of the Kings Canyon National Park. In 1941, Brower assumed the editorship of the University of California Press, and a year later he edited the *Manual of Ski-Mountaineering* to assist in the instruction of mountain battalions. He himself served in France and Italy with the U.S. Army's Tenth Mountain Division.

After returning to the United States, Brower became the editor of the *Sierra Club Bulletin* (now *Sierra* magazine) and edited the *Sierra Club Handbook* (1947). In 1952, he became the Sierra Club's first executive director. Writes historian Susan Schrepfer, "One of the earliest national battles over wilderness, and the first in which he led the club, involved the successful defense of Dinosaur National Monument. During the 1950s and 1960s, the club grew from a moderately activist group chiefly concerned with the Sierra Nevada to a militant, national organization. Much of this change was due to Brower's vision, effective publicity, and aggressive conservation tactics." Under his leadership, club membership went from 7,000 in 1953 to 70,000 when he left the group in 1969. He fought the construction of dams and the cutting of trees, but it was the fight over the placement of a nuclear power plant at Diablo Canyon in California that estranged him from his Sierra Club colleagues, whose opposition to the plant was not as absolute as his. When he tried to take over the group, he lost and resigned. He went on to found the Friends of the Earth (FOE) and the John Muir Institute, which one source says "agitated to save the whales, kill the supersonic transport, kill nuclear power, kill ... leases of coal and oil fields on federal lands and lock up all of Alaska in no-use designations." Brower was editor of the FOE's *The Earth's Wild Places* (10 volumes, 1970–1977) and *Celebrating the Earth* (1972–1973). In November 1979, citing differences with the rest of the group, Brower again departed, setting up the Earth Island Institute. Presently, as head of the Institute, he remains one of the most respected environmentalists alive.

See also Earth Island Institute; Sierra Club.

Browning, Orville Hickman (1806–1881)

A lawyer from Kentucky, Orville Hickman Browning rose to become a U.S. senator, confidant to Presidents Abraham Lincoln and Andrew Johnson, and Johnson's secretary of the interior from 1866 to 1869. Browning was born near Cynthiana in Harrison County, Kentucky, on 10 February 1806, the son of Micaijah Browning, a farmer, and Sarah (nee Brown) Browning. Orville Browning attended local schools and Augusta College before reading law with several area attorneys. In 1831, he was admitted to the Kentucky bar but moved to Quincy, Illinois. Five years later, having become a trusted citizen, he was elected to the Illinois state senate. Here he started a friendship with another lawyer and former Kentuckian, Abraham Lincoln, who was serving in the lower house of the Illinois legislature. Browning worked with Lincoln on legislation to move the state capital from Vandalia to Springfield. Browning refused a second term, but two years later in 1842, he was elected to a seat in the state assembly. A year later, he ran unsuccessfully against Stephen A. Douglas for a seat in the U.S. House. Subsequent congressional election attempts in 1850 and 1852 met with similar failure.

In 1856, Browning was instrumental in the founding of the Illinois State Republican Party and setting up its rules and resolutions. Although a friend of Lincoln's, Browning backed Edward Bates for the Republican presidential nomination. Browning became a key member of Lincoln's election team upon Bates's defeat. With Lincoln's election, Browning, out of elective office since 1843, desired a government post. Rumors flew that Browning would get the first open seat on the U.S. Supreme Court, but Stephen A. Douglas's death in 1861 led Illinois

Gov. Richard Yates to name Browning to fill the vacancy in the U.S. Senate. Once a staunch antislavery Republican, Browning became disenchanted with Lincoln's harsh policies toward the South and his signing of the Emancipation Proclamation. Browning's vote against the second Confiscation Act drove the two men to become enemies. Browning refused to run for reelection on a Lincoln ticket, and he left the Senate.

After Lincoln's assassination, Browning got close to President Andrew Johnson, backing the president's stand against the radical Republicans in Congress. By 1866, Browning was a key advisor to Johnson. On 26 July, wanting to move Interior Secretary James Harlan out of the cabinet, Johnson asked Browning to take over the post. On 1 September, Harlan resigned, making way for Browning. His tenure was most noted for his dismissals of radical Republicans in the Interior Department and his liberal policies toward the Indians. When Johnson faced impeachment, Browning stepped in and served as attorney general while the occupant of that office, Henry Stanbery, defended Johnson in the Senate.

With the inauguration of Ulysses S. Grant in 1869, Browning returned to Illinois, where he was elected as a Democrat to the state Constitutional Convention of 1869–1870. During this period of Reconstruction, he spoke out against black suffrage and minority rights. Browning also served as a railroad attorney and represented the Chicago, Burlington, and Quincy Railroad in the landmark Supreme Court litigation known as the *Granger Cases*. Browning died in Quincy on 10 August 1881.

See also Department of the Interior.

Buckley, Samuel Botsford (1809–1884)

Today, the works of naturalist and botanist Samuel Botsford Buckley are known only slightly outside scientific circles. Buckley was born in Torrey, New York, on 9 May

1809, although nothing is known of his parents. Buckley attended Wesleyan University in Middletown, Connecticut, graduating with a bachelor of arts degree in 1836. He subsequently attended the College of Physicians and Surgeons in New York from 1842 to 1843, but never received a degree from that institution.

Following his graduation from Wesleyan, Buckley spent two years collecting botanical specimens in Virginia and Illinois. In 1839 and 1840, he was the principal at an academy in Allenton, Alabama. While in the South, he toured the southwestern region of the Appalachian Mountains and discovered 24 new species of plants and a shrub, which was later named *Buckleya distichophylla* by botanist John Torrey. He described some of his findings in "Description of Some New Species of Plants" in the *American Journal of Science* in 1843.

In 1858, Buckley studied the mountains of North Carolina and Tennessee; Buckley's Peak in Tennessee is named after him. In 1859, at the height of the slavery controversy, he traveled to the South to collect specimens at the request of naturalists François André Michaux and Thomas Nuttall, then collaborating on their *North American Sylva*. He served under Francis Moore of the Texas Geological Survey from 1860 to 1861 and with the U.S. Sanitary Commission for four years collecting statistical information on wounded soldiers fighting in the Civil War. Later, he returned to Texas, where he prepared two geologic maps of the state. Later, in Austin, Buckley was editor of the *State Gazette*. He spent 1877 to 1881 penning several works on the trees and shrubs of the United States, and particularly the natural history of the state of Texas, but none of these was ever published. He died at Austin on 18 February 1884 at the age of 74.

Buford v. Houtz (133 U.S. 618 [1890])

This Supreme Court case involved the issue of grazing on public lands. M. B. Buford was part-owner with J. W. Taylor and Charles and George Crocker of the Promontory Stock Ranch Company in Utah. Buford sued sheepherder John S. Houtz and his partner, Edward Conant, for straying onto his property. The Third Judicial Court of Utah in Salt Lake City ruled in favor of Houtz. Buford sued to the U.S. Supreme Court. On 3 February 1890, the Court ruled in favor of Houtz and his right to graze unimpeded on public lands. The decision stated:

> We are of the opinion that there is an implied license, growing out of the custom of nearly a hundred years that the public lands of the United States, especially those in which the native grasses are adapted to the growth and fattening of domestic animals, shall be free to the people who seek to use them where they are left open and unenclosed and no act of government forbids this use.... The government of the United States, in all its branches, has known of this use, has never forbidden it, nor taken any steps to arrest it. No doubt it may be safely stated that this has been done with the consent of all branches of the government and, as we shall atempt to show, with its direct encouragement.

In essence, the Court held that there was a fundamental freedom for persons to use the federal lands, a freedom that was bound into American society. There is some evidence that conservationists used the wording of this decision to help pass section 24 of the General Revision Act the following year and in so doing set aside millions of acres of federal land. Wrote author Wayne Hage, "It is doubtful if the forest reserve clause, Section 24 of the 1891 General Revision Act (also known as the Payson Act), would have emerged had it not been for one important catalyst: the United States Supreme Court ruling in the case of *Buford v. Houtz*."

See also Payson Act.

Bureau of Fisheries
See United States Fish and Wildlife Service.

Bureau of Land Management (BLM)
The Bureau of Land Management is the governmental agency that oversees what used to be the separate offices of the General Land Office and the United States Grazing Service. The agency was created under section 403 of Congressional Reorganization Plan No. 3 of 16 May 1946. Previously, all land policy and grazing procedures had been handled by the General Land Office, which was created in 1812, and the U.S. Grazing Service, which came into being in 1934 as the Division of Grazing, founded under the Taylor Grazing Act. At first, the new agency's mandate was the same as that of the old General Land Office: continue to administer the lands and oversee grazing policy. Secretary of the Interior Julius A. Krug chose the last head of the General Land Office, Fred W. Johnson, as temporary BLM commissioner. One of Johnson's first acts was to establish seven BLM districts nationwide where closer administration of the lands could occur. Angry at the apparent lack of congressional oversight such a plan would create, Congress overruled the Johnson plan in the Interior Department appropriations for 1947. Further, to reassure western senators and congressmen that the creation of the Bureau of Land Management meant an end to the continual sale of public land, Congress demanded that the BLM dispose of as much public land as it could in one sale. In a final insult, total BLM staff was cut to 85 people nationwide, whose duty was now to administer 150 million acres of grazing land.

To ease the situation, Krug named Rex L. Nicholson, a California rancher, as a bureau advisor to come up with a land and grazing procedure that was good for public policy while meeting the demands of ranchers. Nicholson's plan called for an increase in BLM personnel to 242, as

well as a small increase in the grazing fees, part of which would be used to administer the lands. Congress passed Nicholson's plan in 1947.

In 1948, Krug replaced Johnson with Marion Clawson, who had served in the BLM's San Francisco office. At once, Clawson asked Congress to decentralize the bureau, allowing it to be closer to the lands it had to administer; later, he reorganized the entire bureau and introduced "new blood" into the programs of soil and moisture conservation, oil and gas leasing, and range management. In 1953, Edward Woozley, the commissioner for state lands in Idaho, replaced Clawson. Key legislation enacted during his tenure included the Outer Continental Shelf Act and the Multiple Mineral Development Act. He also initiated Project 2012, a 50-year program for improving the management of the public domain. Woozley's successors, among them Karl Landstrom (1961–1963) and Curt Berklund (1973–1977), have continued to administer the public lands and grazing areas.

See also General Land Office; Sparks, William Andrew Jackson; Taylor Grazing Act; Tiffin, Edward.

Bureau of Land Management Organic Act (BLMOA)
See Federal Land Policy and Management Act.

Bureau of Mines
This agency within the Department of the Interior was originally established by the Organic Act of 16 May 1910 (30 U.S.C. 1, 3, 5–7). It is, according to the *United States Government Manual 1993–1994*, essentially "a research and factfinding agency." Charged with the mission of guaranteeing that the nation has an adequate supply of important minerals for economic and security needs, it is "a bureau of mining, metallurgy, and mineral technology." It also

gathers, amasses, scrutinizes, and publishes factual and statistical information on "all phases of nonfuel mineral resource development, including exploration, production, shipments, demand, stocks, prices, imports, and exports." The bureau also examines nonfuel alternatives and what impact proposed regulations and legislation might have on mineral resources.

Bureau of Reclamation

This government agency, a part of the Interior Department, was created in 1902 to address the demands of reclamation and irrigation concerns, mainly in the western United States. Before the presidency of Theodore Roosevelt, presidential leadership for the support of federal legislation for reclamation projects was lacking. With Roosevelt's election to the White House, a bill nurtured for years by Sen. Francis G. Newlands of Nevada passed both houses of Congress. On 17 June 1902, Roosevelt signed it into law. This Newlands Reclamation Act created the Reclamation Service, whose first director, known as the chief engineer, was Frederick H. Newell, formerly head of the Hydrographic Branch of the U.S. Geological Survey. Newell's assistant was Arthur Powell Davis, nephew of John Wesley Powell and himself a noted reclamation and irrigation expert. The Reclamation Act authorized the secretary of the interior to survey the western arid lands, draw up programs that would first reclaim and then irrigate these lands, and then provide water for year-round irrigation. The new service worked hand in hand with the Geological Survey, mapping the 17 western states under the program and reporting back to Washington. However, the service was not given sufficient appropriations to do a complete job. By 1907, although 79 sites for irrigation had been surveyed and mapped, there was only enough money to start 25 projects. That year, the Reclamation Service was separated from the Geological Survey as a government entity.

In its first decade, the Reclamation Service was responsible for the creation of several major projects, including the Pathfinder Dam on the Platte River in Wyoming, the Elephant Butte Dam on the Rio Grande, and the Roosevelt Dam on the Salt River in Arizona, as well as the construction of numerous water-storage facilities. Newell, however, came into conflict with Secretary of the Interior Franklin K. Lane over the repayment schedules of farmers who were irrigating their own land. In a showdown, Lane cut Newell's responsibilities, and Newell resigned in May 1915.

Arthur Powell Davis succeeded Newell, but the repayment-schedule controversy soon touched him as well. The 1922–1923 Teapot Dome Affair, which Davis was not involved in, nonetheless claimed his job in a grand housecleaning of the Interior Department. In 1923, Congress renamed the service the Bureau of Reclamation. Davis was replaced first by businessman David William Davis (no relation), then Elwood Mead, a water expert from Wyoming. Before his death in 1936, Mead oversaw the bureau during the throes of the Depression by writing off repayment schedules for farmers whose lands were not profitable. The passage of the Reclamation Project Act of 1939 allowed for a grace period of ten years before payment of reclamation projects was to begin; the enactment of the Wheeler-Case Act of 1939 permitted the employment of Civilian Conservation Corps (CCC) and Works Progress Administration (WPA) laborers on reclamation projects, with a waiver of payment from settlers who were unable to pay.

Today the Bureau of Reclamation is the leading government agency in the business of developing and managing water resources. According to the *United States Government Manual 1993–1994*, the mission of the bureau "is to manage, develop, and protect, for the public welfare, water and related resources in an environmentally and economically sound manner." Further, it "draws up water conservation plans, designs and constructs water resources projects, conducts research into water resource management, and operates facilities for the maximum usage of water resources, namely through its operation of 355 storage reservoirs, 69,400 miles of canals and other water conveyances and distribution facilities, and 52 hydroelectric power plants."

See also Irrigation and Reclamation Policy, National; Newell, Frederick Haynes; Newlands, Francis Griffith.

Burnham, John Bird (1869–1939)
Noted conservationist and environmental author John Bird Burnham is one of the least-known conservationists of the late nineteenth and early twentieth centuries. Born on 16 March 1869 in Newcastle, Delaware, Burnham received his bachelor's degree from Trinity College in Hartford, Connecticut, in 1891. That year, he began working as the business manager for *Field and Stream* magazine and became the protégé of editor George Bird Grinnell. In 1897, he resigned his position at the magazine to head for Alaska and join the Klondike gold rush. When he returned from Alaska, he purchased a farm in Willsboro, New York, which he managed as the Highland Game Preserve. His work on the preserve led to his appointment in 1905 as chief game protector for New York, a post he held for three years. In 1908, he was named to the office of the state commission of fish and game, becoming head of that office in 1911.

On 11 September 1911, Burnham, Frederic Collin Wolcott of Connecticut, and others interested in a sportsmen's group that would lobby for wildlife refuges formed the American Game Protective and Propagation Association (AGPPA). Writes Stephen Fox, "Beginning in 1920, the eastern wildlife establishment annually tried to push through

Congress a game-refuge bill drafted by John Burnham of the American Game Protective Association (AGPA) and Dr. E. W. Nelson of the U.S. Biological Survey." This bill, called the Weeks Act, was enacted by Congress in 1913. As part of the new law, Burnham was named to the Advisory Committee to the Department of Agriculture on Migratory Bird Law; Burnham also served on a three-man commission that worked hand in hand with the state of New York to codify the state's fish and game laws in 1915. He also lobbied for passage of the 1916 treaty between the United States and Canada on migratory birds. He chronicled a 1921 journey to Siberia to collect endangered sheep specimens in *The Rim of Mystery* (1921). John Bird Burnham died on 24 September 1939 at the age of 70.

See also American Game Protective and Propagation Association.

Burroughs, John (1837–1921)
John Burroughs's works are considered classics in nature and conservation study. He was born of old New England stock on 3 April 1837 near Roxbury (once called Beaver Dam), New York. Burroughs spent much of his early life roaming the scenic trails and valleys of New York, thus establishing his love of nature. He later wrote:

> From childhood I was familiar with the homely facts of the barn, and of cattle and horses; the sugar-making in the maple woods in early spring; the work of the corn-field; the delicious fall months, with their pigeon and squirrel shooting, [the] threshing of buckwheat, [the] gathering of apples and burning of fallows; in short, everything that smacked of, and led to, the open air and its exhilarations. I belonged, as I may say, to them; and my substance and taste, as they grew, assimilated them as truly as my body did its food. I loved a few books much, but

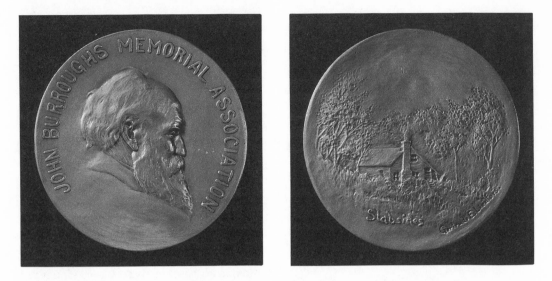

The John Burroughs medal, showing his portrait and his New York home, Slabsides, has been awarded annually since 1926 to honor natural history writers. Winners include Rachel Carson for The Sea around Us (1952).

I loved Nature, in all those material examples and subtle expressions, with a love passing all of the books of the world.

Burroughs's education was spread out among various institutions; he returned to one, Tongore Academy in Ulster County, New York, to teach for eight years. He later held positions as a clerk in the U.S. Department of the Treasury and as a bank examiner. Yet it was his love of nature that led him to declare at age 20, "I shall be an author."

Burroughs soon realized his dream. Scratching out unsigned articles for magazines such as the *Atlantic Monthly* and *Knickerbocker's*, Burroughs eventually combined several works into the novel *Wake Robin* (1871). His work *Winter Sunshine* (1875) dealt with nature he observed while touring England. *Birds and Poets* came out in 1877, followed by *Locusts and Wild Honey* (1879). Burroughs would compose over his lifetime uncounted articles and 27 books.

In 1874, he acquired a small fruit farm in West Park on the Hudson River, where he spent the rest of his life. Noted for his long beard, he was a fixture in Washington, D.C., during the presi-

dency of Theodore Roosevelt. Although after 1908 his output of literary works dwindled to a trickle, Burroughs remained, to some, the quintessential authority on nature writing. Writes one source, "He seem[ed] to be, by turns, an artist, a poet, a naturalist, and a sportsman, always without the least pretense or passion." Five years after his death on 29 March 1921 at the age of 83, the American Museum of Natural History in New York established the John Burroughs Medal for nature writing.

Butler, Ovid McQuat (1880–1960)

Ovid Butler was a noted naturalist and supporter of the conservation of forests. He was born 14 July 1880 in Indianapolis, Indiana, and later attended Butler University (founded by his grandfather, Ovid Butler). After graduation, Butler worked as a newspaperman for three years before earning a master's degree from the Yale University Forest School in 1907.

That same year, Butler became a member of the United States Forest Service, where he was assigned to the Boise National Forest in Idaho. Later, he was sent to the Ogden, Utah, district office, where he advanced to chief of the Division of Forest Management. During World War I, he was instrumental in preparing reports for the federal government dealing with the use of lumber materials for the war effort. His 1917 and 1918 reports, *The Distribution of Softwood Lumber in the Middle West*, brought him wide acclaim in the forestry industry.

In 1922, Butler was named forester of the American Forestry Association; he later served from 1923 to 1948 as executive secretary of the organization and editor of its *American Forests*. In 1924, working with his friend William Buckhout Greeley, Butler lobbied Congress for federal legislation in the area of fire control and reforestation on nonfederal lands. The resulting legislation became the Clarke-McNary Reforestation Act of 1924. At the beginning of the Depression, Butler called for a federal jobs program that would employ the nation's jobless while supplying manpower in forestry and other internal conservationist improvements. President Franklin D. Roosevelt took Butler's suggestion, sending to Congress the idea that became the Civilian Conservation Corps.

A familiar name in the area of conservation for much of his life, Ovid Butler was named executive director emeritus of the American Forestry Association in 1948. He died on 20 February 1960 at the age of 79.

See also Civilian Conservation Corps; Clarke-McNary Reforestation Act; Greeley, William Buckhout.

Cabinet Committee on the Environment
See Executive Order 11472 of 29 May 1969.

Cameron v. United States (252 U.S. 450 [1920])
This Supreme Court case settled the issue of whether a president could sign a proclamation that designated private land as public land reserved within the boundaries of a national park, monument, or forest reserve. Ralph H. Cameron, whom one author called a "canny speculator," took out a mining claim on a rim of the Grand Canyon so that he could charge tourists for the view. The matter was brought to the White House's attention, and in 1908, Theodore Roosevelt, without waiting for congressional authority, established the Grand Canyon National Monument, effectively canceling Cameron's claim. Cameron sued the United States government on the grounds that the government had stepped on his rights by making his claim invalid. First a district court in Arizona, and then the Ninth Circuit Court of Appeals, found for the government. Cameron sued to the U.S. Supreme Court. On 19 April 1920, the Court held unanimously that the president had the right to establish the national monument under the authority granted to him by the Creative Act of 3 March 1891, thereby invalidating Cameron's claim. Justice Willis Van Devanter wrote the Court's opinion, in which he approved of Roosevelt's action in protecting an area that was "an object of unusual scientific interest."

Capper Report (1920)
This government report was sponsored in the Senate by Arthur Capper and pre-pared by Assistant Chief Forester Earle Hart Clapp. Capper's resolution no. 311 called for a detailed report to Congress on the nation's forestry reserves. Clapp's report, officially called *Timber Depletion, Lumber Prices, Lumber Exports, and Concentration of Timber Ownership*, came to three basic conclusions: (1) that three-fifths of the original timber of the United States was gone (in 1920) and that the nation was using timber four times faster than it was being reforested; (2) that the depletion of timber was not the sole cause of post–World War I price rises, but was an important factor; and (3) that the solution to the overall problem was to increase the production of lumber stockpiles while looking for an end to deforestation. With aid from forestry expert Raphael Zon, Clapp called for the passage of legislation that would achieve these ends as well as the production of an all-encompassing governmental report on the nation's wild resources. His latter suggestion was not heeded until 1933, with the issuance of the Copeland Report.

See also Clapp, Earle Hart; Copeland Report; Zon, Raphael.

Carey, Joseph Maull (1845–1924)
Joseph M. Carey, who served as governor and U.S. senator from Wyoming, was the sponsor of the Carey Irrigation Act, one of the most important land-use acts in U.S. history. Carey was born on 19 January 1845 in Milton, Delaware, the son of Robert Hood Carey, a merchant, and Susan (nee Davis) Carey. After graduating from Union College in Schenectady, New York, and the University of Pennsylvania Law School, Carey was admitted to the Pennsylvania bar.

After practicing law for two years, Carey was appointed in 1869 by President

Judge Joseph Maull Carey

Ulysses S. Grant as the first U.S. attorney for the newly established territory of Wyoming. Two years later, he was selected as an associate justice of the Wyoming Territorial Supreme Court. He served in this latter position until 1876. Carey also served as mayor of Cheyenne from 1881 to 1885. In 1885, he organized the Wyoming Development Company, which attempted to use diverted waters for irrigation in order to create cities to which railroads would be connected. He was also the founder in 1891 of the National Irrigation Congress.

In 1885, Carey was elected Wyoming's territorial delegate to Congress and served as such until 1890. He made his mark in the House with the passage of the Wyoming Statehood Act, which made Wyoming the forty-fourth state on 10 July 1890. The Wyoming legislature then elected Carey as the state's first U.S. senator. It was in the Senate that Carey pushed for the establishment of an irriga-

tion law that could help western states like Wyoming prosper and grow. His bill, introduced in 1894, easily passed the House and Senate and was signed into law on 18 August 1894 as the Carey Irrigation Act. It was Carey's key accomplishment as senator. The following year, the legislature refused to reelect him, and he left public office to resume business and law practices in Wyoming. His final public service, however, came in 1910, when he was elected governor of Wyoming, serving until 1915. In that office, he lobbied for the direct election of senators and increased agricultural development as the way to state prosperity.

Although a staunch Republican, Carey joined the Progressive walkout of the party in 1912 and supported Theodore Roosevelt's bid for the presidency that year. Refusing to run for a second term as governor, Carey returned to private life. His son later served as governor from 1919 until 1923. Joseph M. Carey died at his home in Cheyenne, Wyoming, on 5 February 1924.

See also Carey Irrigation Act; National Irrigation Congress.

Carey Irrigation Act (28 Stat. 422)

This federal legislation was enacted on 18 August 1894 to give the president of the United States the power to allow each state to sell no more than 1 million acres of public lands in each state for settlement and irrigation. Named after its chief sponsor, Sen. Joseph Maull Carey of Wyoming, this law was enacted with the goal "that to aid the public land, States in the reclamation of the desert lands within, and the settlement, cultivation and sale thereof in small tracts to actual settlers, the Secretary of the Interior with the approval of the President ... hereby is ... authorized and empowered to contract and agree ... to donate, grant and patent to the State ... not exceeding one million acres ... to be irrigated, reclaimed [and] occupied, within ten years after the passage of this act."

The shortcomings of this act were apparent almost from the start. Because of an 1890 land speculation scandal in the West, legislation was enacted that cut in half, from 640 to 320 acres, the size of tracts settlers could obtain. By 1902, even though 1 million acres had been mandated, less than 8,000 had been achieved. Although the Carey Act was not as important as the Newlands Reclamation Act of 1902, it did open many western cities to development through the use of irrigation projects.

See also Carey, Joseph Maull.

Carhart, Arthur Hawthorne (1892–1978)

Writes forestry historian Dennis Roth, "In the first half of the twentieth century, Aldo Leopold, Arthur H. Carhart, and Robert Marshall designed and helped implement a wilderness policy for the U.S. Forest Service." Barely known outside of the conservation community, Carhart was a landscape artist when he was hired in 1919 to work on the recreational betterment of national forests. He was born on 28 September 1892 in Mapleton, Iowa, and attended Iowa State College, where he earned a bachelor's degree in landscape engineering in 1916. Three years later, after serving in the U.S. Army during World War I, he was commissioned by the U.S. Forest Service to survey the Trappes Lake area of Colorado's White River National Forest and report whether it was suitable for homesites. His analysis concluded that the area should remain free of all commercial and residential development. His land-use studies of the possibilities of recreational development in national forest areas in Colorado and Minnesota made him an expert in that area. He later served with the Colorado Fish and Game Department as director of the funds derived from the federal Pittman-Robertson Act. Historian Frank Harmon wrote of Carhart, "He was one of the first persons in the United States to formulate and

apply on the ground practical principles to safeguard the wilderness values of forestlands. As the first landscape artist employed by the Forest Service and one of the first by the U.S. government, he drew up plans ... for protecting large outstanding scenic areas of the National Forests in Colorado and Minnesota from commercial development and motorized vehicles." Adds Forest Service historian Terry West, "Carhart saw the wilderness as a recreational experience." After 1960, Carhart was the manager of the Conservation Center Library at the Denver Public Library. He died on 27 November 1978 at the age of 86.

Carr, Archie Fairly, Jr. (1910–1987)

Carr was a noted naturalist and authority on snakes and amphibians who spent the last three decades of his life trying to save the various species of endangered sea turtle from extinction. Born in Mobile, Alabama, in 1910, Carr took his first trip to Florida in 1924 in the backseat of an old Dodge touring car and began a lifelong love affair with nature. During the Depression, he joined the staff of the University of Florida at Gainesville as an English teacher. His love of nature led him to change his subject to biology, and he soon became an expert in that area. He was awarded a bachelor's degree in biology from the university in 1932 and a master's degree in 1934. In 1937, he received the university's first doctorate in biology. Over the years, Carr became renowned as an authority on reptiles and amphibians, becoming known as a "herp," or herpetologist. He authored 11 books, including *Handbook for Turtles* and *High Jungles and Low*.

In the 1950s, Carr became aware of the drastic situation facing the several species of sea turtles that inhabited much of the Caribbean and southern United States. Exploited by humans for hundreds of years for their shells, eggs, and meat, almost all the species were faced with extinction. Many sanctuaries where the turtles had laid their eggs for centuries were becoming slaughtering areas. Following the publication of his *Windward Road*, which described in detail the plight of the sea turtle, Carr began to investigate a beach in Costa Rica where he had heard that the turtles still nested. This site, now known as Tortuguero, is the last major sea turtle nesting area in the Caribbean. Because of Carr's advocacy over the next 30 years, he convinced several governments, including that of the United States, to stop the slaughter of the turtles and to ban their importation and use for jewelry and trinkets. Because of his efforts and the continuing work of his son, Archie Carr, Jr., the outlook for the survival of the sea turtle is brighter than ever.

Archie Carr died at his home on Wewa Pond near the town of Micapony, Florida, on 21 May 1987.

Carson, Rachel Louise (1907–1964)

More than any person in the last 50 years, Rachel Carson, author, conservationist, and scientist, whose landmark work *Silent Spring* (1962) called attention to the dangers of pesticides, especially DDT, is responsible for the modern environmental movement. She was born on 27 May 1907 in Springdale, Pennsylvania, and attended the local public schools. After graduating from Parnassus High School, she attended the Pennsylvania College of Women at Pittsburgh, where she studied to be a writer. Although her major was English composition, she became fascinated with biology. She did her graduate work at Johns Hopkins University in that field, graduating with a master's degree in biology in 1932. The year before, she had become a member of the biology department at the University of Maryland, where she taught until 1936.

In 1936, after completing some postgraduate course work at the Woods Hole Marine Laboratory in Massachusetts, Carson was hired as an aquatic biologist with the U.S. Bureau of Fisheries in Washington, D.C. She stayed on until 1940, when Congress merged the Bureau of Fisheries and the Bureau of Biological Survey to create the U.S. Fish and Wildlife Service. Carson's informative article on the marine world, "Undersea," which appeared in the September 1937 issue of *Atlantic* magazine, led to her first book, *Under the Sea Wind* (1941), which in turn led to her appointment as chief of the Fish and Wildlife Service in 1949. Her time there was short; she resigned in 1952 to devote herself to writing.

In 1951, Carson published *The Sea Around Us*, for which she was highly praised and received the National Book Award. This work remained on the *New York Times* best-seller list for 86 straight weeks—a record. Her 1955 work *The Edge of the Sea* was based on her trip in a fishing trawler off the coast of Massachusetts. Her biggest and most controversial work lay ahead, however.

In the 1940s, scientists began to work with a new chemical called dichlorodiphenyltrichloroethane, known as DDT. At the time, it was hoped that DDT could safely kill lice on people as well as insects on food. Author Edwin Way Teale sounded the first alarm against DDT in an article in the March 1945 *Nature* magazine. Pesticides had become a part of modern society. By 1955, 600 million pounds—about 5 pounds for every man, woman, and child in the United States at that time—were being used on the crops and fields of the nation. By the early 1960s, however, after intense spraying worldwide, particularly of DDT, reports of malformed animals and their eggs began to surface, and the alarm on DDT's hazards was quietly sounded. An activist against all indiscriminate pesticide spraying, Carson began to write her epic, *Silent Spring*. When it was published in 1962, it started a heated battle between conservationists, who hailed her book, and the chemical industry, which denounced it. As the landmark work *Man and Nature* by

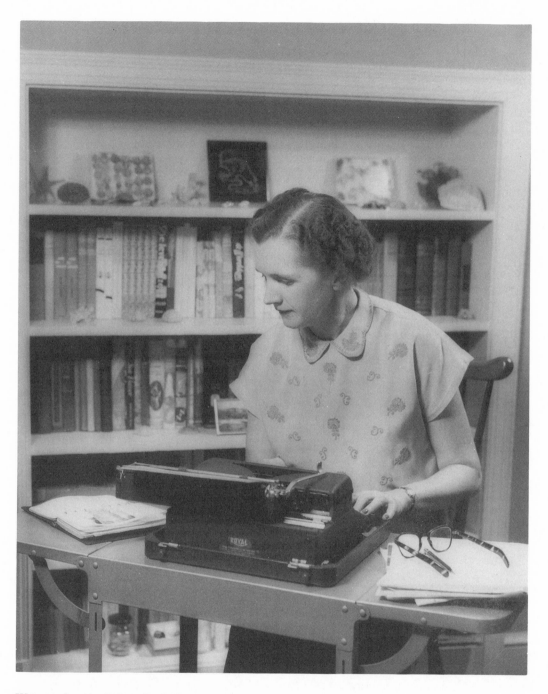

Writer and marine biologist Rachel Louise Carson

George Perkins Marsh a hundred years earlier had shocked the nation into the conservation movement, Carson's particular warning against 12 of the most harmful chemical poisons gave rise to the modern environmental movement. Hers was not the only book of this genre, however. In 1962, Murray Bookchin's *Our Synthetic Environment* exposed the growing danger humans faced from the use of pesticides. But Bookchin's book was not written for the layperson, and it did not receive the attention it should have. Carson's flowery style combined with down-to-earth writing gave *Silent Spring* the push *Our Synthetic Environment* never had. In her work, Carson discussed what she called not pesticides, killers of pests, but biocides, killers of the earth, which had to be carefully regulated and in some cases banned altogther.

On 3 April 1963, CBS broadcast "CBS Reports: The Silent Spring of Rachel Carson," in which the famed writer said, "Now, I truly believe that we in this generation must come to terms with nature, and I think we're challenged as mankind has never been challenged before to prove our maturity and our mastery, not of nature, but of ourselves." At the time she spoke these words, Rachel Carson knew that she was dying of cancer. She succumbed to the disease on 14 April 1964, at the age of 56. Said former Interior Secretary Stewart L. Udall, "A great book has a flow to it, and it changes people's minds; it changes their outlook, and it has a long reach. Rachel Carson's *Silent Spring* is still affecting our thinking and policymaking today. The essence of Rachel's message was that we had to come to terms with nature—to work with it, and not against it. In a complex, modern society, of course that was a radical concept. But in a sense of a change in our thought, I think she was a revolutionary."

See also DDT; Teale, Edwin Way.

Carter, Thomas Henry (1854–1911)

Carter was a U.S. senator from Montana and served as commissioner of the General Land Office. He was born on 30 October 1854 in Scioto County, Ohio, of Irish parents, Edward C. Carter and Margaret Byrnes, who had both emigrated from Ireland and met in Virginia. In 1865, the Carter family moved from Ohio to Christian County, Illinois, where Thomas grew up on a farm. According to his biography, Carter "engaged in railroading and school teaching" as professions. In 1878, he left Illinois and went to Burlington, Iowa, where he began the study of the law. To pay expenses, he sold a book, *The Footprints of Time: A Complete Analysis of Our System of Government.* Four years later, he went to Helena, in the Montana Territory, where he started up a law practice.

In 1888, Carter was nominated by the Republicans of the Montana Territory to be their territorial representative in the U.S. House of Representatives. After being elected, he went to Washington and oversaw Montana's entry into the Union as the forty-first state on 8 November 1889. Carter was thus Montana's last territorial representative and its first state representative. In 1890, he was swept out of office by the Democratic landslide that year, but in appreciation of his party loyalty, President Benjamin Harrison named him commissioner of the General Land Office. Carter's appointment was welcome in Montana, where land policy had been dictated by politicians in far-off Washington who had little contact with the West. Carter secured the opening of the western section of the Crow Indian reservation for development, which opened 100,000 acres of territory for homesteaders. For these new homesteaders, Carter was able to get an exemption of the $1.50 an acre payment, thus saving the settlers a total of $2 million. He was able to convince the Indians of Fort Belknap to lease 20,000 acres to these new settlers for the production of sugar beets and other crops. Further, he instituted new regulations in the areas of forest harvesting and strip mining. In 1892, Carter left the General Land Office to head the Republican National Committee. In 1893, he retired to Montana to take up his old law practice but was elected by the state legislature to the U.S. Senate in 1895. Although he lost the seat in 1901, he was elected again in 1905, replacing the conservation-minded Paris Gibson, and served in the Senate until 1911. As a senator, he pushed for improvements in Yellowstone

National Park and the creation of Glacier National Park. In 1911, President William Howard Taft named Carter chairman of the U.S. section of the Joint International Committee, created to settle water disputes between the United States and Canada. Carter was the head of the commission for only a few months when he died on 17 September 1911 at the age of 56.

Catesby, Mark (1679? or 1683?–1749)

Although naturalist Mark Catesby spent virtually his entire life in his native England, the few years that he did spend in the colonies, from about 1710 to 1719 and again from 1722 to 1726, were used to assemble his landmark *Natural History of Carolina, Florida, and the Bahama Islands, with Observations on the Soil, Air, and Water* (two volumes with illustrations and text in French [1731–1743] and English [1729–1747]), possibly the first in-depth work on the nature of the New World. Catesby was born in Castle Hedington, Essex, England. One source, *The Biographical Dictionary of American Science*, claims his birth date as 3 April 1683, but *The Dictionary of National Biography* gives it as 1679 with a question mark and asserts London as his probable birthplace. His parents were John Catesby, a lawyer and mayor of nearby Sudbury, Suffolk, and Anne (nee Jekyll) Catesby. The former source reports that Mark Catesby's education was probably limited to private tutoring at Castle Hedington; however, he was schooled in Latin.

His father's wealth allowed Catesby to travel to the colonies to study botany in about 1710 (it may have been as late as 1712). Until 1719, he traveled in Virginia, the Carolinas, Florida, the Bahamas, and Bermuda, where he collected plant specimens and had them sent back to England for further study. These plants have been characterized as "the most perfect which had ever been brought" to England. Catesby returned to his homeland in 1719 to arrange his

specimens, which included ornithological samples. With the aid and encouragement of British scientists such as Sir Hans Sloane, Catesby returned to the colonies in 1722 for a four-year sojourn. In 1726, upon his return to England, he settled in Hoxton and dedicated the rest of his life to assembling his collection into a printed and illustrated manuscript. The result was the aforementioned *Natural History*, which contained 220 illustrations. A German edition, with an introduction by someone identified as "M. Edwards du College Royal des Medecins de Londres," was released in 1756. New English editions were released in 1754 and 1771; Catesby's biographers, George F. Frick and Raymond P. Stearns, rereleased it in 1974 with a new introduction. The first volume "was largely devoted to ornithology with botanical backgrounds, while the second volume had elements of cartography [mapmaking], geology, anthropology, and zoology, as well as botany," reports editor Clark Elliott in his *Biographical Dictionary of American Science*, one of the few American sources on Catesby.

Although best known for this single work, Catesby also produced *Hortus Britanno-Americanus, or a Collection of 85 Curious Trees and Shrubs, the Production of North America, Adapted to the Climate and Soil of Great Britain* (1737?), and, after a period of living on the Isle of Providence, *Piscium, Serpentum, Insectorum aliorumque nonnullorum Animalum, nec non Plantarum quarundam, Imagines,* which was published in Nuremberg, Germany, in 1777. For his lifetime of work, the West Indian shrub *Catesbaea* was named in his honor. Catesby died at his home in London on 23 December 1749. His biographers, Frick and Stearns, have called the mostly unknown Catesby a "colonial Audubon."

Catlin, George (1796–1872)

George Catlin is one of the best-known artists, along with Frederic Remington,

Nineteenth-century artist George Catlin documented a plains Indian hunting a bison.

who preserved the history of the American West in its pristine state. Born in Wilkes-Barre, Pennsylvania, on 26 July 1796, Catlin was the fifth of 14 children of Putnam and Polly (nee Sutton) Catlin. George Catlin's great-great-great-grandfather, Thomas Catlin, emigrated from England in 1644 and settled in Hartford, Connecticut. Putnam Catlin was born in Litchfield, Connecticut, in 1764 and served in the Revolutionary War (he enlisted at age 13) as a fife-major of the Second Connecticut Regiment; his father, Eli Catlin, was a lieutenant in the same company. Polly Sutton Catlin had been captured by the Indians as a small girl, and she regaled her son George with wild tales of the Native Americans. It was these stories and a growing love of nature that marked George's life. He spent his childhood, as he wrote later, with "books reluctantly held in one hand, and a rifle or fishing pole firmly and affectionately grasped in the other."

George Catlin's education was ob-

tained largely at home. According to one source, "In 1817–1818 he read law in the office of Reeves and Gould in Litchfield, Connecticut, where he became celebrated as an amateur artist." Although it is not known when he was admitted to the bar, he did begin to practice law in Pennsylvania in about 1820. This profession, however, was not to be his life's vocation. Instead, he turned to painting. As he wrote years later, "During this time, another and stronger passion was getting the advantage of me, that of painting, to which all my pleading soon gave way." He slowly converted his law office into an art center. Writes Peter Wild, "Finally, he indulged [in] his new passion. Selling lawbooks and other possessions— but keeping his rifle and fishing pole— Catlin set off for Philadelphia, at the time the nation's cultural center."

For the next several years, Catlin painted mostly political figures, including New York Gov. DeWitt Clinton and First Lady Dolley Madison. It was not until his attention was drawn to a delegation of Indians visiting New York that he decided to make it his life's work to capture the lives and portraits of the Native Americans on canvas. From 1829, when he began his forays into Indian territory, until 1838, he painted over 600 Indians, as well as collecting Indian artifacts such as costumes and jewelry. As part of his cataloging of Indian life, he wrote a series of articles, *Notes of Eight Years' Travel amongst the North American Indians*, which appeared in the *New York Commercial Advertiser* from 1830 to 1839. He also wrote *Letters and Notes on the Manners, Customs, and Condition of the North American Indians* (1841), *Catlin's North American Indian Portfolio: Hunting, Rocky Mountains and Prairies of America* (1845), *Catlin's Notes of Eight Years' Travels and Residence in Europe* (1848), *Life among the Indians* (1867), and *Last Rambles amongst the Indians of the Rocky Mountains and the Andes* (1867).

George Catlin was interested not only in the preservation of Indian life but also in the conservation of the American West as he saw and documented its people and landscape. He expressed this interest in his articles in the *Commercial Advertiser*, in which he called for the creation of a "nation's Park, containing man and beast, in all the wild and freshness of their nature's beauty." This concern was also exhibited in his 1841 work, *Letters and Notes on the Manners, Customs, and Condition of the North American Indians*, a series of diary entries on his travels. In one such entry, he wrote:

Many are the rudenesses and wilds in Nature's works, which are destined to fall before the deadly axe and desolating hands of cultivating men; and so amongst her ranks of *living*, of beast and human, we often find noble stamps, or beautiful colours, to which our admiration clings; and even in the overwhelming march of civilised improvements and refinements do we love to cherish

their existence, and lend our efforts to preserve them in their primitive rudeness. Such of Nature's works are always worthy of our preservation and protection; and the further we become separated (and the face of the country) from that pristine wildness and beauty, the more pleasure does the mind of enlightened man feel in recurring to those scenes, when he can have them preserved for his eyes and his mind to dwell upon.

"For all his early adventures, the last two decades of the artist's life were the most bizarre as Catlin seesawed between bouts of depression and bursts of enthusiasm," wrote Peter Wild. The deaths of his wife and several children from pneumonia, as well as the destruction of Native American culture, slowly drove him insane. The failure of the scientific community to recognize his documentation of Indian culture was also a blow. Forced to sell his collection of Indian artifacts, Catlin lived a solitary life until his death at his daughter's home in Jersey City, New Jersey, on 23 December 1872 at the age of 76. Following his death, many of his paintings were sent to the Smithsonian Institution, where Thomas Donaldson, a member of the Public Lands Commission of 1879, collated them and reported on them in the 1886 annual report for the Smithsonian. George Catlin is now considered one of the greatest documentarians of the American West.

Chandler, Zachariah (1813–1879)

One of Michigan's most distinguished politicians of the nineteenth century, Zachariah Chandler went on to become a U.S. senator and secretary of the interior. The son of Samuel and Margaret (nee Orr) Chandler, Zachariah Chandler was born in Bedford, New Hampshire, on 10 December 1813. He received a common school education, including instruction at an academy, but he never attended college. At age 20, he moved to Detroit, where he began a dry-goods retail business. During the next several years, he branched out into what one source termed "trade, banking, and land speculation"; within a few years, Chandler was a wealthy man.

Chandler's wealth brought calls for him to enter politics, but it was not until 1848, when he spoke out in favor of Zachary Taylor for president, that he entered the political arena. A delegate to the Whig state convention in 1850, he was elected mayor of Detroit the following year. In 1852, he was the Whig candidate for governor but was defeated. Two years later, disgusted by his party's stand on slavery, he was one of several Michigan political leaders to form the state Republican Party. A delegate to the first Republican National Convention in 1856, Chandler was elected to the party's national committee.

In 1857, the Michigan state legislature elected Chandler to the U.S. Senate to fill the seat of Lewis Cass. He held the seat until 1875. In those years, he used the Senate as a bully pulpit to become one of the leading politicians in the nation. An outspoken foe of slavery, he nonetheless was a personal adversary of Sen. Charles Sumner, who was most noted for authorship of the Civil Rights Act of 1875. As chairman of the Senate Committee on Commerce, Chandler favored internal improvements, such as roads and canals. During the Civil War, he was a radical Republican, decrying any lackluster performance by a Union general and calling for the execution of former rebels. His ill temper during the war earned him the sobriquet "Xanthippe in pants," a reference to a shrew. His domineering attitude toward the Republican Party and the Michigan political apparatus led to his defeat by Democrats in the 1875 elections. He left the Senate on 3 March of that year and returned to Michigan.

On 30 September 1875, Secretary of the Interior Columbus Delano resigned.

On 19 October, President Ulysses S. Grant named Chandler as his replacement. Hamilton Fish, Grant's secretary of state, thought that the Chandler nomination was a serious mistake. He believed, according to his biographer, that although Chandler was "incorruptible and generous ... [he] was violent of temper, hasty in judgment, fond of excitement, adverse to labor, and surrounded by the worst type of politicans." Chandler served only until 4 March 1877. He spent much of his tenure working behind the scenes as Republican national chairman to elect Rutherford B. Hayes president in 1876. Chandler had little or no impact on Interior Department policies, and with the change in administrations, he left office. He did not remain out of politics for long, however; just four years later, the resignation of Sen. Isaac P. Christiancy led the Michigan legislature to once again send Zachariah Chandler to Washington and the U.S. Senate. Chandler had served just nine months of this term when he suddenly died on 1 November 1879 in Washington at the age of 65.

See also Department of the Interior.

Chapman, Frank Michler (1864–1945)

Frank Chapman was a noted ornithologist and one of the founders of the National Audubon Society. He was born in Englewood, New Jersey, on 12 June 1864, the son of a Wall Street attorney. The wilds of the New Jersey countryside beckoned Frank Chapman in his youth, and he soon began to study nature more closely. His education was limited to 10 years at the Englewood Academy and one year at a public school in Baltimore, where the family settled following his father's death in 1876. Chapman never earned a degree; he was awarded an honorary degree in 1913 from Brown University.

Although his first job was as a debt collector for the American Exchange Bank in

New York City, Chapman spent his off-hours collecting and cataloging ornithological material. In 1884, he presented his first report to the American Ornithologists' Union, and a year later he became an associate member of that group. An 1886 trip to Florida, where he collected bird specimens that later made their way into the collections of the American Museum of Natural History in New York, gave him the knowledge to apply for a job as a staff member at the museum in 1888. This began a 54-year period during which Chapman worked exclusively for the museum. He was made a member of the ornithology and mammology department, eventually becoming assistant curator of the department in 1901, curator of birds in 1908, and chairman of the bird department in 1920, a post he held until his retirement in 1942. In 1899, Chapman started, and served for 36 years as editor of, the ornithological journal *Bird-Lore*, using the pages of the magazine to call for the protection of wild birds, which were being slaughtered by the millions to supply the ladies' hat industry with plummage. A Republican, Chapman lobbied President Theodore Roosevelt in 1903 to set aside Pelican Island in Florida as the nation's first federally protected refuge for wild birds. Pelican Island later became the first such refuge in the federal government's national wildlife refuge system. In 1905, Chapman was one of the founding members of the National Audubon Society, and in 1935 he gave *Bird-Lore* to the society, which now publishes it as *Audubon* magazine.

Various honors, including the John Burroughs Medal for his writing and the Brewster Medal of the American Ornithologists' Union for his life's work, were bestowed on Chapman. His published works include *Bird Studies with a Camera* (1900) and *Camps and Cruises with an Ornithologist* (1908). Chapman retired from active work in 1942 and died three years later, on 15 November 1945, of uremic poisoning.

See also Roosevelt, Theodore.

Chapman, Oscar Littleton
(1896–1978)

Oscar Chapman was secretary of the interior, serving from 1949 to 1953 in the Truman administration. He was born, according to his own writing, "under a tobacco plant" on 22 October 1896 in the town of Omega, Virginia. He attended the one-room schoolhouse in Omega, where his idol was Abraham Lincoln. In the former Confederate state, such devotion cost him a suspension from school.

Chapman graduated in 1918 from Randolph-Macon Academy in Bedford, Virginia. Immediately, he enlisted as a seaman in the U.S. Navy. He spent 18 months in the service before contracting tuberculosis; he was then sent to a veterans' hospital in Denver, Colorado, to recover. Following his recovery, Chapman took up law classes at the University of Denver while working as an officer in a juvenile court. Although he studied law at the University of New Mexico for a time, Chapman returned to Denver to earn his law degree in 1929 and subsequently joined the law office of Edward P. Costigan, a top Denver attorney. In 1930, Chapman managed Costigan's successful run for the U.S. Senate. It was Chapman's first taste of politics.

As he became more well known in Democratic Party circles, Chapman founded the Spanish-American League to fight the exploitation of Mexican labor on U.S. beet farms. In 1932, he managed the successful U.S. Senate run of Alva B. Adams and served as a delegate to the Democratic National Convention. With the election of Franklin D. Roosevelt to the White House, Costigan and Adams pushed to have their young protégé named to a government post. Roosevelt obliged and appointed Chapman assistant secretary of the interior under Harold L. Ickes. Chapman would serve in this post for 13 years—the longest on record. He served for a time as acting secretary (February–March 1946), then undersecretary to Julius A. Krug. In 1949, upon Krug's resignation, Chapman was elevated in his place to become interior secretary. As secretary, he was instrumental in pushing conservation programs and calling for the protection of the nation's natural resources. In 1953, with the election of Republican Dwight D. Eisenhower to the presidency, Chapman left the government. He founded the Washington, D.C., law firm of Chapman, Duff and Paul, which specialized in energy and trade law. Chapman worked right up until the summer of 1977, when his health began to fail. He died in Washington on 8 February 1978 at the age of 81.

See also Department of the Interior.

Chemical Emergency Preparedness Program (CEPP)

See Emergency Planning and Community Right-to-Know Act.

Chemical Waste Management, Inc. v. Hunt et al. (119 L Ed 2d 121 [1992])

In this litigation before the U.S. Supreme Court, it was held that a state may not impose higher fees on hazardous waste coming from outside as opposed to inside the state; such higher fees were found to be unconstitutional. The Alabama legislature enacted a law that called for a fee of $26.50 per ton for the disposal of all hazardous waste in the state, with an added assessment of $72 for waste generated out of state. Chemical Waste Management, Inc., a commercial landfill facility, sued Gov. Guy Hunt and others to overturn the law. The original trial court found that such an extra fee violated Article 1, section 8, clause 3 of the U.S. Constitution, which is known as the commerce clause. When the state appealed, the Alabama Supreme Court reversed the judgment, holding that "the additional fee advanced legitimate local purposes that could not be adequately served by reasonable nondiscriminatory purposes." On certiorari, the Supreme Court struck down the law on 1 June

1992 by a vote of 8–1 (Chief Justice William Rehnquist dissenting), finding that the fee did violate the commerce clause. In an opinion written by Justice Byron White, the Court ruled that "no state may attempt to isolate itself from a problem (the generation of waste) common to the several states by raising barriers to the free flow of interstate trade."

Civilian Conservation Corps (CCC)

Known as the CCC, the Civilian Conservation Corps was a New Deal agency set up under the Unemployment Relief Act of 31 March 1933 to cure the nation's unemployment problem. On 21 March 1933, newly inaugurated President Franklin D. Roosevelt asked Congress to enact legislation to use unemployed workers to improve the natural infrastructure of the nation. Although originally called "Emergency Conservation Work," it was soon dubbed the Civilian Conservation Corps (CCC) as part of Roosevelt's "alphabet soup" of New Deal agencies. The name was officially changed as part of the Conservation Corps Act of 28 June 1937. Two men headed the agency from its inception until its demise in 1942. The first was Robert Fechner (1876–1939), a labor union official who steadfastly held that although blacks would be allowed to work in the Corps, they must serve in segregated units. Many historians cite this as the reason that so few blacks, suffering as did all Americans from the Great Depression, enrolled in the CCC. With Fechner's death on 31 December 1939, James J. McEntee (1884–1957), a machinists' union official who had served as Fechner's assistant, became the agency head.

Housed in camps across the United States, CCC workers, numbering some 2.5 million (including 225,000 World War I veterans and 15,000 American Indians) over nine years, did conservation work for $30 a month. In total, 2.25 billion trees were planted, 6 million erosion

Members of the Civilian Conservation Corps, organized in 1933 during President Franklin Roosevelt's administration, worked on forests, parks, and reclamation projects until 1942. Two men roll a log in a timber clearing operation near Shasta Reservoir, California, in 1940.

dams were constructed, countless reservoirs were cleaned and repaired, and states were assisted in topsoil recovery, fire fighting in forests, and reforestation.

With the end of the Depression nearing and World War II beginning in Europe, Congress became wary of investing any more funds in the CCC. In 1942, it allocated only enough money to close down the camps and disperse the workers. That June, the CCC ceased to exist.

Clagett-Pomeroy Act
See Yellowstone National Park Act.

Clapp, Earle Hart (1877–1970)
Earle Clapp was the forestry expert whose 1933 report, *A National Plan for American Forestry*, became the basis for federal forestry policy for many years.

Clapp was born in North Rush, New York, on 15 October 1877. He attended local schools, then decided on a career in forestry when he attended Cornell University from 1902 to 1903. He finished his education at the University of Michigan, where he received his bachelor's degree in forestry in 1905.

Clapp was hired right after graduation to serve the U.S. Forest Service as a timber surveyor in the Medicine Bow Forest Reserve. Later, he was sent to the service's Washington State office. From 1908 until 1942, when he retired, Clapp moved up the ladder in the southwest district office from forest assistant to chief. In 1920, he prepared *Timber Depletion, Lumber Prices, Lumber Exports, and Concentration of Timber Ownership*, a report known as the Capper Report. His 1926 report, *A National Program of Forest Research*, was the basis for the passage of the McSweeney-McNary Act of 1928.

51

His landmark 1933 work, *A National Plan for American Forestry*, also known as the Copeland Report, was used by the federal government for many years in the area of forestry planning. Among his many honors, Clapp was awarded the Gifford Pinchot Medal by the Society of American Foresters. He died on 2 July 1970 at the age of 92.

See also Capper Report; Copeland Report; Mc-Sweeney-McNary Forest Research Act.

Clark, William Patrick (1931–)

A key member of the Reagan administration in its early years, William P. Clark served as undersecretary of state and secretary of the interior from 1982 to 1985. He was born in Oxnard, California, on 23 October 1931, the son of William Pettit Clark, a cattleman, and Bernice (nee Gregory) Clark. William Clark attended Villanova Preparatory School in Ojai, California, before entering Stanford University. Although he left Stanford to enroll at the Augustinian Novitiate in New Hamburg, New York, in 1950 to study for the priesthood, he eventually returned to Stanford and finally wound up at the University of Santa Clara and Loyola Law School. He never earned a law degree but did take the bar examination and was admitted to the bar in 1958.

In December 1953, Clark was drafted by the U.S. Army and spent two years in Germany as a counterintelligence agent. There he met Joan Brauner, a Czechoslovakian refugee. In 1955, the two were married in Switzerland. When he returned to the United States, Clark took night classes for three years before passing the bar exam in 1958. Soon after, he became the senior partner of the firm of Clark, Cole & Fairfield, where he stayed until 1966. At the time, William Clark was a Democrat. However, after becoming disenchanted with the administration of California Gov. Edmund G. Brown, he met former actor Ronald Reagan and joined the Republican Party.

According to Clark's mother, her son was "so far right we can't even discuss politics."

Clark's career then rose with Reagan's. In 1966, he served as a close advisor in Reagan's successful gubernatorial campaign and was named the new governor's chief of staff. Beginning in 1969, Reagan named Clark to a succession of court seats, starting with the California Superior Court, the state Court of Appeals, and, finally, in 1972, associate justice of the California Supreme Court. A conservative, Clark clashed with the more liberal members of that court, but according to one newspaper, his "honesty and integrity" made him a leading member of Reagan's entourage.

With Reagan's election to the White House in 1980, Clark was selected as deputy secretary of the state, the number-two position in that department. Although he conceded at his confirmation hearings that he "had no formal training in foreign policy," he was confirmed. Clark served in this post for less than a year. After the resignation of National Security Advisor Richard V. Allen on 4 January 1982, Reagan named Clark assistant to the president for national security affairs. As a member of the National Security Council, Clark was at the core of U.S. foreign policy decisions. The *Wall Street Journal* described his "ol' boy combination of sagacity and affability," and *Newsweek* reported several months after his appointment that "after a year of disarray ... the National Security Council is quietly being restored to the lofty status it enjoyed under Henry Kissinger and Zbigniew Brzezinski."

On 8 November 1983, Secretary of the Interior James G. Watt resigned after nearly three years as perhaps the most controversial interior secretary since Albert B. Fall. Ten days later, Reagan named Clark as Watt's replacement. He served a little more than 15 months. Among his accomplishments were, according to the *New York Times*, "restoring communications with Congress, conservationists, and other outside groups ... changes in coal and offshore oil leasing programs ... [and] a return to [the] acquisition of National Park and wildlife refuge land after a moratorium imposed by Watt." On 1 January 1985, Clark announced his resignation to return to private life and his ranch near San Luis Obispo, California, adding that he intended to stay on until his replacement was sworn in. A month later, he left when he was succeeded by Donald Hodel.

See also Department of the Interior.

Clarke-McNary Reforestation Act (16 U.S.C. 567 [repealed], 568)

This federal legislation was enacted into law on 7 June 1924 and called for increased cooperation between the secretary of agriculture and the states to procure forest tree seeds and plants to regrow areas of forest that had been cut down or burned. Officially called "an Act to provide for the protection of forestlands, for the reforestation of denuded areas, for the extension of national forests, and for other purposes, in order to promote the continuous production of timber on lands chiefly suitable therefor," the bill was cosponsored by Rep. John D. Clarke of New York and Sen. Charles L. McNary of Oregon. Part of the act was repealed on 1 October 1978.

See also McNary, Charles Linza; McSweeney-McNary Forest Research Act; Woodruff-McNary Act.

Classification and Multiple Use Act (78 Stat. 986)

This federal legislation, enacted on 19 September 1964, directed the Bureau of Land Management (BLM) to manage the public lands under its control according to the rules of sustained yield and multiple use while classifying these lands in a survey for the aim of selling the lands or managing them as federal resources. Modeled from the Multiple Use–Sustained Yield Act of 1960, this act differed

in that it instructed the BLM to do the survey, whereas the 1960 act requested the study from the Forest Service.

See also Multiple Use–Sustained Yield Act.

Clean Air Act (77 Stat. 392)

This landmark legislation was enacted on 17 December 1963. Although it set national policy as well as uniform standards for air pollution control, the Clean Air Act was in fact an amendment to a previous law. In 1955, Congress passed an Act to Provide Research and Technical Assistance Relating to Air Pollution Control (Public Law 84-159), which only supplied funds for the investigation into air pollution issues. The 1963 Clean Air Act, as one source notes, "amended the 1955 act to provide for encouragement of uniform laws among states and local governments, research, enforcement, automobile emission control, and cooperation among federal agencies to control air pollution from federal facilities." The 1966 Amendments to the Clean Air Act (Public Law 89-675) gave states appropriations for the establishment and continuance of such air pollution control programs. Further supplementing this line of legislation was the Air Quality Act of 1967 and the Clean Air Act Amendments of 1990.

See also Air Quality Act; Clean Air Act Amendments of 1990.

Clean Air Act Amendments of 1990 (Public Law 101-549)

These revisions of the Clean Air Act of 1963 dealt with the commission of investigations into an ethanol substitute for diesel fuel, which pollutes the air; the expansion of states' authority to regulate the emissions of oxides of nitrogen from test cells; and the determination of the effect of "fugitive dust" (residue from surface mining) on the air we breathe. They also established a Risk Assessment and Management Commission, "which shall make a full investigation of the policy

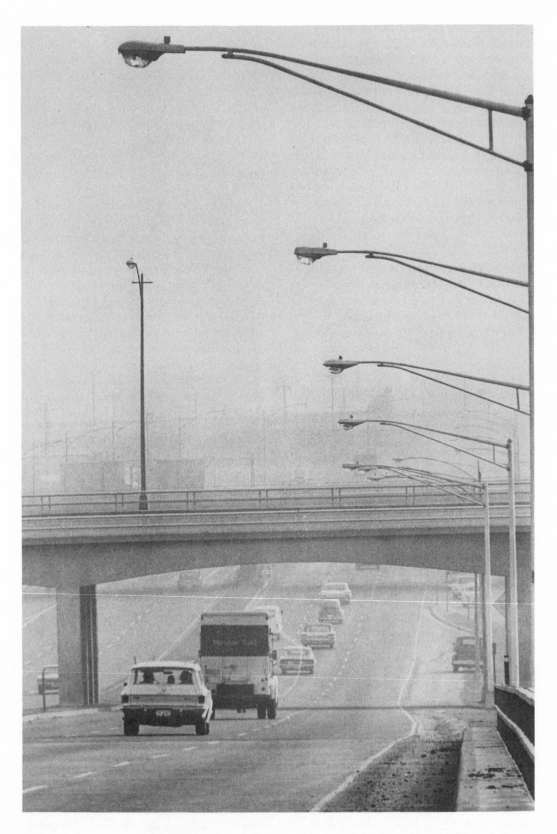

The Clean Air Act of 1963 encouraged federal, state, and local automobile emission controls and standards. Smog obliterates the Denver, Colorado, skyline three years after the act's passage.

implications and appropriate uses of risk assessment and risk management in regulatory programs under various federal laws to prevent cancer and other chronic human health effects which may result from esposure to hazardous substances."

See also Air Quality Act; Clean Air Act.

Clean Water Act
See Federal Water Pollution Control Act; Federal Water Pollution Control Act Amendments of 1972; Federal Water Pollution Control Act Amendments of 1977.

Clean Water Act Amendments of 1977 (Public Law 95-217)
See Federal Water Pollution Control Act Amendments of 1977.

Clean Water Restoration Act (80 Stat. 1246)
This federal legislation, enacted on 3 November 1966, was a response to the worsening condition of the nation's water and the failure of previous legislation to fix the problem. Under this act, increased funds were appropriated to the states for city waste-treatment facilities, a "clean rivers" restoration program was created, and research funds were doled out to find new ways to end water pollution.

Clements, Frederic Edward (1874–1945)
Pioneer botanist Frederic Clements "was the most influential ecologist of the first half of the twentieth century," according to biographer Joseph Ewan. Clements, however, has received little recognition for his part in the study of vegetation and its impact on human populations. Born in Lincoln, Nebraska, Clements was the only son and one of three children of Ephraim George Clements and his first wife, Mary Angeline (nee Scoggin) Clements. Ephraim Clements's father, George Clements, emigrated from Somerset-shire, England, in about 1842; Ephraim Clements himself was a Civil War veteran (he served with the 122d Regiment of the New York Infantry) before moving to Nebraska to set up a photography studio. Frederic Clements grew up in the Great Plains of Nebraska and, after attending local schools, entered the University of Nebraska in 1890 at age 16. In his junior year, he published his first botany-related work, a paper on a previously undiscovered species of fungi. The next year, 1894, the 20-year-old Clements received his bachelor's degree in botany and was immediately named an assistant in that field. He received a master's degree in 1896 and a Ph.D. two years later.

While at the University of Nebraska, Clements rose through the academic rank and file: botany instructor in 1897, adjunct professor two years later, assistant professor in 1903, and full professor of plant physiology in 1906. At the same time, he was participating in studies of Nebraska's vegetation with Professor Charles E. Bessey, a renowned botany expert. Clements and botany student (and later dean of Harvard Law School) Roscoe Pound coauthored *The Phytogeography of Nebraska* (1898). In 1899, Clements married one of his students, Edith Gertrude Schwartz, who earned a doctorate in botany and would later participate in many of his experiments.

Clements's work was not limited to the academic world. Twice he served the U.S. government: first from 1893 to 1896 as a special agent of the Department of Agriculture, and then from 1907 to 1910 as a consultant to the U.S. Forestry Service. In 1907, Clements was named head of the botany department of the University of Minnesota, a post he held for ten years. In 1917, he became a research associate at the Carnegie Institution in Washington, and eight years later he became a member of the institution's Coastal Laboratory in California. He was also a founder of the institution's Alpine Laboratory on Pike's Peak in Colorado. There he investigated how the plant world was affected by environmental factors. Wrote one source, "With adequate funds from the institution, Clements planned and directed researches at the two laboratories along the main project of determining the origin of species in the plant world by means of the impact of the physical factors in their environment." His wife's doctoral dissertation was on how leaves reacted to pollution.

Clements's greatest work is considered to be *Plant Succession: An Analysis of the Development of Vegetation* (1916), in which he discussed how plant creation was influenced by climatic and environmental conditions, not genetics. This "Clementsian theory" has been debated by scientists since its espousal but is widely accepted. Among Clements's other works and collaborations are *Histogenesis of the Caryophyllales* (1899), *Laboratory Manual of High School Botany* (1900), *Greek and Latin in Biological Nomenclature* (1902), *Development and Structure of Vegetation* (1904), *Research Methods in Ecology* (1905), *Genera of Fungi* (1909), *Minnesota Mushrooms* (1910), *Plant Indicators* (1920), *Phylogenetic Method in Taxonomy: The North American Species of Artemisia, Chrysothamnus, and Atriplex* (1923), and *Climatic Cycles and Human Populations on the Great Plains* (1938). In 1939, one of his final works, *Bio-Ecology*, was written with Victor E. Shelford. Coming on the heels of the great drought in the grasslands region of the country, it demonstrated how ecology could be used to restore the affected areas. Clements retired in 1941 and died four years later on 26 July 1945 at the age of 70.

Coastal Barrier Resources Act (16 U.S.C. 3501)

Enacted on 18 October 1982, this law attempted to set national policy for the protection and management of coastal barriers. In its statement of findings, Congress wrote, "Coastal barriers along the Atlantic and Gulf coasts of the United States and the adjacent wetlands, marshes, estuaries, inlets and near shore waters

provide (A) habitats for migratory birds and other wildlife; and (B) habitats which are essential for spawning, nursery, nesting, and feeding areas for commercially and recreationally important species of finfish and shellfish, as well as other aquatic organisms such as sea turtles." It continued, "Coastal barriers contain resources of extraordinary scenic, scientific, recreational, natural, historic, archeological, cultural, and economic importance, which are being irretrievably damaged and lost due to development on, among, and adjacent to, such barriers." Finding that "coastal barriers serve as natural storm protective buffers and are generally unsuitable for development because they are vulnerable to hurricane and other storm damage and because natural shoreline recession and the movement of unstable sediments undermine manmade structure," this act authorized the establishment of the Coastal Barriers Resource System, a sort of "reserve" for coastal barriers, to be overseen by the secretary of the interior.

Coastal Wetlands Planning, Protection, and Restoration Act (16 U.S.C. 3956)

This federal legislation, passed on 29 November 1990, authorizes the secretary of the interior to establish projects for "the protection, restoration, or enhancement of aquatic and associated ecosystems, including projects for the protection, restoration, or creation of wetlands and coastal ecosystems." Further, studies were to be conducted under the act to test the feasibility of "changing flood control projects on the Mississippi River to increase water and sediment flow down the Aychafalaya River for the purpose of land building and wetlands nourishment."

Coastal Zone Management Act (16 U.S.C. 1451)

This act, enacted on 27 October 1972 and amended in 1976 and 1980 (the latter as the Coastal Zone Management Improvement Act), attempted to formulate policy on the management of coastal zones. According to the act, such areas are defined as "the coastal waters (including the lands therein and thereunder) and the adjacent shorelands (including the waters therein and thereunder), strongly influenced by each other and in proximity to the shorelines of the several coastal states, and includes islands, transitional and intertidal areas, salt marshes, wetlands, and beaches." The degradation of these areas and the importance they play in the environment prompted Congress to enact this law. In it, Congress found that "there is a national interest in the effective management, beneficial use, protection, and development of the coastal zone." It further established that "the coastal zone is rich in a variety of natural, commercial, recreational, ecological, industrial, and esthetic resources of immediate and potential value to the present and future well-being of the Nation." The act ordered the secretary of commerce to bestow program development grants to states for the creation of plans to safeguard their respective coastal zones. The two amendments to this act added wording that expanded the definition, policy, and mandates of the original act.

Colorado River Compact (1922)

This landmark agreement, signed 24 November 1922, was the result of a meeting of representatives of several western states for the purpose of finding a solution to the impending western water crisis. Commissioners from the states of Arizona, California, Colorado, Nevada, New Mexico, Utah, and Wyoming, appointed by the governors of these states, met to dole out equally the waters of the Colorado River. Herbert Hoover, secretary of commerce in the Harding administration and acting as a representative of the federal government, oversaw the gathering, which took place at Bishop's Lodge, a retreat outside of Santa Fe, New Mexico. The meeting's main objective was outlined in Article 1 of the compact: "The major purposes of this compact are to provide for the equitable division and apportionment of the use of the waters of the Colorado River System; to establish the relative importance of different beneficial uses of water; to promote interstate comity (courtesy); to remove causes of present and future controversies; and to secure the expeditious agricultural and industrial development of the Colorado River Basin, the storage of its waters, and the protection of life and property from floods." Dividing the Colorado at the arbitrary boundary of Lee's Ferry, Arizona, the compact created two artificial water basins: California, Arizona, and Nevada were to be in the lower basin, and the rest of the states were in the upper basin. Hoover signed the pact for the federal government. After passage, the representatives went home to get their respective state legislatures to ratify the pact. Many of the states objected to various portions, and the matter lingered for six years. In 1928, Congress stepped forward and passed the Boulder Dam Act on the condition that six of the seven Colorado Compact states ratify the treaty. Arizona refused, but because its vote was not needed, the Boulder Dam Act became law.

During the compact conference, Herbert Hoover became interested in ways to harness the waters of the Colorado River—so interested, in fact, that upon his return to Washington, he began to lobby Congress to build a dam on the river for electric power. Although the bill for the dam was sponsored in the Senate by Hiram Johnson of California, Hoover was its chief lobbyist. On 21 December 1928, just after his election as president, Hoover saw outgoing President Calvin Coolidge sign the Boulder Dam bill, which included provisions for passage of the Colorado River Compact, into law. Today the dam is called Hoover Dam in honor of its most ardent advocate.

See also Arizona v. California (1963); Arizona v. California (1983); Colorado River Storage Project.

Colorado River Storage Project
(43 U.S.C. 620)

This landmark legislation, enacted by Congress on 11 April 1956, established a series of storehouses for Colorado River waters for public use. Pronounced the act:

> In order to initiate the comprehensive development of the water resources of the Upper Colorado River Basin, for the purposes, among others, of regulating the flow of the Colorado River, storing water for beneficial consumptive use, making it possible for the States of the Upper Basin to utilize, consistently with the provisions of the Colorado River Compact ... the Secretary of the Interior is authorized (1) to construct, operate, and maintain the following initial units of the Colorado River storage project, consisting of dams, reservoirs, power plants, transmission facilities and appurtenant (additional) works: Wayne N. Aspinall, Flaming Gorge, Navajo (dam and reservoir only), and Glen Canyon.

The act also called for the construction of reclamation projects in various parts of the country.

Commission on Country Life

Intrigued by the issue of life on American farms and in rural areas, Gifford Pinchot, chief forester of the United States, recommended in 1908 that President Theodore Roosevelt appoint a Commission on Country Life "as a means for directing the attention of the Nation to the problems of the farmer, and for securing the necessary knowledge of the actual conditions of life in the open country." Establishing the commission in August 1908, Roosevelt said, "It is at least as important that the farmer should get the largest possible return in money, comfort, and social advantages from the crops he grows, as that he should get the largest possible return in life. The great rural interests are human interests, and good crops are of little value to the farmer unless they open the door to a good kind of life on the farm." Appointed to the commission as chairman was Liberty Hyde Bailey of Cornell University, an expert on agricultural matters. Other members included Henry Wallace, editor of *Wallace's Farmer*, whose grandson, Henry Wallace, was Franklin D. Roosevelt's second vice president; Kenyon L. Butterfield, president of the Massachusetts Agricultural College; Walter Hines Page, editor of the influential journal *World's Work*; Charles S. Barrett, president of the Farmers' Cooperative and Educational Union of America; W. A. Beard, a writer for *Great Western Magazine*; and Pinchot. Congress, led by westerners fighting Roosevelt's conservation policies, denied the commission any appropriations; funds were eventually accepted from the Russell Sage Foundation, a private organization. Thirty hearings were held with farmers from forty states, and questionnaires were mailed that elicited some 120,000 responses. Wrote Pinchot in his autobiography, *Breaking New Ground*:

> The Commission found that the general level of American country life was high compared with any preceding time or with any other land, but that nevertheless farming did not yield either the profit or the satisfaction that it ought to yield.... [T]he remedies suggested by the Commssion included better organization among farmers, greater influence by farmers on legislation, and above all a more creative social life. It declared that, while the question of better life in the country intimately concerns both the state and national Governments, the farmers themselves must decide whether or not country life is to become more dignified, better thought of, with larger rewards in comfort, income, and social advantages.

When Roosevelt delivered the commission's report to Congress, it was refused. In fact, under an amendment sponsored by Rep. James A. Tawney of Minnesota, Congress declared that no president could appoint future commissions without congressional consent.

See also Tawney, James Albertus.

Committee on the Conservation and Administration of the Public Domain
See Garfield, James Rudolph.

Comprehensive Environmental Response, Compensation, and Liability Act (CERCLA)
See Superfund.

Comstock, Anna Botsford
(1854–1930)
Anna Comstock was a naturalist and conservation advocate of the late nineteenth and early twentieth centuries. Born Anna Botsford in the small village of Otto, New York, on 1 September 1854, she was encouraged to love and respect nature by her Quaker mother, Phebe Irish Botsford. Anna attended local schools and a "select" school in Otto before enrolling at a Methodist women's academy. In 1874, she entered Cornell University. In her first year at Cornell, she was courted by Dr. John Henry Comstock (1849–1931), the chief entomologist at Cornell. His 1895 work, *A Manual for the Study of Insects*, remains to this day one of the most notable in its field. Although Anna left Cornell after two years, she continued her studies under Comstock's tutelage and married him in 1878.

John and Anna Comstock became a team that broke new ground in the field of entomology. From 1879 to 1881, John Comstock was the chief entomologist for the U.S. Department of Agriculture, and Anna was his assistant. Her detailed drawings of insects made their way into the 1880 work *Report of the Entomologist*.

Anna Botsford Comstock

Over the years, Anna Comstock published several works of her own, including *How to Keep Bees*, *How to Know the Butterflies*, and *The Pet Book*. She became known as the mother of the American nature study. Soon after the turn of the century, the Comstocks founded their own printing interest, the Comstock Publishing Company, in Ithaca, New York. In 1911, they jointly published the *Handbook of Nature Study*, which was later issued in eight languages.

Anna Comstock was awarded her bachelor's degree in natural history from Cornell in 1885 and received instruction in environmental studies at New York's Cooper Union. In 1888, she was made a member of Sigma Xi, the national honor society of the sciences. In 1923, she was chosen by the National League of Women Voters as one of the twelve greatest living women. Finally, in 1988, she was posthumously inducted into the National Wildlife Federation's Conservation Hall of Fame. A teacher in the natural sciences until just before her death on 24 August 1930, Anna Botsford Comstock remains one of America's most important early conservationists.

Conservation Easements

This increasingly used policy is defined by author Margaret Haapoja as "a legal agreement by which a landowner voluntarily restricts or limits the type and amount of development that may take place on his or her property. Each easement's restrictions are tailored to the particular piece of property, the interests of the individual owner, and the resource being protected. These restrictions bind the present and future owners of the property." The plan works by having an owner of a piece of property by a river, for instance, donate the land to the state with the promise that certain types of development, or none at all, may be allowed. Hunting and logging may be prohibited as part of the easement. By law, the state agency overseeing the easement must comply with its restrictions in totality.

Conservation easements have existed as private contracts for more than 100 years. Recently, however, with the increased awareness of the possible uses of private land after the death of an owner, states have begun to pass easement laws to facilitate land donations. By 1984, some 44 states had easement laws on the books. The federal government established a federal easement policy with the passage of the Forest Legacy Program contained in the Food, Agriculture, Conservation and Trade Act of 1990 (better known as the 1990 farm bill). Labeled the Forest Stewardship Act of 1990, it "authorizes the ... Forest Service to acquire permanent conservation easements on private forestlands that are at risk of being converted to non-forest uses."

See also Forest Legacy Program.

Conservation Movement

Prior to 1963, the movement to protect the nation's natural resources was known as the conservation movement. The term *conservation* has been defined by author J. Leonard Bates as "the prevention of waste" of our natural resources. Bates reports that Herbert Hoover, secretary of commerce and president of the United States, defined conservation as having "an emphasis on the efficiency of use." The conservation movement, whose goals were to lobby for reform of the nation's laws and attitudes toward the environment and protect the nation's varied natural resources, can be naturally broken down into four epochs: the first from 1731 to 1812, the second from 1812 to 1864, the third from 1864 to 1907, and the fourth from 1907 to 1963.

The first epoch, although not considered by historians to be part of the conservation movement per se, began with the delineation and discovery of the North American wilderness by men who documented their travels, such as John and William Bartram, Thomas Nuttall, John Bradbury (1768–1823), Charles Christopher Parry (1823–1890), George Vancouver, George Featherstonhaugh (1780–1866), and the father-son team of André and François André Michaux, as well as with the explorations and discoveries of eminent botanists and scientists such as John Clayton (c. 1685–1773), Dr. Alexander Garden (c. 1730–1791), and Cotton Mather (1662–1727), an ancestor of another conservationist, Stephen Tyng Mather, founder and first director of the National Park Service. The landmark expedition of Meriwether Lewis (1774–1809) and William Clark (1770–1838) mapped much of the land ceded to the United States in the Louisiana Purchase, and Zebulon Pike (1779–1813) explored the Rio Grande and what is now Colorado. As early as 1787, John Taylor, a wealthy planter from Virginia, spoke out on soil erosion as a member of the Philadelphia Society for Promoting Agriculture. Other early advocates of soil conservation were

Samuel Deane, Solomon and William Drown, Isaac Hill (editor of *The Farmer's Monthly Visitor*), John Lorain, and Nicholas Sorsby. Journals such as Hill's *Visitor*, John S. Skinner's *American Farmer*, and Edmund Ruffin's *Farmers' Register* advocated soil-conservation techniques.

The second era of the movement to identify, conserve, and protect the nation's natural resources began just before the War of 1812. In the years surrounding the turn of the nineteenth century, Congress began to dictate land policy through the passage of several laws, including the Federal Timber Purchases Act (1799) and the Land Acts of 1800, 1804, 1820, and 1821. With the settlement of the western lands, new states began to be admitted to the Union. Trailblazers such as Jim Bridger, Jedediah Strong Smith, and John Charles Frémont led the way in exploring these new areas, including California and Utah. The subsequent establishment of the Oregon Trail and similar paths to the far reaches of the American West, which beckoned to tens of thousands of pioneers, led to the passage of the Preemption Act (1841) and the Homestead Act (1862).

Conservation historian Donald Fleming believes that the roots of the conservation movement of the mid-nineteenth century began with the transcendentalist thought of such writers as Ralph Waldo Emerson (1803–1882) and Henry David Thoreau. Science led many into this area, and those involved in the scientific shaping of this era of conservation included John Muir, John James Audubon, Spencer Fullerton Baird, William Henry Brewer, and George Barrell Emerson. With the creation of the Department of the Interior in 1849, the federal government moved from a passive to an active role in shaping the nation's conservation and environmental policies. The 1860s were, according to one source, "like the 1960s, a decade of American upheaval." While the Civil War raged in the East and parts of the West, Congress passed the Homestead Act, which offered ownership of 160 acres of untouched land to any settler who worked on it for five years. First towns, then villages, and finally small cities sprang up as pioneers moved west. While railroads were shaping the borders of new population centers, Congress was appropriating funds to establish agricultural and mechanical (land grant) colleges. The California gold rush in 1849 and the land rush following the Civil War made the western United States prime territory for irrigation, reclamation, and conservation. Between 1860 and 1890, the West grew in population from less than 1 million to over 9 million.

With the publication in 1864 of George Perkins Marsh's landmark work, *Man and Nature; or, the Earth as Modified by Human Action*, those who sought to preserve and conserve the environment, and those who endeavored to establish restrictions on the use of the lands and natural resources of the country, stepped forward in ever-increasing numbers. This third epoch of new thinking is considered the beginning of the conservation movement as most historians know it. The impetus behind the movement led to the prominence of many people and groups in various fields of study, such as wildlife and forestry preservation, irrigation, reclamation, and the conservation of natural resources. Even artists did their part in documenting what was left of the untouched parts of America. Historian Tom Turner reports that one of the first organizations founded in the name of conservation was the Williamstown Alpine Club, established in Massachusetts in 1863. The White Mountain Club was also created in 1873, followed by the Rocky Mountain Club in Colorado Springs, Colorado, in 1875 and the Appalachian Mountain Club of Boston in 1876. The first official wildlife refuge in America, Lake Merritt in California, was created by a bill introduced in the California legislature by state Sen. Edward Thompson and enacted on 18 March 1870.

In the late 1860s and early 1870s, various voices—those of men like the Rev. Frederick Starr, Jr., and J. William Dawson—were raised to protest the plight of forestry reserves in the United States. Starr's warning was published in a report by the secretary of agriculture in 1866. The American Forestry Association was formed in 1875, and people such as Carl Schurz, secretary of the interior in the Hayes administration; James A. Williamson and William A. J. Sparks, both commissioners of the General Land Office; Bernhard Fernow; Gifford Pinchot; Robert Underwood Johnson; John Muir; and Rep. William Holman of Indiana used their influence to direct conservation policy. Meanwhile, scientists such as Othniel Charles Marsh, John Wesley Powell, and William Brewer used their scientific knowledge and informative explorations to advance the cause of conservation. Artists such as George Catlin, Albert Bierstadt, Camillus S. Fly, Eadweard Muybridge, Thomas Moran, and William Henry Jackson used paint and camera to preserve the noteworthy but disappearing scenes of land and water. Muir used his power to get Congress to establish Yellowstone National Park in 1872. Twenty years later, he was one of the founding members of the Sierra Club. Author John F. Reiger argues, however, that it was sportsmen, including the founders of the Boone and Crockett Club, who were the real spearheads of the conservation movement in the 1870s. Among these men was George Bird Grinnell, editor of *Forest and Stream* magazine, and the first conservationist president, Theodore Roosevelt. Such advocates of irrigation as George Hebard Maxwell, Sen. Francis G. Newlands of Nevada, and Sen. Henry Clay Hansbrough of North Dakota led the way in shaping and formulating congressional policy on land reclamation and water irrigation. Sen. William Morris Stewart of Nevada was a key player in the enactment of mining legislation. Organizations that led the way in wildlife conservation included the American Ornithological Union of Boston, founded in 1883, the first Audubon "societies" in 1886, and the Boone and Crockett Club in 1887.

In about 1875, the conservation movement broke into two sections—those who were called "conservationists" and those who were called "preservationists." Those in the latter group believed that the preservation of natural areas and resources, not just conserving them, was necessary. The goal of preservationism was, according to author Stephen Fox, "the absolute exclusion [of these resources and lands] from human use." Conservation, however, meant "only a more prudent, more efficient use by humans" of these resources, not exclusion. By 1907, following the massive program of environmental regulation through Theodore Roosevelt's presidential directives and federal legislation, the groups reemerged as one, and the term *conservation* was conveniently affixed to everything having to do with the environment.

The last era of the conservation movement, 1907–1963, was initially imprinted by the Governors' Conference on the Environment (1908), the first such gathering in the history of the country aimed at identifying problems in the environment and earmarking government policies and funds to combat them. The principal players in the early years of this era were from the progressive Republican fold: Theodore Roosevelt, W J McGee, James R. Garfield, and Sens. Irvine L. Lenroot of Wisconsin, Charles L. McNary of Oregon, and George Norris of Nebraska, among others. McNary was the impassioned leader of farm reform in the 1920s, sponsoring numerous acts such as the Clarke-McNary Reforestation Act of 1924, the McSweeney-McNary Forest Research Act of 1928, and the Woodruff-McNary Act of 1928. Norris's idea to dam the waters of the Tennessee Valley led the way in the 1930s to the establishment of the Tennessee Valley Authority. The Civilian Conservation

Corps (CCC), a New Deal agency, was created to give jobs to the unemployed while preserving the environment.

The years from 1920 to 1962 were marked by a drought in conservationist thought. Although such people as Jay Norwood Darling and Ira N. Gabrielson advocated the increased protection of the wild, their work mainly involved formulating state policy and establishing environmental groups, such as Darling's National Wildlife Federation. Until the 1960s, the conservation movement had no leader and no real program to adhere to. It took Rachel Carson's landmark 1962 work, *Silent Spring*, to rejuvenate the movement into action. The publication of her book launched the environmental movement, with its marches, celebrations of Earth Day, recycling, and awakened awareness of the health of the earth.

See also Audubon, John James; Baird, Spencer Fullerton; Bartram, John; Boone and Crockett Club; Brewer, William Henry; Department of the Interior; Emerson, George Barrell; Environmental Movement; Fernow, Bernhard Eduard; Garfield, James Rudolph; Governors' Conference on the Environment; Grinnell, George Bird; Holman, William Steele; Lenroot, Irvine Luther; McGee, William John; Marsh, Othniel Charles; Muir, John; Norris, George William; Nuttall, Thomas; Pinchot, Gifford; Roosevelt, Theodore; Schurz, Carl; Sierra Club; Sparks, William Andrew Jackson; Thoreau, Henry David.

Convention for the Protection of Fur Seals in Alaska

This agreement for the protection of the Alaskan fur seal populations that were being decimated at the close of the nineteenth century was signed by U.S. Secretary of State James G. Blaine and British Foreign Secretary Julian Pauncefote on 29 February 1892.

With the purchase of Alaska in 1867, the United States became the "possessor" of the 1.5 million fur seals that settled on the Pribilof Islands in the Bering Sea to roost, rest, give birth, rest some more, moult, and then go back to the sea and repeat the cycle. In the first

two years after the purchase of Alaska, some 200,000 seals were killed illegally, leading the government to sell a contract to the Alaska Commercial Company to hunt the seals (about 100,000 a year) and slaughter them for their fur.

In 1872, a 26-year-old naturalist with the Smithsonian Institution, Henry Wood Elliott, was hired by the Treasury Department to go to Alaska and study the sealing operations there. After extensive research, he concluded that the seals were in no danger. Unfortunately, soon after his visit, the natives increased their take in pelagic hunting, that is, hunting in the seas and oceans. In 1886, the U.S. government apprehended three British ships in the Bering Sea transporting sealskins back to Great Britain. A controversy arose as to whether the United States could declare the Bering Sea a "closed" sea. The contention lasted for several years. Meanwhile, hunting that amounted to two and three times the quota continued. On 2 March 1889, Congress authorized the president to work to protect U.S. rights and interests in the area. That year, Special Agent James Goff, reporting on his observations of the situation in the Pribilofs, sent a message back to Washington: "The seals are about gone." Elliott went back to Alaska and discovered that the natives and the sealing company had conspired to wipe out the entire seal herd. Angered, he returned to Washington, where he pleaded for an end to sealing. Ignored, he went to the newspapers, which printed his charges and a subsequent report. Under pressure, the United States asked for a panel of arbitration to decide whether it had the right to ban pelagic hunting—and thus freeze Britain out of the slaughter. An excerpted transcript of the agreement follows:

A Convention for the Settlement of the Disputed Right of the United States to Protect the Seals in Behring Sea
Concluded, February 29, 1892.

The United States of America and Her Majesty the Queen of the United Kingdom of Great Britain and Ireland, being desirous to provide for an amicable settlement of the questions which have arisen between their respective governments concerning the jurisdictional rights of the United States in the water of Behring's Sea, and concerning also the preservation of the fur-seal in, or habitually resorting to, said Sea, and the rights of the citizens and subjects of either country as regards the taking of fur-seal in, or habitually resorting to, the said waters, have resolved to submit to arbitration the questions involved. . . .

Article I—The questions which have arisen between the Government of the United States and the Government of Her Britannic Majesty concerning the jurisdictional rights of the United States in the waters of Behring's Sea, and concerning also the preservation of the fur-seal in, or habitually resorting to, the said Sea, and the rights of the citizens and subjects of either country as regards the taking of fur-seal in, or habitually resorting to, the said waters, shall be submitted to a tribunal of Arbitration, to be composed of seven Arbitrators, who shall be appointed in the following manner, that is to say: Two shall be named by the President of the United States; two shall be named by Her Britannic Majesty; His Excellency the President of the French Republic shall be requested by the High Contracting Parties to name one; His Majesty the King of Italy shall be so requested to name one; and His Majesty the King of Sweden and Norway shall be so requested to name one. The seven arbitrators to be so named shall be jurists of distinguished reputation in their respective countries. . . .

Article II—The arbitrators shall meet at Paris within twenty days after the delivery of the counter-case mentioned in Article IV and shall proceed impartially and carefully to examine and decide the questions that have been or shall be laid before them as herein provided on the part of the Governments of the United States and Her Britannic Majesty respectively. All questions considered by the tribunal including the final decision, shall be determined by a majority of all the Arbitrators.

Each of the High Contracting Parties shall also name one person to attend the tribunal as its Agent to represent it generally in all matter connected with the arbitration. . . .

Article VI—In deciding the matters submitted to the Arbitrators, it is agreed that the following five points shall be submitted to them, in order that their award shall embrace a distinct decision upon each of said five points, to wit:

1. What exclusive jurisdiction in the sea now known as the Behring's Sea, and what exclusive rights in the seal fisheries therein, did Russia assert and exercise prior and up to the time of the cession of Alaska to the United States?
2. How far were these claims of jurisdiction as to the seal fisheries recognized and conceded by Great Britain?
3. Was the body of water now known as the Behrings Sea included in the phrase "Pacific Ocean," as used in the Treaty of 1825 between Great Britain and Russia; and what rights, if any, in the Behrings Sea were held and exclusively exercised by Russia after said Treaty?
4. Did not all the rights of Russia as to jurisdiction, and as to the seal fisheries in Behrings Sea east of the water boundary, in the

Treaty between the United States and Russia of the 30th March, 1867, pass unimpaired to the United States under that treaty?

5. Has the United States any right, and if so, what right of protection or property in the fur-seals frequenting the islands of the United States in Behring Sea when such seals are found outside the ordinary three-mile limit?

Article VII—If the determination of the foregoing questions as to the exclusive jurisdiction of the United States shall leave the subject in such position that the concurrence of Great Britain is necessary to the establishment of Regulations for the proper protection and preservation of the fur-seal in, or habitually resorting to, the Behring Sea, the Arbitrators shall then determine what concurrent Regulations should extend, and to aid them in that determination the report of a Joint Commission to be appointed by the respective governments shall be laid before them, with such other evidence as either Government may submit.

The High Contracting Parties furthermore agree to cooperate in securing the adhesion of other Powers to such regulations.

Article VIII—The High Contracting Parties have found themselves unable to agree upon a reference which shall include the question of the liability of each for the injuries alleged to have been sustained by the other, or by its citizens, in connection with the claims presented and urged by it; and, being solicitous that this subordinate question should not interrupt or longer delay the submission and determination of the main questions, do agree that either may submit to the Arbitrators any question of fact involved in said claims and ask for a finding thereon, the question of the liability of either Government upon the facts found to be the subject of further negotiation. . . .

The arbitration panel held on 15 August 1893 that because the United States was supervising the killing on land, it could not order an end to hunting in the water. Further, it ordered damages to be held against the United States; the sum of $473,151 was finally paid on 16 June 1898. Wrote author Jeanne Van Norstrand, "The decision of the tribunal unloosed a carnival of slaughter. Between 1893 and 1896, pelagic sealers took not less than 500,000 seals." Elliott tried during the next several years to confront the problem, but to no avail. An Anglo-American Commission to the Pribilofs in 1896 found nothing wrong with the seal populations. After trying for so long to get the United States to admit wrongdoing, Elliott finally wrote to author William Temple Hornaday in 1907: "Do something to save those seals." Working with Sen. Joseph Moore Dixon (1867–1934) of Montana, chairman of the Senate Natural Resources Committee, Hornaday fashioned a bill that ended the leasing system and put a five- to ten-year moratorium on all hunting in the Pribilofs. Even after the passage of the bill, Secretary of Commerce and Labor Charles Nagel ordered the 1911 hunting season to proceed. Elliott called for a congressional investigation of the Bureau of Fisheries and its handling of the seal rookeries.

Elliott's hard work led to the Four Power Conference in Washington, which opened on 7 July 1911. Attended by the United States, Russia, Great Britain, and Japan, the nations agreed on a treaty that outlawed pelagic sealing north of the 30th parallel for a period of 15 years, permitted the United States to have a monopoly on the seal catch, and allowed for sharing of the profits from the cull. Japan nullified its signature on the treaty on 23 October 1940.

Convention on International Trade in Endangered Species of Wild Flora and Fauna (CITES)

This international treaty was signed in 1973 by 80 nations, including the United States, to end the selling, trafficking, and exploitation of endangered species of animals and plants, their parts, or any products resulting from such exploitation. It set up a system of regulations that all countries would abide by to end such traffic, although the failure of many countries to sign has weakened the act's effectiveness worldwide. Explains one source, "Species listed in Appendix 1 are considered rare and endangered, and their commercial trade is generally prohibited. Species listed in Appendix 2, not currently endangered, are considered at risk of overexploitation if their trade is not regulated."

Convention on Nature Protection and Wildlife Preservation in the Western Hemisphere

This treaty was entered into by the United States and other Central and South American nations in 1940; it was ratified by the U.S. Senate on 15 April 1941. The statement of purpose notes that the treaty's intention is "to protect and preserve in their natural habitat, representatives of all species in general of their native flora and fauna, including migratory birds, and in sufficient numbers in areas extensive enough to prevent them from becoming extinct through any agency within man's control; to protect and preserve scenery of extraordinary beauty, unusual and striking geologic formations, regions or natural objects of aesthetic, historic or scientific value in areas characterized by primitive conditions in those cases covered by the Convention, and to conclude a Convention on the protection of nature and preservation of flora and fauna." The Convention calls on the signing nations to establish national parks, monuments, and forest reserves that would protect endangered nature and wildlife.

Convention on the Conservation of Polar Bears

This 1976 agreement, signed by the United States, Canada, Denmark, Norway, and the USSR, established national policies among all these nations to protect and preserve polar bear populations; it outlaws hunting, killing, and capturing bears. It does, however, allow the killing of bears for scientific purposes, for conservation purposes, by native and local peoples in the carrying out of their cultural heritage, and when such polar bears are a threat to other areas of wildlife or nature. The act expressly forbids the commercial sale of skins or other items taken from bears allowed to be killed under this treaty.

Cooper, James Graham (1830–1902)

Pioneer naturalist James Graham Cooper was among the trailblazers of the conservation movement. Born in New York City on 19 June 1830, he was the eldest of six children of William Cooper, one of America's first trained naturalists, and Frances (nee Graham) Cooper. James Cooper learned about his environment while growing up in Hoboken, New Jersey, where the family moved in 1837. He was influenced by his father's friends, namely John James Audubon, Thomas Nuttall, the botanist John Torrey, and Spencer Fullerton Baird. William Cooper was a founding member of the Lyceum of Natural History of New York, the precursor to the New York Academy of Sciences.

James Cooper graduated from the New York College of Physicians and Surgeons, but after two years of practice in New York City he asked his father's friend Spencer Baird to get him a job as a naturalist. Baird convinced Washington Territory Gov. Isaac I. Stevens to sign Cooper on as a physician in the U.S. Army, which was then surveying the 47th and 49th parallels for the transcontinental railroad. For two years Cooper examined the territory in the area of Puget Sound in Washington State while amassing numerous plant samples. When he returned to the East Coast in 1855, his reports on the specimens he collected were published in a report on the railroad's activities. Other trips taken between 1857 and 1860 included voyages to eastern Florida, the White Mountains of New England, and the Great Plains. In 1855, he wrote to Baird from California, "I have lately been considering the subject of the botanical and zoological regions on this coast, which are known to be strongly marked and limited by natural boundaries much more than on the Atlantic side. I send a very rough sketch to show my views on the subject and to suggest a complete map, such as I think you now have the means of constructing and which would be of great assistance in pursuing scientific explorations out here." Taking an interest in the idea that 40 years later evolved into forestry reserves, Cooper composed a summary of his discoveries in *On the Distribution of Forests and Trees of North America, With Notes on Its Physical Geography*, which was included in the Smithsonian Institution's annual report for 1858. Wrote Cooper biographer Eugene Coan, "The thirty-four page article contained a catalog of the native trees of North America, a series of observations about the 'extreme points to which each species extends in every direction,' and a call for new examples and field notes, with a view to 'extending or correcting the list of trees herewith appended.'"

In 1860, Cooper joined an expedition that traveled along the Columbia River until reaching California, where Cooper spent the rest of his life. Cooper worked with scientist Josiah D. Whitney, head of the California Geological Survey, on survey projects as a zoologist. His reports for the survey on avian sightings were distinguished by Spencer Baird in his work *History of North American Birds* (1874) as "by far the most valuable contribution to the biography of American birds that has appeared since the time of Audubon." Further investigations by

Cooper into forestry resources were published in the essay *The Forests and Trees of North America, as Connected with Climate and Agriculture* in the *Agricultural Report for 1860* (1861).

Cooper's career as a naturalist ended in 1865 when he married Rosa Wells in Oakland, California, and settled down to life as a physician. He moved around California, eventually settling in Hayward, where he spent the last 27 years of his life. James Graham Cooper died there on 19 July 1902 at age 72.

Cooperative Farm Forestry Act of 1937
See Norris-Doxey Act.

Cooperative Forestry Assistance Act (16 U.S.C. 2101)
This piece of federal legislation was enacted into law on 1 July 1978. In it, Congress declared that "most of the Nation's productive forest land is in private, State, and local governmental ownership, and the Nation's capacity to produce renewable forest resources is significantly dependent on these non-Federal forest lands." Problems arising from this situation were discussed in the act. To meet these head-on, several solutions were proposed, namely, that the secretary of agriculture would (1) assist in the advancement of forest resources management, (2) encourage the production of new sources of timber, (3) utilize programs to prevent and control insects and other diseases affecting trees, (4) prevent and control the spread of forest fires, (5) formulate ideas for the recycling of wood and wood residues and fibers, (6) improve and maintain the fish and wildlife habitat in forests, and (7) plan and conduct urban forestry programs. This act supplemented, but was not an amendment to, the Forest and Rangeland Renewable Resources Planning Act of 1974.

See also Forest and Rangeland Renewable Resources Planning Act.

Coordination Act Amendment (Public Law 732)

Enacted on 14 August 1946, this legislation mandated that the Fish and Wildlife Service be allowed to participate in the conception and construction of all water utilization projects, such as dams and water- and flood-control projects, so as to anticipate or alleviate environmental damage to fish and wildlife resources, or to exploit the projects for wildlife management uses.

Cope, Edward Drinker (1840–1897)

A leading scientific authority in the area of zoology and paleontology, Edward Cope was a member of two early surveys of the American West that led to the founding of the U.S. Geological Survey. Cope was born on 28 July 1840 in Philadelphia, Pennsylvania. His parents were Alfred Cope, a wealthy Quaker merchant, and Hannah (nee Edge) Cope, who died when her son was three. Because of his father's Quakerism, Cope attended the Friends Select School and the West-town Boarding School in and around Philadelphia before attending the University of Pennsylvania and completing his education through private tutoring. Cope was urged by his father to take up agriculture, but his time spent on the farms of family members bred a love not of the land but of what was in it—the materials of paleontology.

In 1859, Cope went to Washington and came under the influence of Spencer F. Baird, then assistant secretary of the Smithsonian Institution and a noted fisheries expert. Here the budding scientist wrote his first paper, "On the Primary Divisions of the Salamandridae, With a Description of Two New Species," which was published in the *Proceedings of the Academy of Natural Sciences* that same year. Cope later returned to Philadelphia, where he worked for a time at the Academy of Natural Sciences, studying the herpetological collections there. Much of his early work was on reptiles.

In 1863 and 1864, he toured the natural history museums of Europe. In 1866, he was named chairman of the department of comparative zoology at Haverford College in Haverford, Pennsylvania, but he had to withdraw after one year because of ill health. For the rest of his life, Cope researched paleontology and became an expert and writer in that field.

From 1871 to 1874, Cope was a paleontologic consultant to the Hayden and Wheeler surveys of the West. Wrote biographer Joseph Maline, "Cope spent eight months of each year with one of the . . . surveys. During this time he visited or discovered fossil fields in Colorado, Kansas, Montana, New Mexico, South Dakota, Texas, and Wyoming." As part of the surveys' reports, Cope contributed *The Vertebrata of the Cretaceous Formations of the West* (1875) and *The Vertebrata of the Tertiary Formations of the West, Book 1* (1884), which is referred to as "Cope's Bible" or "Cope's Primer."

Edward Cope's research clashed with that of Othniel Charles Marsh, the paleontologist attached to the Clarence King survey. Their caustic dispute originated when Cope questioned some of Marsh's samples. When Marsh advocated the consolidation of the surveys into one federal survey in 1878, Cope bitterly objected. His protestations came to naught, and when the U.S. Geological Survey was established, Marsh was named the survey's chief paleontologist, and Cope was shut out.

Cope's final years were marked by financial collapse caused by a bad mining investment. Forced to sell his personal collection of fossils, he went on a lecture tour, edited the biology-oriented journal *American Naturalist*, and attempted to publish his writings to support his family. Financial stability returned in 1895, when he was named chairman of the zoology and comparative anatomy department at the University of Pennsylvania. He had been in this post only two years when he died on 12 April 1897 at the age of 56.

See also Marsh, Othniel Charles.

Copeland Report (1933)

This government analysis, which was released by the U.S. Forest Service to address the condition of the nation's forests, was issued under the direction of Earle Hart Clapp, assistant chief of the Forest Service. One of the workers on the report was Robert Marshall, a forestry and wilderness expert who had just returned from Alaska. The report was named for Sen. Royal Copeland of New York, who introduced a resolution in the U.S. Senate to have the study done. Officially titled *A National Plan for American Forestry*, the 1,677-page document, released on 13 March 1933, discussed forestry issues; the condition of timber, water, and range areas; wildlife; and federal involvement in these areas. The most controversial aspect of the report was the official rebuke of the forestry industry for overlogging private lumber reserves, and the report called for the federal government to buy these lands at a rate of $50 million a year under the Clarke-McNary Reforestation Act, which Congress had passed in 1924. The recommendations of the Copeland Report were used for a number of years as a basis for federal and private industry policy on logging and timber cutting.

See also Clapp, Earle Hart; Clarke-McNary Reforestation Act; Marshall, Robert.

Coues, Elliott (1842–1899)

Considered one of the greatest ornithologists and scientists in the field of bird study, Elliott Coues is barely known today. Born in Portsmouth, New Hampshire, on 9 September 1842, Coues (pronounced "cows") was the son of Samuel Elliott Coues, a merchant and patent officer, and Charlotte Haven (nee Ladd) Coues, a descendant of John Mason, whom biographer Witmer Stone called "the original grantee of New Hampshire." When his son was 11, Samuel Coues moved the family to Washington, D.C., where he worked in the Patent Office. Elliott Coues attended Gonzaga College, a Jesuit institution, and Columbian University (now George Washington University), where he earned an A.B. degree in 1861 and a medical degree two years later. Shortly before becoming a physician, he enlisted with the Union Army as a medical cadet and saw limited action as a surgeon in a Washington, D.C., hospital. In 1864, he was named assistant surgeon and was immediately sent to Arizona.

Coues's time in Washington was spent cultivating his love of nature, and he soon became an expert in birds. His years in Arizona were, according to one source, "utilized in investigating the natural history of the region, respecting which he published various scientific papers." He served at various military outposts, including Fort Whipple, Arizona; Fort Macon, North Carolina; and Fort Randall, Dakota Territory. In 1873, he was named to the United States Northern Boundary Commission and served a three-year term. During these assignments, he spent his time collecting and cataloging specimens of animal and plant matter. His early investigations led to the publication of *Key to North American Birds* (1872), which was reprinted five times. *Birds of the Northwest* followed in 1874. In 1876, Coues was named a naturalist for the geological survey of the western states under Ferdinand V. Hayden and served for four years. Coues's work there resulted in his *Fur-Bearing Mammals* (1877), *Monographs of North American Rodentia* (1877), and *Birds of the Colorado Valley: A Repository of Scientific and Popular Information Concerning North American Ornithology*, which were published as part of the Hayden survey's reports. In all, over his lifetime, Coues published nearly 1,000 scientific papers.

In 1881, Coues resigned from the army and spent the rest of his life writing on scientific (as well as religious) subjects. On the advice of Spencer Fullerton Baird, head of the Smithsonian Institution, Coues joined the Smithsonian, where he wrote *New England Bird-Life* (1881) and

Dictionary and Check List of North American Birds (1882), as well as a revised edition of *Key to North American Birds*, which one source said was "recognized as the standard textbook on ornithology."

During a trip to Europe in 1884, he became fascinated with religion, particularly theosophy. (Coues's biographers, Paul Russell Cutright and Michael J. Brodhead, wrote of this faith: "We can characterize theosophy as a ... cult which attempts to combine and synthesize mystical and occult teachings from all the major religions, but actually leaning heavily on Buddhism and Hinduism.") When he returned to the United States, Coues founded the Gnostic branch of the Theosophistic Society. Although he was later involved in a controversy with members of the society, he was partly responsible for getting an English version of the Buddhist catechism released in the United States. In his final years, Coues spent much of his time writing. From 1884 to 1891, he was biology editor of *Century* magazine, and he wrote for *Osprey* magazine from 1897 to 1899. Most of his works, however, dealt with the American West. They include *History of the Expedition of Lewis and Clark* (1893), *Expeditions of Zebulon Montgomery Pike* (1895), *Journal of Major Jacob Fowler* (1898), and *Forty Years a Fur Trader on the Upper Mississippi by Charles Larpentuer* (1898).

From 1893 on, Coues was a sickly man. By 1899, he was having serious bowel trouble and returned to the East to have doctors examine him. They found that he was suffering from inoperable rectal cancer. Coues died in a hospital in Baltimore, Maryland, on Christmas Day, 1899, at the age of 57. He was buried in Arlington National Cemetery. His name appears in the Latin names of at least eight species he discovered, including the pink-footed shearwater (*Puffinus creatopus Coues*), the ashy petrel (*Oceanodroma homochroa Coues*), Baird's sandpiper (*Erola bairii Coues*), and Bendire's thrasher (*Taxostomabendirei Coues*).

Coville, Frederick Vernon (1867–1937)

Author Samuel P. Hays wrote, "In 1897, Frederick V. Coville, botanist of the Department of Agriculture, carried out in the Western reserves the first scientific range investigation of the United States." Barely known outside the scientific community, Coville contributed greatly toward the establishment of a fundamental national range conservation policy. He was born in the village of Preston in Chenango County, New York, on 23 March 1867, the son of Joseph Addison Coville, a farmer and Civil War veteran, and Lydia Smith (nee More) Coville. Frederick Coville studied at the Oxford (New York) Academy before entering Hamilton College in Clinton, New York. After only one year at Hamilton, he transferred to Cornell University in Ithaca, New York, where he majored in botany. In 1887, he was awarded a B.A. degree.

The summer following his graduation, Coville accompanied explorer John C. Branner as a botanist on a geological survey of Arkansas. Although Coville returned to Cornell to teach botany, he was hired by the Department of Agriculture as an assistant botanist in 1888. In 1893, upon the death of Dr. George Vasey, Coville became chief botanist, a position he held for the rest of his life. Reported one source, "Coville's most important field work was as a botanist on the Death Valley expedition in 1891, the results of which were published in 1893 as 'Botany of the Death Valley Expedition.'" He soon became an acknowledged expert on the flora called *Juncaceae* and published an 1896 article on *Grossulariaceae* in the work *Illustrated Flora of the Northern States and Canada*. During a trip west in 1897, he began to take an interest in grazing problems on the western range. After careful study, he concluded that grazing permits should be allowed in the national forests in the West on a limited scale. In 1898, upon his recommendation, the General Land Office released permits for cattle grazing across the West but limited sheep grazing permits to Oregon and Washington. Further, it called grazing a privilege, not a right—one that could be taken away if abused. Coville's suggestion pleased neither Secretary of the Interior Ethan Allan Hitchcock, who wanted no grazing at all, nor Chief Forester Gifford Pinchot, who believed that grazing was a right, not a privilege. Such a policy was upheld by the Supreme Court in the case of *United States v. Grimaud* in 1911.

Frederick Coville helped establish the Desert Botanical Laboratory near Tucson, Arizona, and developed the strain of blueberries called Coville hybrids. He was the coauthor with landscape architect Frederick Law Olmsted, Jr. (son of the landscape pioneer) and Harlan P. Kelsey of *Standardized Plant Names* (1923), as well as the writer of some 175 scientific papers. President of the Botanical Society of America (1903–1904) and of the Washington Academy of Sciences (1912–1913), Coville was awarded the George Robert White Medal of the Massachusetts Horticultural Society in 1931 for his lifetime of work in botany and horticulture. He died in Washington, D.C., on 9 January 1937, less than three months shy of his seventieth birthday.

Cox, Jacob Dolson, Jr. (1828–1900)

Jacob Dolson Cox served as secretary of the interior from 1869 to 1870 during the Grant administration. Born in Montreal, Ontario, Canada, on 27 October 1828, Cox was the son of Jacob Dolson Cox, Sr., a building contractor (working on a project in Montreal at the time), and Thedia Redelia (nee Kenyon) Cox. The family's original name, Koch, was changed when Michael Koch, a German immigrant, came to New York City in about 1705. Soon after Jacob Cox's birth, the family returned to New York. The economic collapse of 1837 forced his family into poverty and deprived Cox of

a college education. Instead, Cox clerked in a law office and studied the law; in 1844, he changed his mind and clerked as a broker and banker, a profession that allowed him to save enough money to attend Oberlin College in Ohio, where he earned a degree in theology in 1851. While at Oberlin, he courted and married Helen Finney, the daughter of Dr. Charles Grandison Finney, theologian and first president of Oberlin College.

For the next two years, Cox served as the superintendent of schools and principal of the high school in Warren, Ohio, while again reading the law. In 1853, he was admitted to the Ohio bar and began a law practice. Although a Whig, Cox became a radical abolitionist and attended a convention in Columbus in 1855, which led to the founding of the state Republican Party. In 1859, he was elected to the Ohio state senate where, with future president James A. Garfield and future Supreme Court justice Salmon P. Chase, he formed the core of a radical antislavery group. With the outbreak of the Civil War, he volunteered for military service and was made the head of the Ohio militia. He was involved in the battles at Antietam, Atlanta, Franklin, and Nashville and was promoted in 1864 to major general. He ended the war in North Carolina, overseeing the parole of Confederate captives.

With his return to Ohio, Cox was nominated for governor. Ironically, he ran on a platform that opposed Negro suffrage and called for the segregation of former slaves into a separate state. He defeated his Democratic rival, George Mayan, and served from 1866 to 1868. His support of a separate Negro entity and his backing of President Andrew Johnson against the radical Republicans in Congress lost him the support of his party, and he was not renominated. After leaving office, he returned to his law practice, this time in Cincinnati.

In 1869, newly elected President Ulysses S. Grant tried to reach out to the more liberal wing of the Republican

Party and chose Cox as his secretary of the interior. Cox treated the Indians in a benevolent fashion and reformed the civil service area of the Interior Department, but for the duration of his tenure he was a political thorn in Grant's side. Wrote the president, "General Cox thought the Interior Department was the whole government, and that Cox was the Interior Department. I had to point out to him that there were three controlling branches of the Government, and that I was the head of one of these and would so like to be considered by the Secretary of the Interior." Eighteen months after taking office, on 5 October 1870, Cox resigned.

Jacob Cox spent the last three decades of his life in various political and business pursuits. A member of the Liberal Republican Party, he was elected to a seat in the U.S. House of Representatives and served a single term (1877–1879). He then served as dean of the Cincinnati Law School (1881–1897), retiring to write his *Military Reminiscences of the Civil War* (1900). Shortly after its completion, he died while on vacation in Magnolia, Massachusetts, on 4 August 1900.

See also Department of the Interior.

Dana, Samuel Trask (1883–1978)

Samuel Dana is considered one of the greatest forestry experts in the history of the United States. He was born in Portland, Maine, on 21 April 1883. After attending local schools, he enrolled at Bowdoin College in Brunswick, Maine, where he received his bachelor's degree in 1904. In 1907, he was awarded a master of forestry degree from Yale University.

In 1907, Dana entered the U.S. Forest Service. He rapidly advanced from forest assistant to assistant chief of the Office of Silvics (the study of trees and the environment) and Forest Investigations. During World War I, Dana served as a captain in the U.S. Army, advising the military on the efficient uses of wood products. After the war, he returned to the Forest Service and was named assistant chief of the Branch of Research under the direction of Earle Hart Clapp, Forest Service chief. Dana served as the forest commissioner for the state of Maine from 1921 to 1923, during which time he spoke out for the creation of a federal forest research center in the Northeast. When the Northeastern Forest Experiment Station was created in 1923, Dana was made its first director. He also served as the dean of the University of Michigan's School of Forestry until his retirement in 1953.

After retirement, Dana became a prolific writer, penning such works as *Forest and Range Policy: Its Development in the United States* (1956), *California Lands* (1958), and *Minnesota Lands* (1960). He was at one time editor of the *Journal of Forestry*. In his final years, Dana was awarded various honors for his lifetime of work, including the John Astor Warder Medal of the American Forestry Association. He died on 8 May 1978 at the age of 95.

Darling, Jay Norwood (1876–1962)

Jay Darling, known as "Ding," was a satirist and cartoonist as well as one of the founders of the National Wildlife Federation. He was born on 21 October 1876 in Norwood, Michigan (the source of his middle name), the son of a Congregational minister and his wife. When Darling was a young boy, the family moved to Sioux City, Iowa, and Darling considered himself an Iowan for the rest of his life. It was in Iowa and on his uncle's farm that he grew to love the outdoors and nature.

Darling studied at Yankton College in South Dakota in 1894, but the following year he was expelled because of a college prank. He worked for a time as a cub reporter for the *Sioux City Tribune*. Eventually he went to Beloit College in Wisconsin, where he earned a Ph.D. in biology in 1900. After graduation from Beloit, Darling joined the *Sioux City Journal*, where he started his career penning humorous and comical cartoon sketches. In 1906, he went to work for the *Des Moines Register and Leader*, where he remained (except for a short period, 1911–1913, when he worked for the *New York Globe*) until his retirement in 1949. In 1917, the *New York Tribune* syndicated his cartoon column in 130 newspapers nationwide. In 1923 and 1942, he was awarded the Pulitzer Prize for cartooning.

As his following grew, "Ding" Darling, as he signed all his cartoons, began to speak out in favor of conservation policies. He was instrumental in getting the Iowa legislature to create a state fish and game commission in 1931, and he was one of the first commissioners. Although a staunch Republican who counted Presidents Calvin Coolidge and Herbert Hoover among his friends, Darling accepted an appointment from Democrat

Cartoonist and conservationist Jay Norwood Darling

Franklin D. Roosevelt as chief of the Bureau of Biological Survey in 1934. As chief, he was responsible for enlarging the national system of federally protected wildlife refuges. With his backing and influence, environmentalists and sportsmen at the North American Wildlife Conference in 1935 founded the National Wildlife Federation. Darling served as the federation's first president from 1936 to 1939. Further, he lobbied for congressional passage of the Migratory Bird Hunting Stamp Act of 1934, which established the conservation stamp program to finance wildlife refuge management through the sale of stamps to hunters. Darling himself drew the first duck stamp.

Near the end of his life, Darling was named cochairman, with Walt Disney, of National Wildlife Week 1962, but before he could serve in this capacity he succumbed to heart failure in Iowa on 12 February 1962. He was 85 years old. The J. N. "Ding" Darling Wildlife Refuge on Sanibel Island, on the west coast of Florida, is named after him.

See also Migratory Bird Hunting Stamp Act; North American Wildlife Conference.

Dawes General Severalty Act
See General Allotment Act.

DDT
DDT (short for dichlorodiphenyltrichloroethane) is the insect-killing chemical that was used worldwide from the late 1940s until its banning in the United States by the Environmental Protection Agency (EPA) in 1972.

In 1874, German chemist Othmar Zeidler mixed the compounds that compose DDT. At first, he had no idea what he had discovered. In fact, his findings did not attract any interest until shortly before World War II, when it was found that the chemical could control lice and other pests. Its use picked up after the war, when it was utilized to kill lice on refugees in Europe. Its efficient properties and ease of application allowed it to

Only one mallard duckling hatched successfully in this clutch of twelve eggs. DDE, a breakdown product of DDT, caused eleven eggs to develop incompletely. Published in the December 1970 National Geographic, the photograph dramatized the cumulative effect of pesticide concentrations among wildfowl.

be used worldwide as a pesticide. Its hidden property, that it remains in the environment for a long time and produces massive damage, was its downfall. By the 1960s, animal malformations (such as cracked eggs and animals with birth defects not related to genetics) caused voices to be raised against DDT's continued use. Author Edwin Way Teale was the first to raise an alarm against the pesticide in an article in *Nature* magazine in March 1945, but his outcry was ignored. In 1962, author Rachel Carson sounded the first real alarm against DDT in her monumental work *Silent Spring*. It took the EPA ten years of study to finally conclude that DDT was deadly to all parts of the food chain, and its use was banned in the United States. However, to this day, some Third World countries, too poor to afford costlier pesticides, continue some use of DDT.

See also Carson, Rachel Louise.

De Brahm, William Gerard
(1717?–c.1799)

Dutchman William De Brahm served as surveyor general of the southern district of North America. He was an important early narrator of the fight to survey and preserve the environment in the colonial United States. Originally named John Gerar William De Brahm, he was probably born in Holland in 1717. Nothing is known of his background except that he received military training. All sources agree that he served as a captain of engineers in the army of Emperor Charles VI (1711–1740) of the Holy Roman Empire. His whereabouts after his service are unknown; he reappeared in England in 1751, leading a band of 160 Protestant worshippers from Salzburg to the English colonies. The group landed in Georgia, and it was there that De Brahm helped found the colony of Bethany. The settlement was later noted for its applications of farming and agriculture.

In 1754, British King George II appointed De Brahm as surveyor of Georgia.

In his first year in the position, De Brahm made a survey of eastern Georgia, joining it two years later with one prepared for South Carolina by that colony's Lt. Gov. William Bull (1710–1791). De Brahm also constructed fortifications at Charleston and Savannah and established Fort London and Fort George in Georgia. In 1764, he was appointed surveyor of the southern district. From 1765 until 1772, De Brahm conducted a survey of the southern Atlantic and Gulf coasts. His work in Florida was aided by his assistant, Bernard Romans, who later became a naturalist of some note himself. A biographer, Wilbert H. Siebert, wrote, De Brahm "prefaced his reports on his surveys of South Carolina, Georgia and East Florida with summaries of the early history of those provinces. His reports fill four folio volumes, are copiously illustrated with local and general maps, and contain his observations on practicable inlets and their rivers, their surroundings, his directions for mariners, his notes on soils, useful trees and shrubs, game, fish, Indians, Spaniards, etc. These volumes show an excellent command of the English language, some classical language, and much special and technical information." During a 1771 trip to England, according to one source, he described the Gulf Stream by means of a chart, which he publicized in his *The Atlantic Pilot* (London, 1772).

In 1770, De Brahm was removed from his position by Gov. James Grant under an order from the king for "overcharges, incivility and obstructing gentlemen in acquiring lands." De Brahm returned to England, where he wrote and published *The Atlantic Pilot* as well as *The Leveling Balance and Counter-Balance; or, The Method of Observing, by the Weight and Height of Mercury* (1774) and *De Brahm's Zonical Tables for the Twenty-Five Northern and Southern Climates* (1774). During a trip to Charleston in 1775, he described what he called the George Stream. In 1791, he returned to what was now the United States, where he settled in

Philadelphia and published *Time and Apparition of Eternity* (1791) and *Apocalyptic Gnomon Points Out Eternity's Divisibility* (1795). His work *History of the Province of Georgia* was later edited and published posthumously in 1849. De Brahm died in Philadelphia, although the date, 1799, is in some doubt. If so, he was about 82 years old.

See also Romans, Bernard.

Dead and Down Timber Act
(25 Stat. 673 [25 U.S.C. 196])

This legislation was enacted on 16 February 1899 to give the president of the United States the power to dispose of dead or downed timber on Indian reservations. The act reads: "Be it enacted ... that the President of the United States may from year to year in his discretion under such regulations as he may prescribe authorize the Indians residing on reservations or allotments, the fee to which remains in the United States, to fell, cut, remove, sell or otherwise dispose of the dead timber standing or fallen, on such reservation or allotment for the sole benefit of such Indian or Indians."

Delano, Columbus (1809–1896)

Columbus Delano was secretary of the interior during the Grant administration, serving from 1870 to 1875. Delano was born in Shoreham, Vermont, on 5 June 1809. His father died when he was six, and in 1817, he and his mother moved to Mount Vernon, Ohio, a suburb of Cleveland. There he received a limited education. He began the study of law in 1825 and was admitted to the Ohio bar in 1831. For the next several years, he was the prosecuting attorney of Knox County, Ohio.

In 1844, Delano was elected to a seat in the U.S. House of Representatives, where he served a single term in the 29th Congress. Although a Whig, he vehemently objected to the slave trade. In 1846, he attempted to be his party's candidate for governor of Ohio but failed. Instead, he moved to New York, where he engaged in the banking business as a member of the firm of Delano, Dunlevy & Company. In 1855, he quit the business and returned to Ohio, where he built a farm outside of Mount Vernon and raised sheep. Although he attended the 1860 and 1864 Republican National Conventions as a delegate and served a single term (1863) in the Ohio state legislature, it seemed that his political career was over.

With the end of the Civil War, however, Delano reentered politics. He was elected to his old seat in the U.S. House of Representatives. Now a radical Republican, he spoke out for the harsh reconstruction of the South. In 1869, President Ulysses S. Grant named him commissioner of internal revenue, a post he held for one year. During that time, Delano was notorious for cracking down on whiskey revenue fraud; his reorganization of the bureau resulted in a 100 percent increase in revenue returns in just eight months. In 1870, Grant asked him to replace Jacob Dolson Cox as secretary of the interior. Delano oversaw the reorganization of the Bureau of Indian Affairs, in which massive fraud was discovered. Blamed by his many critics for the fraud, Delano resigned. He returned to his sheep farm and served as president of the National Wool Growers' Association. Columbus Delano died at his home on 23 October 1896 of what one newspaper called "paralysis of the heart."

See also Department of the Interior.

Department of the Interior

The Department of the Interior is the cabinet-level agency responsible for land preservation and development in the United States and its possessions. Incredibly, the U.S. government spent the first half century of its existence without a central agency to manage the nation's land and water resources. In 1812, President James Madison named Edward Tiffin to head the new General Land Office, but the office was limited in its capability and scope and did little to cure the lack of cohesiveness among the various government agencies that had powers over land and water supplies. During the 1820s, the idea of a "Home Department" arose but was never legislated by Congress. The land and gold rushes, territories taken in the Mexican War, and explorations that accompanied the expansion of the United States during the 1840s made the need for such a department all the greater.

In the mid-1840s, Secretary of the Treasury Robert J. Walker thought that the work of the General Land Office, which was part of the Treasury Department, was taking up too much of his time. He approached Rep. Samuel Finley Vinton (1792–1862) of Ohio, chairman of the House Ways and Means Committee and a member of the House Public Lands Committee, to draft legislation creating a separate department to handle natural resources matters and have jurisdiction over the General Land Office. Vinton's legislation included the Office of Indian Affairs, Patent Office, Census Bureau, and other offices that existing agencies wanted to unload. The land department bill came to the floor of the House in early 1849, with one member objecting, according to one source, that the name "Interior Department" sounded too French. The bill finally passed both houses after vigorous debate, with dissenters claiming that a new government was being created.

President James K. Polk signed the Interior Department bill into law (9 Stat. 395) on 3 March 1849, consolidating many of the government's land programs. A day later, President Zachary Taylor took office. He named Thomas Ewing (1789–1871), a former U.S. senator, as the first secretary of the interior. Ewing served only one year before resigning in August 1850, but he recommended in his only annual report that a passage to the Pacific Ocean be constructed. Ewing was succeeded by Thomas M. T. McKennan, who served only several months until Alexander Hugh Holmes Stuart (1807–1891), a former congressman from Virginia, took office. During Stuart's term (1850–1853), the department was organized into efficient units that began to deal with the country's more serious land and water problems. President Franklin Pierce's interior secretary, Robert McClelland (1807–1880), a former governor of Michigan, was, according to one source, "personally low-keyed and unobtrusive ... brusque and demanding in his official capacity." Overseeing the offices dealing with land, Indians, patents, and pensions, he attempted major reforms, introduced rules to make the procedures of the offices more orderly, and tightened work discipline, but land and pension frauds and land speculators doomed his endeavors. One of the more colorful of the early interior secretaries was Jacob Thompson of Mississippi, appointed by James Buchanan because of his southern loyalties. Thompson served all but four months of his four-year term (1857–1861), resigning over objections to Fort Sumter being resupplied just before the outbreak of the Civil War. Thompson later served as a Confederate agent, and after the war he was forced into exile in Europe over reports that he was involved in the assassination of Abraham Lincoln.

Following the election of 1860, the interior secretary became an important member of the cabinet, with increased control over the railroads, Indians, and, with the passage of the Homestead Act in 1862, homesteaders and their land claims. It was during the tenure of Caleb Blood Smith (1808–1864), the first of Abraham Lincoln's two secretaries of the interior, that the landmark Homestead Act was passed, opening the door of opportunity to thousands of families moving west. Yet it was under Smith's successor, John Palmer Usher (1816–1889), that the full ramifications of the Homestead Act were seen, and larger appropriations were requested for Indian reservation management and protection. Usher's successor, James Harlan (1820–1899), served admirably as secretary. His tenure was noted for the clearing out of unnecessary workers in the department, although he resigned over differences with President Andrew Johnson (later, as a senator, he even voted to impeach Johnson) and was later accused of corruption and graft in Cherokee Indian sales. Orville Hickman Browning (1806–1881) spent a considerable amount of time liberalizing government policy toward the Indians.

In the latter part of the nineteenth century, the department became responsible for national parks, forest reserves, and, in the first years of the twentieth century, wildlife refuges. Under the leadership of German expatriate Carl Schurz (1829–1906), Lucius Q. C. Lamar (1825–1893), Hoke Smith (1855–1931), Ethan Allen Hitchcock (1835–1909), and Walter Fisher (1862–1955), the Interior Department became a leading promoter of responsible conservation policy and ethics. Although the Teapot Dome scandal during the tenure of Albert B. Fall (1861–1944) gave the department a black eye, later secretaries such as Harold L. Ickes (1874–1952), Oscar Chapman (1896–1978), Fred A. Seaton (1909–1974), and Stewart Udall (1920–) made the post one of the leading cabinet positions. The department again became controversial during the tenure of James Watt (1938–) but seemed to revert to its conservationist tendencies under Donald Hodel (1935–) and Bruce Babbitt (1938–).

Desert Land Act (19 Stat. 377)

This federal legislation, enacted on 3 March 1877, authorized the sale of 640-acre lots of "nontimber, nonmineral, uncultivable" land at 25 cents an acre to any person who would settle on it and irrigate it according to law. The act affected only the states of California,

Oregon, and Nevada and the territories of Washington, Idaho, Montana, Utah, Wyoming, Arizona, New Mexico, and Dakota. The downside of this act was its requirement that such settlers bring the irrigation waters to the lands by themselves. This requirement sounded the death knell for reclamation in the West and led to the formation of the National Irrigation Congress in 1891 to lobby for federal aid in irrigation, paving the way for passage of the Carey Irrigation Act in 1894 and the Newlands Reclamation Act in 1902.

See also National Irrigation Congress.

DeVoto, Bernard Augustine (1897–1955)

As a well-known Pulitzer Prize–winning writer, critic, and essayist, Bernard DeVoto used his reputation and national standing to become what the *New York Times* called "an emphatic proponent of [the] preservation of the nation's natural resources." He was born on 11 January 1897 in Ogden, Utah, the only child of Florian Bernard DeVoto, an Italian immigrant, and Rhoda (nee Dye) DeVoto. "I was the child of an apostate Catholic and an apostate Mormon," he later wrote. His maternal grandfather, Samuel Dye, had immigrated to the United States in 1856 after converting to Mormonism. DeVoto later chronicled Dye's life in his essay "The Life of Jonathan Dyer." He attended a convent school but, becoming disenchanted with religion, transferred to public schools. His turn away from religion became complete when, at the University of Utah, he founded a campus socialist society and proudly boasted of owning a copy of Karl Marx's *Das Kapital.*

After transferring to Harvard, DeVoto was encouraged by his professors to become a writer. Although a pacifist, he served a short stint in the U.S. Army during World War I as a weapons instructor. After the war, he returned Harvard and was awarded a bachelor's degree in 1920. In 1921, he taught history at Ogden High School, and from 1922 to 1927 he was a professor of English at Northwestern University in Evanston, Illinois. It was during these years that his first two novels were published. *The Crooked Mile* (1924) lambasted the Western way of life in which he had been raised, and *The Chariot of Fire* (1926) questioned his religious roots in a story reminiscent of the tale of Elmer Gantry. His 1928 work, *The House of Sun-Goes-Down*, was the first in what was supposed to be a trilogy about life in Utah.

In 1926, DeVoto began writing a series of essays that appeared in social critic H. L. Mencken's *American Mercury*. After moving with his family to Cambridge, Massachusetts, so that he could teach at Harvard, he penned *Mark Twain's America* (1932), a biting social criticism of U.S. history. Articles that appeared in *Collier's* magazine were eventually published as novels under the pseudonym John August; these were *Troubled Star* (1939), *Ran before Seven* (1940), *Advance Agent* (1942), and *The Woman in the Picture* (1944). Historian Wallace Stegner, who edited *The Letters of Bernard DeVoto* (1975), noted that DeVoto's essays in literary criticism and articles on conservation, civil liberties, and censorship continued to appear in a wide range of magazines. His conservation philosophy is evident in the works *Year of Decision: 1846* (1943), which discussed the impetus behind Manifest Destiny, the great push to expand the American frontier, and *Across the Wide Missouri* (1947), a narrative of the fur-trade industry. This latter work won DeVoto the Pulitzer and Bancroft Prizes. His best-known work, *The Course of Empire* (1952), appeared shortly before his death and is considered one of the great histories of the American West.

For the last 20 years of his life, from 1935 to 1955, DeVoto wrote a monthly column for *Harper's Magazine* called "The Easy Chair," in which he was free to voice social criticism. Wrote biographer Glenn Q. Snyder, "Constantly re-minding his readers that 'this is *your* land we are talking about,' he frequently used the 'Easy Chair' as an educational forum on the public domain and the threats to it from stockmen, loggers and power companies. In the late 1940s, his exposé effectively stopped an attempted landgrab by western stockmen." A number of his essays appeared under the title "Treatise on a Function of Journalism" in the work *The Easy Chair* (1955), published shortly before his death. His conservation ethic is best illustrated in an essay he wrote in 1955: "I have got to have the sight of clean water and the sound of running water. I have got to get to places where the skyshine of cities do not dim the stars, where you can smell land and foliage, grasses and marshes, forest duff and aromatic plants and hot underbrush turning cool. Most of all, I have to learn again what quiet is. I believe that our culture is more likely to perish from noise than from radioactive fallout." DeVoto was also the editor of the *Saturday Review of Literature* for a number of years. He suffered a heart attack in New York City and died on 13 November 1955 at the age of 58. Wallace Stegner profiled him in 1974's *The Uneasy Chair: A Biography of Bernard DeVoto*.

Dilg, Will (1867–1927)

Will Dilg was a sportsman who founded the Izaak Walton League, a conservation group composed mainly of sportsmen. Little information is available about Dilg's early life. Born in Milwaukee, Wisconsin, in 1867, he later became a publicist and advertising salesman. His favorite pastime, fly-fishing, led him to write several well-known articles, including "The Cork-Bodied Black-Bass Bug," and his column, "Here and There With Will Dilg."

On 14 January 1922, Dilg invited fellow fishermen as well as outdoor writers and publishers to a luncheon at the Chicago Athletic Club. That afternoon, the Izaak Walton League was

founded. Named for the English sportsman whose 1653 work, *The Compleat An-
gler*, is a minor classic of English
romanticism, the league intended to call
for conservation from a sportsmen's perspective. Among its first drives, wrote author Robert Herbst, "was to fight to
preserve clean waterways and combat
stream pollution." The group led the way
in lobbying the governor of Illinois to
outlaw the sale of black bass because of
its low numbers. Expanded interests later
led the league to call for the protection of
the Superior National Forest (which received a wilderness designation in 1926
and was incorporated into the Wilderness Act of 1964) and to help win (with
Dilg's personal assistance) the establishment of the Upper Mississippi River
Wild Life and Fish Refuge in 1924.

In April 1926, Dilg was removed as
head of the league after his opponents
made allegations against him. At the
time, Dilg was battling throat cancer and
was unable to fight his dismissal. Instead,
he went to Washington to finish his personal fight to establish a national department of conservation as a cabinet agency.
Although legislation was introduced in
Congress, it failed to pass. Will Dilg died
in a Washington, D.C., hospital on 8
March 1927. Wrote fellow sportsman
William Temple Hornaday, "The death
of Will Dilg, inspired leader in national
wild life protection, is a great loss to the
American people and to wild life. He it
was who accomplished in unbelievably
quick time the theoretically impossible
task of inducing Congress to make the
vast and far reaching Upper Mississippi
Wild Life Sanctuary, thereby amazing
his friends and confounding his enemies." E. C. Bassett, a field representative of the Izaak Walton League, once
said of his boss, "Will Dilg is a visionist, a
dreamer of dreams, impractical in some
things but cold and calculating and keen
and amazingly efficent in others, one of
the greatest publicists of a decade and
whether you like to admit it or not it is
his finger that points the way to aggres-

sive militant conservation activities. He
does not deal with consideration of possible conservation problems; he deals
with performance of definite conservation activities. He is willing to start doing
the things that cannot be done." In 1952,
the Izaak Walton League established a
monument to Dilg in Prairie Island Park
in Minnesota. It reads: "Years of untiring
and unselfish effort were devoted by Will
Dilg to the cause dearest to his heart: the
restoration and conservation of America's
wild life resources."

See also Izaak Walton League of America.

Dingell-Johnson Act (64 Stat. 430)

This act, also known as the Fish Restoration and Management Act, was passed on
9 August 1950 to fund wildlife refuges
and fish management through an excise
tax on fishermen. It is, according to one
source, an "extension of the Pittman-
Robertson Act of 1937." Sponsored in
the House by Rep. John David Dingell
of Michigan (1894–1955), this act placed
a 10 percent tax on the manufacturer's
price of "fishing rods, creels, reels, and
artificial lures, baits, and flies."

See also Pittman-Robertson Act.

Divison of Biological Survey

See Merriam, Clinton Hart.

Division of Ornithology and Mammalogy

See Merriam, Clinton Hart.

Dolan v. City of Tigard (U.S. 1994)

This landmark Supreme Court decision,
handed down at the end of the 1994
Court term, established that government
may not take private property without
just compensation, even when the purpose is to protect the environment. John
and Florence Dolan owned A-Boy Electric & Plumbing Supply, a store in the
city of Tigard, Oregon. The Dolans

asked the city for permission to expand their 9,700-square-foot store to 17,600 square feet by ripping up the existing parking lot and paving a new one next to the store on their own land, which included Fanno Creek. The city granted the Dolans' request with the proviso that they "donate" a portion of the property for use as a storm drainage system and another strip for a bicycle path. The Dolans refused and appealed to the Land Use Board of Appeals (LUBA), arguing that the regulations amounted to an unfair and uncompensated taking of their property, which was unconstitutional under the Fifth Amendment. LUBA found that because the expansion of the Dolans' business would cause increased traffic and lessen the amount of land available for drainage, it was "reasonable" for them to donate some land for a drainage system as well as a strip for a bicycle path to stimulate "alternative means of transportation." The Oregon Court of Appeals affirmed the decision, and so did the Oregon Supreme Court on appeal. The Dolans took the case to the U.S. Supreme Court, but John Dolan died of leukemia before the case could be heard. His son and widow carried on the case.

On 24 June 1994, in a 5–4 decision (Justices John Paul Stevens, Harry Blackmun, Ruth Bader Ginsburg, and David Souter dissented), the Court held that, under the Fifth Amendment, such restriction of the Dolans' land use was unconstitutional and sent the case back for a rehearing before the Oregon Supreme Court. Speaking for the majority, Chief Justice William Rehnquist wrote:

The Takings Clause of the Fifth Amendment of the United States Constitution, made applicable to the States through the Fourteenth Amendment ... provides: "[N]or shall private property be taken for public use, without just compensation." One of the principal purposes of the Takings Clause is "to bar Government from forcing some people alone to bear public burdens which, in all fairness and justice, should be borne by the public as a whole." Without question, had the city simply required petitioner to dedicate a strip of land along Fanno Creek for public use, rather than conditioning the grant of her permit to redevelop her property on such a dedication, a taking would have occurred. Such public access would deprive petitioner of the right to exclude others, "one of the most essential sticks in the bundle of rights that are commonly referred to as property."

Douglas, Marjory Stoneman (1890–)

Author and environmental advocate Marjory Stoneman Douglas has spent most of her life defending Florida's Everglades. Born to Frank Bryant Stoneman, an attorney who later became a judge in Miami, Florida, and Lillian (nee Trefethen) Stoneman in Minneapolis, Minnesota, on 7 April 1890, Marjory Stoneman spent much of her childhood with her grandmother in Taunton, Massachusetts, following her mother's death. She attended the public schools in Taunton, then enrolled at Wellesley College where she earned a bachelor's degree in English in 1912. Her father moved to Miami, where he founded the *Miami Herald*. In 1914, Marjory followed him to Miami and joined the staff of his paper. That year, she married Kenneth Douglas (a writer 30 years her senior), but the two were divorced in 1917 and Marjory never remarried. She served as the *Herald*'s society editor (with her column "The Galley") before serving for a time in Europe as a member of the American Red Cross. In 1920, she became the *Herald*'s assistant editor. In 1923, she left the *Herald* to begin a career of freelance writing.

Frank Stoneman had been interested in the welfare of the nearby Everglades

Marjory Stoneman Douglas receives the Coe Award from Everglades National Park Superintendent Bob Chandler in 1990. By 1994 there were fewer than 60 Florida panthers in existence.

since his arrival in Florida. The lush wet-lands of the Everglades, the runoff point of much of the water table for southern Florida, was considered by many land speculators as worthless swampland to be drained and exploited. Frank's daughter took up his cause and became its best-known advocate. Several of her articles in the *Saturday Evening Post* on the Ever-glades marshes near Miami brought her to the attention of a group that ap-pointed her in 1927 to lobby for the es-tablishment of Everglades National Park. Twenty years later, her efforts and the history of the park were examined in her landmark *The Everglades: River of Grass.* This work, according to the *Miami Herald,* "established Douglas as the god-mother of the Everglades." Her work led the *Herald* to say that "hers is the voice of authority about Florida."

Following *River of Grass* was *Road to the Sun* (1951), Douglas's first novel. Her other works include a series of stories about the colorful people and places that make up southern Florida, which were published in the *Saturday Evening Post* in the 1930s and 1940s. These were com-piled as *Nine Florida Stories by Marjory Stoneman Douglas* (1990). Her autobiog-raphy, *Voice of the River,* appeared in 1987. Douglas also wrote book reviews for the *New York Herald-Tribune.* In 1993, she was awarded the Medal of Freedom by President Bill Clinton. She continues to live in the small masonry house near the Everglades that she built in the 1920s.

Duke Power Company v. Carolina Environmental Study Group, Inc.
(438 U.S. 59 [1978])

In this case, the U.S. Supreme Court held that the defendant, Carolina Environ-mental Study Group, Inc., had standing to sue a nuclear power plant, although the Court upheld a congressional act that set the maximum liability of the plaintiff in such an action.

Attempting to halt the construction of a nuclear power plant to be built by Duke Power Company, a group of environmentalists, a labor union, and persons residing near the planned facility sued the company and the Nuclear Regulatory Commission to enjoin its construction, alleging that (1) the plant could cause them injury from radiation and (2) under the Price-Anderson Act (42 U.S.C. 2210), in case of an accident, the facility was limited to a liability of $560 million, which restricted the people's right to adequate compensation. A district court found the Price-Anderson Act unconstitutional because it denied people the right of due process. The Duke Power Company asked the Supreme Court for a hearing on the case. Four questions needed to be answered: Did the district court have jurisdiction over the lawsuit? Did the original plaintiffs have the right to sue? Should the Court review the case? Was the Price-Anderson Act an unconstitutional burden on the right of due process? The Court answered yes to the first three questions and no to the fourth. In a sweeping decision written by Chief Justice Warren Burger, it was held that under 28 U.S.C. 1331(a), district courts have jurisdiction over decisions of the Nuclear Regulatory Commission; the plaintiffs had standing to sue because, as the district court found, the nuclear power plant was a threat to the health of the people around it; and the Price-Anderson Act was constitutional because, as an economic regulation, it met the standard of constitutionality if it "advances the nation's legitimate interest in fostering the development of nuclear power."

Dutcher, William E. (1846–1920)

William Dutcher was a noted ornithologist during the latter part of the nineteenth century and was one of the founders of the National Association of Audubon Societies. Born in Stelton, New Jersey, on 20 January 1846, he received a public school education. For many years he was an insurance agent for the Prudential Insurance Company.

In 1883, Dutcher joined the American Ornithologists' Union and, over the next year, worked on its Committee on Bird Protection. In 1896, he became the chairman of the committee and began to encourage the founding of state clubs to advance the cause of wildlife protection, particularly bird preservation. These clubs became Audubon Societies. In 1905, when more than 40 such clubs existed, Dutcher founded the National Association of Audubon Societies (now the National Audubon Society). He served as the association's president from 1905 until his death.

Dutcher was not just the head of an organization; he was also an investigator and lobbyist. Following the passage of the landmark Lacey Act in 1900, which gave federal protection to wild birds and their eggs, he and ornithologist Frank M. Chapman surreptitiously visited pet shops in New York to find violations of the law. In 1910, he helped lobby for the passage of the Plumage Act in New York State, which outlawed the sale of plumage of domestic and foreign birds. Such legislation was made national by the passage of the Weeks-McLean Act in 1913 and international by the signing of the Migratory Bird Treaty Act between the United States and Great Britain in 1916.

In his final decade, Dutcher was afflicted by the effects of a stroke that left him speechless. He lost control of his beloved Audubon Society. His daughter's death in 1910 was also devastating. William Dutcher died at his son's home in Plainfield, New Jersey, on 1 July 1920.

See also National Audubon Society.

Dutton, Clarence Edward (1841–1912)

Although not well known for his work during the John Wesley Powell survey of

Clarence Edward Dutton

the arid West, Clarence Dutton was one of the leading geologists of his time and contributed to the significant early discoveries of the U.S. Geological Survey. Born in Wallingford, Connecticut, on 15 May 1841, he was the son of Samuel and Emily (nee Curtis) Dutton. After attending local schools in nearby Ellington, he was ready to enter Yale University at age 13 but was held back because his parents thought him too young for collegiate life. Two years later, however, he enrolled at Yale and graduated in 1860 with a bachelor's degree. He then attended the Yale Theological Seminary.

When the Civil War broke out, Dutton wanted to enlist but waited until September 1862, when he was commissioned adjutant of the 21st Connecticut Volunteers. Specializing in ordnance, Dutton was promoted in March 1864 to captain. He served much of the war at the Watervliet, New York, arsenal, where a nearby Bessemer steel factory secured his interest. In 1870, he was transferred to the Frankford Arsenal in Philadelphia, and later to Washington, D.C., where he became acquainted with scientists such as Spencer Fullerton Baird and Joseph Henry of the Smithsonian Institution, Ferdinand V. Hayden and John Wesley Powell of the geological surveys of the West, and Simon Newcomb of the Naval Observatory. Dutton's interest in geology led Powell to seek a congressional mandate in 1875 to allow the young officer to leave the military and travel west as a member of his geological survey.

Wrote Dutton's biographer, Wallace Stegner, "From the beginning of his geological studies, Dutton was interested in orogenic problems, and during his years of work in the plateau region of Utah, Arizona, and New Mexico he had the opportunity to study not only the faults and monoclines along which uplift and subsidence had taken place but also the extensive volcanism that had accompanied these earth movements." Part of his work was with the Powell survey; when the great surveys of the West were consolidated into the U.S. Geological Survey in 1879, Dutton continued his labors as a valued aid to Clarence King and Powell, the first two survey directors. His monographs on his discoveries, the 307-page *Report on the Geology of the High Plateaus of Utah* (1880), volume 32 of the *U.S. Geographical and Geological Survey of the Rocky Mountain Region*, and the 264-page *Tertiary History of the Grand Canyon District*, U.S. Geological Survey Monograph no. 2 (1882), confirmed Dutton as a leading scientific mind in the study of the earth's physical features. Later works included "Mount Taylor and the Zuñi Plateau" in *Report of the United States Geological Survey* no. 6 (1885). He later studied Hawaiian volcanoes, documented in his *Hawaiian Volcanoes* (1884), and the route to be taken by the proposed Nicaraguan canal. From 1888 to 1890, he was a leading member of the Powell irrigation survey. In 1890, Dutton left the survey and returned to military duty, resigning in 1901. He died in Englewood, New Jersey, on 4 January 1912 at the age of 70.

See also United States Geological Survey.

E. I. Du Pont de Nemours and Company v. Train
(430 U.S. 112 [1977])

This Supreme Court case hinged on the question of whether the Federal Water Pollution Control Act Amendments of 1972 authorized the administrator of the Environmental Protection Agency (EPA) to establish effluent restrictions for existing, but not new, chemical manufacturing plants nationwide. The 1972 amendments called for the EPA administrator to implement a series of measures that would eliminate effluent discharge from chemical plants by 1985. Under section 301 of the 1972 amendments, existing plants must apply "the best available technology" to achieve this goal. In 1977, Congress amended the 1972 amendments and instituted a 1977 deadline for applying this mandate industrywide.

In 1977, Russell Train, the EPA administrator, issued guidelines that were consistent for both existing and new chemical plants. E. I. Du Pont de Nemours, one of the world's largest chemical companies, joined eight other smaller companies in demanding that the regulations apply only to existing plants and that such rules be employed on a plant-by-plant basis. Although the court of appeals hearing the case sided with Train and the EPA, it recommended that the agency give new plants "variances" in allowing effluent discharges. Unimpressed by the recommendation, Du Pont appealed to the U.S. Supreme Court. The Court ruled 8–0 (Justice Lewis Powell did not participate) that the 1972 amendments allowed the EPA to establish such industrywide regulations regardless of the age of the plant. Justice John Paul Stevens wrote the Court's unanimous opinion. He argued that if the EPA were forced to consider effluent discharge permits on a plant-by-plant basis, it would force the administrator "to give individual consideration to the circumstances of each of more than 42,000 dischargers who have applied for permits, and to issue or to approve all these permits well in advance of the 1977 deadline in order to give industry time to install the necessary pollution control equipment."

Earth Day

Celebrated originally on 22 April 1970 and subsequently every year on that day, Earth Day was the idea of Sen. Gaylord Nelson, a Democrat from Wisconsin. Demonstrators numbering about 20 million brought environmental awareness to the forefront of the nation's psyche with marches and lobbying efforts to influence Congress to enact laws that would improve the nation's environment. Wrote Nelson, "It was on that day that Americans made it clear that they understood and were deeply concerned over the deterioration of our environment and the mindless dissipation of our resources. That day left a permanent impact on the politics of America. It forcibly thrust the issue of environmental quality and resource conservation into the political dialogue of the nation. That was the important objective and achievement of Earth Day. It showed the political and opinon leadership of the country that the people cared, that they were ready for political action, that the politicans had better get ready, too. In short, Earth Day launched the Environmental Decade with a bang." Due to that undertaking, such legislation as the Clean Air Act, the Clean Water Act, the Endangered Species Act, the Federal Land Policy and Management Act, the Toxic Substances Control Act, and the Resource Recovery Act, among others, has been enacted.

Earth First!

This group is considered the most radical in thought and action in the environmental movement today. Founded in the early 1980s in the northwestern United States to protest the growing destruction of the forests in that area, Earth First! began to use "monkey-wrenching"—destructive techniques known as ecotage (environmental sabotage). These techniques include pouring sugar and sawdust in the gas tanks of bulldozers clearing areas for logging roads and booby-trapping trees with metal spikes to destroy tree-cutting equipment in the forests or the lumber mills. Recently, some of the group's members have been tried for allegedly attempting to cut the power lines to nuclear power plants. As of 1990, the group claims to have some 15,000 members nationwide.

Earth Island Institute

Founded in 1982 by Sierra Club member David R. Brower, this environmental group was responsible for the passage in 1992 of the "dolphin safe" law, which prohibits U.S. tuna fishing fleets from catching tuna at the expense of the lives of dolphins that swim nearby. The cause of this action was an undercover film made by institute member Sam LaBudde, which depicted the senseless slaughter of dolphins by tuna fishermen. With a 1993 membership of about 35,000, this group ranks among the smaller environmental organizations. According to the group, its members "sponsor and develop innovative projects to conserve, preserve, and restore our natural environment." Its official magazine is the *Earth Island Journal*. Current programs include the Sea Turtle Restoration Project, Baikal Watch (devoted to saving Russia's Lake Baikal and the surrounding Siberian area), and the Brower Fund, which provides "short-term seed funding for innovative approaches to environmental problem solving."

See also Brower, David Ross.

Eastern Wilderness Act

See National Wilderness Preservation System.

Edge, Mabel Rosalie Barrow (1877–1962)

A socialite who had little interest in conservation until midlife, Mabel Edge was a member of the National Audubon Society and a founding member of the Emergency Conservation Committee, an anti-Audubon group. Born Mabel Barrow on 3 November 1877 into a rich New York family (her father was a cousin of Charles Dickens), she attended a prestigious Manhattan finishing school before marrying wealthy engineer Charles Noel Edge in 1909. For the next four years, she lived in Europe. In 1913, as she was returning to the United States, she met a British suffragette who gave her a cause worth fighting for: the right of American women to vote. When she moved to a home on Long Island, she began to watch the birds on the seashore. She later said of her bird-watching activities, "When we who have a feeling for birds observe a mighty eagle, or the perfection of a tiny warbler, we see, not the inspiration of God filtered through the human agency, but the very handiwork of the Creator himself."

In 1929, while in the process of divorcing her husband, she took up a new cause: reforming the Audubon Society. Having read William G. Van Name's scathing *A Crisis in Conservation*, which accused the society of receiving funds from firearms manufacturers, Edge set out to remedy these abuses. In 1930, she, Van Name, and writer Irving Brant formed the Emergency Conservation Committee (ECC), with the intention of forcing changes in the society. Its greatest effect came in 1934 with the resignation of Audubon president T. Gilbert Pearson. After this initial victory, the ECC turned to the environment as a whole, working for the creation of the Hawk Mountain Wildlife Sanctuary in

Pennsylvania, the establishment of Olympic National Park in Washington State, and, in conjunction with the Sierra Club and other groups, the creation of the Kings Canyon National Park in 1940. Because of her inflexible refusal to consider any development, Edge made enemies of many environmentalists, eventually driving even Brant and Van Name away. She worked on alone for a number of years, financing her activities from her private fortune. When she died on 30 November 1962 at the age of 85, the ECC went out of business.

Egleston, Nathaniel Hillyer (1822–1912)

Nathaniel Egleston was the chief of the U.S. Division of Forestry who called for governmental protection for federal forestlands as well as their limited use by the general public. He was born in Hartford, Connecticut, on 7 May 1822, the son of Nathaniel and Emily (nee Hillyer) Egleston. The family's origins date back to Bagot Egleston, an Englishman who left Exeter in 1630 and landed at Nantasket, Massachusetts. Nathaniel Egleston attended local schools in Hartford, then enrolled at Yale College, from which he graduated in 1840, and Yale Divinity School, which awarded him a degree in theology in 1844. From his graduation until 1869, he served in various pastorates from Brooklyn, New York, to Chicago, Illinois. In 1853, he was a founding member of the American Congregational Union. He also served as a professor at Williams College in Williamstown, Massachusetts.

While at Williams, Egleston developed a love of forestry matters. He began to compose articles for national magazines that laid out his concern about the condition of the nation's forestry resources. When the American Forestry Association and the American Forestry Congress consolidated operations in 1882, Egleston, a delegate to the Montreal convention where the two merged, was named

the new organization's vice president. In 1883, association president George B. Loring, who was also the commissioner of agriculture in the Chester A. Arthur administration, named Egleston chief of the U.S. Division of Forestry, replacing Franklin B. Hough. Egleston served until 1886. His three years in the forestry post were marked by his sole report to Loring, a 462-page treatise calling for action "to ensure that the extensive forestlands owned by the federal government were properly cared for and were used for the general welfare." Egleston, working under Grover Cleveland's new Agriculture Commissioner Norman J. Colman, found his work irritating and refused to issue another annual report. On 15 March 1886, Cleveland fired Egleston and replaced him with Bernhard E. Fernow. Egleston remained with the division, amassing information for future forestry reports. He resigned in 1898.

During the last two decades of his life, Egleston lived at his home in Jamaica Plain, Massachusetts. The author of *Villages and Village Life* (1878), *Hand-Book of Tree-Planting* (1883), and *Arbor Day Leaves* (1893), he died on 24 August 1912 at the age of 90.

Emergency Conservation Committee (ECC)

See Edge, Mabel Rosalie Barrow.

Emergency Planning and Community Right-to-Know Act (42 U.S.C. 11001)

This legislation, also known as the Superfund Amendment and Reauthorization Act, was enacted by Congress on 17 October 1986 in response to the emergency handling of chemicals across the nation. Following the release of toxic gas in Bhopal, India, in 1984, which killed more than 2,500 people, the Environmental Protection Agency (EPA) established the Chemical Emergency Preparedness Program (CEPP) to increase the awareness of state and local authorities of the probability of accidents involving hazardous chemicals and advance the creation of plans to deal with such accidents. Chemical companies constituted the Community Awareness and Emergency Response (CAER) program, which called for greater public awareness of chemical creation by chemical plants. The right-to-know movement came into the picture when some 30 states passed laws giving citizens the right to know what chemicals were being produced in their communities.

There are several important sections of the Emergency Planning and Community Right-to-Know Act (EPCRA). They deal with emergency planning (sections 301–303), emergency release notification (section 304), hazardous chemical reporting (sections 311–312), toxic chemical release reporting (section 313), chemical company trade secrecy (section 322), and the allowance of penalties and citizen lawsuits (sections 325–326). Under the emergency planning section, governors would appoint state emergency response commissions (SERCs), which in turn would form local emergency planning committees (LEPCs). Facilities dealing in chemicals would then inform these SERCs and LEPCs of any accidental releases of chemicals. Material safety data sheets (MSDSs) would be required to inform communities what chemicals were on-site. These data would be compiled in the EPA's Toxic Chemical Release Inventory (TCRI), which would be open for public inspection. Companies could claim that certain chemicals were subject to trade secrecy, but by law, they would have to verify this claim. In the final provision, EPCRA allows for penalties of $10,000 to $75,000 a day for failure to comply with the law. SERCs and LEPCs could initiate lawsuits on behalf of citizens injured by chemical spills.

Emergency Wetlands Resources Act (16 U.S.C. 3901)

This law was enacted on 10 November 1986 to protect what remains of the nation's wetlands. According to the act, Congress discovered that

1) wetlands play an integral role in maintaining the quality of life through material contributions to our national economy, food supply, water supply and quality, flood control, and fish, wildlife, and plant resources, and thus to the health, safety, recreation, and economic well-being of all of the citizens of the nation, and 2) that wetlands provide habitat essential for the breeding, spawning, nesting, migration, wintering and ultimate survival of a major portion of the migratory and resident fish and wildlife of the Nation; including migratory birds, endangered species, commercially and recreationally important finfish, shellfish and other aquatic organisms, and contain many unique species and communities of wild plants.

The act allowed for the collection of fees for admission permits to refuges, with 30 percent of such fees going toward offsetting the cost of operations and for administration and maintenance of the refuges, and 70 percent going into the migratory bird conservation fund established in the act.

This act was amended by the North American Wetlands Conservation Act of 1989 to request a comparison of U.S. wetlands existing in 1780 with those existing in 1990.

See also North American Wetlands Conservation Act.

Emerson, George Barrell (1797–1881)

Known more as an educator than as a botanist, George B. Emerson was an American scientist who contributed to the early conservation movement. One of nine children of Dr. Samuel Emerson, a Revolutionary War veteran and physician, and Sarah (nee Barrell) Emerson,

Emerson was born on 12 September 1797, in Wells, Maine, then part of Massachusetts. He went to local schools and attended to his family's farm before attending the Dummer School in Byfield, Massachusetts. Partially tutored by his father, he entered Harvard University and graduated in 1817 with a B.A. degree. After graduation, he was the principal of a private school in Lancaster, Massachusetts, a tutor in mathematics at Harvard, and principal of the Boston English Classical School before founding a private school for girls in Boston. In the area of education, his written works include *A Lecture on the Education of Females* (1831), *The Massachusetts Common School System* (1841), and *Education in Massachusetts: Early Legislation and History* (1869).

George Emerson's other work was in the area of botany and arboriculture. His early interest in the shrubs and trees of Massachusetts led him to join the Boston Society of Natural History, which elected him its president in 1837. When the state legislature established the Zoological and Botanical Survey of Massachusetts, it named Emerson as its administrator. His nine years of study were documented in his *Report on the Trees and Shrubs Growing Naturally in the Forests of Massachusetts* (1846), which naturalist Asa Gray called a classic of New England botany, according to one source. Emerson subsequently reissued the report in two volumes in 1875, with a fifth and final edition published posthumously in 1903. In 1873, he and Franklin Benjamin Hough were members of a committee formed by the American Association for the Advancement of Science to prepare a report detailing the need for a national forestry policy that would be submitted to Congress. One source even credits Emerson with obtaining the bequest that evolved into the Arnold Arboretum, a conservatory of plant, flower, and shrub culture in Boston. Emerson died in Newton, Massachusetts, on 4 March 1881 at the age of 83.

See also Hough, Franklin Benjamin.

Emmons, Samuel Franklin (1841–1911)

A pioneer mining and geology expert, Samuel Emmons was a key member of the Clarence King survey during the 1860s and 1870s. He was born in Boston on 29 March 1841, the son of Nathaniel Henry Emmons, a prosperous Boston merchant, and Elizabeth (nee Wales) Emmons. Emmons was a descendant of Thomas Emmons, whom one source notes was "one of the founders of the Rhode Island Colony" and a citizen of Boston in about 1648. On his mother's side, he was descended from Nathaniel Wales, who emigrated from Yorkshire, England, to Boston in 1635. Samuel Emmons was named after a great-grandfather, Samuel Franklin, a cousin of Benjamin Franklin. His father's wealth allowed Emmons to attend private Boston schools, including the Dixwell Latin School, where he was educated in geology and physical geography, areas of study in which he later majored. In 1858, he entered Harvard University and was awarded a bachelor's degree three years later.

Following graduation, Emmons wanted to enlist in the Union Army, but, bowing to his parents' wishes, he instead headed off to Europe, where he spent five years exploring the Alps and getting an extended private education in Paris before enrolling at the prestigious École Impériale de Mines (Imperial Mining School) in France and the Royal School of Mines, known as the Bergakademie, in Freiberg, Germany. At the latter institution, he became friends with Arnold Hague, who would, like Emmons, become one of America's leading geological authorities. In 1866, Emmons returned to the United States.

With the establishment of the Geological Surveys of the Fortieth Parallel, known more commonly as the King survey, Arnold Hague and his older brother James were hired as geologists. On Arnold Hague's recommendation, King hired Emmons as an assistant geologist, and he stayed with the survey until its completion in 1877. Because of his work, a peak in the Uinta Mountains of Utah is called Mount Emmons. As a member of the survey, he was the author of volume 3 of the survey reports, *Geology of Toyabe Range* (1870), and the coauthor with Hague on volume 2, *Descriptive Geology of the 40th Parallel* (1877). When the U.S. Geological Survey was created in 1879, with King as its director, Emmons was appointed the geologist in charge of the survey's Rocky Mountain division headquartered in Denver, Colorado. He directed that a compilation of minerals of the region be made and wrote, "The study of the origin of ore deposits is one upon which comparatively little systematic scientific work has been done; and which presents peculiar difficulties in its prosecution. Its results may not be of immediate practical value for any particular district, but for the advancement of the interests of mining, in general, they are of the utmost importance." Emmons's results were published in conjunction with geologist George F. Becker in *Statistics and Technology of the Precious Metals* (1885), volume 13 of the Tenth Census.

Toward the end of his life, Emmons served as head of the Geological Survey's Division of Economic Geology. He was in Washington, D.C., on 28 March 1911 when he died unexpectedly, the day before his seventieth birthday. His friend and coauthor George Becker remarked that "there was not a geological society or even a mining camp from Arctic Finland to the Transvaal, or from Alaska to Australia, where his name was not known and his authority recognized."

See also Hague, Arnold; King, Clarence Rivers; United States Geological Survey.

Endangered Species Conservation Act (16 U.S.C. 1530)

Known popularly as the Endangered Species Act, this law, enacted on 28 December 1973, was a milestone in envi-ronmental conservation. In it, Congress declared that "various species of fish, wildlife, and plants in the United States have been rendered extinct as a consequence of economic growth and development untempered by adequate concern and conservation; [that] other species of fish, wildlife, and plants have been so depleted in numbers that they are in danger of or threatened with extinction; [that] these species of fish, wildlife, and plants are of esthetic, ecological, educational, historical, recreational, and scientific value to the Nation and its people; [that] the United States has pledged itself as a sovereign state in the international community to conserve to the extent practicable the various species of fish or wildlife or plants facing extinction." It asserted that "it is the policy of Congress that all Federal departments and agencies shall seek to conserve endangered species and threatened species and shall utilize their authorities in furtherance of these purposes."

Endangered Species List

This catalog of endangered and threatened species of plants, mammals, birds, reptiles, amphibians, fish, mussels, crustaceans, insects and arachnids, and snails was mandated as part of the Endangered Species Conservation Act of 1973. "A species achieves protection when the U.S. Fish and Wildlife Service, after scientific review, puts it on the list. Listing is followed by a protection and recovery plan, and since such plans limit ongoing habitat destruction (for example, the decimation of ancient forests in the case of the northern spotted owl), they generate opposition," wrote Peter A. A. Berle in *Audubon* magazine. Presently, there are more than 600 species on the Endangered Species List, with the U.S. Fish and Wildlife Service adding about 65 a year. In 1994, however, the gray whale and the bald eagle, having sufficiently recovered in population, were removed from the list.

Whooping cranes, shown here at the Aransas National Wildlife Refuge, Texas, were listed as a threatened and endangered species by the U.S. Fish and Wildlife Service, after a study mandated by the Endangered Species Conservation Act of 1973.

Endangered Species Preservation Act (16 U.S.C. 842)

According to one source, this legislation was the first federal act to protect endangered species. Enacted on 15 October 1966, it instructed the secretary of the interior to establish an agenda of preserving, reinstating, and replenishing chosen species of fish and wildlife that were threatened. The act declares that "for the purpose of consolidating the authorities relating to the various categories of areas that are administered by the Secretary of the Interior for the conservation of fish and wildlife, including species that are threatened with extinction, all land, waters, and interests therein administered by the Secretary as wildlife refuges, areas for the protection and conservation of fish and wildlife that are threatened with extinction, wildlife ranges, game ranges, wildlife management areas, or waterfowl production areas, are hereby designated as the 'National Wildlife Refuge System,' which shall be ... administered by the Secretary through the United States Fish and Wildlife Service." To pay for the establishment of this system, Congress authorized appropriations that had been collected under the Land and Water Conservation Fund Act of 1965. This act was later supplemented by the Endangered Species Conservation Act of 1973.

See also Endangered Species Conservation Act; Land and Water Conservation Fund Act.

Environmental Impact Statements

These summaries are mandated by the National Environmental Policy Act of

1969 (NEPA) to assess the possible effects on the environment of proposed improvements or developments, such as highways, building construction, dams, or canals. Under NEPA, all federal agencies are required to file an environmental impact statement (EIS) before any project proceeds. In addition, many developers and the federal government must comply with local zoning laws, which favor the environment, and the Endangered Species Act, which prohibits development that may disturb wildlife in affected areas.

See also Andrus, Secretary of the Interior v. Sierra Club; Kleppe, Thomas Savig; National Environmental Policy Act of 1969.

Environmental Justice Movement

The environmental justice movement has developed into a growing force over the last decade. An extension of the civil rights movement of the 1960s, it arose from the perceived notion that companies that deal in refuse dumping, chemical manufacturing, and sewage treatment have located their factories and plants (called "environmental risk centers") in areas heavily populated by minorities, including blacks, Hispanics, and Native Americans. The originator of the movement was Dr. Benjamin F. Chavis, Jr., who subsequently became executive director of the National Association for the Advancement of Colored People (NAACP). In 1982, Chavis, then head of the United Church of Christ Commission on Racial Justice, came to the aid of a group of blacks who were fighting the placement of a dump for PCB-saturated soil in their small community. (PCBs—polychlorinated biphenyls—have been found to cause cancer in humans.) Chavis led an unsuccessful fight against the dump, resulting in several hundred arrests. In the aftermath of the protest, Chavis coined the term environmental racism and sought to fight it with what he called environmental justice, which meant not an equal distribution of environmental risk centers among races but an end to them altogether. Chavis and others brought together a vast network of groups in this new movement at the People of Color Environmental Leadership Summit in Washington, D.C., in October 1991.

Environmental Movement

The recent impetus of the reform and heightened awareness of the nation's environmental laws is rooted in what is known as the conservation movement, but its beginnings are rather modern. Prior to 1962, those who sought to conserve or preserve distinct areas of the environment were considered conservationists. Writers such as Edwin Way Teale and Joseph Wood Krutch and politicians such as Sens. Henry ("Scoop") Jackson and Warren G. Magnuson, both of Washington State, spoke out on environmental issues with little effect during the years after World War II and before the 1960s. In 1962, Murray Bookchin's Our Synthetic Environment became the first major work on pesticides. Stephen Fox, writing about Bookchin's book, noted that it "set forth the full range of modern technology's incidental effects: polluted air, food with pesticide residues, milk contaminated by strontium 90, intolerable living conditions in cities, water not fit to drink, diets of chemical additives, and so on." Bookchin's work, however, was generally not understood and poorly received. That same year came the publication of Rachel Carson's groundbreaking work, Silent Spring, which in a more simplistic way detailed the abuse of pesticides, most notably DDT, and their effect on people and animals.

Recreation became a more important matter in the middle and late 1960s, with the formation of the Outdoor Recreation Resources Review Commission and the President's Council on Recreation and Natural Beauty. Both these groups found that making recreational resources available to the American people increased their awareness and appreciation of the environment. In the 1970s, the environmental movement went international, as the United Nations sponsored conferences on environmental problems in foreign nations. The establishment of the Environmental Protection Agency (EPA) was the U.S. government's immediate response to the growing movement. The first Earth Day, held in 1970, highlighted the issue of pollution and other problems that people saw as urgent and worthy of government and individual attention. The National Environmental Policy Act of 1969 seemed headed in the right direction but was soon overwhelmed by calls for action. The horrors of whaling and the killing of fur seals in Canada were visually documented for the first time, bringing pictures of unending slaughter and abuse into people's living rooms. The discovery of pollution at Love Canal in upstate New York in the 1970s and at Times Beach, Missouri, in the 1980s and the disaster at the Three Mile Island nuclear power plant in 1979 made the environment front-page news and led to the passage of the Superfund Act in 1980, but the problem continued.

In the 1980s, the environment became upscale as "yuppies" led the way in recycling and the marketing of dolphin-safe tuna. But a backlash developed in the form of the wise-use movement, springing from the Sagebrush Rebellion, which called for the government to dispose of public lands; forestry advocates who battled to use trees in the ancient growth forests of the Northwest; and mining supporters who tried to keep mining fees from being increased by the government. Fights over the *Exxon Valdez* oil spill in Alaska in 1989 and the battle over the habitat of the northern spotted owl in the Pacific Northwest have brought the subject of the environment directly to the people.

See also Carson, Rachel Louise; Conservation Movement; DDT; *Exxon Valdez* Oil Spill; Jackson, Henry Martin; Krutch, Joseph Wood; Magnuson, Warren Grant; Teale, Edwin Way; Wilderness Act.

Environmental Protection Agency (EPA)

The Environmental Protection Agency is the federal bureau designed to handle the nation's environmental problems. Established by Reorganization Plan no. 3 of 2 December 1970, which was signed by President Richard M. Nixon, it brought together the National Air Pollution Control Administration, the Bureau of Water Hygiene, the Bureau of Solid Waste Management, the Bureau of Radiological Health, and the Office of Pesticides in the Department of Health, Education, and Welfare; the Federal Water Quality Administration and the Pesticides, Fish, and Wildlife Office in the Department of the Interior; and the Pesticide Regulation Office in the Department of Agriculture. In his message to Congress establishing the EPA, Nixon wrote, "As concern with the condition of our physical environment has intensified, it has become increasingly clear that we need to know more about the total environment—land, water and air. It also has become increasingly clear that only by reorganizing our Federal efforts can we develop that knowledge, and effectively ensure the protection, development and enhancement of the total

The logo of the Environmental Protection Agency, established in 1970

environment itself." William Ruck-elshaus, a former Justice Department official, was named the first administrator of the new agency. He served until 1973 and then served a second time in the 1980s during the Times Beach dioxin scare. Two other noted administrators are William K. Reilly, who served during the Bush administration, and Carol Browner, administrator under the Clinton administration.

The EPA's responsibilities for the nation's air, water, and ground quality are enormous. The passage of a number of laws during the 1970s and 1980s expanded this obligation: the Clean Air Act; the Clean Water Act; the Safe Drinking Water Act; the Comprehensive Environmental Response, Compensation, and Liability Act (CERCLA, or Superfund); the Federal Insecticide, Fungicide, and Rodenticide Act (FIFRA); the Toxic Substances Control Act; the Marine Protection, Research, and Sanctuaries Act; the Uranium Mill Tailings Radiation Control Act (UMTRCA); and the Pollution Prevention Control Act.

See also Reilly, William Kane; Ruckelshaus, William Doyle.

Environmental Quality Act (EQA)

This legislation, passed by Congress on 1 April 1970, recognized the degree to which the environment had been altered in the United States. The act declared that "man has caused changes in the environment," that "population increases and urban concentration contribute directly to pollution and degradation in our environment," and "that many of these changes may affect the relationship between man and his environment." The act called for the implementation, by all pertinent federal agencies involved in environmental and public works programs, of all environmental laws previously passed by Congress in achieving the aims of this act, and it provided administrative personnel to the Council on Environmental Quality set up under the National Environmental Policy Act of 1969.

See also National Environmental Policy Act of 1969.

Estuary Protection Act (16 U.S.C. 1221)

This law was enacted on 3 August 1968 to provide federal protection for estuaries, which are tributary waterways teeming with wildlife. "Congress finds and declares that many estuaries in the United States are rich in a variety of natural, commercial, and other resources, including environmental natural beauty, and are of immediate and potential value to the present and future generations of Americans," reads the legislation. "It is therefore the purpose of this chapter to provide a means for considering the need to protect, conserve, and restore these estuaries in a manner that adequately and reasonably maintains a balance between the national need for such protection in the interest of conserving the natural resources and natural beauty of the Nation and the need to develop these estuaries to further the growth and development of the Nation. In connection with the exercise of jurisdiction over the estuaries of the Nation and in consequence of the benefits resulting to the public, it is declared to be the policy of Congress to recognize, preserve, and protect the responsibilities of the States in protecting, conserving, and restoring the estuaries in the United States." The act called for the secretary of the interior to conduct a study and inventory of the nation's estuaries, including "coastal marshlands, bays, sounds, seaward areas, lagoons, and land and waters of the Great Lakes." Appropriations for the project, which was completed in 1970, were $500,000.

Ewing, Thomas (1789–1871)

Thomas Ewing was the first secretary of the interior, serving from 1849 to 1850 during the Zachary Taylor administration. Born on 28 December 1789 on a small farm near the Ohio village of West

Liberty, Ewing was the son of George Ewing, a Revolutionary War veteran, and Rachel (nee Harris) Ewing. About April 1792, the family moved, according to one source, "to the mouth of Olive-green Creek on the Muskingum River," or, according to another source, to Marietta, Ohio. Six years later, because of Indian attacks, Thomas spent time in West Liberty attending school. A year later, the family moved again, this time to Ames Township in Athens County, Ohio. Ewing expanded his education through reading books, a self-imposed curriculum that included the Bible.

Ewing's three years working at the Kanawha Salt Mines in Virginia to pay a debt on what had once been his father's land taught him to value hard work and independence. In 1812, he finished paying off the debt, returned home, and attended Ohio University at Athens for three years, earning his bachelor of arts degree in 1815.

Starting in 1815, Ewing began the study of law, which was to bring him to the steps of the U.S. Senate. He read law in the Lancaster, Ohio, office of former congressman Philemon Beecher, a War of 1812 veteran, and was admitted to the bar in 1816. From about 1817 to 1822, Ewing served as the prosecuting attorney of Fairfield County, Ohio. In 1829, just before his election to the U.S. Senate, he adopted an orphan who was to make a major impact on U.S. history—William Tecumseh Sherman. His term as senator (1830–1836) was marked by his opposition to the administration of Andrew Jackson and by his appointment of Sherman to West Point. In 1836, the Ohio legislature refused to reelect Ewing, and he returned to his law practice in Ohio.

In 1841, Ewing was named to be William Henry Harrison's secretary of the treasury, but Harrison's death a month later left Ewing in conflict with the new president, John Tyler. After resigning, Ewing again returned to Ohio. On 5 March 1849 came his appointment

by President Zachary Taylor as secretary of the newly created Interior Department, and he took on the responsibilities of organizing the brand-new government agency. In his only report as secretary, Ewing called for the construction of a passage to the Pacific and the establishment of a government mint near the California gold mines. Following Taylor's death in August 1850, Ewing resigned his cabinet post to fill the unexpired Senate term of Thomas Corwin, who had been named secretary of the treasury. Ewing served only until 1851. In the last 20 years of his life, he practiced law in Ohio and was an aide to President Abraham Lincoln during the Peace Convention in 1861, which tried to head off the Civil War. Thomas Ewing died at his home in Lancaster, Ohio, on 26 October 1871 at the age of 81.

See also Department of the Interior.

Executive Order 1014 of 14 March 1903

Signed by Teddy Roosevelt, Executive Order 1014 created the Pelican Island National Refuge, the first government-sanctioned sanctuary in the United States for marine wildlife. Wrote James Trefethen, "Pelican Island, in size, was a relatively insignificant bit of real estate lying in the Indian River near Sebastian, Florida. But its mangrove thickets were crowded with the nests of brown pelicans, egrets, great blue herons, and roseate spoonbills. All of these species were targets of poachers exploiting the millinery feather trade in violation of a state law, enacted in 1901, that accorded full protection to nongame birds."

Frank M. Chapman, later one of the founding members of the National Audubon Society, had been a frequent visitor to Pelican Island and saw its value as a refuge for the many birds of the region; at the time, it was the last known sanctuary for the brown pelican on the eastern coast of the United States. With the help of an employee of the General

Land Office, Frank Bond, Chapman drafted a proposal that would allow President Roosevelt to sign an executive order to create a refuge there. There was precedent for such an action. In 1892, President Benjamin Harrison, upon the recommendation of Secretary of Interior John Willock Noble, had established the Afognak Forest and Fish Culture Reserve in Alaska. With little hesitation, Roosevelt signed Executive Order 1014 on 14 March 1903. Less than two years later, he created the Breton Island Wildlife Reserve in Louisiana, and by the end of his first term in 1905, he had created 51 wildlife refuges in 17 states.

See also Afognak Forest and Fish Culture Reserve; Chapman, Frank Michler; Roosevelt, Theodore.

Executive Order 8000 of 22 July 1936
See Great Plains Drought Area Committee.

Executive Order 9634 of 28 September 1945
President Harry S Truman signed this decree to establish fishery conservation zones. In the order, he wrote, "By virtue of and pursuant to the authority vested in me as President of the United States, it is hereby ordered that the Secretary of State and the Secretary of the Interior shall from time to time jointly recommend the establishment by Executive orders of fishery conservation zones in areas of the high seas contiguous to the coasts of the United States, pursuant to the proclamation entitled 'Policy of the United States With Respect to Coastal Fisheries in Certain Areas of the High Seas' this day signed by me, and said Secretaries shall in each case recommend provisions to be incorporated in such orders relating to the administration, regulation and control of the fishery resources of and fishing activities in such zones, pursuant to authority of law heretofore or hereafter provided."

Executive Order 11278 of 4 May 1966
This administrative dictate was signed by President Lyndon B. Johnson on 4 May 1966 to establish the President's Council on Recreation and Natural Beauty, a commission headed by Vice President Hubert H. Humphrey. Said Johnson in the order:

Because the Federal Government administers massive programs that affect the natural beauty of our land, it must pursue a course that will enhance and protect that beauty. It must stimulate action in behalf of natural beauty and outdoor recreation on the part of others—of State and local governments, of private organizations and individual citizens.

If it is to do this well, its own house must be in order. Its programs must be wise, and they must be coordinated. Its organization must reflect its responsibilities.

Therefore, by virtue of the authority vested in me as President of the United States ... there is hereby established the President's Council on Recreation and Natural Beauty ... and the Citizens' Advisory Committee on Recreation and Natural Beauty.

See also President's Council on Recreation and Natural Beauty.

Executive Order 11472 of 29 May 1969
This presidential mandate was signed by President Richard M. Nixon on 29 May 1969. It created the Cabinet Committee on the Environment, a short-lived government scheme to get cabinet officers, such as the secretaries of agriculture; commerce; health, education, and welfare; housing and urban development; interior; and transportation, to "coordinate" governmental work on environmental matters. The committee would have been chaired by the president himself. However, the burden of the secretaries' already prescribed work, as well as the dispersion

of environmental business throughout the government, left the committee with few if any tangible functions, and it was terminated under Executive Order 11541 of 1 July 1970.

Executive Order 11523 of 9 April 1970

This presidential directive established the National Industrial Pollution Control Council, which would "advise the President and the Chairman of the Council on Environmental Quality, through the Secretary [of Commerce], on programs of industry relating to the quality of the environment." The council's duties included (1) surveying and evaluating the plans and actions of industry in the field of environmental quality; (2) identifying and examining the effects of industrial practices on the environment and recommending solutions to problems; (3) providing liaison among members of the business and industrial communities to improve the quality of the environment; (4) encouraging the business and industrial communities to improve the quality of the environment; and (5) advising on plans and actions of federal, state, and local agencies involving environmental quality policies.

Executive Order 11574 of 23 December 1970

Signed by President Richard M. Nixon, this order established a permit schedule to manage the nation's refuse emissions. Wrote Nixon:

By virtue of the authority vested in me as President of the United States, it is hereby ordered as follows: the executive branch of the Federal Government shall implement a permit program ... to regulate the discharge of pollutants and other refuse matter into the navigable waters of the United States or their tributaries and the placing of such matter upon their banks ...

The Secretary [of the Army] shall be responsible for granting, denying, conditioning, revoking, or suspending Refuse Act permits. In so doing: (A) He shall accept findings, determinations, and interpretations which the Administrator [of the Environmental Protection Agency] shall make respecting applicable water quality standards and compliance with those standards in particular circumstances, including findings, determinations, and interpretations arising from the Administrator's review of State or interstate water quality certifications .... A permit shall be denied ... where issuance would be inconsistent with any finding, determination, or interpretation of the Administrator pertaining to applicable water quality standards and considerations. (B) In addition, he shall consider factors, other than water quality, which are prescribed by or may be lawfully considered under the Act or any pertinent laws.

See also Federal Water Pollution Control Act Amendments of 1972; Refuse Act; Rivers and Harbors Act of 1899.

Executive Order 11644 of 8 February 1972

Signed by President Richard Nixon, this order dealt with the use of off-road vehicles on public lands. According to the order, "an estimated five million off-road vehicles—motorcycles, minibikes, trail bikes, snowmobiles, dune-buggies, all-terrain vehicles, and others—were in use in the United States," and "the widespread use of such vehicles on the public lands—often for legitimate purposes but also in frequent conflict with wise land and resource management practices and environmental values, has demonstrated the need for a unified Federal policy toward the use of such vehicles on the public lands." The order called for all concerned federal agencies to formulate plans that would balance the need for the recreational use of the public lands by off-road vehicles with environmental policies.

See also Roadless Area Review and Evaluation.

Executive Order 11870 of 18 July 1975

This mandate, signed by President Gerald Ford, established safeguards for operations dealing with animal damage control on federal lands. Wrote Ford:

It is the policy of the Federal Government, consistent with ... the policies of the National Environmental Policy Act of 1969 and the Endangered Species Conservation Act of 1969 ... to (1) manage the public lands to protect all animal resources thereon in the manner most consistent with the public trust in which such lands are held; (2) conduct all mammal or bird damage control programs in a manner which contributes to the maintenance of environmental quality, and to the conservation and protection of the nation's wildlife resources, including predatory animals; (3) restrict the use on public lands and in federal predator control programs of any chemical toxicant for the purpose of killing predatory animals or birds which would have secondary poisoning effects; (4) restrict the use of chemical toxicants for the purpose of killing predatory or other animals or birds in Federal programs and on Federal lands in a manner which will balance the need for a responsible animal damage control program consistent with the other policies set forth in this Order; and (5) assure that where chemical toxicants or devices are used pursuant to section 3, only those combinations of toxicants and techniques will be used which best serve human health and safety and which minimize the use of toxicants and best protect nontarget wildlife species and those individual predatory animals and birds which

do not cause damage, consistent with the policies of this Order.

Exxon Valdez Oil Spill

This environmental accident was perhaps the worst oil-spill catastrophe to occur in U.S. waters. The *Exxon Valdez*, a 987-foot oil tanker, hit a reef in Alaska's Prince William Sound on 24 March 1989. Over the next several days, before a hole in the ship could be sealed, more than 11 million gallons of oil spilled across 1,200 miles of the environmentally sensitive bay, killing an estimated 270,000 marine birds (even the government admits between 100,000 and 300,000 dead birds) and thousands of sea otters, destroying the fish stocks of the bay, and annihilating the bald eagle population in the area. Eventually, some 12,000 workers were hired to clean up the oil from the beaches of the sound (parts of which are still polluted by oil), at a cost of over $1 billion. The captain of the ship, Joseph Hazelwood, was later found not guilty for his part in the accident. The General Accounting Office (GAO) found that the Coast Guard's "response to the *Exxon Valdez* grounding was clearly inadequate to contain and recover the spilled oil." A GAO study concluded that "experts had estimated that the more than 36,000 dead seabirds recovered after the spill represented only a small portion of the total number killed. The estimate generally believed was a range of 100,000 to 300,000 birds killed. Preliminary results from the study indicated that the total number of seabirds from the spill ranged from 260,000 to 580,000, with a best approximation of between 350,000 and 390,000 seabirds."

Five years after the spill, the effects of this tragic accident can still be seen in oil-stained beaches and a vacuum in the diversity of life in the bay. The spill's one good effect has been to increase the incentive for oil and shipping companies to use double-hulled tankers; the *Exxon Valdez* was a single-hulled ship.

Fall, Albert Bacon (1861–1944)

Exemplifying the rugged individualist of the West, Albert B. Fall served as a U.S. senator from New Mexico. He is best remembered, however, for his tenure as secretary of the interior, during which he was caught up in one of the greatest scandals in the history of the federal government. Born in Frankfort, Kentucky, on 26 November 1861, Fall was the eldest of the three children of Williamson Ware Robertson Fall and Edmonia (nee Taylor) Fall, both schoolteachers. When Williamson Fall joined the Confederate army during the Civil War, Albert was sent to Nashville, Tennessee, to live with his grandfather, Philip Slater Fall, a former Baptist minister who instilled in his grandson a love of reading. Remaining in Nashville after the war, Fall took a job in a cotton mill when he was 12. He later returned to Kentucky, where the only public education he received was in classes taught by his father. Ironically, Fall later was a teacher during the time he read law before moving westward in 1881.

Starting first in Clarksville, on the Red River in Texas, continuing into Mexico, and finally ending up in the territory of New Mexico, Fall worked at odd jobs such as bookkeeping, ranch handing, and mining. It was the latter occupation that drove him to settle in the Black Range of New Mexico. There he met Edward Doheny, who would later be linked with Fall and Teapot Dome. In 1887, Fall moved on to Las Cruces, New Mexico, where, after a time as a frontier attorney, he represented the area during a single term in the New Mexico Territorial House of Representatives (1890–1892) and later served three disjointed terms on the Territorial Council (1892–1893, 1896–1897, and 1902–1904), two years as an associate justice of the Territorial Supreme Court (1893–1895), and two terms (1897, 1907) as territorial attorney general.

Although early on Fall considered himself a Democrat, he began an ideological move toward the Republican Party that culminated in his serving as a member of the state delegation to the 1908 Republican National Convention. While serving during the Spanish-American War (1898) as a captain of the New Mexican infantry, Fall met and befriended Theodore Roosevelt. It was this friendship that led Fall in the early 1900s to become a leading Republican in the New Mexico Territory.

In 1912, with New Mexico's entry into the Union as the forty-seventh state, the new state legislature elected Fall one of its two U.S. senators. In the Senate, Fall became a leading member of the Republican Party. Said the *New York Times*, "He was a formidable debater, quick-witted and with a biting tongue." Known for his harsh criticism of Woodrow WIlson's Mexican policy, Fall became good friends with Ohio Sen. Warren G. Harding. Serving with Fall on the Senate Foreign Relations Committee, the card-playing Harding fell under Fall's spell; as one historian wrote, "Fall had a special talent for being a crony and played poker with a passion."

In 1920, Harding was elected president of the United States; meeting with Gifford Pinchot, the new president seemed to convey a sense of commitment to the conservation movement. He promised the former chief forester that the new secretary of the interior would meet Pinchot's standards. Unfortunately, while on vacation, Harding bumped into Fall and, after that one chance meeting, decided that the New Mexican should get the Interior post. Fall is considered Harding's worst cabinet appointment. Historian

Albert Bacon Fall in 1910. President Warren G. Harding appointed him the twenty-seventh secretary of the interior in 1921.

Page Smith reported that Woodrow Wilson's Secretary of the Treasury David Houston wrote on hearing of Fall's appointment, "He [Fall] had about the same interest in the conservation of the natural resources of the nation that a tiger has in a lamb." Writer William Allen White called Fall "a tall, gaunt, unkempt, ill visaged" man marked by "a disheveled spirit behind restless eyes"; he described him as "a patent medicine vendor" and "a cheap, obvious faker." Houston and White were not far off. Almost as soon as he took control of the Interior Department, Fall began to lease control of the federal oil reserves at Teapot Dome in Wyoming and Elk Hills in California to cronies who rewarded him with kickbacks.

Fall also dealt with other matters as interior secretary. A westerner who desired the full use of government lands, he once said, "I stand for opening up every resource." He advocated a reservoir on Lake Yellowstone and urged relaxed regulation of water power concerns. He is best known, however, for his role in the Teapot Dome scandal. Teapot Dome was leased to the Mammoth Oil Company, owned by Harry F. Sinclair, and Elk Hills was turned over to Fall's old friend Edward Doheny, who owned the Pan-American Petroleum Company. When the Senate got wind of these dealings, Fall quickly resigned and returned to New Mexico. Sens. Robert LaFollette of Wisconsin and Thomas Walsh of Montana of the Senate Public Lands Committee began full-scale hearings in 1923, which embarrassed the Harding administration. In the midst of the inquiry, in August 1923, Harding died while traveling. Fall and Doheny were indicted in 1926 for conspiracy, but both were eventually acquitted. In 1929, however, Fall was convicted of accepting a bribe from Doheny. After a year in prison, Fall went home, the only secretary of the interior ever to serve time.

Suffering from ill health (including a heart ailment he had had when he entered prison) and his reputation ruined, the now-destitute Fall lost his ranch in Three Rivers, New Mexico, in 1936. He spent the last eight years of his life at the Hotel Dieu Hospital in El Paso, Texas, where he died on 30 November 1944, just four days after his eighty-third birthday. Author David Stratton called Fall a "New Mexican Machiavellian."

See also Department of the Interior.

Federal Aid in Fish Restoration and Management Act
See Dingell-Johnson Act.

Federal Aid in Wildlife Restoration Act of 1937
See Pittman-Robertson Act.

Federal Food, Drug, and Cosmetic Act (Public Law 717)
Although not important as an environmental act, this 1938 law contains section

406, which calls for the setting of "residue limits" on food if it is shown that the residue of any chemical appears on that food. The Food and Drug Administration (FDA) must set a "residue tolerance limit" on such foods before they are sold. Under the Federal Food, Drug, and Cosmetic Act Amendments of 1954 (Public Law 518), all pesticides used on food have to be registered with the FDA and then have limits approved by the Department of Agriculture. These two laws were the first important steps in the regulation of pesticides on foods.

Federal Insecticide, Fungicide, and Rodenticide Act (FIFRA)

This law was passed on 25 June 1947 to address the problem of toxic chemicals being used to kill such pests as insects, rodents, and weeds. The act labeled such chemicals "economic poisons" and stated that they were "intended for preventing, destroying, repelling, or mitigating any insects, rodents, nematodes, fungi, weeds," or any other environmental pests. The act was also intended to "protect consumer(s) from labeling fraud" on such chemicals, with a "requirement of a warning or caution statement" to warn of chemical effects.

See also Federal Food, Drug, and Cosmetic Act.

Federal Land Policy and Management Act (FLPMA) (90 Stat. 2744)

Also known as the BLM (Bureau of Land Management) Organic Act, this legislation was enacted by Congress on 21 October 1976 to reiterate the federal policy of allowing for multiple-use management of public lands under the care of the Bureau of Land Management. The act gave the BLM authority over land totaling about 450 million acres in the 11 western states and Alaska, all of which is to officially remain public land (not to be sold for private use). Grazing fees are to be divided equally with the U.S. Forest Service, and half of those fees are to be used for range management. The act also abolished the National Forest Reservation Commission, a governmental body set up under the Weeks Act of 1911, and transferred its powers to the Department of Agriculture.

See also National Forest Reservation Commission.

Federal Water Pollution Control Act (33 U.S.C. 1151–1175)

Known as the Taft-Barkley Act, this legislation was enacted on 30 June 1948 as a way to authorize government policy on clean water standards. Jon Luoma wrote that the act "established a relatively small, $50 million construction-grant program to assist communities with treatment-system needs." An amendment in 1965 added new appropriations, and with the 1972 amendments (which established a federal permit program for pollution dumping), the program topped $1 billion a year. By 1981, the government had spent $33 billion to construct wastewater treatment and sewage treatment plants nationwide.

See also Federal Water Pollution Control Act Amendments of 1972.

Federal Water Pollution Control Act Amendments of 1972 (86 Stat. 816)

Known as the Clean Water Act Amendment, this legislation, passed on 18 October 1972, was enacted "to limit effluent discharges and set water quality standards." According to one source, this act provided the United States "with a new, comprehensive, and forceful strategy not only to stop [water] pollution but also to restore and maintain the chemical, physical, and biological integrity of lakes, streams, and surface waters." Added a historian of the Senate committee where the legislation originated, "The 1972 amendments established two ambitious water quality goals: total elimination of pollutant discharges by 1985, and, by

1981, an interim goal of water quality adequate to provide for the protection of aquatic life and wildlife and for recreation in and on the water." Unfortunately, these goals were not met and had to be amended by the Federal Water Pollution Control Act Amendments of 1977.

Established under this act was the National Water Quality Commission (NWQC), which was to evaluate the development of water quality programs. Section 402, enacted because of President Richard Nixon's Executive Order 11574 of 23 December 1970, authorized the granting of refuse and discharge emission permits. Under section 404 of the act, the U.S. Army Corps of Engineers was to "regulate the discharge of dredge or fill material into the nation's waters." In a 1975 Supreme Court case, *Natural Resources Defense Council v. Callaway*, wetlands and headwaters were included in the areas under the Corps of Engineers' jurisdiction.

See also Executive Order 11574 of 23 December 1970; Refuse Act; Rivers and Harbors Act of 1899.

Federal Water Pollution Control Act Amendments of 1977 (33 U.S.C. 1251)

"The objective of this act is to restore and maintain the chemical, physical, and biological integrity of the nation's waters," reads the act passed on 27 December 1977, amending the Clean Water Act of 1972. Because the 1972 act had failed to end the dumping of refuse and sewage into navigble waters, this act again called for that ambitious goal:

In order to achieve this objective it is hereby declared that, consistent with the provisions of this act, (1) it is the national goal that the discharge of pollutants into the navigable waters be eliminated by 1985; (2) it is the national goal that wherever attainable, an interim goal of water quality which provides for the

protection and propagation of fish, shellfish, and wildlife and provides for recreation in and on the water be achieved by July 1, 1983; (3) it is the national policy that the discharge of toxic pollutants in toxic amounts be prohibited; (4) it is the national policy that Federal financial assistance be provided to construct publicly owned waste treatment works; (5) it is the national policy that areawide waste treatment management planning processes be developed and implemented to assure adequate control of sources of pollutants in each State.

A historian of the committee that enacted this law, the Senate Environment and Public Works Committee, claims that this legislation improved the three previous acts in three ways. "First, it extended authorizations for the municipal sewage treatment construction grants program, providing $24.5 billion over five years while extending the municipal compliance deadline from 1981 to 1983. Second, it included numerous 'fine-tuning' provisions to adjust municipal and industrial source deadlines, improve administrative flexibility, and enhance implementation by the Environmental Protection Agency (EPA) and the States. And third, it provided new authorities for controlling toxic pollutant discharges into the Nation's waters."

See also Federal Water Pollution Control Act; Federal Water Pollution Control Act Amendments of 1972.

Federal Water Power Commission

This agency was created by the Water Power Act (41 Stat. 1063) on 10 June 1920. The act, which repealed the power of the Newlands Waterways Commission (which had been passed in 1917), was officially "an Act to create a Federal Power Commission; to provide for the improvement of navigation; the development of water power; the use of the public lands in relation thereto, and to repeal section 18 of the River and Harbor Appropriation Act, approved 8 August 1917, and for other purposes." The act, sponsored by Sen. Wesley L. Jones (1863–1932) of Washington State, set up the commission to be run by the secretaries of war, interior, and agriculture, although a 1930 amendment broadening the commission's powers changed it to a five-man, non-cabinet delegation. It was later transferred to the jurisdiction of the Department of Energy.

See also Lenroot, Irvine Luther.

Federal Water Project Recreation Act (79 Stat. 213)

This legislation was intended "to provide uniform policies with respect to recreation and fish and wildlife benefits and costs of Federal multiple-purpose water resource projects." It was enacted on 9 July 1965 to restate the goals of the Fish and Wildlife Coordination Act of 1958 as well as to create an environment for the establishment of programs that not only benefited wildlife but had recreational uses as well, such as national parks, dams, and wildlife refuges that could be accessible to the public.

Fernow, Bernhard Eduard (1851–1923)

Bernhard Fernow was a noted forestry expert and author on the subject. He was born in Inowrazlaw, in the province of Posen, Prussia (now Germany), on 7 January 1851 into a family of aristocrats. Fernow's grandfather, Graf Karl Frederick Leopold von Ferno, owned a large estate. Bernhard's father, Eduard Ernst Leopold Fernow (there is no known reason for the change in spelling), was a member of the Prussian government. Bernhard Fernow received his early education at Bromberg and then started studies in forestry management. In 1870, he enlisted in the Prussian army to fight in the Franco-Prussian War and served

for one year in France. Following the war, he completed his education with a year of law studies at the University of Königsberg. His studies in forestry management were completed at the Hanover-Müenden Forest Academy. Fernow eventually joined the Prussian Forest Service.

Soon after this, Fernow met and fell in love with Olivia Reynolds, a tourist from Brooklyn, New York. At once, they became engaged, and when his family expressed their disapproval, Fernow immigrated to the United States in 1876. The two were married in 1879. Fernow found that the field of forestry research and management was virtually unknown in the United States, so he joined the American Institute of Mining Engineers, the closest position available. He also took a job as manager of a 15,000-acre range of forestland in Pennsylvania that belonged to the New York firm of Cooper Hewitt and Company, which used the wood for charcoal.

Fernow's activity in the area of forest management led him to found in 1882 the American Forestry Congress, and he served as the group's secretary from 1883 to 1895. In 1886, President Grover Cleveland named Fernow as the third chief of the Division of Forestry of the Department of Agriculture, replacing Nathaniel H. Egleston. Immediately, the new forestry chief convened a cadre of experts in forestry management, trees, and other lumber fields, including silviculture. It was up to Fernow to improve the nation's wood management resources, distribute self-written articles on forestry management, and lobby the government for legislation on wood and forestry preservation. After his resignation as Forestry Division chief in 1898, Fernow started the Cornell School of Forestry, which had to close in 1903 over a controversy involving the Adirondack Mountains.

For the last two decades of his life, Fernow was involved in private business and teaching forestry at Yale University.

In 1907, the University of Toronto (Canada) gave him a faculty seat as a professor in forestry, where he taught until his retirement in 1919. He died in Toronto on 6 February 1923.

See also Egleston, Nathaniel Hillyer.

Fish and Wildlife Act (70 Stat. 1119)

This act, signed into law on 8 August 1956 by President Dwight D. Eisenhower, created the U.S. Fish and Wildlife Service. Under Reorganization Plan no. III of 1 July 1940, Congress had combined the Bureau of Fisheries and the Bureau of Biological Survey, two Interior Department bureaus, into the Fish and Wildlife Service. Under the Fish and Wildlife Act of 1956, the service was renamed the *U.S.* Fish and Wildlife Service, with the status of a federal bureau and the responsibility of two new departments: the Bureau of Commercial Fisheries and the Bureau of Sport Fisheries and Wildlife. The act was officially called "an Act to establish a sound and comprehensive national policy with respect to fish and wildlife; to strengthen the fish and wildlife segments of the national economy; to establish within the Department of the Interior the position of Assistant Secretary for Fish and Wildlife; to establish a United States Fish and Wildlife Service; and other purposes."

See also United States Fish and Wildlife Service.

Fish and Wildlife Act Airborne Hunting Amendment (16 U.S.C. 742j–l)

This amendment to the Fish and Wildlife Act of 1956 was enacted on 18 November 1971 specifically to outlaw the airborne killing of protected fish and wildlife. The act reads, "Any person who (1) while airborne in an aircraft shoots or attempts to shoot for the purpose of capturing or killing any bird, fish, or other animal; or (2) uses an aircraft to harass any bird, fish, or other animal; or (3) knowingly participates in using an aircraft for any purpose referred to in paragraph (1) or (2); shall be fined not more than $5,000 or imprisoned not more than one year, or both."

Fish and Wildlife Conservation Act (49 U.S.C. 2901)

Enacted on 29 September 1980, this act was intended "to provide financial and technical assistance to the States for the development, revision, and implementation of conservation plans and programs for nongame fish and wildlife, and to encourage all Federal departments and agencies to utilize their statutory and administrative authority, to the maximum extent practicable and consistent with each agency's statutory responsibilities, to conserve and to promote conservation of nongame fish and wildlife and their habitats." Congressional findings that led to the passage of this law included the following: that "fish and wildlife are of ecological, educational, esthetic, cultural, recreational, economic, and scientific value to the nation"; that "the improved conservation and management of fish and wildlife, particularly nongame fish and wildlife, will assist in restoring and maintaining fish and wildlife and in assuring a productive and more esthetically pleasing environment for all citizens"; and that "many citizens, particularly those residing in urban areas, have insufficient opportunity to participate in recreational and other programs designed to foster human interaction with fish and wildlife and thereby are unable to have a greater appreciation and awareness of the environment." It mandated states to conceive conservation plans that provide "for an inventory of the nongame fish and wildlife, and such other fish and wildlife as the designated State agency deems appropriate ... determine the size, range, and distribution of their populations ... identify the extent, condition, and location of their significant habitats ... identify the significant problems which may adversely affect the plan species and their significant habitats ... determine those actions which should be taken to conserve the plan species and their significant habitats ... establish priorities for implementing [these actions] ... provide for the monitoring, on a regular basis, of the plan species and the effectiveness of the conservation actions [that have been implemented]."

Fish and Wildlife Coordination Act (48 Stat. 401)

This legislation was passed by Congress on 10 March 1934. It authorized the secretaries of commerce and agriculture to cooperate with federal, state, and other agencies in establishing a national program of "wildlife conservation and rehabilitation." The act also called for studies of the effects of water pollution on animals and studies of the use of impounded water by the Bureau of Fisheries or the Bureau of Biological Survey (later combined into the Fish and Wildlife Service).

Fish and Wildlife Coordination Act (72 Stat. 563)

This legislation amended the 1934 Fish and Wildlife Coordination Act to authorize the secretary of the interior and other pertinent agencies to work with the Fish and Wildlife Service to create conservation programs that "better protect" fish and wildlife resources. The act states that "for the purpose of recognizing the vital contribution of our wildlife resources to the Nation, the increasing public interest and significance thereof due to expansion of our national economy and other factors, and to provide that wildlife conservation shall receive equal consideration and be coordinated with other features of water-resource development programs, through the effectual and harmonious planning, development, maintenance, and coordination of wildlife conservation and rehabilitation," programs would be authorized through the

Fish and Wildlife Service to protect fish and wildlife resources.

Fish and Wildlife Service

See United States Fish and Wildlife Service.

Fish Restoration and Management Act

See Dingell-Johnson Act.

Fisher, Walter Lowrie (1862–1935)

Attorney, conservation advisor, and civic leader Walter Fisher also served as secretary of the interior. Born on 4 July 1862 in Wheeling, Virginia (now West Virginia), Fisher was the son of Daniel Webster Fisher, a Presbyterian clergyman, and Amanda (nee Kouns) Fisher. Walter Fisher was educated in the public schools of Wheeling, after which he attended Marietta College in Marietta, Ohio, and Hanover College in Hanover, Indiana, where his father was the president (1879–1908). Fisher received his bachelor of arts degree from Hanover in 1883.

With his admittance to the Illinois bar in 1888, Fisher began practicing law in Chicago. According to one source, as an attorney "he specialized in railroad and transportation litigation." Before the end of 1888, he was named by the mayor as special assessment attorney for the city, a position he held for a year (he was succeeded by another up-and-coming attorney, Clarence Darrow). Fisher remained in private practice until 1899.

Known as a reformer, Fisher joined and became the secretary of the Chicago Municipal Voters' League, a reform-minded organization, in 1899. He also helped found the City Club of Chicago, which, according to Fisher, was a place where people could "meet informally and confer about municipal affairs and have luncheon or dinner served if this was desired." Chicago Mayor Edward F. Dunne had been elected on a platform of constructing municipal railways, and in 1906, he named Fisher as special traction (railway specialist) counselor for the city. In that position, Fisher drew up the legislation that became the Street Railway Ordinance of 1907.

Fisher's abilities also drove him in the direction of conservation. An intimate of Gifford Pinchot (and, as can be seen in Pinchot's papers, a frequent correspondent with the chief forester), Fisher was one of the organizers of the Conservation League of America (serving as its president from 1908 to 1909) and the National Conservation Association.

As an attorney, Fisher became involved politically with William Howard Taft when Taft was secretary of war and Fisher represented the city of Chicago before the War Department. Taft's election to the presidency brought a promise of a cabinet post for Fisher, but instead he was named to the Federal Railroad Securities Commission, a post he held until 1911.

Following the controversy between Secretary of the Interior Richard A. Ballinger and Pinchot, and Ballinger's subsequent resignation, Taft looked for a replacement who could placate the Theodore Roosevelt–Gifford Pinchot wing of the Republican Party. He looked no farther than his friend Fisher, who was acceptable to both conservationists and conservatives. One author called the Interior post when Fisher assumed it the "Trouble Portfolio." Fisher took the oath of office on 13 March 1911 and served until the end of Taft's term on 4 March 1913. In his two years, Fisher is remembered for endorsing the leasing of Alaskan coal lands and the construction of an Alaskan railroad, recommendations taken up by his Democratic successor, Franklin K. Lane. Most importantly, his tenure reinvigorated the morale of the Interior Department, whose spirit had suffered during the anxious months of fighting between Ballinger and Pinchot and Ballinger's investigation by Congress. Convinced that he would be re-

placed after the election, Fisher made way for Lane with aplomb and courtesy.

Fisher returned to Chicago and picked up his law practice. For the rest of his life he participated in the civic and legal affairs of that city. He was counsel for federal Judge James H. Wilkerson in front of a Senate committee, and in 1933, he was appointed by a federal court to coordinate the merger of surface and elevated train lines in Chicago. Walter Fisher died at his home in the Chicago suburb of Winnetka on 9 November 1935 at the age of 73.

See also Department of the Interior.

Fishery Conservation and Management Act
See Magnuson Fishery Conservation and Management Act.

Fishery Conservation Zones
See Executive Order 9634 of September 1945.

Flood Control Act, First (39 Stat. 948–951)
This legislation, enacted on 1 March 1917, was the first of its kind to deal with the matter of flood control of public lands. Known as the Ransdell-Humphreys Act, it is officially "an Act to provide for the control of the floods of the Mississippi River and of the Sacramento River, California, and for other purposes." The act empowered the secretary of war (now the secretary of defense) to spend some $45 million (but not more than $10 million in any one fiscal year) to control the floodwaters of the Mississippi and Sacramento Rivers. The act basically failed in its approach, and there were terrible floods along the Mississippi River in 1927, leading to passage of the Mississippi Flood Control Act of 1928 as well as national flood control acts in 1944, 1954, and beyond.

See also Mississippi Flood Control Act.

Flood Control Act of 1936
(49 Stat. 1570)

Enacted on 22 June 1936, this was the first federal legislation designed to control floodwaters nationally rather than target such programs to specific areas. In the name of federal-state cooperation, the law authorized the construction of multipurpose flood control projects. In its declaration of policy, Congress wrote:

It is hereby recognized that destructive floods upon the rivers of the United States, upsetting orderly processes and causing loss of life and property, including the erosion of lands, and impairing and obstructing navigation, highways, railroads, and other channels of commerce between the States, constitute a menace to national welfare; that it is the sense of Congress that flood control on navigable waters or their tributaries is a proper activity of the Federal Government in cooperation with the States, their political subdivisions, and localities thereof; that investigations and improvements of rivers and other waterways, including watersheds, thereof, for flood-control purposes are in the interest of the general welfare; that the Federal Government should improve or participate in the improvement of navigable waters or their tributaries, including watersheds thereof, for flood-control purposes if the benefits to whomsoever they may accrue are in excess of the estimated costs, and if the lives and social security of people are otherwise affected.

Flood Control Act of 1944
(58 Stat. 890)

Enacted on 22 December 1944, this flood-control legislation was the first to include a congressional dictate that water-control areas, such as reservoirs, were to be open to public recreation such as "boating, swimming, bathing, fishing, and other recreational purposes."

See also Flood Control Act of 1954.

Flood Control Act of 1954
(68 Stat. 1266)

This legislation, enacted on 3 September 1954 as an amendment to the Flood Control Act of 1944, expanded the definition of public water-control areas, allowing for greater public access to these areas.

Flood Control Act of 1960
(74 Stat. 480)

The third congressional attempt to control floodwaters nationwide (and the fifth such act to control any floodwaters) resulted in the passage of this act on 14 July 1960. Reads the act:

In recognition of the increasing use and development of the flood plains of the rivers of the United States and of the need for the information on flood hazards to serve as a guide to such development, and as a basis for avoiding future flood hazards by regulation of use by States and political subdivisions thereof, and to assure that Federal departments and agencies may take proper cognizance of flood hazards, the Secretary of the Army, through the Chief of Engineers [of the Army Corps of Engineers], is hereby authorized to compile and disseminate information on floods and flood damages, including identification of areas subject to inundation by floods of various magnitudes and frequencies, and general criteria for guidance of Federal and non-Federal interests and agencies in the use of flood plain areas; and to provide advice to other Federal agencies and local interests for their use in planning to ameliorate the flood hazard. Surveys and guides will be made for States and political subdivisions thereof only upon the request of a State or a political subdivision thereof, and upon approval by the Chief of Engineers, and such information and advice provided them only upon such request and approval.

Section 206 of this act, which ordered the Army Corps of Engineers to assist the states and local governments with the data necessary to manage the use of floodplain lands by way of a $15 million a year appropriation, led to the establishment of the Flood Plain Management Service (FPMS).

Flood Control Act of 1968 (82 Stat. 731)

Enacted by Congress on 13 August 1968, this act authorized the acquisition of land for use for flood-control projects and policies. Amending the earlier flood-control acts passed in the previous 50 years, it allowed for appropriations to purchase land so that people and businesses could be moved when floods threatened them.

Flood Plain Management Service (FPMS)

See Flood Control Act of 1960.

Forest and Rangeland Renewable Resources Planning Act (FRRRPA) (92 Stat. 353)

This law was enacted on 17 August 1974 to ease long-range planning for the use of natural resources in the National Park System. With the congressional finding that "the management of the Nation's renewable resources is highly complex and the uses, demand for, and supply of the various resources are subject to change over time," this act mandated the Forest Service to establish programs that would conserve these resources. One of the laws recommended for passage by the Public Land Law Review Commission, it further directed the secretary of agriculture to "assess all lands, and prepare a management program" for the utilization, protection, management, and development of natural resources in national parks, with these plans to be updated every five years.

Forest Congress

See American Forest Congress.

Forest Experiment Stations

See Zon, Raphael.

Forest Homestead Act (34 Stat. 233)

This act, which became law on 11 June 1906, is also known as the Agricultural Settlement Act. It allowed farmers to use agricultural lands on forestry reserves that had previously been closed by the government. Passed on the recommendation of the Public Land Commission of 1903, the act was the first attempt to settle people on federal lands. Wrote Gifford Pinchot, "Its passage brought a notable increase of good will in many parts of the West. Moreover, its effect was to settle settlers on nonforest lands—permanent settlers, with a real interest in helping our men against fire and trespass, to the marked advantage of National Forests."

Forest Legacy Program

This program was established as part of the Food, Agriculture, Conservation, and Trade Act of 1990 (better known as the 1990 farm bill). The act "authorizes the ... Forest Service to acquire permanent conservation easements on private forestlands that are at risk of being converted to non-forest uses." Under the program, sponsored by Sen. Patrick Leahy of Vermont, the Forest Service purchases land sections, known as easements, at fair-market prices from willing owners who wish to see their lands preserved as nature reserves of sorts. The Forest Service is the sole holder of easements acquired with federal funds. (Easements purchased with state monies are overseen by the particular state agency concerned with land preservation.)

See also Conservation Easements.

Forest Management Act of 1897
See Pettigrew Act.

Forest Products Laboratory

The Forest Products Laboratory (FPL) is a government–private industry project set up in 1908 by Overton W. Price, assistant forester of the National Forest Service, and leaders of the forestry industry to promote research into forestry matters. As early as 1887, the Department of Agriculture had been conducting experiments to see what kind of products could be made from wood. These experiments, as one source noted, "proved relatively ineffectual and difficult to correlate and manage." Through the efforts of Price and Gifford Pinchot, chief forester of the United States during the Teddy Roosevelt administration, the Forest Products Laboratory was established at the University of Wisconsin. Charles Richard Van Hise, a geologist and author of a conservation textbook, *The Conservation of Natural Resources in the United States*, was president of the university at the time. The Forest Products Laboratory opened its doors on 4 June 1910.

In its 80-plus years of operation, according to the FPL's public-relations department, "the Laboratory developed the semichemical pulping process that is now used to produce corrugating medium used in corrugated containers. . . . It constructed the first all-wood prefabricated house and first large-scale glued-laminated wooden arches in the United States, tested all-weather wood adhesives, and improved production techniques for softwood plywood, tested waferboard and particleboard—concepts

that are now used in a variety of products. Current design standards for wood construction drew heavily on the research conducted on wood properties . . . [E]qually important was the Lab's development of the lumber dry kiln which revolutionized lumber seasoning." The FPL conducts research into six areas: wood products; pulp, paper, and composites; wood protection and marketing; microbial and biochemical conversion; energy resources from wood; and the use of forest products on an international scale. The lab employs about 300 people, including more than 100 scientists and other professionals.

Forest Reserve Act of 1891
See Payson Act.

Forest Service
See United States Forest Service.

Fort Gratiot Sanitary Landfill, Inc. v. Michigan Department of Natural Resources et al. (119 L Ed 2d 139, 112 S Ct 2019 [1992])

Although decided on the basis of the commerce clause of the U.S. Constitution, *Fort Gratiot* was one of a series of recent Supreme Court cases dealing with the states' ability to legislate the difference between intrastate and interstate garbage and wastes. Under a Michigan state law, persons were prohibited from accepting wastes not generated in the county they were to be disposed in. The Fort Gratiot Sanitary Landfill, a private waste-disposal firm, requested a waiver from the state so that it could accept some out-of-state waste. When it was turned down, the company sued the Michigan Department of Natural Resources, the state agency in charge of enforcing the statute, on the ground that the statute violated the commerce clause of the Constitution. The District Court for the Eastern District of Michigan denied the landfill's assertions, and the U.S. Court of Appeals for the Sixth District agreed. The landfill owners appealed to the U.S. Supreme Court. By a 7–2 vote (Chief Justice William Rehnquist and Justice Harry Blackmun dissenting), the Court held that the Michigan statute was an unconstitutional burden on the commerce clause. In an opinion written by Justice Anthony Kennedy, the Court ruled that the state had shown no clear reason why waste from outside the county should be treated differently from waste from inside the county. Further, the Court explained that if the regulations had been set up in an effort to further public health and safety, and had clearly spelled this out, the Court would have ruled in favor of this approach.

Fortieth Parallel Survey

See King, Clarence Rivers; United States Geological Survey.

Francis, David Rowland (1850–1927)

David R. Francis's tenure as the chief watchdog of the nation's natural resources is considered an interregnum between those of his predecessors in the post of interior secretary, John Willock Noble and Hoke Smith, and those of his successors, Cornelius Bliss and Ethan Allen Hitchcock. Francis was born in Richmond, Kentucky, on 1 October 1850, the son of John Broaddus Francis, the sheriff of Madison County, and Eliza Caldwell (nee Rowland) Francis. David Francis attended a local girls' academy in Richmond run by Rev. Robert Breck. From 1866 to 1870, he attended Washington University in St. Louis, Missouri, graduating with a bachelor of arts degree. Too poor to study law, he returned home.

Francis began his career as a clerk in his uncle's Kentucky merchant house, Shyrock & Rowland. He studied commissions and, in 1876, opened his own merchant house, D. R. Francis & Brother,

which dealt in grain commissions in St. Louis. Gradually he built the company into a prosperous business, making his name in the industry. In 1884, he was elected president of the St. Louis Merchant's Exchange. That year, he was a delegate to the Democratic National Convention in Chicago. In 1885, Francis was elected mayor of St. Louis. Advocating a reform platform, he led the way in internal improvements, collecting tax debts from corporations and fighting corrupt legislation with vetoes. His administration was such a success that the Democrats nominated him for governor at the end of his first term in 1888. He was elected over Republican E. E. Kimball by 13,000 votes. In a single four-year term, Francis established a Board of Mediation and Arbitration to manage labor strikes, created a Geological Survey Commission to map the state, and instituted the secret ballot, modeled on the Australian system. In 1892, he was a major supporter of Grover Cleveland's attempt to win back the White House.

Three years after Francis left office, Cleveland asked him to fill the vacancy created by the resignation of Interior Secretary Hoke Smith. Francis served only six months, from 3 September 1896 to 5 March 1897. Wrote Interior Department historian Eugene Trani, "During his tenure, he defended hotly contested forest reserves, recommended that the President withdraw twenty-one million acres, and refused to support a bill to modify any executive oder creating forest reserves." When Cleveland left office on 4 March 1897, Francis was out of a job. His opposition to free silver and Democrat William Jennings Bryan's runs for the presidency in 1896 and 1900 left him isolated from his party. Instead, he was named head of the Louisiana Purchase Exposition, which was held in 1904, and his travels in Europe to entice European participation resulted in his work *A Tour of Europe in Nineteen Days* (1903). He told the story of the exposition in *The Universal Exposition of 1904* (1913).

By 1908, Francis had been accepted again by both his party and Bryan, and for a time he was considered vice-presidential material. Two years later, he was an unsuccessful candidate for the U.S. Senate. His career reached its zenith when President Woodrow Wilson named him U.S. ambassador to Russia in 1916. He was ambassador during the highly crucial period when Czar Nicholas II was dethroned, the interim Kerensky government was toppled, and the Bolshevik movement seized power. During this time, Francis was forced to move the official ambassador's residence to various cities to avoid bloodshed. On 6 November 1918, he was carried on a stretcher to a U.S. warship in Archangel, Russia, to be taken to England for an operation, from which he never fully recovered. After the surgery, he resigned as ambassador and retired. David R. Francis died at his home in St. Louis on 15 January 1927 at the age of 76.

See also Department of the Interior.

Fur Seal Act (16 U.S.C. 1151–1175)

This act of 2 November 1966 outlawed the killing and harvesting of north Pacific fur seals in the Pribilof Islands of Alaska. It was passed to give strength to the Interim Convention on the Conservation of North Pacific Seals signed in Washington on 9 October 1957 and amended by the protocol signed on 8 October 1963. The act reads:

It is unlawful, except as provided in this chapter or by regulation of the Secretary [of Commerce], for any person or vessel subject to the jurisdiction of the United States to engage in the taking of fur seals in the North Pacific Ocean or on lands or waters under the jurisdiction of the United States, or to use any port or harbor or other place under the jurisdiction of the United States for any purpose connected in any way with such taking, or for any person to transport, import, offer for sale, or possess at any port or place or on any vessel, subject to the jurisdiction of the United States, fur seals or the parts thereof, including, but not limited to, raw, dressed, or dyed fur seal skins, taken contrary to the provisions of this chapter or the Convention.

The act permitted sealing by Aleuts, Eskimos, and Indian tribes. Subsequent amendments to the convention agreement were signed on 3 September 1969, 7 May 1976, and 14 October 1980.

Gabrielson, Ira Noel (1889–1978)
See North American Wildlife Conference.

Garden, Alexander (1730–1791)
A Scottish naturalist, Dr. Alexander Garden was one of the first careful observers of the natural wonders of North America. He was born on 20 January 1730 in Edinburgh, Scotland, the son of a clergyman whose Birse Parish was located in Aberdeenshire. Alexander Garden's earliest schooling is unknown, but from about 1743 to 1746 he was apprenticed to Dr. James Gordon (or Gregory). He qualified to join the British navy as a surgeon's second mate and served from 1748 to 1750. Subsequently he went on to study medicine at Marischal College in Aberdeen and the University of Edinburgh. The former institution awarded him a medical degree in 1753.

In 1752, Garden traveled to the colonies, where he settled in South Carolina. He formed a medical partnership with a Dr. Rose in Prince William's Parish and later practiced by himself in Charleston. Garden biographer Donald C. Peattie wrote, "From the first he took an interest in the fauna and flora of South Carolina, partly as an adjunct to the practice of medicine." His interest was stirred by a trip north in 1754, where he met naturalists Cadwallader Colden (1688–1776) and John Bartram. He later corresponded with British naturalist John Ellis (1710?–1776) and Swedish scientist Carolus Linnæus (1707–1778). Ellis named the flower *Gardenia* after Garden. Eventually, Garden sent all these men plant and animal specimens. Although he eventually came to disagree with Ellis and Linnæus over certain matters, Garden was installed by Linnæus as a member of the prestigious Royal Society of Uppsala. In 1755, during a journey inland with the governor of South Carolina, Garden observed a clay that "was declared by English pottery makers to be equal to the finest porcelain clay from India," but he did not mark the spot, and it remains unknown to this day.

A practicing physician, Garden was noted for his humanitarian work during the 1760 outbreak of smallpox in Charleston. He was, however, an unrepentant Loyalist, and when the American Revolution began, he left his property behind and fled to England. During the voyage, wracked by tuberculosis, he came down with a serious case of seasickness that made his health worse. American authorities confiscated his property, but his son, Maj. Alexander Garden, who had sided with the colonists, had it restored to him, although the Scottish naturalist never returned to America. For a time after his return to Britain, he served as vice president of the Royal Society of London. His health soon took a turn for the worse, however, and he died at his home in London on 15 April 1791 at the age of 61.

Dr. Alexander Garden remains largely unknown in early American naturalist history. His numerous publications, including *Medical Properties of the Virginia Pink Root, With a Botanical Description* (1764) and *Contributions to Essential Physics and Literature* (1771), and his articles "An Account of an Amphibious Biped; the Mud Inguince or Syren of South Carolina" and "An Account of Two New Tortoises," both written in 1771, linger untouched by historians.

Garfield, James Rudolph (1865–1950)
James R. Garfield was the twenty-third secretary of the interior, serving in the

James Rudolph Garfield, appointed the twenty-third secretary of the interior by President Theodore Roosevelt in 1907

Theodore Roosevelt administration from 1907 to 1909. Born in Hiram, Ohio, on 17 October 1865, he was the second son of James A. Garfield—a Civil War general, congressman, and twentieth president of the United States (1881)—and Lucretia (nee Rudolph) Garfield. While Garfield was growing up, his father was in Congress, and the family spent much of the time shuttling between Ohio and Washington, D.C. James Garfield received much of his education in the schools of Washington and at St. Paul's School in Concord, New Hampshire. He was being privately tutored when his father was elected president but gave up the privilege when his father was shot and died of his wounds after two months of suffering. Garfield entered Williams College in Williamstown, Massachusetts, with his older brother Harry Augustus, who later became president of Williams

as well as serving as national fuel administrator during World War I. James Garfield graduated with a bachelor's degree in 1885. After contemplating a course in medicine, he took up the law instead, studying simultaneously at Columbia University's School of Law and the law firm of Bangs and Stetson in New York City. In 1888, he returned to Ohio, where he was admitted to the bar and opened a law office in Cleveland with his brother Harry.

Except for the periods when Garfield served in state and federal government, his law practice lasted for 60 years. Originally the firm of Garfield and Garfield, it expanded to Garfield, Howe & Westenhaver when Harry Garfield left for other professional pursuits. In 1895, James Garfield was nominated for the Ohio state senate and was elected the following year. Among his activities there was what one source called his sponsoring of a bill "requiring sworn statements of expenditures by candidates for public office." Although he was defeated in runs for Congress in 1898 and 1900, his reform-minded stance brought him to the attention of President Theodore Roosevelt, who named Garfield to the U.S. Civil Service Commission in 1902. The next year, Roosevelt named him the first commissioner of corporations in the Department of Commerce and Labor (now separate government agencies). "Is He a Trust Buster?" questioned the *New York Tribune* when Garfield began his work probing Standard Oil Company's monopoly over oil supplies. Added the *Tribune*, "Next to the Attorney General and Senator [Henry Cabot] Lodge, Mr. Garfield is probably the man most frequently consulted by the President." Garfield's reports on abuses in the meat-packing industry led to the passage of the Pure Food Act of 1904; other investigations focused on railroads and licenses for companies involved in interstate commerce. Garfield's work made him a member of Roosevelt's "Tennis Cabinet."

On 4 March 1907, Secretary of the Interior Ethan Allen Hitchcock resigned. The following day, Roosevelt named Garfield to succeed him. With only two years until Roosevelt left office in March 1909, the new interior secretary had little time to establish much of an imprint on the department. However, according to Interior Department historian Eugene Trani:

> Garfield charged into battle with the fervor of a Rooseveltian Progressive . . . from the first [he] saw inefficiency, waste, and duplication. . . . His first change was to abolish the clerical divisions serving as screening agencies. . . . He gradually became more interested in the policies of the Department, and came to believe that the Federal Government alone had the power to preserve the country's resources, and with Roosevelt's support and encouragement he went out to battle the interests threatening these resources. He withdrew timber reserves from public sales, and refused to relent when the cattle and logging interests objected . . . . His efforts to reform land laws were not so successful. Here opposition was too strong; he could not get needed legislation. He did succeed, if briefly, in abolishing administrative duplication concerning the forest preserve of the Indians. . . . His term as Secretary . . . was one of the most important in the middle period of the Department, and his organizational reforms proved some of the most far-reaching.

When Roosevelt left office in 1909, Garfield left the Interior Department. Roosevelt wrote to a friend of Garfield's, "[He] is a great comfort to me. . . . He has such poise and sanity—he is so fearless, and yet possesses such common sense, that he is a real support to me. He tells me that his father, President Garfield, when he and his brother were at school and only 15 and 13 years of age, used to write them in full of his plans—why he went to the

Senate instead of staying in the House, etc., etc. The man has the sound body, sound mind, and above all the sound character, about which I intend to preach until I become prosy!"

After practicing law for several years, Garfield joined Roosevelt's breakaway Progressive Party, running in 1914 for governor of Ohio. After his loss, he returned to the Republican fold. One of his last political acts was to serve as chairman of the Committee on the Conservation and Administration of the Public Domain, or, as it is better known, the Garfield Public Land Commission. This council was appointed by President Herbert Hoover in 1929 to recommend options for the use and management of those lands still in the public domain. Other members of the commission included William Buckhout Greeley, chief forester of the U.S. Forest Service from 1920 to 1928; former U.S. Sen. Holm Olaf Bursum (1867–1953) of New Mexico; Gardner Cowles, publisher of the *Des Moines Register*; George Horace Lorimer, editor of the *Saturday Evening Post*; Commissioner of Reclamation Elwood Mead; and writer Mary Roberts Rinehart. The commission ultimately proposed ceding all federally controlled lands back to the states; Secretary of the Interior Ray Lyman Wilbur supported the suggestion, but Congress balked, and nothing came of the commission's work.

Garfield's last public service was as a member of the platform committee at the Republican National Convention in 1932. In 1940, he spoke out against what he perceived as an assault by Franklin Roosevelt on the nation's judiciary. James R. Garfield died of pneumonia in a nursing home in Cleveland on 24 March 1950. He was 84 years old.

See also Department of the Interior.

Geer v. Connecticut
(161 U.S. 519 [1896])

In this landmark case, the Supreme Court held that states had the right under the Tenth Amendment to pass game and hunting laws. A man named Geer shot several birds in Connecticut and wanted to ship them out of state. When he was arrested for interfering with the state's game laws, he sued in district court, which ruled against him. Geer appealed the case to the U.S. Supreme Court. Justice Edward White wrote the Court's opinion, holding that a state may pass laws dealing with wildlife. A state has the right "to control and regulate the common property in game," he wrote, which is to be applied "as a trust for the benefit of the people." Wrote Dyan Zaslowsky, "States could name whatever conditions were justified in taking game, and these conditions also applied after the game had been killed. White asserted that because of the 'peculiar nature' of game and the state's ownership of it, it was doubtful whether interstate commerce was even created in the case."

See also Weeks-McLean Act.

General Allotment Act (24 Stat. 388)

This act, also called the Dawes General Severalty Act after its principal sponsor and advocate in the U.S. Senate, Henry L. Dawes of Massachusetts, became law on 8 February 1887. It broke up Indian tribal lands and authorized the president of the United States to distribute reservation lands to Indians by allotting 160 acres to family heads, 80 acres to single persons younger than 18, and 40 acres to other persons under the age of 18. The titles to the land were to be held in trust by the government for 25 years, after which time the Indians would receive the titles and be freed of any government supervision. Section 6 of the act granted all Indians U.S. citizenship. The Dawes Act is still considered one of the landmark land laws passed in the history of the country; for the first time the Indians were recognized as citizens, and the tribal lands concept was created. But it failed to achieve the goal of making In-

dians into independent farmers; instead it made it easy for whites to buy or lease Indian lands and resulted in a reduction in Indian land holdings from 138 million acres to 52 million acres by 1934, when the act was repealed. At that time two-thirds of Indians had become landless or lacked enough land to be self-sufficient.

General Land Office

The General Land Office was established by Congress on 25 April 1812 to "superintend, execute, and perform all such acts and things touching or respecting the public lands of the United States." Before its inception, land policy was dictated by the states. In some cases, various claims to parcels of land had nearly caused states to pull out of the Union. In one such case, a cession of land led Maryland to sign the Articles of Confederation.

The first land act passed by Congress was the Land Ordinance of 1785, which surveyed and plotted townships in what is now the middle United States into plots six miles square. The ratification of the Constitution in 1788 made the Land Ordinance moot, leading to the passage of the Land Law of 1796, which created townships one-quarter the size of the original plans, with 640-acre sections to be sold. This law was supplemented by the Land Law of 1800, which cut the size of plots to be sold to 320 acres.

The sale of public lands led to the creation of the General Land Office in 1812. Faced with mounting problems that diverted him from his other duties, the secretary of the treasury asked for an office to deal directly with land sales. Edward Tiffin, a British-born physician, was named the first Land Office commissioner. After two years, he exchanged his post in Washington for the surveyor general's job held by his friend, Josiah Meigs. Among the more noted commissioners was John McLean, who later became an associate justice on the U.S. Supreme Court; Thomas Hendricks, vice president of the United States under Grover Cleveland; William Andrew Jackson Sparks, considered the most effective commissioner; James A. Williamson, whose work with Interior Secretary Carl Schurz led to the most efficient two-man conservation team in the U.S. government; and Fred Johnson, the last commissioner and first director of the Bureau of Land Management. The General Land Office became part of the Interior Department in 1849 and was melded with the U.S. Grazing Service into the Bureau of Land Management in 1946.

See also Bureau of Land Management; Carter, Thomas Henry; Sparks, William Andrew Jackson; Tiffin, Edward.

General Mining Law (17 Stat. 91)

This landmark legislation of 10 May 1872 was authored by Sen. William Morris Stewart of Nevada. Stewart, a former miner and mine law attorney, wanted the rights of miners protected against federal land laws. In 1866, he sponsored the Lode Law, which secured patent rights for lode miners. This 1872 legislation assured the same rights for placer mines. Defined simply, placer mines are those in the "ground with defined boundaries containing loose deposits of minerals in the earth, sand or gravel on or near the surface," as opposed to lode mines, which contain metal (such as gold and silver) embedded in rock (*United States v. Iron Silver Mining Company*, 128 U.S. 673 [1888]). The 1872 act reads, in part, "All valuable mineral deposits in lands belonging to the United States, both surveyed and unsurveyed, shall be free and open to exploration and purchase, and the lands in which they are found to occupation and purchase, by citizens of the United States and those who have declared their intention to become such, under regulations prescribed by law, and according to the local customs or rules of miners in the several mining districts, so far as the same are applicable and not inconsistent with the laws of the

United States." Essentially, this law opened the West for free mining by any person or company.

See also Lode Mining Law; Stewart, William Morris.

General Motors Corporation v. United States (110 L Ed 2d 480, 110 S Ct 2528 [1990])

In *General Motors*, the Supreme Court held that the Clean Air Act Amendment of 1970 did not hold the Environmental Protection Agency (EPA) to reviewing state implementation plans (SIPs) on air quality within a set time frame and did not prevent the enforcement of existing SIPs when revision was not "timely reviewed." In 1970, Congress amended the Clean Air Act to require the EPA administrator to set national air standards, known as national ambient air quality standards (NAAQS). Section 7410(a)(1) of the act required each state to submit to the government a state implementation plan (SIP) on how it would implement, preserve, and enforce NAAQS. Section 7410(a)(2) instructed the EPA administrator to approve or disapprove a submitted SIP "within four months." Section 7410(a)(3)(A) allowed the administrator to "approve any revision of an implementation plan applicable to an air quality control region if he determines that it meets the requirements" set out in section 7410(2). In 1980, the EPA granted a request from the state of Massachusetts to permit it to revise its previously approved SIP, allowing for a higher degree of emissions from automobile-painting companies, provided that such standards be reinstituted by 31 December 1985. A paint factory owned by General Motors asked for a waiver for full compliance until 1987; when the state tried to get such a deferment, the federal government sued for full enforcement by the 1985 date. On 4 September 1988, the EPA rejected Massachusetts' revised SIP. A district court struck down the original SIP because the EPA had not evaluated the revision within four months. The U.S. Court of Appeals for the First Circuit reversed, holding that although the EPA had failed to meet the four-month deadline, this did not preclude it from enforcing the original plan. On certiorari, the Supreme Court agreed with the court of appeals by a unanimous vote on 14 June 1990. Writing for the Court, Justice Harry Blackmun ruled that section 7410(a)(2) applied only to the original SIP, not to any revision. Since the original SIP had been approved within the four-month deadline, the plaintiff, General Motors, had no recourse.

General Revision Act of 1891
See Payson Act; Timber Culture Act.

Gentry, Alwyn Howard (1945–1993)

Before his untimely death in the jungles of Ecuador in August 1993, Alwyn Gentry was one of the most eminent U.S. biologists and botanists. Born on 6 January 1945 in Clay Center, Kansas, he was exposed to the sweeping majesty of the plains of the Midwest. As he later wrote, "I have always been fascinated by nature and the outdoors. As a boy growing up on the Kansas plains, I devoted every free moment to studying nature—butterflies, birds, wildflowers, snakes—but had no idea that that fascination could be turned into a career." He attended local schools before enrolling at Kansas State University, where he earned a bachelor of arts degree in physical science and a bachelor of science degree in botany and zoology, both in 1967. He was awarded a master's degree in botany from the University of Wisconsin in 1969 and a doctorate in biology from Washington University in St. Louis. His doctoral dissertation was "An Ecoevolutionary Study of the Bignoniaceae of Southern Central America."

While an undergraduate, Gentry was accepted into the Organization of Tropical Studies introductory course, which conducted plant studies in Costa Rica.

Here, Gentry noted, "I immediately fell in love with the tropics and consequently decided to study tropical plants in graduate school." Over the next several years, he worked for the U.S. Army doing vegetational analyses in the Panama Canal Zone; was curator of the Summit Herbarium in the Canal Zone; and served as assistant curator, associate curator, and finally curator of the Missouri Botanical Garden in St. Louis, where he worked with his wife, Rosa Ortiz de Gentry. Over a 20-year period, he personally collected and identified some 70,000 plant specimens—more than any other living botanist. According to one source, "his greatest achievement [as curator of the Missouri Botanical Garden] was to develop a methodology for identifying plants by common traits like bark and leaves—a great leap forward for tropical botanists, whose previous guideposts were seasonal characteristics like flowers and fruit." Editor of *Four Neotropical Rainforests* (1990) and author of *A Field Guide to the Families and Genera of Woody Plants of Northwest South America* (1993), which was published posthumously, as well as hundreds of botany articles, Gentry was considered one of the greatest living botanists. In 1991, Gentry joined the Rapid Assessment Program of Conservation International, a small environmental group in Washington, D.C. The program's goal is to use the talents of a dizzying array of scientists specializing in rainforest science and send them into affected areas to assess what needs to be done to secure the forest's survival. As part of the program, Gentry joined noted ornithologist Theodore Parker III in exploring several areas of Latin America, including the Alto Madidi region in northwest Bolivia. While in Ecuador surveying a rain forest to be preserved, Gentry and Parker were killed in a plane crash. Gentry was 48.

Gentry's feelings about his work preserving the environment were expressed in a paper he wrote:

Conservation never crossed my mind when I fell in love with tropical forests. Instead I found them fascinating from a purely selfish intellectual viewpoint. I was lucky enough to begin graduate school knowing I wanted to study tropical plants and to find support to be able to return repeatedly to the Tropics. Of course I soon came face to face with the rapid disappearance of the rainforest. This was long before rainforest destruction achieved public awareness and for a while I tried hard to ignore how fast the tropical forests were disappearing all around me, and concentrate instead on learning as much as I could about them. But as my favorite collecting areas and study sites literally disappeared in front of my eyes, I became increasingly appalled at what was happening, and began worrying about what could be done. I began to seek out opportunities to talk about the beauty and significance of rainforests and the catastrophe that seemed imminent; I also began spending more time thinking about the practicalities of how rainforest conservation could be encouraged and how the biodiversity that so enchants me could be turned into a tool for conservation, after much soul-searching coming up with the idea that the most effective rainforest conservation might come from careful harvesting of forest-grown products [the concept that has come to be called extractive reserves]. Today, much of my time is spent trying to sell the importance of rainforest biodiversity and conservation to the general public, both in this country and in Latin America. Indeed, I feel that perhaps the single most important contribution I have made to the world is in helping to stimulate a significant number of young Latin Americans to share my love of and concern for tropical forests.

See also Parker, Theodore A., III.

Geological Surveys
See United States Geological Survey.

Gibson, Paris (1830–1920)

Paris Gibson of Montana was an advocate in Congress for conservation. Born on his family's farm in Brownsville, Maine, on 1 July 1830, he was the son of Abel and Ann (nee Howard) Gibson, both farmers. He attended Bowdoin College in Brunswick, Maine, and graduated in 1851. A member of the Maine state legislature, he quit upon his father's death to tend to the family's homestead. In 1858, he moved to Minneapolis, where he began a woolen mill that earned him a small fortune, which he lost in the financial panic of 1873. He then relocated to Montana, where he engaged in sheep ranching. He was the planner and first mayor of the city of Great Falls, Montana, which he came upon while sheep grazing.

A follower of the Democratic Party, Gibson was a member of the constitutional convention that drafted Montana's first constitution, and he served in the first state senate session (1891–1893). In 1900, upon the resignation of Sen. William Andrews Clark, the Montana state legislature elected Gibson to serve out the remaining four years of Clark's term. Gibson soon became an advocate of the repeal of the Desert Land Act and the Timber Culture Act. In a 1903 article, "The Repeal of Our Objectionable Land Laws: A Clear and Concise Statement of the Dangerous Manner in Which the Remaining Public Domain Is Being Absorbed," Gibson wrote, "We now find that the public lands are being absorbed with a rapidity never equaled in our country's history, and that only a small part of these lands is becoming property of the actual settlers." In 1904, when the Republicans took control of the state senate, Gibson was replaced by Thomas Henry Carter, a former senator and one-time commissioner of the General Land Office.

A noted politician whose finances were tied up in mining, ranching, and other pursuits, Gibson ran his business until he was 85 years old. He died on 16 December 1920 at the age of 90.

Gilbert, Grove Karl (1843–1918)

Geologist Grove Gilbert was an important member of the Wheeler survey west of the 100th meridian and John Wesley Powell's Rocky Mountain survey. He was born on 6 May 1843 in Rochester, New York, the son of Grove Sheldon Gilbert, a painter, and Eliza (nee Stanley) Gilbert. Grove Karl Gilbert's ancestors were of English origin, immigrating to America in 1630. Gilbert attended the public schools of Rochester before he enrolled at the University of Rochester, where he earned a bachelor's degree in 1862. While at the university, he took a course in geology taught by Henry A. Ward, owner of the Ward Natural Science and History Establishment in Rochester, which distributed scientific material to schools and was also a research institution. Gilbert later worked for Ward's company from 1863 to 1868.

In 1869, Gilbert went to work for Professor John Strong Newberry on a geological survey conducted by the state of Ohio. In 1871, he was hired by Lt. George Montegue Wheeler to be a geologist on his geographical and geological survey west of the 100th meridian. Gilbert found Wheeler's military expedition, done more for mapping purposes than for geological research, to be restricting. After a chance meeting with explorer John Wesley Powell, he joined Powell's survey of the Rocky Mountain region in 1874. Wrote Powell biographer Lindsey Morris, "During the next two years, Powell's survey made a special study of the Uinta Mountains and adjacent portions of the Green River Basin" in Utah, as well as the Henry Mountains of that state. Gilbert's first work for the Powell survey was a monograph, *Report on the Geology of the Henry Mountains*

Geologist Grove Karl Gilbert was a member of the Wheeler survey (1871), the Powell survey (1874), and the Harriman expedition to Alaska (1899). He served as chief geologist of the U.S. Geological Survey from 1889 to 1892.

(1877), a detailed analysis of the laccolithic configuration of that particular mountain range. In 1879, when all the western surveys were consolidated into the U.S. Geological Survey, Gilbert published what is considered his greatest work, *Lake Bonneville*, and described what biographer Ronald DeFord called "the gigantic Pleistocene ancestor of the Great Salt Lake." Gilbert's other works include *The Transportation of Debris by Running Water* (1914), *Hydraulic-Mining Debris in the Sierra Nevada* (1917), and the posthumously published *Studies of Basin-Range Structure* (1928). Named chief geologist of the U.S. Geological Survey in 1889, Gilbert held this position until 1892. His later studies included those of the Great Lakes and the Niagara River. He was a member of the National Academy of Sciences, the Geological Society of America and its counterpart in London, and the Bavarian Royal Academy of Sciences in Germany and participated in the 1899 Harriman expedition to Alaska. He died in Jackson, Michigan, on 1 May 1918, five days short of his seventy-fifth birthday.

See also Powell, John Wesley; Wheeler, George Montegue.

Global 2000 Report to the President

This analysis from the Council on Environmental Quality (CEQ) was conducted to study the environmental situation in the world as the twenty-first century approaches. On 23 May 1977, in his environmental message to the nation, President Jimmy Carter asked the CEQ, working with other federal agencies, including the Department of State, to consider the "probable changes in the world's population, natural resources, and environment through the end of the century." The subsequent report would act as "the foundation of our longer-term [policy] planning." Heading the reporting commission was Gus Speth, chairman of the CEQ, and Thomas R. Pickering, assistant secretary of oceans and international environmental and scientific affairs at the Department of State, who later served as U.S. ambassador to the United Nations in the Bush administration. The director of the study was Dr. Gerald O. Barney. The Global 2000 team looked at the three pressing issues of population, resources, and the environment and their growing interdependence. The report, released in 1978, concluded that if nothing was done to curb population growth and environmental destruction worldwide, by the year 2000, total world population would approach 6.35 billion, with 100 million people being added each year (90 percent of them in poor, Third World countries); food production would increase 90 percent from 1970 to 2000, but that would be 15 percent less than needed, with most of the food being produced in developed nations and lacking in less developed countries (LDCs); arable land would increase by only 4 percent, leaving LDCs to produce higher yields or suffer starvation; the energy needs of LDCs would suffer as the world stretches the maximum allowable yields of oil and other energy sources to be removed from the earth; forests would be cut down in greater numbers, particularly in LDCs, as the cost of energy and the need for land for farming and living increases; the elimination of trees would lead to soil erosion, increasing environmental destruction; and animal and plant species found only in LDCs would be threatened, endangered, and perhaps rendered extinct. The Global 2000 report estimated that at present birth rates, the world population could reach 10 billion by 2030 and perhaps 30 billion by 2100. It called for immediate policy changes worldwide to avoid this catastrophe.

Governors' Conference on the Environment

This 1908 gathering brought together the governors and administrators from 52 states, territories, and dependencies under the U.S. flag. In addition, Supreme Court justices, members of the House and Senate, and other invited guests swelled the crowd that met on the White House lawn to over 500 people. The conference, which was originally recommended by the Inland Waterways Commission, lasted from 13 to 15 May 1908 and was organized by Gifford Pinchot, W J McGee, and Frederick Haynes Newell. It was divided into lecture sessions on minerals, water resources, and land and timber conservation. Andrew Carnegie opened the meeting with an address called "The Conservation of Ores and Related Materials"; railroad magnate James J. Hill opened the third session with a talk on wealth invested in land, and Theodore Roosevelt made a stirring speech in which he called for congressional funding to make the Inland Waterways Commission a permanent entity.

By the second day, the conference was in full swing. Robert A. Long, head of the Long-Bell Lumber Company of Kansas City, Missouri, used his speech, "Forest Conservation," to denounce those who called for a cutback in logging. He did, however, call for a "national inventory" of the nation's resources, including timber reserves. Gov. Newton C. Blanchard of Louisiana and McGee cooperated on the conference's declaration of views in a 1,000-word statement.

The Governors' Conference was historic in several ways. It was the first time a call had been issued for greater awareness of the nation's environment; it led to the creation of a National Conservation Commission, which in 1909 released a lengthy report on the nation's environmental resources; and it gathered persons from the entire political and business spectrum to discuss government and public policy on the environment. If anything, it demonstrated that the White House could be used as a bully pulpit. Although several conservationists did not appear or were not invited—among them John Muir, whose personal battle with Pinchot over development had strained their relationship—the conference is considered by historians to have been a major success. From 1908 on, the policy of the federal government, through federal legislation and presidential directives, was one of conservation and increased protection of the environment.
See also National Conservation Commission.

Governors' Conferences on the West
See Western Governors' Conference.

Graves, Henry Solon (1871–1951)
A leading forestry expert, Henry S. Graves served as chief forester from 1910 to 1920. Born in Marietta, Ohio, on 3 May 1871, he was the son of William Blair Graves, a professor of natural sciences at the prestigious Phillips Academy in Andover, Massachusetts, and Luranah

(nee Hodges) Graves. Henry Graves attended Phillips as well as Yale University, from which he graduated with a bachelor's degree in 1892. One of his classmates at Yale was Gifford Pinchot, who urged Graves to take up professional forestry as a career. Graves took Pinchot's advice and went overseas to attend the University of Munich (Germany). He returned to the United States in 1896 and joined Pinchot as a professional forester in a consulting position. (Pinchot referred to Graves as "Harry.")

In 1896, the two men coauthored *The White Pine*, an investigation into white pine trees in the United States, particularly those on the Dodge reservation in Pennsylvania. As Pinchot described it, the work "kept in mind both the forest and its owner—[it] did not forget interest, taxes, and other expenses in relation to forest production." As Pinchot admitted, Graves did most of the fieldwork. Two years later, when Pinchot was named chief of the U.S. Forestry Division, Graves accompanied him to Washington as his assistant. For ten years, starting in 1900, Henry Graves served as dean of the Pinchot-inspired Yale Forestry School, close to Graves's home in Connecticut. The school was funded by Pinchot's family, with students doing extracurricular work on the Pinchot estate in Pennsylvania. While at Yale, Graves authored two works, which were the earliest books on forestry management written by an American, *Forest Mensuration* (1906) and *Principles of Handling Woodlands* (1911). In 1910, when Pinchot was fired in the midst of the Ballinger-Pinchot affair, he was able to use his influence to allow Graves to succeed him. Related Pinchot, "Things being as they were ... the appointment of Harry Graves as Forester was the very best possible. A man of the highest character, the best-trained forester in America, with no little executive experience, Graves had in addition friendly relations with the President which promised extremely well."

Graves served as chief forester from 1910 to 1920. Author Jean Pablo reported that during that time, he "quickly restored orderly relationships between his agency, the secretary of agriculture, and the Department of the Interior." In 1917, Gen. John Pershing sent an urgent message from Europe that U.S. forestry engineers gathering lumber supplies for U.S. troops needed an authority in forestry to aid them. Graves (commissioned a major) immediately went to France with his assistant, William B. Greeley, to give advice about the utilization of wood products for the troops.

In 1920, he returned to the United States and resigned as chief forester to become the Sterling Professor of History at Yale University. In 1922, he was made dean of the Yale Forestry School. In 1932, he collaborated with Cedric H. Guise on the textbook *Forest Education*. From 1923 to 1924, and again from 1934 to 1936, he was president of the American Forestry Association. In 1944, he received the Sir William Schlich Memorial Medal (named after a German forester who trained British foresters for work in India), and in 1950 the Gifford Pinchot Medal from the Society of American Foresters. His last work was *Problems and Progress of Forestry in the United States* (1947). Henry Graves died on 7 March 1951 at the age of 79.

See also Greeley, William Buckhout.

Great Plains Drought Area Committee

This federal commission was created by President Franklin D. Roosevelt's Executive Order 8000 on 22 July 1936 to devise "a long term program for the efficient utilization of the resources of the Great Plains area." Wrote James Olson, "Because of severe droughts in the Midwest in 1934 and then another series of droughts in the Midwest and Southeast in 1936, President Roosevelt established [the committee] on 22 July 1936." Heading the council was Morris L. Cooke, admin-

istrator of the Rural Electrification Administration (REA). Also named to the commission were Professor Harlan H. Barrows, a member of the National Resources Committee in Chicago; Dr. Hugh Hammond Bennett, chief of the Soil Conservation Service; Dr. L. C. Gray, assistant administrator of the Resettlement Administration (RA); Rexford Tugwell of the RA; Col. Francis Clark Harrington, assistant administrator of the Works Progress Administration (WPA), a New Deal agency; Harry Hopkins of the WPA; Col. Richard C. Moore, engineer in the Missouri River Division of the Corps of Engineers; John C. Page, acting director of the Bureau of Reclamation; Dr. Harlow S. Person of the Rural Electrication Administration in Washington, D.C.; and Frederick H. Fowler of the National Resources Committee.

From its inception until a final report was issued, the committee held public hearings in Dalhart, Texas; Bismarck, North Dakota; and Washington, D.C., in an attempt to listen to governors and their representatives, members of state planning boards and farmers' organizations, federal officials of pertinent agencies, and average citizens. Said author Olson, "The committee surveyed drought conditions, identified 1,194 counties in twenty-five states as drought areas, adjusted the soil conservation program to provide for increased grain production, created a Federal Livestock Feed Agency in Kansas City to coordinate feed distribution and stabilize prices, provided work relief to the needy, helped resettle hundreds of families, and generally coordinated federal relief activities." The committee's final report, *The Future of the Great Plains*, advocated the establishment of a federal agency to coordinate all drought-related operations for the federal government.

Greeley, William Buckhout (1879–1955)

Chief of the U.S. Forest Service from 1920 until 1928, William Greeley did much to shape the nation's policies on forestry matters. The son of Frank Greeley, a Congregational minister descended from a long line of religious men, he was born in Oswego, New York, on 6 September 1879. His biographer, George T. Morgan, notes that Greeley was influenced "by the stern Calvinist teachings of his grandfather and father." At the age of 11, Greeley and his family moved to California, where young William attended public schools in San Jose before entering the University of California. There he earned a bachelor's degree in English and history in 1901.

Prior to his college education, Greeley taught high school English, but his real interest was in forestry. After graduating from the University of California, he returned to the East Coast and enrolled at the Yale University Forestry School, where he came under the influence of such forestry pioneers as Bernhard Fernow and Henry S. Graves. In 1904, Greeley graduated from Yale with a master of forestry degree. That same year, he entered the Bureau of Forestry (now the U.S. Forest Service), where he spent a year studying commercial trees in the southern Appalachian Mountains. In 1905, he was assigned to the Sequoia National Forest in California as an inspector of timber sales. He eventually served as the regional forester for the Rocky Mountain region (1908–1911) and as chief of the Forest Service's Branch of Forest Management. He eventually became assistant chief forester.

In 1917, Gen. John J. Pershing, head of the American Expeditionary Force in Europe, sent a request to Henry Graves, chief of the Forest Service, to establish a corps of foresters to supply U.S. troops with wood for "dock, barracks, warehouses, railroad ties, barbed wire entanglement stakes, fuelwood, and other forest products." Graves and Greeley were both commissioned majors, and the two went off to Europe to give advice and create a forestry corps. Greeley was eventually promoted to lieutenant colonel

before returning to the United States in 1919. A year later, Graves resigned as chief forester, and Greeley was named to replace him. He served in this position until 1928. During those years, he battled with former chief forester Gifford Pinchot over the enactment of the Clarke-McNary Reforestation Act of 1924 and the McSweeney-McNary Act of 1928. Wrote historian Henry Clepper, "During his [Greeley's] administration, under the Clarke-McNary Reforestation Act, fundamental forest policy of [the] United States was established, providing for federal-state cooperation in fire control, reforestation, and farm forestry extension. Net area of national forests was enlarged to nearly 160 million acres." Adds historian Dennis Roth, "During his tenure as forester, Greeley also promoted the passage of the McSweeney-McNary Act of 1928, which securely established the Forest Service's research program." In 1928, Greeley resigned from the Forest Service to become the secretary-manager of the West Coast Lumbermen's Association in Seattle, a position he held for 18 years. He also served as chairman of the board of American Forest Products Industries, Inc., where he was a leader in the "tree farms" crusade. He was also the inspiration behind the "Keep America Green" and "More Trees for America" movements. His philosophy was summed up when he said, "True forest conservation can be accomplished more by the voluntary cooperation of individuals, rather than by law." Greeley was honored by the Society of American Foresters with the Sir William Schlich Memorial Medal (named after a noted German forester) and was posthumously recognized by the organization when it named a forestry research foundation the William B. Greeley Memorial Laboratory. He died at his home in Suquamish, Washington, on 20 November 1955, after being ill for several months. The *Seattle Times* editorialized, "One of his pet projects was the forest industry's tree nursery at Nisqually

[Washington]. Established in 1941, it now [1955] has a stock of 20,000,000 seedling trees for the replenishment of depleted stands in the Douglas fir region, supplementing the work of Nature in providing a perpetual growth of new trees.... At ceremonies to be held ... the Nisqually nursery will be christened officially the Col. William B. Greeley Forest Nursery. His long leadership in American forestry could not be accorded a more fitting memorial, not could this important forest enterprise be given a more appropriate designation." A bronze plaque at the Nisqually Nursery has the following inscription: "This Industrial Forest Nursery Dedicated to William B. Greeley, Forester—A man strong in his faith in the land and the ability of informed people to manage it wisely during the transition from old forests to new, he became a builder of industrial forestry in the Northwest and a teacher of forestry to all the world."

See also Clarke-McNary Reforestation Act; Graves, Henry Solon; McSweeney-McNary Forest Research Act.

Greenpeace

This environmental group is considered one of the most radical in its thought and actions. It was established out of the Canadian Don't Make a Wave Committee, which, in 1971, protested the United States' explosions of nuclear weapons in the Aleutian Islands in Alaska. Greenpeace (named for the concept of an ecological peace with the earth) was founded by several leftists in British Columbia: Irving Stowe, a former Philadelphia attorney who had protested the Vietnam War and left the United States for Canada after refusing to pay taxes; James and Marie Bohlen, who had fled the United States when their son became eligible for the draft; Paul Cote, a recently graduated Canadian attorney; Terry Simmons, an antiwar activist from Minnesota; and environmental advocate Bill Darnell.

Greenpeace and its radical agenda were at the forefront of the environmental movement of the 1970s and 1980s. With its program of "Direct Action," the group has confronted communities and governments with disruptions of various industries. Greenpeace literature claims that members have "scaled smokestacks of air polluters and plugged discharge pipes of chemical polluters." In 1985, a Greenpeace ship, the *Rainbow Warrior*, was blown up by the French government for protesting French nuclear tests in the south Pacific, and a Greenpeace photographer was killed. The scandal that ensued nearly toppled the French government. The *Washington Times* called Greenpeace a "1990s Pied Piper trying to lure a whole generation toward a promised land of total political transformation." The group has about 2.3 million members in the United States and 5 million worldwide. Its 1991 budget was $47.6 million.

Grinnell, George Bird (1849–1938)

George Bird Grinnell was a noted naturalist, conservationist, and author. He was born in Brooklyn, New York, on 20 September 1849, the oldest of five children of George Blake Grinnell and his wife Helen. Both parents were of New England ancestry and claimed to be related to five colonial governors and Betty Alden, one of the first white women born in New England. When George was young, the family moved to what was then called Audubon Park in New York, where the widow of famed naturalist and artist John James Audubon conducted educational classes. This is where George Grinnell received his first schooling. Later, he attended Churchill's Military School in Ossining, New York, and Yale University, where he was awarded a bachelor's degree in 1870. That summer, he went with Professor Othniel Charles Marsh of Yale University's Peabody Museum to serve as a naturalist and aide on an expedition to uncover dinosaur fossils

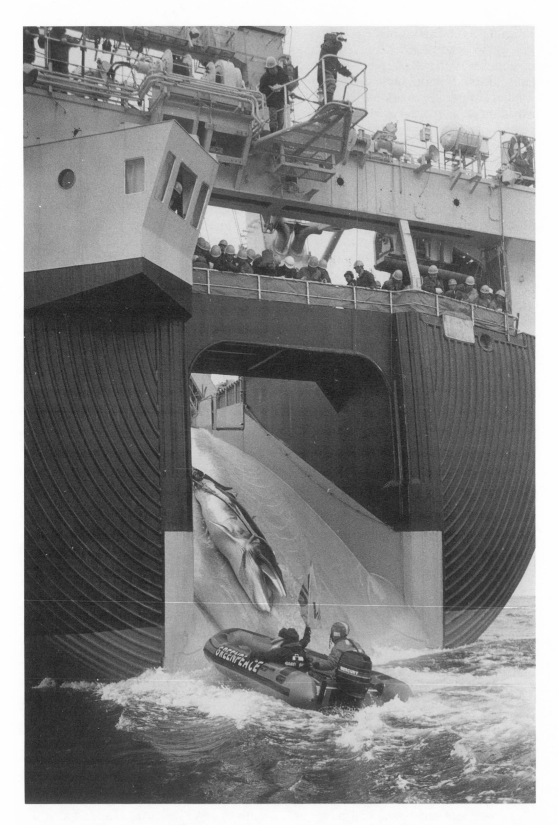

A Greenpeace crew protests whaling in international waters by confronting a Japanese factory ship operating off Antarctica in January 1992.

in the West. In 1874, after a time spent working for his father, Grinnell returned to Yale, where he studied osteology under Marsh and became Marsh's assistant. In 1880, Grinnell was awarded a Ph.D. in paleontology from Yale.

Grinnell's study of the western United States continued in late 1874 when he accompanied Gen. George Armstrong Custer on a military journey to South Dakota's Black Hills. In 1876, Custer asked Grinnell to make a similar trip, but Grinnell had to turn down the offer. History notes that Custer and all his men were wiped out that year at the Battle of the Little Big Horn. In 1875, however, Grinnell accepted an invitation to travel with Col. William Ludlow, chief engineer of the Department of the Dakotas, to Yellowstone National Park, and Grinnell's reports on the birds and other animals of the area remained for many years the authority on the subject.

In 1876, Grinnell became a natural history writer for the sports magazine *Forest and Stream*, which was the forerunner of today's *Field and Stream*. Although he became the periodical's editor and publisher in 1880, Grinnell continued to contribute many articles to its pages. As editor, he lobbied for the protection of Yellowstone National Park as a sanctuary for wildlife. In 1894, Congress passed this legislation in the form of the Yellowstone Animal Protection Act. In 1885, while on a hunting trip in the St. Mary's Lakes region of Montana, Grinnell discovered a glacier that was later named for him. He described the area in the September 1891 issue of the magazine *Century*. He was also responsible for the establishment under the General Revision Act of 1891 (also known as the Payson Act) of the Afognak Forest and Fish Culture Reserve.

Grinnell became intimate with Theodore Roosevelt after writing a favorable review of Roosevelt's work, *Hunting Trips of a Ranchman*. At the end of 1887, Roosevelt returned to New York from the western United States, where he had seen the wanton slaughter of many species. Roosevelt met Grinnell and other sportsmen and called for the formation of an organization that would "work for the preservation of the large game of this country, further legislation for that purpose, and assist in enforcing existing laws." Roosevelt opined that the group should be named after Daniel Boone and Davy Crockett. The following month, January 1888, saw the formation of the Boone and Crockett Club. Grinnell was a leading member of that organization as well as a founder in 1895 of the New York Zoological Society. In the pages of *Forest and Stream* in 1886, he announced the formation of an "Audubon Society," named after his mentor, John James Audubon. He solicited donations to form this new organization, which would lobby for the protection of bird species, but when incoming funds failed to cover mounting expenses, Grinnell closed the books on the group. It was eventually revived by salesman William E. Dutcher in 1905.

In an article called "The Last of the Buffalo" that appeared in *Scribner's Magazine* in September 1892, Grinnell wrote:

Of the millions of buffalo which even in our own time ranged the plains in freedom, none now remain. From the prairies which they used to darken, the wild herds, down to the last straggling bull, have disappeared. In the Yellowstone National Park, protected from destruction by United States troops, are the only wild buffalo which exist within the borders of the United States. These are mountain buffalo, and, from their habit of living in the thick timber and on the rough mountain sides, they are only now and then seen by visitors to the park. . . . On the great plains is still found the buffalo skull half buried in the soil and crumbling to decay. The deep trails once trodden by the marching hosts are grass-grown now, and fast filling up. When these most enduring relics of a vanished race shall have passed away, there will be found, in all the limitless domain once darkened by their feeding herds, not one trace of the American buffalo.

A noted writer on many conservation subjects, Grinnell also authored *Blackfoot Lodge Tales, By Cheyenne Campfires*, and *The Story of the Indian*. He served as chairman of the Council on National Parks, Forests, and Wild Life and was elected president of the National Parks Association in 1925. Although still a leading conservationist after retiring from *Forest and Stream* in 1911, Grinnell was not the major voice he had been in the late nineteenth century. He died at his home in New York City on 11 April 1938. The *New York Times* called him "the Father of American Conservation."

See also Afognak Forest and Fish Culture Reserve; Audubon, John James; Boone and Crockett Club; Burnham, John Bird; Dutcher, William E.; Marsh, Othniel Charles; National Audubon Society.

Hague, Arnold (1840–1917)

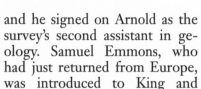

Noted geologist Arnold Hague was an assistant to Clarence King during King's survey of the western territories as well as during the latter's tenure at the U.S. Geological Survey. Born in Boston, Massachusetts, on 3 December 1840, Hague was the son of William Hague, a Baptist minister, and Mary Bowditch (nee Moriarty) Hague and the younger brother of mining engineer and King intimate James Duncan Hague. In about 1852, the family relocated to Albany, New York, where Arnold Hague attended the Albany Boys' Academy. Rejected by the Union army during the Civil War, he instead enrolled at the Sheffield Scientific School at Yale University, graduating in 1863. At the Sheffield School, Hague became intimate with Othniel Charles Marsh, Clarence King, J. Willard Gibbs, George Jarvis Brush, and James Dwight Dana—men who would later form the foundation of American scientific thought and study for the rest of the nineteenth century.

After being rejected a second time for service in the Union army, Hague went to Europe, where he studied at the Universities of Göttingen and Heidelberg and at the prestigious Royal School of Mines at Freiberg (known as the Bergakademie), where his older brother James had studied. There he befriended Samuel Franklin Emmons, another scientist in the making. In 1867, when he returned home to the United States, Hague was approached by Clarence King, who was forming a scientific team to accompany him on his survey of the territory between the Rocky Mountains and the Sierra Nevada, known as the Geological Survey of the Fortieth Parallel. King had recruited Arnold's brother James to head the survey's geology team, and he signed on Arnold as the survey's second assistant in geology. Samuel Emmons, who had just returned from Europe, was introduced to King and placed on the survey's crew as a nonsalaried geologist. The survey lasted until 1877, and the Hagues and Emmons followed it to the end. In fact, in 1877, Hague and Emmons coauthored *Descriptive Geology*, volume 2 of the seven-volume summary, *Report of the Geological Exploration of the Fortieth Parallel*.

In the several years that followed, Hague was hired by the government of Guatemala to study that nation's geologic makeup and by Li Hung Chang of the

Geologist Arnold Hague

Chinese government to investigate the gold and silver mines of northern China. With the establishment of the U.S. Geological Survey in 1879, Hague returned home and was hired as a survey geologist by the survey's first director, Clarence King. He was assigned to study the Eureka mining area and, under the directorship of John Wesley Powell, the geology of Yellowstone National Park. Hague had been one of the leading advocates calling for the establishment of Yellowstone as a national park, which finally occurred in 1872. His work on the Eureka district was published in 1882 in a collaborative effort with geologist Charles Doolittle Walcott.

In 1891, Congress enacted the General Revision Act, known also as the Payson Act. This legislation, which created the nation's forestry reserves and established the rules for their administration, included section 24, which declared that the president could, "from time to time, set apart and reserve, in any State or Territory having public land bearing forests, in any part of the public lands wholly or in part covered with timber or undergrowth, whether commercial value or not, as public reservations, and the President shall, by public proclamation, declare the establishment of such reservations and the limits thereof." Historians have credited this passage as being the work of Secretary of the Interior John Willock Noble. But Arnold Hague's letters in the National Archives illustrate that it was Rep. William Steele Holman of Indiana who crafted section 24, which Gifford Pinchot called "the most important legislation in the history of Forestry in America." The Payson Act led to the creation of 15 forest reserves totaling some 13 million acres of timber by the end of 1893.

Elected to the National Academy of Sciences in 1885, Arnold Hague is perhaps one of the least known men who dominated U.S. science and geology in the nineteenth and early twentieth centuries. In 1892, he authored *The Geology of the Eureka District, with Atlas*, monograph 20 of the Geological Survey. Hague died in Boston on 14 May 1917 after suffering a fall the previous winter. He was 76.

See also Emmons, Samuel Franklin; Holman, William Steele; King, Clarence Rivers; Noble, John Willock; Payson Act; Walcott, Charles Doolittle.

Hallstrom, et ux. v. Tillamook County (493 U.S. 1989, 107 L Ed 2d 237, 110 S Ct 304)

This Supreme Court case dealt with a section of the Resource Conservation and Recovery Act of 1976 (RCRA) that allowed citizens to bring lawsuits against violators of pollution regulations. Olaf A. Hallstrom owned a farm in Oregon that was located next to a landfill. In April 1981, Hallstrom sent Tillamook County officials written notice that he intended to file a lawsuit against the landfill under RCRA's section 6792. Under section 6792(b)(1), RCRA allows for lawsuits by citizens (except under certain circumstances) only within 60 days after the alleged violation was brought to the attention of the Environmental Protection Agency (EPA), the state in which the alleged violation occurred, and the alleged violator. Hallstrom waited a full year before filing suit. On 1 March 1983, the county asked for dismissal of the lawsuit, since the proper officials had not been notified. Hallstrom then notified these officials, and the case went to trial. The district court held that the county, in operating the landfill, had violated the RCRA provisions. On appeal, the U.S. Court of Appeals for the Ninth Circuit in San Francisco ordered the case remanded to the district court for dismissal, finding that Hallstrom had not followed the law. Hallstrom appealed to the U.S. Supreme Court, which held 7–2 that because Hallstrom had not followed the RCRA guidelines, he had no standing. Justice Sandra Day O'Connor, writing for the majority, stated that because RCRA clearly spelled out that all relevant agencies had to be notified 60 days before the filing of a lawsuit, a district court could not "disregard those requirements at its discretion, but must dismiss the action as barred by the terms of the statute." Justices Thurgood Marshall and William Brennan dissented.

Hansbrough, Henry Clay (1848–1933)

Said *Forestry and Irrigation* magazine of Henry Clay Hansbrough in 1902, "He has been one of the most active and efficient advocates of the development of the West through irrigation." A U.S. congressman and senator, Hansbrough was a leader in Congress in the attempt to formulate a national irrigation policy. He was born near the small village of Prairie du Rocher, Illinois, on 30 January 1848, the son of Eliab Hansbrough, a farmer, and Sarah (nee Hagen) Hansbrough. Both his parents were native Kentuckians, with ancestors on his father's side from Virginia. Eliab Hansbrough was an intimate of Henry Clay, the "Great Compromiser," and he named his son after the noted Kentucky lawmaker. Henry Clay Hansbrough's education was limited to the public schools of Randolph County, Illinois; in 1866, however, he moved with his parents to San Jose, California, where he learned the newspaper trade and worked for a time as managing editor of the *San Francisco Chronicle*. When his health failed, he moved back east to Baraboo, Wisconsin. When he recovered in 1881, he again moved, this time to the Dakota Territory.

Upon Hansbrough's arrival in the Dakota Territory, he started the *Grand Forks News*, where he worked for two years. He then moved to the small town of Devils Lake, where he began publication of the *Devils Lake Inter Ocean*. He later served two terms as mayor of Devils Lake, from 1885 to 1888. When North Dakota was admitted to the Union as the thirty-ninth state on 2 November 1889, Hansbrough became the first person to be nominated to represent the state in the U.S. House of Representatives. A conservative Republican, he was elected and served a single term (1889–1891). In 1891, he was elected by the state legislature to the U.S. Senate, where he served until 1909.

Hansbrough has been called a political maverick. Biographer Leonard Schlup said that Hansbrough "earned a reputation for independence that confounded both his supporters and opponents." Although he was a champion of the free coinage of silver, in 1900 he backed President William McKinley, a gold man. Although conservative, he called for the nomination of progressive Theodore Roosevelt for vice president in 1900 and supported a pure food bill in the Senate. But his most important work, as chairman of the Committee on Public Lands, came when he cosponsored in the Senate the landmark legislation that has become known as the Newlands Reclamation Act of 1902. On the need for a national irrigation policy, Hansbrough said in a speech in the Senate on 6 February 1902, "The purpose of this policy is to assist in providing homes for the rapidly increasing population of the country. President Roosevelt stated the case in a few words when he said in his [annual] message that 'successful home-making is but another name for the up-building of the nation.'" Because of his independence, the party machine back in North Dakota aimed its guns at him, and the state legislature did not return him to Washington in 1908. Hansbrough finished out his term and returned home.

In the last 25 years of his life, Hansbrough pursued business interests in North Dakota, Florida, New York, and Washington, D.C., where he died on 16 December 1933, just before his eighty-sixth birthday. According to one source, his remains were cremated and the ashes scattered, according to his wishes, near an elm tree on the grounds of the U.S.

Capitol. *Forestry and Irrigation* magazine said of him, "By his long experience in the far West he has become thoroughly familiar with the opportunities and results of irrigation, and has come to be regarded as the most strenuous advocate in the Senate of the utilization of the vacant public lands through national irrigation works. His position as Chairman of the Senate Committee on Public Lands, and also as a member of the Committee on Agriculture and Forestry, has given him exceptional opportunities for advancing the interests of the West in Congress, and his name will always be connected with the irrigation movement because of the introduction by him of the first bill providing for the construction of national irrigation works." The northern North Dakota city of Hansboro is named after him.

See also Newlands, Francis Griffith; Newlands Irrigation (or Reclamation) Act.

Harlan, James (1820–1899)

A U.S. senator from Iowa, James Harlan also served as the eighth secretary of the interior from 1865 to 1866. Born on 26 August 1820 to pioneer farmers Silas and Mary (nee Conley) Harlan in Clark County, Illinois, James Harlan shared his ancestry with Supreme Court Justice John Marshall Harlan. Silas and Mary Harlan, born in Pennsylvania and Maryland, respectively, had journeyed west, where their son James was born. Four years later they moved again, this time to the settlement of "New Discovery" in Parke County, Indiana, where, according to one source, "the Harlan home became a stopping place for travelling clergy."

James Harlan received a home education, supplemented by a short period at a local seminary and four years at Asbury University (now DePauw University) in Greencastle, Indiana. He taught at a small school in Missouri during his time at Asbury, and his first job after college was as principal of the Iowa City College. He married Ann Eliza Peck in 1845. Two years later, he was elected state superintendent of public instruction on the Whig ticket; unfortunately, the election was declared invalid because of some lost votes, and he lost in the runoff. Harlan began studying law and was admitted to the Iowa bar in 1848.

Two years later, Harlan won the Whig nomination for governor but was disqualified because he had not yet attained the age of 30. He was president of Iowa Conference University (now Iowa Wesleyan) from 1853 to 1855. During this time, Harlan was a leading figure in the establishment of the antislavery Free-Soil Party in Iowa. In 1855, the Iowa state legislature recognized his growing political power and elected him to the U.S. Senate for a full term. He spoke on behalf of the right of Kansans to decide whether to enter the Union as a slave or free state and was instrumental in helping found the Iowa Republican Party in 1856. Harlan was an intimate of Abraham Lincoln, and his daughter Mary married Lincoln's son Robert. Reelected to a second term in 1860, Harlan served as chairman of the Senate Committee on Public Lands, which enacted the legislation that later became the Homestead Act.

When conflict arose concerning the Department of the Interior's handling of the Pacific Railroad, Interior Secretary John Palmer Usher resigned. On 18 May 1865, President Andrew Johnson named Harlan to succeed Usher. The new secretary began by firing en masse those workers in the department that he thought were unnecessary (including a then unknown clerk named Walt Whitman). However, Harlan was soon angered by Johnson's Reconstruction policies and clashed with the president over land and railroad frauds. On 31 August 1866, less than 15 months after accepting the Interior post, Harlan resigned. He returned to Iowa, where in 1867 the state legislature returned him to the U.S. Senate.

Harlan served in the Senate until 1873; he was a leading critic of his former boss and even voted for Johnson's impeachment. Unsuccessful at reelection, Harlan returned to Iowa. Caught up in a scandal involving the famous Credit Mobilier company, Harlan had to defend himself against charges that he had accepted a bribe as secretary to ensure that the railroads received land grants. Although he was cleared, his reputation was ruined. He was nominated for the U.S. Senate in 1881 but was forced to decline the honor. His last public service was from 1882 to 1886 as the presiding judge of the Alabama Claims Commission, set up to define claims against the British government for anti-American acts during the Civil War. His last years were spent in retirement. James Harlan died in Mount Pleasant, Iowa, on 5 October 1899, at the age of 79.

See also Department of the Interior.

Harlan, Richard (1796–1843)

A writer and naturalist, Dr. Richard Harlan was the author of *Fauna Americana*, the first methodical composition on American mammals. Born in Philadelphia on 19 September 1796, he was the eighth child of Joshua Harlan, a Philadelphia merchant, and Sarah (nee Hinchman) Harlan, both of whom were Quakers. Harlan was related to the great jurists John Marshall Harlan and his grandson of the same name. Richard Harlan's brother, Dr. Josiah Harlan (1799–1871), was a soldier of fortune whose exploits were recounted in *A Memoir of India and Afghanistan* (1842). There is no record of Richard Harlan's early education, except that he studied medicine with Dr. Joseph Parrish of Philadelphia. Prior to receiving his medical degree from the University of Pennsylvania in 1818, Harlan served as a ship's surgeon on a voyage to Calcutta, India.

Harlan practiced medicine for a time in the dissecting room of Dr. Parrish's anatomical school. In 1820, he was made a physician in the Philadelphia Dispensary; the following year, he was named professor of comparative anatomy at naturalist Titian Ramsay Peale's Philadelphia Museum. In 1821, he presented a paper to the Academy of Medicine on the properties of animal heat. As biographer Whitfield Bell, Jr., wrote, "Harlan was the first American to devote a major part of his time to vertebrate paleontology." His first publication in this field was "Observations on Fossil Elephant Teeth of North America," which appeared in the *Journal of the Academy of Natural Sciences of Philadelphia* in June 1823. A confidant of such famous naturalists as Titian Peale, John James Audubon, and Thomas Nuttall, Harlan worked closely with these men on several outings and collected his material into the landmark work *Fauna Americana* (1825), the first academic collection of data on North American zoology, which some critics charged too closely resembled the studies of French naturalist Anselme-Gaétan Desmarest (1784–1838). Desmarest's work, *Mammalogie* (1820–1822), grouped both living and deceased fossil forms into distinct families, as did Harlan's. The leading critic of *Fauna Americana* was naturalist John Davidson Godman (1794–1830), author of *Rambler of a Naturalist* (1833). Godman's charges were so widespread that Harlan countered them in his *Refutation of Certain Misrepresentations Issued Against the Author of Fauna Americana* (1826). Harlan's next work was *American Herpetology* (1827), a study of American reptiles. His collected works appeared in 1835's *Medical and Physical Researches*. In that work, he wrote, "We have every reason to conclude, that every distinction of existing species has existed from the earliest periods of the formation of the present world; and has its origin ultimately in the nature of the *soil*; every variety of which is marked by a corresponding variety in its animal and vegetable productions; and many of these are limited by geographical distribution."

Harlan spent some time in Europe in the early 1830s. After returning to the

United States, he settled in New Orleans and began his medical practice once more. It was there on 30 September 1843 that he collapsed and died after an attack of apoplexy (a stroke). Harlan had celebrated his forty-seventh birthday only 11 days before.

Hatch Act (7 U.S.C. 361(b))

This act of 2 March 1887, which was written by Norman Jay Colman (1827–1911), the first secretary of agriculture, granted assistance to the states for the establishment of agricultural experiment stations. The act reads, "It is ... the policy of the Congress to promote the efficient production, marketing, distribution, and utilization of products of the farm as essential to the health and welfare of our peoples and to promote a sound and prosperous agriculture and rural life as indispensable to the maintenance of maximum employment and national prosperity and security.... It shall be the object and duty of the State agricultural experiment stations through the expenditure of the appropriations hereinafter authorized to conduct original and other researches, investigations, and experiments bearing directly on and contributing to the establishment and maintenance of a permanent and effective agricultural industry of the United States."

Hathaway, Stanley K.

See Kleppe, Thomas Savig.

Hayden, Ferdinand Vandeveer (1829–1887)

An engineer who led one of the four great surveys of the American West, Ferdinand Hayden remains one of the least-known geologists and explorers of his time. The son of Asa and Melinda (nee Hawley) Hayden, he was born in Westfield, Massachusetts, on 7 September 1829. Asa Hayden died when his son was

ten, and the boy was sent to live on his uncle's farm near Rochester, New York. What schooling Ferdinand had at this time is unknown, but he did teach school during the winters after he reached age 16. He left the farm at age 18 and walked to Ohio, where he persuaded the president of Oberlin College to allow him entry. He graduated with the class of 1850 with a bachelor of arts degree. He then attended the Albany Medical College in New York and was awarded a medical degree in 1853.

While in Albany, Hayden came under the influence of paleontologist James Hall (1811–1898), who encouraged Hayden to join an expedition led by geologist Fielding Bradford Meek to the Badlands of what is now South Dakota to collect fossils. In 1854, Hayden joined the American Fur Company in exploring the geological formations of the Missouri-Yellowstone Rivers area. In 1856 and 1857, he was attached as a geologist to Lt. Gov. K. Warren's survey of the Yellowstone-Missouri Rivers area and the Black Hills of the Dakotas. Finally, he worked with Meek in an investigation of the Kansas Territory to establish "the age of the lowest Cretaceous stratum" and joined the survey of Capt. William F. Raynolds of the Rocky Mountains (1859–1860). With the outbreak of the Civil War, Hayden volunteered for the Union army and served as a surgeon until his discharge in 1865, leaving with the rank of lieutenant colonel.

In the decade following the Civil War, Congress, in a quest to map and detail the lands of the American West, funded a number of surveys. By an act passed on 2 March 1867, it allocated appropriations for "a geologic survey of Nebraska, said survey to be prosecuted under the direction of the Commissioner of the General Land Office." Known better as the Survey of the Nebraska Territory, it was later expanded by Congress to become the Geological and Geographical Survey of the Territories, or the Hayden survey. It was eventually scheduled to cover Col-

Members of the 1870 western geological survey include leader Ferdinand Vandeveer Hayden (seated center, second row) and photographer William Henry Jackson (far right) at a Wyoming camp.

orado, New Mexico, Wyoming, Idaho, Nebraska, and Montana. Hayden assembled an impressive list of scientists to accompany him on the mission, among them Fielding Meek, Edward Drinker Cope, Leo Lesquereux, William Henry Holmes, and, eventually, engineer James Terry Gardiner (1842–1912), who deserted the rival survey of Clarence King. Hayden's 1870 survey of the Dakota Badlands included Meek, botanist Cyrus Thomas, and photographer William Henry Jackson, whose remarkable pictures of the surveys made him the preeminent photographer of the late nineteenth century. After the release of Jackson's stunning photos of the Hayden survey in 1875, the *New York Times* editorialized, "While only a select few can appreciate discoveries of the geologists or the exact measurements of the topographers, everyone can understand a picture." Another member was scientist Elliot Coues,

whom one historian called "a prolific writer and contributor to ornithology." Coues's report, *Birds of the Colorado Valley: A Repository of Scientific and Popular Information Concerning North American Ornithology*, was eventually published as Publication 11 of the Hayden survey reports. In 1879, because of overlap in the surveys, Congress consolidated them into the U.S. Geological Survey, with Hayden's rival Clarence King as the agency's first director. Hayden was given the position of geologist and sent back to the West, but his health soon failed and he suffered from locomotor ataxia (loss of coordinated movement caused by a disease of the nervous system). He retired in 1886 and died the next year in Philadelphia on 22 December.

See also Cope, Edward Drinker; Coues, Elliott; King, Clarence Rivers; Powell, John Wesley; United States Geological Survey; Wheeler, George Montegue.

135

Hazardous and Solid Waste Amendments (Public Law 98-616)

Enacted on 9 November 1984, this legislation supplemented the Resource Conservation and Recovery Act of 1976. It included new provisions that mandated a study by the Environmental Protection Agency (EPA) of hazardous wastes disposed of in large and small generators. Reads section 221(c):

> The Administrator [of the EPA] in cooperation with the States shall conduct a study of hazardous waste identified or listed under section 30001 of the Solid Waste Disposal Act which is generated by individual generators in total quantities for each generator during any calendar year of less than 1,000 kilograms. The Administrator may require from such generators information as may be necessary to conduct the study. Such study shall include a characterization of the number and type of such generators, the quantity and characteristics of hazardous waste generated by such generators, State requirements applicable to such generators, the individual and industry waste management practices of such generators, the potential costs of modifying those practices and the impact of such modifications on national treatment and disposal facility capacity, and the threat to human health and the environment and the employees of transporters or others involved in solid waste management posed by such hazardous wastes or such management practices.

The act also called for an inventory of all wells in the United States that inject hazardous wastes, as well as the establishment of a National Groundwater Commission.

See also Resource Conservation and Recovery Act.

Hetch-Hetchy Valley Controversy

This was a major battle between the forces of public power and the conservation movement in the early twentieth century. In 1913, Rep. William Kent of California (1864–1928) introduced legislation that would lead to the flooding of the Hetch-Hetchy Valley in Yosemite National Park and the construction of a dam to create electrical power; the water would be pumped 175 miles to San Francisco. Opposing this plan were leading environmentalists, most notably conservationist John Muir. Muir, who had founded the Sierra Club in 1892 and was the chief advocate of the protection and management of the Yosemite Valley, had been asked in 1903 to be Teddy Roosevelt's guide during the president's travels to the valley. Although he was 75 years old in 1913, Muir drew physical and mental strength from protecting the valley. In 1907, he predicted that the power interests would win out in destroying the Hetch-Hetchy. In a 1910 article, Muir described the valley: "The floor of the Hetch-Hetchy Valley is about three and one-half miles long and one-half mile wide. The lower portion is mostly a level meadow about a mile long, with the trees restricted to the sides and partially separated from the upper forested portion by a low bar of glacier-polished granite, across which the [Tuolumne] river breaks in rapids."

The records of the fight against the flooding of the valley are voluminous. The U.S. Senate received many angry letters from conservationists and others concerned about the issue. The King County (California) Chamber of Commerce wrote to Sen. George C. Perkins, "Our sense of justice compels us to protest against the cession to San Francisco of the waters of the Hetch Hetchy." From Visalia, California: "The Tulare County Board of Trade, representing twelve civic organizations, renews its protest against the diversion of the

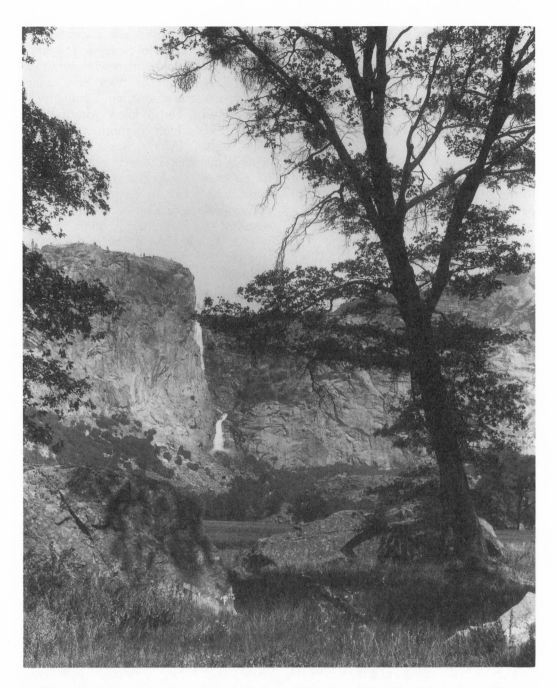

Geologist and Sierra Club member Joseph Le Conte photographed Hetch Hetchy Valley in California's Yosemite National Park about 1908. Controversy raged when California Congressman William Kent proposed in 1913 that the Tuolumne River be dammed to flood the valley and provide electricity and water to San Francisco. John Muir unsuccessfully led the opposition; the dam was completed in 1928.

waters of the rivers of the San Joaquin Valley. The flow of these rivers is absolutely essential to the development and prosperity of this great valley." Muir wrote to Sen. George E. Chamberlain, "In behalf of all the people of the nation we ask your aid in putting an end to these assaults on our great national parks and to prevent this measure from being rushed through before it can be brought to the attention of the ninety millions of people who own this park."

In his 1911 work *Yellowstone*, Muir swore against "these temple destroyers, devotees of ravaging commercialism, [who] seem to have a perfect contempt for Nature." Although Muir was a powerful force in the conservationist community, newly elected President Woodrow Wilson owed many favors to California politicians who had helped his campaign, and he signed the legislation to flood the Hetch-Hetchy on 13 December 1913. Muir never recovered from the shock of the destruction of his beloved Hetch-Hetchy Valley; he lost his zest for life and succumbed to pneumonia in December 1914 at the age of 76.

See also Muir, John.

Hickel, Walter Joseph (1919–)

Elected in 1990 for the second time as governor of Alaska, Walter J. Hickel, a former secretary of the interior, came under fire for a state plan to shoot wolves in the Alaska wilderness. He was born in Ellinwood, Kansas, on 18 August 1919, the eldest of ten children of Robert and Emma (nee Zecha) Hickel, both tenant farmers of German ancestry. He attended public school in Claflin, Kansas, up until he was 16. In 1940, when he was 20, he left home, intending to go to Australia and make his fortune. He worked briefly as a carpenter until he found out that he could not get a passport until he was 21. Undaunted, he purchased a steerage ticket for Alaska and shipped out, arriving in what would become the forty-ninth state with only 37 cents in his pocket. As he related in an interview

years later, "I didn't know a soul when I got off the boat in Seward. I washed dishes in an Anchorage restaurant and at night I slept on the floor of a cabin. When I was in that cabin, I thought 25 years ahead, and I knew where I was going." He added in another interview, "I thought it [Alaska] was a challenging, young, dynamic country. I wanted a frontier. I knew I could be somebody. I like to work and I like to accomplish things." After working at a series of odd jobs, including as a carpenter, he built a model home, sold it for a profit, and repeated the process until, by 1946, he was a developer with a sizable fortune.

A year later, in 1947, he founded the Hickel Construction Company of Anchorage and spent the next two decades building various massive construction projects in and around Anchorage, including Turnagain-by-the-Sea, an area of residential homes. Politics, although secondary, was a part of his program. A lifelong Republican, Hickel served as chairman of the Alaska Republican National Committee in 1964. Two years later, he ran a bitter campaign for governor against incumbent William A. Egan and won. Taking office as the state's first elected Republican governor, he initiated a program of development across the state and opened up the state's arctic areas, where minerals valued in the billions of dollars were thought to be deposited.

On 11 December 1968, less than two years after taking office, Hickel was chosen by President-elect Richard Nixon to be secretary of the interior. Wrote the *New York Times*, "Few men have been nominated for cabinet posts who seemed to have so few qualifications for the job and who aroused so much apprehension and opposition as did Walter J. Hickel. Here was a governor of Alaska, an apparently parochial figure with the self-made millionaire's limitations of national outlook and interest, named to head the Department of the Interior, which has authority over the regulation over much of the nation's natural resources—the public lands, oil and mineral leasing, wildlife and the national parks—and also over the pollution of the nation's waters." Initially opposed by every major conservation group, Hickel was nonetheless confirmed and sworn into office on 20 January 1969.

In his less than two years as interior secretary, Hickel surprised his critics and made several decisions in favor of conservation and the environment, including championing federal legislation obligating oil companies to clean up oil spills at their own expense, changing his mind about the feasibility of a trans-Alaska oil pipeline until safety studies could be conducted, and calling for the expurgation of pollutants from detergents, vegetable oils, and other chemicals used by ordinary people. In May 1970, Hickel got in trouble when a letter he had written to Nixon was leaked to the newspapers. In it, Hickel brought up the student demonstrations over the U.S. invasion of Cambodia and demanded that "youth must be heard." He cautioned the president that the administration was "embracing a philosophy which appears to lack appropriate concern" for America's youth and called for more attention to be paid to that segment of society. The letter was so embarrassing to the White House that Hickel's standing with the administration decreased soon after. It was no surprise when, on 25 November 1970, Hickel was called to the White House and told to resign. When he refused, he was fired and replaced the next day by Rep. Rogers C. B. Morton of Maryland.

Hickel returned to Alaska and his business enterprises. Once a Republican, he became a member of the Alaska Independence Party, founded in 1980 after passage of the Alaska National Interest Lands Conservation Act, which protected selected lands in Alaska from development. The Alaska Independence Party's platform calls for the gutting of the act, which it claims unjustly intrudes on state sovereignty. In his article "Alaska's Rush for the Gold," Ted Williams wrote, "If Hickel and his party get their way, the

blessings of . . . mining will be extended to 205 lakes and rivers, including 12 of 26 wild and scenic rivers [covered by the Wild and Scenic Rivers Act of 1968], 26 rivers and lakes inside national parks, 74 rivers in national wildlife refuges, and 20 rivers and lakes in national wilderness areas." In 1990, Hickel was elected governor of Alaska a second time. In 1993, he and his administration came under fire by environmentalists for a plan to shoot wild wolves and coyotes to save the moose population for hunting season. Although threatened with a nationwide boycott, the state began the killing in late 1993. In 1994, Hickel rejoined the Republican Party, but he decided against seeking a third term as governor.

See also Alaska National Interest Lands Conservation Act; Department of the Interior.

Highway Beautification Act (79 Stat. 1028)

This act of 22 October 1965 was designed to "provide for the scenic development and road beautification of the Federal highway system." The act reads, "The Congress hereby finds that the erection and maintenance of outdoor advertising signs, displays, and devices in areas adjacent to the Interstate System and the primary system should be controlled in order to protect the public investment in such highways, to promote the safety and recreational value of public travel, and to preserve natural beauty." The law, which called for the effective control of such advertising, allotted $20 million in fiscal 1966 and an additional $20 million in fiscal 1967 to carry out this mandate.

Historic Sites and Buildings Act of 1935

See Antiquities Act.

Hitchcock, Ethan Allen (1835–1909)

Ethan Allen Hitchcock was the twenty-second secretary of the interior, serving during both the McKinley and Roosevelt administrations. Born in Mobile, Alabama, on 19 September 1835, Hitchcock was the great-grandson of Revolutionary War hero Ethan Allen (of Green Mountain Boys fame), nephew of Civil War Gen. Ethan Allen Hitchcock, and son of Henry Hitchcock, a noted Vermont attorney and one-time chief justice of the Alabama Supreme Court. Henry Hitchcock died during the financial panic of 1837, leaving his widow Anne (nee Erwin) Hitchcock to care for her two sons, Ethan Allen and Henry, who later became dean of the St. Louis Law School (now the Washington University School of Law). The family moved first to New Orleans and then to Nashville, where Ethan attended public schools. He finished his education at a military academy in New Haven, Connecticut.

Hitchcock became a member of a mercantile house in St. Louis, then went to China and joined the St. Louis concern of Olyphant & Company as the commission enterprise's Hong Kong agent. He was made a partner of the business in 1866 and retired in 1872, having made a small fortune. From 1874 to 1897, he was involved in several capitalist ventures, including the formation of the first successful plate-glass manufacturing factory in the United States. In 1890, he embraced the platform of Rep. William McKinley of Ohio, who was elected president two years later. On 18 August 1897, McKinley appointed Hitchcock first as envoy and then as the first U.S. ambassador extraordinary to Russia and minister plenipotentiary to St. Petersburg. Hitchcock served in this post for little more than a year. In December 1898, Secretary of the Interior Cornelius N. Bliss resigned, and McKinley called Hitchcock home to assume the Interior post. He took over on 20 February 1899.

In this post for eight years, Ethan Allen Hitchcock was destined to become the second longest serving interior secretary in history (Harold Ickes later served just shy of 13 years). Hitchcock's was also one of the most important tenures. A defender of Indians and their land rights, he oversaw passage of the landmark Newlands Irrigation Act of 1902, fought land frauds (and started a bruising fight when he dismissed General Land Office Commissioner Binger Hermann over Hermann's connections with the frauds), and cleared the department of many unqualified employees. The *Boston Evening Transcript* told of the interior secretary's "vigilant services in punishing the thieves of public lands." Historians have cited Hitchcock's important role in a number of Teddy Roosevelt's sweeping presidential orders of 1906–1907, which ended the abuse of mineral mining on public lands. This work earned Hitchcock the enmity of several western senators, who pushed Roosevelt to ask for Hitchcock's resignation. Hounded on all sides, Hitchcock acceded to this demand on 4 March 1907. Just two years later, on 9 April 1909, beset by complications from kidney, heart, and pulmonary disease, Hitchcock died in the Washington, D.C., home of his son-in-law. The *Washington Herald* eulogized him as "An Incorruptible Public Servant." Hitchcock was 73 years old.

See also Department of the Interior.

Hodel, Donald Paul (1935–)

A lawyer and energy expert who would later serve as secretary of energy, Donald Hodel was also the forty-fifth secretary of the interior, from 1985 to 1989. He was born on 23 May 1935 in Portland, Oregon, the son of Philip and Thereia Rose (nee Brodt) Hodel. He attended local schools, then went to Harvard University, where he received a bachelor's degree in government in 1957. He then went back to Oregon, where he obtained a law degree from the University of Oregon Law School in 1960. His final year was spent as editor of the *Oregon Law Review*.

After earning his law degree, Hodel joined the Portland law firm of Davies,

Biggs, Strayer, Stoel & Boley. Three years later he left the firm to become an attorney for Georgia-Pacific, the paper company. During this time, he became involved in politics, serving as chairman of the Oregon Republican State Central Committee from 1966 to 1967. Two years later, he was named deputy administrator of the Bonneville Power Administration (BPA), the federal agency in charge of public power development in the Northwest. In 1972, Hodel was promoted to BPA's top administrative position. He held this post for nearly five years and was a staunch advocate of the establishment of nuclear power plants to supply electricity to the region. As such, he was an arch-enemy of environmentalists. In a speech in 1975, he railed against those who "clamor for a return to more primitive life, to choking off our individual and collective aspirations for ourselves and our children." He claimed that the modern environmental movement had "fallen into the hands of a small, arrogant faction which is dedicated to bringing our society to a halt. I call this faction the Prophets of Shortage. They are the anti-producers, the anti-achievers. The doctrine they preach is that of scarcity and self-denial. By halting the needed expansion of our power system, they can bring this region to its knees." In 1977, when Jimmy Carter became president, Hodel was replaced.

For the next several years, he worked for private industry and as an advisor to a group studying how U.S. energy needs could be met safely. When Ronald Reagan became president in January 1981, Hodel was named undersecretary of the interior under James G. Watt. During Watt's controversial tenure, Hodel backed up his chief in many areas of department policy. His loyalty led Reagan to name Hodel secretary of energy in November 1982, replacing James B. Edwards of South Carolina. Hodel served in this position until 1985. As Ruth Norris reported in *Audubon* magazine, "He saw the eclipse of that agency's planning for a sustainable national energy future while more and more of its budget was consumed by the production of nuclear weapons and the promotion of nuclear energy."

On 2 January 1985, Secretary of the Interior William P. Clark, who had succeeded Watt in 1983, announced his resignation, claiming that his mission was completed. Eight days later, Reagan nominated Hodel to replace Clark. Hodel's confirmation hearings were bitter, with environmental-minded senators quizzing him on his relationship with Watt and demanding to know what views the two men shared. Nonetheless, Hodel was confirmed and took office on 6 February 1985. In his nearly four years as interior secretary, he strove to return respect to the department that had been lost during Watt's tenure. He accepted the nomination with four goals in mind: "preserving the nation's national park, wilderness, and wildlife resources; enhancing America's ability to meet our energy and mineral needs with domestic resources; improving the federal government's relationship with state and local governments; and developing the economic and social resources of native Americans and the people of the U.S. territories." Although by department standards his tenure was virtually an interregnum, Hodel did end Watt's moratorium on the purchase of land for the National Park System (although he did not actually purchase new lands because of fiscal responsibilities), named conservationist William Penn Mott, Jr., to head the National Park Service, and refused to allow contaminated water to be drained into the Kesterton National Wildlife Range in California. However, he announced on 27 April 1987 a five-year plan for the opening of Alaska's Arctic National Wildlife Refuge for oil and gas development, a move that triggered outcries from environmentalists. In 1989, Hodel left the Interior Department for private industry and was replaced by Manuel Lujan of New Mexico.

See also Department of the Interior; Watt, James Gaius.

Hodel, Secretary of the Interior, et al. v. Village of Gambell et al.
(94 L Ed 2d 542, 107 S Ct 1396 [1987])
This Supreme Court case established the legalities of oil and gas leases off the Alaska coast. When Secretary of the Interior Donald P. Hodel attempted to sell leases for oil and gas exploration off the coast of Alaska under the authority of the Outer Continental Shelf Lands Act, two Alaska villages, including Gambell, sued Hodel to enjoin the issuance of the leases, claiming that such exploration would "adversely affect their aboriginal hunting and fishing rights in those areas." Further, they argued, under section 810(a) of the Alaska National Interest Lands Conservation Act (ANILCA) (now 16 U.S.C. 3120), special protections for Alaska natives and other "rural residents" must be considered before such leases could be issued. A district court denied the villages an injunction, deciding that both section 810(a) and their aboriginal rights were irrelevant in this case and that such rights were not spelled out under the Outer Continental Shelf Lands Act. The U.S. District Court of Appeals for the Ninth Circuit in San Francisco affirmed in part and reversed in part, holding that, under the Alaska Native Claims Settlement Act (ANCSA, 43 U.S.C. 1603(b)), their aboriginal rights had been extinguished, but that section 810(a) of ANILCA was binding. The case was remanded for reargument. Hodel reevaluated the leases and, finding them to be satisfactory, approved them again. The original plaintiffs once again sued, and the court of appeals struck down the leases. The defendants, Amoco Production Company and Secretary of the Interior Hodel, appealed to the U.S. Supreme Court. On 24 March 1987, the Court held 7–2 (Justices John Paul Stevens and Antonin Scalia dissented in part and concurred in part) that Hodel had the right to lease the tracts but that ANCSA did not extinguish aboriginal rights. Writing for the Court, Justice Byron White explained that the oil

companies' exploration would not "significantly restrict the subsistence uses protected by ANILCA," but an injunction against the companies would "irreparably harm the national goal of expedited energy exploration."

Holman, William Steele (1822–1897)
William Steele Holman was considered responsible for section 24 of the General Revision Act of 1891, a landmark piece of legislation that set national policy on forestry reserves. He was born at his family's estate near Aurora, Indiana, on 6 September 1822, the son of Jesse Lynch Holman, a noted Indiana figure described by *The Dictionary of American Biography* as an "Indiana legislator, Baptist clergyman, [and] judge," and Elizabeth (nee Masterson) Holman, the daughter of a Kentucky judge. William Holman was educated in the local schools of Aurora before attending the Baptist Manual Labor Institute in Franklin, Indiana, an educational institution established by his father. Holman remained there until 1840, just prior to his father's death. There is no evidence that he received any degree in his lifetime.

After studying law and being admitted to the Indiana bar, Holman was elected probate judge for Dearborn County, and his political career was launched. He remained a judge until 1846, when he was elected prosecuting attorney for the county. In 1850, he was one of 150 members of the state constitutional convention held in Indianapolis that year. In recognition of his work during that meeting, the people of Dearborn County elected Holman their representative in the lower house of the state legislature. After two years there, he was elected to a four-year term as judge of the state court of common pleas. Four years later, he ran the first of 20 campaigns for the U.S. House of Representatives. Holman ultimately served 16 terms (1859–1865, 1867–1877, 1881–1895, 1897). His name became synonymous with his standing as a "war" Democrat (a Democrat who supported Abraham Lincoln's policies on the Civil War), and he was later considered a great friend of veterans.

The most important period in Holman's career covered 1888 to 1891. In 1885, a trip to Yellowstone National Park led him to take an interest in the preservation of the nation's forestry reserves. Three years later, as chairman of the House Public Lands Committee, he introduced "a bill to secure to actual settlers the public lands adapted to agriculture, to protect the forests on the public domain, and for other purposes." It ultimately passed in the House but was killed in the Senate. The matter remained unsettled until 1891. In that year, Holman was assigned to a conference committee—which also included the new chairman of the Public Lands Committee, Lewis Edwin Payson, an Illinois Republican, and Sen. Preston B. Plumb, chairman of the Senate Public Lands Committee—to iron out the differences between the House and Senate versions of what would become the General Revision Act of 1891. In conference, Holman inserted what is now called section 24 into the act, commonly referred to by historians as the Payson Act. Section 24 authorized the president of the United States, "from time to time, [to] set apart and reserve, in any State or Territory having public land bearing forests, in any part of the public lands wholly or in part covered with timber or undergrowth, whether commercial value or not, as public reservations, and the President shall, by public proclamation, declare the establishment of such reservations and the limits thereof." The section was intended to overrule the 1890 Supreme Court case of *Buford v. Houtz*. Although there was eventually limited debate in both houses as to the ramifications of Holman's insertion, the act passed as the conference committee had written it. After the act was signed into law by President Benjamin Harrison, Secretary of the Interior John Willock Noble recommended that the president use it to establish a federal forest reserve in the area next to Yellowstone National Park; before the end of his term in 1893, Harrison had designated protection for 15 reserves totaling some 13 million acres of forest and timberland. Chief Forester Gifford Pinchot later called section 24 "the most important legislation in the history of Forestry in America." The part of the section that completely segregated the reserves from human use was amended by the Pettigrew Amendment of 1897.

Except for this piece of legislation, Holman's most important distinction in Congress was his thrift with appropriations. Known as "the watchdog of the treasury," he was also called "the great objector" because of his frequent protestations against unusual allocation bills. Holman won his last election in 1896. At 74 years old, he was well past his prime. His wife's death in early 1896 presaged his own end. Said one colleague, "He could guide his steps with wisdom along the snares and pitfalls of public life; he could meet the storms and tempests of acrimonious debate; he could bear the 'whips and scorns' of unjust censure," but he could not continue "when his companion of half a century was stricken." His reelection run in 1896 was therapy to him, and when he sat in Congress in March 1897 for his sixteenth term, he held the record for the longest tenure in Congress. Just a few weeks later, he contracted spinal meningitis and died in Washington on 22 April. Said one newspaper about him years later, "Indiana, and especially Dearborn County, was fortunate in having a representative [of] the caliber of William S. Holman."

See also Buford v. Houtz; Hague, Arnold; Noble, John Willock; Pettigrew Act.

Homestead Act (12 Stat. 392)

This landmark legislation, enacted on 20 May 1862, allowed settlers to own 160 acres of land merely by living on it and working it for five years. The act clearly defined those who qualified as "settlers." Officially called "an Act to secure Homesteads to actual settlers on the Public Domain," it reads:

> Be it enacted ... that any person who is the head of a family, or who has arrived at the age of twenty-one years, and is a citizen of the United States, or who shall have filed his declaration of intention to become such, as required by the naturalization laws of the United States, and who has never borne arms against the United States Government or given aid and comfort to its enemies, shall, from and after the first of January, eighteen hundred and sixty three, be entitled to enter one quarter section or a less quantity of unappropriated public lands, upon which said person may have filed a preemption claim, or which may, at the time the application is made, be subject to preemption at one dollar and twenty-five cents, or less, per acre; or eighty acres or less of such unappropriated lands, at two dollars and fifty cents per acre, to be located in a body, in conformity to the legal subdivisions of the public lands, and after the same shall have been surveyed: *Provided*, that any person owning and residing on land may, under the provisions of this act, enter other land lying contiguous to his or her said land, which shall not, with the land so already owned and occupied, exceed in the aggregate one hundred and sixty acres.

Hornaday, William Temple (1854–1937)

Naturalist and taxidermy expert William Temple Hornaday led the fight during the early part of this century for wildlife conservation. Born on 1 December 1854 on a farm in Plainfield, Indiana, the only son of William and Martha (nee Varner)

Hornaday, he was orphaned at age 15 after moving to a farm in Eddyville, Iowa. He grew up with family in Illinois and Indiana and attended Oskaloosa College and Iowa State Agricultural State College in Ames. He left in his sophomore year to work as a taxidermist for Henry A. Ward, owner of Ward's Natural Science and History Establishment in Rochester, New York, which supplied scientific specimens to schools and other institutions and was also a research enterprise. Hornaday wrote of the experience, "It was a place where museums were made and sold! Really, it was wonderful, no less. For me it was a magnificent opportunity and I strove mightily to improve it to the utmost."

At Ward's, according to biographer Charles Callison, "Hornaday became a scientific taxidermist, collected vertebrate animals in different parts of the world, and on one early trip established definitely the previously disputed existence of the Florida crocodile." Related Hornaday, "I roamed to Florida, Cuba, Barbados, Trinidad, Venezuela and British Guiana—and got little for my labors. Then I went to India, Ceylon, the Malay Peninsula and Borneo—and struck such zoological riches that the zoological poverty of South America was forever compensated." Wrote biographer James Dolph, "In 1880, he organized the Society of American Taxidermists and, just two years later, at the age of 27, was named Chief Taxidermist of the U.S. National Museum of Natural History at the Smithsonian Institution." His tenure there was marked by the publication of his pioneering books *Two Years in the Jungle* (1885), *Free Rum on the Congo* (1887), and *The Extermination of the American Bison* (1889). In 1888, he was instrumental in the formation of a Department of Living Animals at the U.S. National Museum. After a disagreement over museum policy, he resigned in 1890.

From 1890 to 1896, Hornaday was involved in the real estate business in Buffalo, New York. During that time he published *Taxidermy and Zoological Collecting* (1891) and *The Man Who Became a Savage* (1896). In 1896, he was named director of the New York Zoological Park (later the Bronx Zoo), where he stayed until 1926. Hornaday wrote that his work at the zoo entailed having to "prepare and submit plans, buy animals and help to keep them alive, form a working force, plan its tasks, train its keepers, keep peace with the city and the public, feed the press, and make all dangerous places fool-proof." He was the leader in U.S. conservation in attempting to save the bison from extinction. In 1905, when the Wichita National Forest and Game Preserve was established in western Oklahoma, Hornaday provided the refuge with 15 buffalo he himself had saved. Two years later, the area became the first official private bison refuge in the United States. That year Hornaday founded the American Bison Society, which sought congressional protection for the nearly extinct animal. In 1908, with Hornaday's backing, Congress passed the National Bison Range Act, in which it appropriated $30,000 for the purchase of land in Montana for the creation of a range for bison. After the secretary of the interior paid the Flathead, Kootenai, and Upper Pond d'Oreille Indian tribes for the land where the range would exist, Hornaday arranged for the release of 37 bison onto the reserve on 17 October 1909. He was also instrumental in the passage of the Weeks-McLean Migratory Bird Act of 1913 and the Migratory Bird Treaty of 1916. His landmark work *Our Vanishing Wild Life* (1913) led to the creation that same year of the Permanent Wild Life Protection Fund, which he helped raise some $100,000 for, to lobby for the preservation of endangered wildlife. In 1926, he left the Zoological Park to write and lobby on his own.

In 1930, Hornaday began what he called "the baiting war." On 1 May of that year, he called for the cessation of game baiting, by which hunters could kill large numbers of birds and other animals. He advocated passage of the Mc-Nary-Haugen bird limit bill, which would have outlawed the practice, but the legislation died in committee because it was opposed by several hunting groups, including the American Wild Fowlers. The next year, Hornaday's *Thirty Years War for Wild Life* appeared. He called it "a chapter-long demand for the stoppage of the use of live decoys to entice birds of their own kinds. Chapter 10 is entitled 'The Bad Ethics of Live Decoys.'" In 1935, with the help of Jay Norwood Darling, Hornaday was successful; President Franklin Roosevelt ordered an end to all baiting of wild game.

Hornaday's other numerous publications include the *Popular Official Guide to the New York Zoological Park* (1899), *The American Natural History* (1904), *Camp-Fires in the Canadian Rockies* (1906), *Camp-Fires on Desert and Lava* (1908), *Wild Life Conservation in Theory and Practice* (1914), *A Searchlight on Germany* (1917), *Awake! America* (1918), *Old-Fashioned Verses* (1919), *The Minds and Manners of Wild Animals* (1922), *Tales From Nature's Wonderlands* (1924), and *Wild Animal Interviews* (1928). He died in Stamford, Connecticut, on 6 March 1937 at the age of 82.

See also Migratory Bird Treaty Act; National Bison Range Act; Weeks-McLean Act.

Hough, Franklin Benjamin (1822–1885)

Initiator of early American policy on forestry, Franklin B. Hough served with the Division of Forestry (forerunner of the modern U.S. Forest Service) upon its creation. He was born Benjamin Franklin Hough in Martinsburg, New York, on 20 July 1822, the son of Horatio G. Hough, a physician, and Martha (nee Pitcher) Hough. When he began to write, he changed the order of his names. He attended Lowville Academy and Black River Institute in Watertown, New York, to prepare for college. While studying, he did geological work that resulted in the discovery of a new mineral named *houghite* in his honor. He attended Union College in Schenectady, New York, graduating with a bachelor's degree in 1843. Following a brief period in which he taught at a small Ohio academy, he entered the Cleveland Medical College, which awarded him a medical degree in 1848.

Hough began a private medical practice in Somerville, New York, which lasted from 1848 until 1852. According to biographer Edna Jacobsen, "Clinical medicine proved uncongenial to Hough." She reports that Hough noted in his diary that he had picked Somerville because its "mineral localities were taken into consideration." He began to write a number of works on the history of the area, and in 1855, he was hired by the state to compile a census much like the national census, but including agricultural and industrial data. In compiling his statistics, Hough noted that the forests of the area were being eliminated at a rapid pace. The Civil War interrupted his studies. He volunteered for duty as an inspector for the U.S. Sanitary Commission, which tried to provide safe conditions for wounded soldiers. Later, Hough served as the regimental surgeon for the 97th New York Volunteers. He was discharged from the army in 1863 after seeing considerable action. He returned to New York to continue his studies. As a result of this work, he was hired to conduct the District of Columbia census of 1867 and was superintendent of the U.S. census of 1870. These numerous investigations gave him a national look at the condition of forests. He was distressed by what he saw. "It did not take much reasoning to reach the inquiry: 'How long will these other supplies last, and what next?'"

At the 1873 conference of the American Association for the Advancement of Science in Portland, Maine, he read a summary of his work in a paper called "On the Duty of Governments in the Preservation of Forests." The Academy formed a committee consisting of Hough and botanist George Barrell Emerson to compose a report to Congress advocating the establishment of a national policy to preserve the forests. Wrote Henry Solon Graves, later chief forester of the United States, "The report of this committee . . . was endorsed by President Grant, who transmitted the plan to Congress in February 1874. Two years later, Congress took action and Hough was chosen to investigate the consumption of timber and the preservation of forests, receiving the appointment as forestry agent in the Department of Agriculture on 30 August 1876." On that date, the commissioner of agriculture (later secretary of agriculture) was requested to appoint an expert in forestry matters to carry out a nationwide study and report back to Congress. The sum of $2,000 was appropriated to hire this expert and pay his salary. Franklin B. Hough got the job. In 1881, his post was given the title Division of Forestry, and Hough was named division chief. Five years later, in 1886, the division became a permanent agency of the government.

Hough worked in this post until 1883, when he was replaced by Nathaniel Hillyer Egleston, who had served as secretary of agriculture during the first Grover Cleveland administration. In his seven years as unofficial chief forester and beyond, Hough's investigations led to the first book on practical forestry in the United States, *Elements of Forestry* (1882); four reports from 1877 to 1885 to Congress; and the first true examination of the condition of the national forests. For a year, 1882–1883, he edited the monthly *American Journal of Forestry*. After being dismissed by the secretary of agriculture in 1883, Hough continued to work in the division, even releasing his fourth report in 1885. His major work, however, was in New York, where he helped write the framework of a bill creating a state forestry commission. His work culminated in the passage of the Forest Commission Act of 15 May 1885, which established the New York State Forestry Commission and laid out the Adirondack and Catskill forest preserves. This was Hough's last accomplishment. A few weeks after its passage, he caught pneumonia and died on 11 June 1885. He was 62.

See also Egleston, Nathaniel Hillyer; Pinchot, Gifford; United States Forest Service.

House Committee on Agriculture

Established on 3 May 1820 "to provide a forum for the interests of the large agricultural population of the country," this committee helped enact many pieces of conservation and environmental legislation. After 1880, forestry matters were added to its jurisdiction; laws enacted in this area included the protection of birds and animals in forest reserves. Other key legislation included the McNary-Haugen Act of 1924, establishment of agricultural colleges under the Morrill Land-Grant College Act of 1862, the Homestead Act of 1862, the Newlands Reclamation Act of 1902, the Soil Conservation and Allotment Act of 1936, and the Agricultural Adjustment Act of 1938. Members of this committee included Charles McNary, who later served in the Senate; Mark Hill Dunnell (1823–1904) of Minnesota; Gilbert N. Haugen (1859–1933) of Iowa; and Owen Lovejoy (1811–1864), brother of the murdered abolitionist the Rev. Elijah Parish Lovejoy.

See also Agricultural Adjustment Act of 1938; McNary, Charles Linza.

House Committee on Interior and Insular Affairs (1951–)

Originally the Committee on Public Lands (1805–1951), this committee oversees land policy issues and embraces all the committees that had jurisdiction over forestry, forest reserve establishment and management, Indian affairs, territories, mines and mining, the irrigation and reclamation of arid lands, and insular affairs. Members of this committee have

included Stewart L. Udall, secretary of the interior in the Kennedy and Johnson administrations; his brother Morris K. Udall, who served as chairman of the committee from 1979 until his retirement in 1990; and Wayne N. Aspinall of Colorado, who served as committee chairman during the crucial negotiations over the Wilderness Act of 1964.

See also Aspinall, Wayne Norviel; House Committee on Public Lands; Udall, Morris King; Udall, Stewart Lee.

House Committee on Irrigation and Reclamation (1924–1946)

See House Committee on Irrigation of Arid Lands.

House Committee on Irrigation of Arid Lands (1893–1924)

This committee was formed to make the previously created House Select Committee on Irrigation of Arid Lands a permanent entity. That council was established to mirror the Senate body of the same name that had jurisdiction over the effort to reclaim and irrigate the arid western states. A historian on congressional committees noted, "The committee exercised jurisdiction over irrigation projects generally, including the preemption and disposition of lands on reclaimed and irrigated projects; authorization of interstate compacts and agreements regarding irrigation projects; and disposal of drainage water from irrigation projects." In 1924, Congress extended the authority of this committee to include the overall reclamation of public land and renamed it the Committee on Irrigation and Reclamation. In 1947, this committee's responsibilities were transferred to the Committee on Public Lands.

House Committee on Public Lands (1805–1951)

This committee had jurisdiction over the national lands. Wrote a historian of

House committees, "Implementation of public land policy was a significant responsibility of the new Federal Government that began functioning in 1789, but the House of Representatives did not create a standing committee to consider land matters during the early Congresses. Instead, the House dealt with land issues in the Committee of the Whole." On 17 December 1805, Rep. William Findlay (1741 or 1742–1821) of Pennsylvania suggested the establishment in the House of "a committee respecting the lands of the United States." A majority of Congress agreed to form the House Committee on Public Lands. Matters involving private land claims, Indian affairs, territories (following the Civil War), mines and mining, and irrigation of arid lands were handled by other committees. In 1946, under the Legislative Reorganization Act, all these concerns were put under the jurisdiction of the Public Lands Committee. Five years later, the House voted to change the name of the committee to the Committee on Interior and Insular Affairs to reflect its new duties and to mirror the name of its counterpart committee in the Senate.

While it existed, the Public Lands Committee oversaw national park creation and management, railroad construction in and near these parks, early geologic surveys, and private land claims. The Committee on Mines and Mining (1865–1946) had control over the U.S. Geological Survey, the Bureau of Mines, and the management of mineral land laws; the Committee on Insular Affairs (1899–1946) oversaw matters involving those land possessions that were received from the Treaty of Spain in 1899. Among the Public Lands Committee's members over the years were Louis E. Payson (1840–1909) of Illinois (author of the General Revision Act of 1891), William Steele Holman of Indiana, Mark Hill Dunnell (1823–1904) of Minnesota, Samuel Finley Vinton (1792–1862) of Ohio, John Fletcher Lacey of Iowa, and Wayne N. Aspinall of Colorado.

See also Aspinall, Wayne N.; Holman, William Steele; Lacey, John Fletcher.

House Select Committee on Irrigation of Arid Lands

See House Committee on Irrigation of Arid Lands.

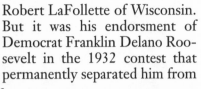

Ickes, Harold LeClaire (1874–1952)

Harold L. Ickes served as secretary of the interior for 13 years (1933–1946) during the Franklin Roosevelt and Truman administrations, the longest-serving cabinet official ever. Born on his grandfather's farm near Hollidaysburg, Pennsylvania, on 15 March 1874, he was the son of Jesse Boone Williams Ickes, a tobacconist, and Martha (nee McCune) Ickes. The Ickeses and McCunes were of German-Scottish stock. Harold's grandfather, Johan Nicholaus Ickes II, was the founder of Ickesburg, a small Pennsylvania village. Martha Ickes died when her son was 16, and Harold and his sister Mary were sent to Chicago to live with an aunt. He enrolled at the University of Chicago and graduated cum laude in 1897.

As a cub reporter for the *Chicago Record*, Ickes covered the Republican National Convention in 1900, then returned to the University of Chicago to study law. He was awarded an LL.D. degree in 1907 and practiced law in Chicago for the next 26 years while dabbling in politics. A liberal Republican, he worked for the mayoral campaign of Republican John Maynard Harlan (son of Supreme Court Justice John Marshall Harlan) in 1897 and was a leading Illinois supporter of Theodore Roosevelt's Progressive movement during the 1912 presidential campaign. He served as a member of the Cook County (Chicago) Progressive Committee and on the Progressive Party National Committee. After the election and the demise of the Progressive Party, Ickes returned to the regular Republican fold. In 1920, he stood against the selection of Warren G. Harding as the Republican Party's presidential candidate, and he left the party again in 1924 to support the third-party campaign of Sen. Robert LaFollette of Wisconsin. But it was his endorsement of Democrat Franklin Delano Roosevelt in the 1932 contest that permanently separated him from his party.

Following Roosevelt's election, the new president selected the inexperienced Ickes for the Interior post. The reasons behind this are murky, but one source speculated that Roosevelt was attempting to "generate the support of midwestern progressive Republicans." Ickes retained this position for the entire 12 years of the three Roosevelt administrations and for one year of the Truman administration. As secretary, he acquired the nickname "Honest Harold." He assembled a first-class staff: Oscar Chapman, a Denver attorney who later served as secretary of the interior under Truman; Louis Glavis, whose charges of fraud in Alaska had been the catalyst of the Ballinger-Pinchot affair; and John Collier, American Indian advocate. Six months into his tenure, Ickes was also named director of the Works Progress Administration (WPA), with a budget of $6 billion to spend on public projects, including construction of the Boulder and Grand Coulee Dams. His support of the Taylor Grazing Act led to its passage in 1934.

According to writer Barry Mackintosh, "Insofar as Ickes had a conservation philosophy before becoming Secretary of the Interior, it was the utilitarian view of natural resource conservation for use promoted by his old friend [and later enemy] Gifford Pinchot." Wrote Interior Department historian Eugene Trani, "Ickes' greatest accomplishments as Secretary occurred in Roosevelt's third and fourth terms. Throughout World War II, he served as custodian of [the nation's] natural resources. He had refused to sell helium to Nazi Germany in 1938, and

Secretary of the Interior Harold Ickes (right) and Colorado Congressman Edward T. Taylor, author of the 1934 Taylor Grazing Act, in 1942. The map behind them is of federal grazing districts.

had headed the Federal Oil Administration in the days of the NRA [National Recovery Administration]. So the President named him Petroleum Coordinator for National Defense in May 1941, making him czar of American Oil. . . . His powers increased to cover coal mines, minerals, and gas and water resources." Adds Ickes's biographer T. H. Watkins, "He also threw his support behind the efforts of . . . an ardent wildlife protector, William T. Hornaday, head of the Bronx Zoo and president of the Permanent Wild Life Protection Fund," to get federal funding for migratory bird refuges passed in Congress. In fact, no person during the New Deal did more to influence and change the United States in the area of conservation. In 1973, Horace Albright, formerly of the National Park Service, proclaimed in an interview that "Harold Ickes was the best secretary of the interior who ever held that office."

Ickes's move to the political left was marked by an acidic confrontation with President Harry S Truman that culminated in Ickes's resignation on 15 February 1946. He continued to blast his critics in his column "Man to Man" in the *New York Post* and in another in the *New Republic*, even writing a sarcastic editorial on the resignation of his successor, Julius A. Krug, entitled "Farewell, Secretary Krug." He was heartened by his protégé Oscar Chapman's rise to interior secretary in 1949. Ickes practiced law and dabbled in political commentary, but his connection with the Independent Citizens Committee of the Arts, Sciences, and Professions, which became a Communist front, stained his career. In the last years of his life, he wrote several books, including *The Autobiography of a Curmudgeon* (1943), in which he began, "If, in these pages, I have hurled an insult at anyone, be it known that such was my deliberate intent." He also wrote the three-volume *Secret Diary of Harold L. Ickes*, published posthumously. Ickes died on 3 February 1952 at the age of 77.

See also Department of the Interior.

Inland Waterways Commission
See Governors' Conference on the Environment; Newlands, Francis Griffith.

Interim Convention on the Conservation of North Pacific Seals
See Fur Seal Act.

International Paper Company v. Ouellette et al. (93 L Ed 2d 883, 107 S Ct 805 [1987])
This Supreme Court case was resolved on the narrow issue of whether the Clean Water Act allowed private suits against companies that polluted water resources. Harmel Ouellette was one of a group of lakeshore property owners on the Vermont section of Lake Champlain. The group filed suit against the International Paper Company for allowing one of its mills in New York to discharge effluent into the water, creating a pollution hazard and lowering the value of the owners' property. International Paper argued that the Clean Water Act did not allow citizens of one state to sue a company in another state. The district court in Vermont held that the Clean Water Act did allow such suits and, on appeal, the U.S. Court of Appeals for the Second Circuit upheld the judgment. International Paper sued to the U.S. Supreme Court. In a 5–4 vote, Justices Lewis Powell, Byron White, Sandra Day O'Connor, and Antonin Scalia and Chief Justice William Rehnquist held that although the group could not sue under the common-law statute of nuisance from one state to another, nothing in the Clean Water Act preempted a lawsuit against the mill when the suit was based on the law of the state in which the pollution originated. Justices William Brennan, Thurgood Marshall, and Harry Blackmun concurred in part and dissented in part in the judgment, which sent the lawsuit back for further argument to the Vermont district court where it had originated.

Irrigation Acts of Congress
See Carey Irrigation Act; Newlands Irrigation (or Reclamation) Act; Right-of-Way Act.

Irrigation and Reclamation Policy, National
The irrigation and reclamation of the lands that today constitute the western

Built by the Reclamation Service, predecessor of the Bureau of Reclamation, a diversion dam in Colorado enabled 53,000 acres to be irrigated in the Grand Valley in the 1900s.

United States have been in motion for several thousand years. Along the Salt River Valley in Arizona, remnants of irrigation ditches built by prehistoric peoples are evident. When first the Spanish and then the Mexicans ruled what is now the western United States, their settlers tried primitive forms of irrigation.

With the establishment of European settlements speading from the eastern to the western coasts of North America, irrigation became a pressing issue. Congress attempted early action with the Swamp Land Acts of 1849, 1850, and 1860, but these had little if any effect. Wrote Samuel Hayes, "In 1824, Congress instructed the Army Corps of Engineers to improve the navigable streams, and since that date the Corps have carried out frequent hydrographic investigations such as

the extensive Humphreys and Abbot surveys of the Mississippi River completed in 1866." Surveys such as those by John Wesley Powell and Clarence King and semisuccessful legislation enacted by Sen. Joseph M. Carey of Wyoming (who also happened to head the National Irrigation Congress) in 1894 made the need for a national reclamation and irrigation policy more pressing.

In the last 100 years, there has been a tug-of-war between those advocating private reclamation and irrigation projects and those supporting government (state, but mainly federal) action in these fields. In the former camp were state officials such as Arizona Territorial Gov. Nathan O. Murphy, who spoke out against federal involvement in irrigation policy. Those who wanted congressional and federal action included Rep. (later Sen.) Francis G. Newlands of Nevada; Sen. Henry Clay Hansbrough of North Dakota; Sen. Francis E. Warren of Wyoming; George H. Maxwell, irrigation advocate from Arizona; and William Ellsworth Smythe, Maxwell's counterpart from Nebraska. In a speech in the House of Representatives on 14 May 1902, Newlands said:

The question is frequently asked: "Why should the Government undertake reclamation?" And the question is generally accompanied by the suggestion that the work should be left to individual enterprise. There are conclusive answers to both the question and the suggestion. First, the limit of individual effort has already been practically reached. Small tracts, favorably located with reference to water supply may continue to reclaimed by settlers, but nothing in the way of a general system can be successfully inaugurated and carried out by individual effort. Nearly every enterprise of the kind heretofore undertaken has proved a financial loss to the projectors, in fact has resulted in bankruptcy, although, as shown by the statements of the Geological Survey, the bankrupt enterprises have created large values and conferred great public benefits. The ill success of these private ventures has been owing mainly to the circumstance that neither individuals nor companies can control both the water supply and the land to be irrigated, which is a prime essential. This the United States Government can do, for it is the owner in fee of the land, and the water can be conserved by it and utilized.

It took the machinations of Newlands in the House and Hansbrough in the Senate and the lobbying powers of Maxwell's National Irrigation Association

(later called the National Reclamation Association) to get the Newlands Reclamation Act of 1902 enacted. This opened the door to the irrigation of the arid West. The act established the Bureau of Reclamation and set firm federal irrigation and reclamation policy for the first time.

See also Bureau of Reclamation; Carey, Joseph Maull; Carey Irrigation Act; Maxwell, George Hebard; National Irrigation Association; National Irrigation Congress; Newlands, Francis Griffith; Newlands Irrigation (or Reclamation) Act; Smythe, William Ellsworth; Swamp Land Acts of 1849, 1850, and 1860.

Irrigation Committees, U.S. Congress

See House Select Committee on Irrigation of Arid Lands; Senate Committee on Irrigation and Reclamation of Arid Lands.

Irrigation Congresses

Under the direction of irrigation advocates such as John Wesley Powell, William Ellsworth Smythe, Richard J. Hinton, Francis G. Newlands, and George H. Maxwell, groups promoting both private and governmental solutions to the problem of irrigating the arid western states sprang up, but they met at "irrigation congresses" held under the auspices of irrigation adherent Joseph M. Carey of Wyoming. Although about a dozen congresses met between 1891 and 1905, they were unable to capture the legislative strength of George Maxwell's National Irrigation Association, which helped pass the Newlands Reclamation Act of 1902.

See also Maxwell, George Hebard; National Irrigation Association; National Irrigation Congress; Newlands, Francis Griffith; Powell, John Wesley; Smythe, William Ellsworth.

Izaak Walton League of America

The Izaak Walton League of America (IWLA) is a nationally known sportsmen's organization that lobbies for increased protection of natural resources in the areas of fish habitat, clean water, and forestry. William Voigt, Jr., historian of the league, wrote, "Through legislative, judicial, and other means, the IWLA sought to correct problems of water pollution, wetland drainage, and cut-and-run logging. At the start, the IWLA declared itself the 'Defender of Woods, Waters, and Wildlife,' but in time, as its leaders learned more about habitat influences and ecological principles, the league also campaigned for soil conservation and air quality. Divided into state groups and local chapters, the IWLA introduced three-level organization and unified action to the American conservation movement."

Founded in Chicago in 1922 by salesman and publicist Will Dilg, the league initially had 54 members but eventually claimed upwards of 100,000 nationwide. One of these was Herbert Hoover, who served as honorary league president in 1926. The league's official journal, which began publication in 1922, is *Outdoor America*. Many conservationists have written for this publication, including writers Zane Grey, Theodore Dreiser, and Gene Stratton Porter.

Although not considered in the top tier of environmental groups, the IWLA has made several contributions to the movement. Reports Stephen Fox, "The ... League was the first conservation group to campaign against water pollution with some persistence.... In 1926, the League's 2700 chapters were asked by President Coolidge, a fisherman, to take samples of local streams and have them analyzed. The results were chastening: even at that early point, 85 percent of the inland waters were found to be polluted." These conclusions, as well as intense lobbying by the group, led to the enactment of the Water Pollution Control Act of 1948. In the 1960s, the league supported the Public Land Law Review Commission and its findings, as well as the enactment of the Wilderness Act of 1964. Headquartered in Arlington, Virginia, since 1971, it remains a minor voice in the environmental movement.

See also Dilg, Will.

Jackson, Henry Martin (1912–1983)

One of the most important legislators of the twentieth century, Henry "Scoop" Jackson "fathered" the National Environmental Policy Act of 1969, among other conservation acts. He was born on 31 May 1912 in Everett, Washington. His parents were Norwegian: Peter Jackson (original name, Gresseth), a concrete mason and building contractor, and Marie (nee Anderson) Jackson. Henry Jackson was given his middle name to honor the German theologian Martin Luther. His sister gave him the nickname "Scoop" after a comic strip character he reminded her of. Henry Jackson sold newspapers while attending school, thereby establishing his reputation as a tireless worker. He attended Stanford University for a short period before enrolling at the University of Washington Law School in Seattle, where he earned a law degree in 1935. After graduating, he worked briefly in the Everett, Washington, office of the Federal Emergency Relief Agency, but he soon joined the law firm of Black & Rucker, where he stayed until 1938.

In that year, he ran for and was elected prosecuting attorney of Snohomish County. Serving until 1940, he was known as a reformer who cracked down on vice and gambling. At the end of his term, Jackson was elected to the U.S. House of Representatives, taking the seat vacated when incumbent Mon Wallgren ran for the U.S. Senate. Jackson held this seat for six consecutive terms (1941–1953), during which time he worked hand in hand with Sen. Warren Magnuson of Washington on several pieces of legislation. Interested in public power and flood-control issues, he was a member of the Merchant Marines and Fisheries Committee; because of his work on that committee, President Truman offered him the interior secretary post in 1950, but the overture was refused.

In 1952, Jackson ran for the Senate and unseated conservative Republican Harry P. Cain in a bitter contest. For the last 31 years of his life, Jackson held that seat. Wrote the *New York Times*, "Jackson made national headlines and television screens in 1954 when he joined two other Democratic Senators [John L. McClellan of Arkansas and Stuart Symington of Missouri] in temporarily resigning from the Senate's permanent investigations subcommittee in protest against the methods of the chairman, Senator Joseph R. McCarthy of Wisconsin." Reported another source, "Magnuson and Jackson worked to diversify Washington state's economic base during the Eisenhower years, when the administration sought to limit federal spending through its partnership programs with private industry and local governments. Their actions included dryland port and irrigation districts. They used their influence in Congress to produce more easily available bank loans, price supports, and import/export legislation responsive to the fishermen, orchardists, cattlemen, wheat ranchers, and lumbermen of Washington state." As author of the Jackson-Vanik Amendment, Jackson sought to deny U.S. funds to countries that prohibited their citizens from emigrating. Although a liberal on social and civil rights issues, Jackson, in his role as the chairman of and later ranking Democrat on the Senate Armed Services Committee, showed himself to be a hawk on the buildup of arms and conservative in his ideas on U.S.-Soviet relations.

It is in the area of conservation, however, that Jackson made his mark. As chairman of the Senate Committee on Interior and Insular Affairs from 1964 to 1981, he was responsible for guiding through the Senate the Wilderness Act of 1964, the Land and Water Conservation Fund in 1965, the National Wild and Scenic River Act and the National Trails System Act in 1968, and the Alaska National Interest Lands Conservation Act of 1980, as well as sponsoring the landmark National Environmental Policy Act of 1969.

Jackson, twice (1972 and 1976) a candidate for the presidential nomination of his party, suffered a massive heart attack and died on 1 September 1983. He was 71 years old.

See also Magnuson, Warren Grant; National Environmental Policy Act of 1969.

Johnson, Robert Underwood (1853–1937)

As author, editor of the prestigious journal *Century*, conservationist, and lobbyist for conservation laws, Robert Underwood Johnson was one of the most important friends of the conservation movement in the late nineteenth century. Born on 12 January 1853 in his grandparents' home in Washington, D.C., he was the second of two sons of Nimrod Hoge Johnson—a Quaker described by one source as "prominent in eastern Indiana as a lawyer and jurist ... noted for his wide and exact knowledge of history, poetry, and general literature"—and Catherine Coyle (nee Underwood) Johnson, a descendent of Calvinists. At an early age, Robert Johnson was taken to Indiana, where he spent his childhood. He attended local schools and Earlham College, an institution run by the Society of Friends (Quakers) in Richmond, Indiana, from which he earned a bachelor of science degree in 1871.

Almost at once, Johnson became a clerk for Scribner's Educational Books Division in Chicago, where he spent two years. In 1873, he was hired as an editorial clerk for *Scribner's Monthly Magazine*, and he advanced under the tutelage of Richard Watson Gilder, who became *Scribner's* editor in 1881 and made Johnson his assistant. That year, the name of the magazine was changed to *Century*. In essence, Johnson took over much of the journal's day-to-day operations, overseeing the production of articles and editorials. When Gilder died in 1909, Johnson became the magazine's editor, serving until 1913.

In 1889, after publishing many of conservationist John Muir's scholarly articles, Johnson went west to meet Muir. Wrote author Tom Turner, "Johnson's object was to persuade Muir to resume writing for the magazine, and his timing was perfect. Muir had his business in order and had decided to devote the balance of his life to preaching the virtues of the Sierra in particular and wilderness in general to the public, in order to save them from the mercenary interests that were fast ruining both. In response to Johnson's entreaty, Muir suggested that they spend a night or two in Tuolumne Meadows, in the high country above Yosemite Valley. Muir and Johnson inspected the meadows and woods and concluded that something drastic had to be done immediately to stop the ravaging of the area by sheep and cattle and loggers and miners." Johnson made a proposal to Muir. He later wrote of this proposition to a friend:

I told him [Muir] that if he would agree to write for the *Century* two articles—the first on "The Treasures of the Yosemite," to attract general attention, and the second on "The Proposed Yosemite National Park," which he and I should propose, and the boundaries of which he should outline—I would ask Mr. Gilder by telegraph on our return to the Valley if I might engage the articles. We would illustrate them with pictures of the wonderful natural features of the Government lands proposed tp be taken for the park, and with these pictures and the proofs of Muir's articles I would go to Congress (where I was to be much engaged in the international copyright campaign) and advocate its establishment before the committees [*sic*] on Public Lands. One of these members was Judge Holman of Indiana, whose circuit had adjoined that of my father when he was Judge of Court. The two, though political opponents—Holman being a Democrat and my father a radical Republican—had nevertheless been friends, and I felt sure that Holman would be predisposed to the scheme. I also knew Mr. Plumb [Sen. Preston Plumb] of Kansas, who was chairman of the Senate Committee, and I believed he would help. I told Muir that I thought there would be no serious objection to such a measure and that in my judgment it would go through.

With Johnson's help, Congress enacted the Yosemite National Park Act on 1 October 1890. Johnson's suggestion to Muir that he should found a national group to lobby for the protection of the Yosemite Valley resulted in the establishment of the Sierra Club in 1892.

As part of his editorial duties, Johnson mixed a zeal for environmentalism with other matters. In 1895, the *Century* sponsored a conference where conservation advocates Muir and Charles Sprague Sargent called for troops to thwart "all types of trespass," in addition to grazing, on federal lands. Johnson also served as secretary of the National Institute of Arts and Letters and helped establish the American Academy of Arts and Letters in 1904. His work for war relief during World War I led President Woodrow Wilson to name him ambassador to Italy, a post he retained from 1919 to 1921.

In the last two decades of his life, Johnson became a prestigious writer, composing his autobiography, *Remembered Yesterdays* (1923), as well as *Poems of Fifty Years* (1931), *Aftermath* (1933), and *Heroes, Children and Fun* (1934). He also authored *The Winter Hour and Other Poems* (1891), *Songs of Liberty* (1897), *Saint-Gaudens, an Ode* (1910), and *Poems of War and Peace* (1916). Robert Underwood Johnson died in New York City on 14 October 1937 at the age of 84.

See also Muir, John; Sierra Club.

Joint Fish Commission of 1892
See United States–Canada Joint Fish Commission of 1892.

King, Clarence Rivers (1842–1901)

A noted geologist and explorer, Clarence King was instrumental in persuading Congress to create the U.S. Geological Survey in 1877. King was born on 6 January 1842 in Newport, Rhode Island, the son of James Rivers King, a merchant, and Caroline Florence (nee Little) King. James was the brother of Charles William King, a noted trader in China and Japan, and a descendant of one Daniel Kinge, who immigrated to this country in 1637 from England and settled in Lynn, Massachusetts. Reported one source, "Clarence King's grandfather, Samuel Vernon King, was a successful East Indian merchant of the firm of Talbot, Olyphant & King." Clarence King received his education at the Hopkins Grammar School in Hartford, Connecticut. Upon the death of his father, his mother moved to New Haven, where Clarence attended the Sheffield Scientific School at Yale University and earned a bachelor of science degree in 1862.

At Yale, King was influenced by the teachings of Professor George Jarvis Brush (1831–1912), a noted geologist. After graduation, he worked across the west on horseback with engineer James Terry Gardiner (1842–1912). They stopped in Nevada to investigate the Comstock lode and then moved on to the Sierra Nevada, which they explored on foot. In 1863, he joined the scientific geological survey of California being conducted under Professor Josiah Dwight Whitney (1819–1896). During the winter of 1865–1866, King served as a scientific aide to the survey of the southern California desert region being conducted by Gen. Irwin McDowell.

In 1867, King returned to Washington and pitched a plan to Congress in which the western region of the country would be explored, mapped, and geologically plotted. On 2 March 1867, Congress mandated a "geological and topographical exploration of the territory between the Rocky Mountains and the Sierra Nevada Mountains, including the route or routes of the Pacific Railroad." Known more commonly as the Survey of the Fortieth Parallel, or the Clarence King survey, it competed for talent and discoveries with three other surveys: those of Ferdinand V. Hayden, John Wesley Powell, and George M. Wheeler. The King exploration, which lasted from 1867 until 1873 (reports continued to be written until 1877), "surveyed ... a belt 105 miles in width extending from longitude 104° 30– to longitude 120°—that is, from Cheyenne, Wyoming, to the eastern border of California," wrote geological survey historians John and Mary Rabbitt. "The survey was primarily geological in character, but included also the topography of the region. The results of the survey were published in 1870–1880 in seven volumes and an atlas. The total cost of the survey and its publications was $383,711." Among the scientists involved in this exploration were geologists Arnold Hague, Samuel Emmons, and Charles Doolittle Walcott.

In 1871, writer Henry Adams (1838–1918), son of diplomat Charles Francis Adams, went to visit King's survey on an invitation from Samuel Emmons. After meeting King, Adams wrote in his third-person style, "King had everything to interest and delight Adams. He knew more than Adams did of art and poetry; he knew America, especially west of the hundredth meridian, better than any one.... Incidentally, he knew more practical geology than was good for him, and saw ahead at least one generation further than the text-books.... His wit and

159

Clarence King (far right), with (left to right) James T. Gardiner, Richard D. Cotter, and William H. Brewer in the 1860s. King became the U.S. Geological Survey's first director in 1879.

humor; his bubbling energy which swept every one into the current of his interest; his personal charm of youth and manners; his faculty of giving and taking, profusely, lavishly, whether in thought or money as though he were nature herself, marked him almost alone among Americans. He had in him something of the Greek—a touch of Alcibiades or Alexander. One Clarence King only existed in the world."

In 1879, King asked Congress to consolidate the four surveys of the West under one federal agency. As a result, the U.S. Geological Survey was established, with King as the first director. He served until 1881, establishing a firm footing for the new bureau. After his resignation, he dedicated the rest of his life to special geological research in the West, particularly California and Arizona. An attack of tuberculosis in 1901 was his undoing. He died on 24 December of that year in Phoenix, Arizona.

See also Emmons, Samuel Franklin; Hague, Arnold; Hayden, Ferdinand Vandeveer; Powell, John Wesley; United States Geological Survey; Walcott, Charles Doolittle; Wheeler, George Montegue.

Kinney, Jay P (1875–1975)

Conservationist Jay P Kinney is best known for his advocacy of the preservation of forests on Indian reserves. He was born on 18 September 1875 in the village of Snowdon in northern New York, the son of Joseph Page Kinney and Isabel Nasbeth (nee Stanhouse) Kinney. According to his Cornell alumni biography, Kinney's middle name was not a name at all but a single letter.

His official biography, *The Organization of the Indian Forest Service*, states that Kinney "received his education in rural district No. 5 of Otsego township ... [and] after several terms of teaching in the public schools, prepared for college at Cooperstown, New York." He won a state scholarship and entered Cornell University in 1898; he was awarded a bachelor's degree in 1902. His master's thesis, "Forest Legislation in America Prior to March 4, 1789," dealt with colonial and state enactments to protect forests and timber reserves. Kinney entered Cornell's forestry school in 1902, but the school was shut down in a clash of state politics in 1903. Instead of transferring to Yale's forestry school, Kinney decided to enter private business. He worked for a Cooperstown publishing company and as a high school principal before taking a position as assistant examiner in the U.S. Patent Office in Washington, D.C., in 1906. At the same time, he attended National University and earned a law degree in 1908.

In 1910, he was hired by the Agriculture Department to head the newly established Indian Forestry Service, which would oversee forest reserves on Indian lands. Alan Newell, Richmond Clow, and Richard Ellis, in a history of the Bureau

of Indian Affairs Forestry Service, wrote, "Kinney assumed responsibility for the Washington, D.C., office as well as all field offices.... [His] position gave him the opportunity to restructure the forestry division."

For the next 23 years, Kinney headed the Indian Forestry Service. In 1933, he was named general production supervisor of the Civilian Conservation Corps' operations on Indian reservations. In 1942, he was appointed associate director of soil conservation in the Interior Department's Office of Land Utilization. Kinney was given a special waiver because he had passed the age of retirement so that he could be named an advisor in forestry for the Department of Agriculture in 1945. He retired to work part time as an advisor to the Department of Justice on Indian land claims in 1954 and retired completely in 1968, after 61 years of working for the federal government in one capacity or another.

Kinney was a prodigious writer, turning out a number of books: *The Essentials of American Timber Law* (1917), *The Development of Forest Law in America* (1917), *A Continent Lost—A Civilization Won: The Indian Land Tenure in America* (1937), and *Indian Forest and Range: A History of the Administration and Conservation of the Redmen's Heritage* (1950). Toward the end of his life, he wrote that he was "thoroughly disgusted as to countenancing of the sexes [living together] ... prior to marriage" and concluded, "since I have now passed the 100 year mark, I leave the future to the rest of you to rule or ruin." Just two and a half months past his hundredth birthday, he died in Hartwick, New York.

See also United States Forest Service.

Kirkwood, Samuel Jordan (1813–1894)

Senator and "war governor" of Iowa, Samuel J. Kirkwood also served as secretary of the interior from March 1881 to April 1882. Born on his father's farm in

Harford County, Maryland, on 20 December 1813, Kirkwood was the son of Jabez Kirkwood, a farmer and blacksmith, and his second wife, Mary (nee Alexander) Wallace Kirkwood, who had been widowed. In 1823, when he was ten, Kirkwood was sent to Washington, D.C., to supplement his limited education by attending the academy of a family friend, John McLeod. He left after four years and clerked in a drugstore in the nation's capital; he then taught for a time in York County, Pennsylvania. When he reached 18, he moved to his father's new farm in Richland County, Ohio, where he read the law in the office of a local attorney. In 1845, he was admitted to the Ohio bar and soon thereafter began a private law practice. Later in 1845, he was appointed prosecuting attorney for Richland County, a post he held until 1849. The following year, he was elected as a Democrat to the state constitutional convention. He then resumed his law practice.

In 1855, Kirkwood, disgusted with his party's stand on slavery, moved to Iowa City and began a flour and sawmill business. In 1856, he was called upon to join other antislavery advocates and serve as a delegate to the 1856 state convention of the Iowa Republican Party. At that convention, he was nominated for a seat in the state senate. He served a single term, during which time he was the Republican Party state chairman. In 1859, the Republicans nominated him for governor. He was elected and then reelected in 1861. Kirkwood served during the bleak early days of the Civil War. He made key decisions in calling up volunteers for the Union army and was forced at one point to call out the home guard to defend the capital from slavery interests. In 1863, President Lincoln nominated Kirkwood to be ambassador to Denmark, but Kirkwood declined the honor. After refusing a third term as governor, he went back to private business. In 1866, he was elected to the U.S. Senate to fill the seat of James Harlan, who had been chosen as Andrew

Johnson's secretary of the interior. Kirkwood served until March 1877 and then returned again to private business.

In 1876, Kirkwood was nominated for governor against his wishes, but he was elected anyway. His term was uneventful, and in February 1877, the state legislature elected him to a full term in the U.S. Senate. Kirkwood was chosen secretary of the interior by newly elected President James A. Garfield in March 1881. Because he was an old-line Republican, Kirkwood immediately abolished the civil service program that had been instituted in the Interior Department by his predecessors. Kirkwood's tenure as interior secretary is considered mostly ineffectual. He served through Garfield's term and into that of Chester A. Arthur, who became president after Garfield's assassination. Seven months into Arthur's term, on 17 April 1882, Kirkwood resigned in favor of Sen. Henry Moore Teller of Colorado.

Kirkwood's last try at political office came in 1886, when he ran an unsuccessful campaign for the U.S. House of Representatives. He spent his last years in Iowa, where he died at his home in Iowa City on 1 September 1894.

See also Department of the Interior.

Kleppe, Thomas Savig (1919–)

Thomas S. Kleppe was mayor of Bismarck, North Dakota, a Republican congressman from that state, administrator of the Small Business Administration, and secretary of the interior from late 1975 until early 1977. Born in the small town of Kintyre, North Dakota, on 1 July 1919, he was the son of Lars O. Kleppe, a homesteader, and Hannah (nee Savig) Kleppe. Lars Kleppe owned Farmers Company, which operated a grain elevator. While attending local schools, Thomas Kleppe essentially became his father's right-hand man. He enrolled at Valley City State College in North Dakota but left after only a year;

instead, he became assistant manager at Farmers Company.

After working as the bookkeeper of a small bank in nearby Napoleon, Kleppe moved to Bismarck in 1941 and took a job as assistant cashier at the Dakota National Bank. He joined the army in 1942 but became a finance officer and never saw combat. When World War II ended, he returned to Bismarck and became the bookkeeper of the Gold Seal Company, which marketed a nationally known glass cleaner. As sales of the product grew, Kleppe moved up the corporate ladder and eventually became the company's president in 1958. He left the company in 1964 because of a disagreement over business tactics and joined an investment banking firm. In the meantime, he had become involved in the North Dakota political scene. In 1950, in his first political race and just 31 years old, he was elected mayor of Bismarck and served until 1954.

After one year (1963) as the North Dakota Republican Party's treasurer, Kleppe decided to run for national office. His 1964 campaign against incumbent Sen. Quentin R. Burdick was unsuccessful, however. Burdick was helped that year by Lyndon Johnson's landslide victory over Barry Goldwater in the presidential contest. Two years later, Kleppe took on incumbent Rep. Rolland Redlin and beat him. Although considered a conservative, Kleppe voted for the Civil Rights Act in 1968 and for the creation of wildlife refuges in the Missouri River basin. When Richard Nixon took office, Kleppe supported him on almost all issues except the Vietnam War, which Kleppe voted to end at the earliest date. In 1970, Kleppe left the House to take on Burdick again. Although Kleppe started the campaign with a large lead, Burdick won in the end with 62 percent of the vote, and Kleppe was out of office. Because of Kleppe's loyalty, President Nixon named him administrator of the Small Business Administration (SBA), which furnishes loans to small enter-

prises. During Kleppe's tenure at the SBA during the Nixon and Ford administrations, credit to minority businesses was up, but the administration was scandalized by accusations of partisanship on the part of Kleppe.

On 25 July 1975, Secretary of the Interior Stanley K. Hathaway resigned after only six weeks on the job, claiming "emotional distress." It took nearly three months before President Ford named Kleppe to replace Hathaway. The reviews on him were mixed. "I think he's qualified," said his former political enemy, Quentin Burdick, "but I think I'm a pretty charitable fellow to support him." He added, "He knows something about water and other Federal resources in the West." Rep. Morris Udall of Arizona commented, "Tom Kleppe is qualified for a lot of jobs, but not Secretary of the Interior." Confirmed easily, Kleppe took office on 17 October 1975 and served until 20 January 1977—less than 15 months. In that short time, he had little chance to put an imprint on the department. However, he oversaw the investigation of the *Argo Merchant* oil tanker wreck off Nantucket Island in Massachusetts in 1976 and tried to proceed with the sale of offshore oil and gas leases in Alaska, although he cut the area available from 1.8 million to 1.1 million acres. The *New York Times* called the latter proposal "Alaska Gamble." In the 1976 case *Kleppe v. Sierra Club*, the Supreme Court held that the Department of the Interior "did not have to prepare an immediate environmental impact statement (EIS) for coal field development in the northern Great Plains area because no federal program was involved." Wrote author Ron Arnold, "Environmentalists viewed this case as a defeat in their drive to push environmental impact statement requirements into private commercial developments." In 1977, Kleppe left the Interior Department to become a private businessman.

See also Department of the Interior.

Krug, Julius Albert (1907–1970)

Julius A. Krug, self-described business-man and entrepreneur, served as secretary of the interior during the Truman administration. He was born on 23 November 1907 in Madison, Wisconsin, the third of seven children of Julius John Krug, a Madison city detective, sheriff, and deputy state fire marshal, and Emma (nee Korfmacher) Krug. At birth, Julius Krug was so large that his father purportedly exclaimed, "Julius is no name for him! We ought to call him Captain Kidd!"—so Julius was nicknamed "Cap."

Krug attended local schools and was athletically inclined; he played football and went sailing on Madison's Lake Mendota. He enrolled at the University of Wisconsin, earning a bachelor's degree in 1929 and a master's degree in public utilities management and regulation the following year. Krug got married when he was a sophomore in college and worked at a series of odd jobs to support his new family, including as a filling station manager and ditch digger. After graduation, he worked as a research statistician for the Wisconsin Telephone Company and as chief of the depreciation section of the Wisconsin Public Service Commission. In this latter post, he worked under the direction of David Lilienthal, who was later director of the Tennessee Valley Authority (TVA). As a member of Lilienthal's staff, Krug's star began to rise. Called to Washington, D.C., in 1935 to serve as a public utilities expert for the Federal Communications Commission, Krug led an investigation into the Long Lines Department of the American Telephone and Telegraph Company, which resulted in lower phone rates in 1936.

Following a stint as director of the Kentucky Public Service Commission, Krug was hired by Lilienthal as a power manager at the TVA, where he worked until 1941. According to Krug, at TVA he "supervised the development of the area-wide program for the acquisition of the electric properties of the privately owned [power] companies, and negoti-ated purchases from Electric Bond and Share, Commonwealth and Southern, Associated Gas and Electric, and other utilities during 1939 and 1940."

In 1941, President Franklin D. Roosevelt named Krug program chief of the War Production Board (WPB), a federal agency whose job was to oversee the establishment of priorities for, and the allocation of, the power of U.S. industry. Krug became the enemy of liberals in the Roosevelt administration when he pushed for an end to price and wage controls following the end of World War II. Krug also served as a defense power consultant to the Office of Production Management (OPM), another New Deal agency. Krug resigned at the end of 1945 when the two agencies were dissolved.

In February 1946, after the resignation of Secretary of the Interior Harold L. Ickes, President Harry S Truman asked Krug to take the Interior post. Krug accepted and took office on 28 February. In his nearly three years as interior secretary, Krug faced down labor leader John L. Lewis on a coal strike ultimatum, advocated the establishment of a national synthetic oil industry, and conducted a survey of natural resources that was published as *National Resources and Foreign Aid: Report of J. A. Krug, Secretary of the Interior, October 9, 1947.* Unfortunately for Krug, his distant relationship with Truman, lack of campaigning for the Democrats in 1948, and controversial relationship with billionaire Howard Hughes cost him his job. He resigned on 1 December 1949.

In the last 20 years of his life, Krug served as president of the Brookside Mill and of the Volunteer Asphalt Company, both of Knoxville, Tennessee. Julius Krug died at his home in Knoxville on 26 March 1970 at the age of 62.

See also Department of the Interior.

Krutch, Joseph Wood (1893–1970)

Stated the *Arizona Republic* upon Joseph Wood Krutch's death, "Describing

Krutch as a conservationist is like describing Henry David Thoreau as an ex-con." Another author called him a "quiet voice for the Devil's Domain." A noted drama critic, scholar, and writer, Krutch was an important essayist on matters relating to the environment. He was born in Knoxville, Tennessee, on 25 November 1893, the son of Edward and Adelaine (nee Wood) Krutch. Joseph Wood Krutch attended local schools and the University of Tennessee, where he received his bachelor's degree in mathematics. He earned a master's degree and a Ph.D., both in the humanities, from Columbia University in New York. He began to teach English at Columbia in 1917 and eventually became the Brander Matthews Professor of Drama Literature in 1943. He was the drama critic for the liberal magazine *The Nation* from 1924 to 1951 and served as president of the New York Drama Critics Circle from 1940 to 1941.

In the 1920s, Krutch wrote several scholarly works, including *Comedy and Conscience after the Restoration* (1924) and *Edgar Allen Poe: A Study in Genius* (1926); he edited *The Comedies of William Congreve* (1927). His most important work of the period was *The Modern Temper: A Study and a Confession* (1929), a series of compositions that exposed his feelings for humanism and nature, which he believed were inextricably mixed. "These series of essays," wrote author Peter Wild, "rejected the wild optimism based on blind faith in science [that was] fashionable in the 1920s. Except for this book, and a sensitive study of Henry David Thoreau in 1948, there was hardly

a hint of the future environmentalist in Krutch."

When Krutch abruptly moved to the Sonoran Desert of Arizona in 1950, even his closest friends were shocked. He settled outside of Tucson, where he took up writing about nature and contributing to the conservation groups Defenders of Wildlife and the Sierra Club. In 1955, he published *The Voice of the Desert: A Naturalist's Interpretation*, followed three years later by *Grand Canyon: Today and All Its Yesterdays*. This work, wrote biographer John Margolis, "provides a potpourri of information for the curious visitor: explanations of the ecology of the area; descriptions of the flora and fauna; discussion of the geological formation of the Canyon and of fossil evidence of early forms of life there; an account of its exploration and of more recent human habitation; suggestions as to how best to relate oneself to the immensity of the spectacle; even reports on tourist amenities and tips concerning spots of particular beauty or interest which 'should not be missed.'"

In his final years, Krutch wrote numerous essays on the environment, including *The Forgotten Peninsula: A Naturalist in Baja, California* (1961). His other works include *Five Masters: A Study in the Mutations of the Novel* (1930), *Henry David Thoreau* (1948), *Great American Nature Writing* (1950), and *The World of Animals: A Treasury of Lore, Legend, and Literature by Great Writers and Naturalists from 5th Century B.C. to the Present* (1961). Joseph Wood Krutch died at his home in Tucson on 22 May 1970 at the age of 76.

Lacey, John Fletcher (1841–1913)

Rep. John Lacey of Iowa was called by some the father of American conservation and the father of federal game laws. Born in a one-room log cabin on the Ohio River near New Martinsville, Virginia (now West Virginia), on 30 May 1841, Lacey was the son of John Mills Fletcher and Eleanor (nee Patten) Lacey, both farmers. In 1855, the family moved to Oskaloosa, Iowa. John Lacey's education was disrupted by farm duties, employment as a brick mason, and the Civil War. With the outbreak of the war, Lacey enlisted as a private in the Federal army but was soon captured at the Battle of Blue Mills (Missouri). During his imprisonment, he read the law with another captured soldier, Samuel A. Rice, who was the attorney general of Iowa. Following his exchange in 1862, Lacey rejoined the army, enlisting in the 33d Iowa Volunteers, with Samuel Rice as his commanding officer. Lacey was promoted to brevet major for his gallantry at the Battle of Mobile in 1865.

When the war ended, Lacey returned home, married his sweetheart Martha Newell, and was admitted to the Iowa bar. He became a giant in Iowa legal affairs. His publication of the *Third Iowa Digest* (1870) and *A Digest of Railway Decisions* (1875 and 1884) earned him a reputation as an expert in legal matters involving railraods. In 1869, he was elected to the Iowa General Assembly and served with distinction on the judiciary committee. In 1888, he was elected to the U.S. House of Representatives and served from 1889 to 1893 and again from 1895 to 1907. He began to lobby for conservation laws starting in 1892, when his bird and game bill was defeated in the House. After his return to Congress in 1895, he pushed for the law until it was finally passed in 1900 as the Lacey Game and Wild Birds Preservation and Disposition Act, or the Lacey Act. From 1893, he was a member of the House Public Lands Committee, and he introduced and fought for such conservation measures as the Yellowstone Protection Act of 1894 (which laid the groundwork for the establishment of the National Park Service in 1916), the Forest Transfer Act of 1905 (which created the U.S. Forest Service), and the American Antiquities Act of 1906.

In 1906, Lacey was defeated for reelection to the House, and in 1908, he lost a bruising battle for the U.S. Senate. With this final defeat he quit politics, but

John Fletcher Lacey

he continued to practice law in Oskaloosa, Iowa, and even joined the League of American Sportsmen's Committee on Conservation to lobby Congress for passage of a migratory bird law, which was passed in 1913, shortly before his death. Lacey died at his home in Oskaloosa on 29 September 1913.

See also Antiquities Act; Lacey Game and Wild Birds Preservation and Disposition Act; Yellowstone National Park Protection Act.

Lacey Act Amendments of 1981 (16 U.S.C. 3371)

These revisions to the original Lacey Act of 1900 were designed to strengthen the laws banning the capturing or killing of certain forms of wildlife. Included in the amendments to the act was the following declaration:

[It is] unlawful for any person (1) to import, export, transport, sell, receive, acquire, or purchase any fish or wildlife or plant taken or possessed in violation of any law, treaty, or regulation of the United States or in violation of any Indian tribal law; (2) to import, export, transport, sell, receive, acquire, or purchase in interstate or foreign commerce (A) any fish or wildlife taken, possessed, transported, or sold in violation of any law or regulation of any State or in violation of any foreign law; or (B) any plant taken, possessed, transported, or sold in violation of any law or regulation of any State; (3) within the the special maritime and territorial jurisdiction of the United States (A) to possess any fish or wildlife taken, possessed, transported, or sold in violation of any foreign law or Indian tribal law; or (B) to possess any plant taken, possessed, transported, or sold in violation of any law or regulation of any State; (4) having imported, exported, transported, sold, purchased, or received any fish or

wildlife or plant imported from any foreign country or transported in interstate or foreign commerce, to make or submit any false record, account, label, or identification thereof; or (5) to attempt to commit any act described in paragraphs (1) through (4).

See also Lacey, John Fletcher; Lacey Game and Wild Birds Preservation and Disposition Act.

Lacey Game and Wild Birds Preservation and Disposition Act (31 Stat. 187)

This law, also known as the Lacey Act, was passed on 25 May 1900 in response to the Supreme Court's declaration in the 1896 case *Geer v. Connecticut* that states had the right to pass game and hunting laws under the powers of the Tenth Amendment. Sponsored in the House by John Fletcher Lacey of Iowa, the act, officially called "an Act to enlarge the powers of the Department of Agriculture, prohibit the transportation by interstate commerce of game killed in violation of local laws, and for other purposes," reads in part, "The object and purpose of this Act is to aid in the restoration of birds in those parts of the United States adapted thereto where the same have become scarce and extinct, and also to regulate the introduction of American or foreign birds or animals where they have not heretofore existed."

See also Geer v. Connecticut; Lacey, John Fletcher.

Lamar, Lucius Quintus Cincinnatus (1825–1893)

Lucius Q. C. Lamar, the first southerner to sit on the Supreme Court following the Civil War, was also the first southerner to hold a post–Civil War cabinet position when he served as secretary of the interior from 1885 to 1887. Born in Putnam County, Georgia, on 17 September 1825, he was the fourth of eight children of Lucius Quintus Cincinnatus

Market hunters on Maryland's Chesapeake Bay display a weekend's work in the 1920s. Such hunting to supply restaurants depleted wildfowl numbers. Congressman John Fletcher Lacey and others eventually passed legislation that affected migratory birds such as Canada geese.

Lamar, Sr., a distinguished Georgia attorney, and Sarah Williamson (nee Bird) Lamar. The Lamars, of French Huguenot ancestry, produced some of the most notable politicians from the South: Lucius, who served as Confederate envoy to Russia, interior secretary, and associate justice on the U.S. Supreme Court; Lucius's uncle Mirabeau, who was the second president of the Republic of Texas; and Lucius's nephew Joseph Rucker Lamar, a Georgia attorney who, like Lucius, served on the U.S. Supreme Court. His mother's family included the noted Bird and Williamson families.

After preparation in the local schools of Baldwin and Newton Counties in Georgia, Lamar graduated from Emory College in Oxford, Georgia, in 1845. It was at this institution that he came under the influence of Judge Augustus Longstreet, a disciple of Vice President John Calhoun and an avid states' rights advocate. Lamar studied law at Macon College in Georgia as well as privately and was admitted to the Georgia bar in 1847. He eventually married Longstreet's daughter, Virginia. In 1849, when Judge Longstreet became president of the University of Mississippi, Lamar followed him there and began a law practice; he also taught mathematics at the university.

In 1852, Lamar returned to Georgia. After a notable debate with Sen. Henry S. Foote over the admission of California as a free state, Lamar was elected to the Georgia state legislature, where he served for three years. In 1855, he returned to

Mississippi; two years later he was elected to the U.S. House of Representatives, a seat he held for the two terms of the 36th Congress (1857–1861). In Congress, Lamar was a radical secessionist, out to preserve the rights of slave owners and the South. When Sen. Jefferson Davis of Mississippi left the Senate over slave policy in 1860, Lamar joined him. He was a key figure in drafting Mississippi's decree of secession. Lamar volunteered for the Confederate army and helped enlist soldiers for the 19th Mississippi Regiment, a group he led until his health declined in 1862 and he was removed from command. He was then named Confederate envoy to Russia, his goal being to secure recognition in Europe of the Confederate government. By the time he reached Europe in late 1863, the war had turned against the South and his mission became futile. He returned home and was named judge advocate of the Third Army Corps of the Army of Northern Virginia. At the end of the war, Lamar returned to Mississippi, where he practiced law and taught. He seems to have escaped the harsh punishment doled out to other leaders of the southern rebellion. With the help of liberal Republicans, he was elected again to the U.S. House of Representatives in 1872 and to the U.S. Senate in 1876. His tenure in the Senate was marked by his attempts to bridge the chasm between the North and the South.

With the election of Grover Cleveland to the presidency in 1884, Lamar was named secretary of the interior. His administration is noted for his interest in Indian land policy. Wrote Interior Department historian Eugene Trani, "A typical 'genteel liberal,' Lamar believed the key to the Indian problem was the elimination of corruption among Indian agents. He instituted proceedings to expel ranchers from lands they had leased from the tribes at minimal cost. He recommended tighter control over land allotment, the disposition of tribal trust funds, and the improvement of Indian schools. A champion of the policy of di-minished reservations and land allotment, Lamar believed that land ownership would teach the Indians individual responsibility and the habit of thrift. Thus he endeared his administration to 'Boomers' and 'Sooners' by agreeing with them that much of the Oklahoma territory was 'surplus' to the Indians, and should be opened to settlement."

On 14 May 1887, Justice William Burnham Woods of the U.S. Supreme Court died. President Cleveland took his time naming a replacement. In August, Postmaster General William Vilas notified Lamar that the president was considering Lamar for the seat on the Court. On 6 December, Lamar was nominated for the Supreme Court. Although Lamar's notable record as interior secretary is considered the reason for his selection, the *Atlanta Constitution* reported upon Lamar's death that the real reason was that Cleveland wanted to give Lamar a more permanent post—one that would not end with Cleveland's administration. "Too old to return to and rebuild a long deserted law practice, too far removed from Mississippi affairs, he would be out of work," wrote historian Willie Halsell. "The place on the Court would be a reward for his long years of service in public life and a secure position for his declining years."

Lamar served a little more than five years on the Court and compiled a lackluster record at best, with no outstanding opinions. His death on 23 January 1893, however, caused universal mourning. Said Chief Justice Melville Weston Fuller, "He rendered few decisions, but [he] was invaluable in consultation. His was the most suggestive mind that I ever knew, and not one of us but has drawn from its inexhaustible store."

See also Department of the Interior.

Land Act of 1800 (2 Stat. 73)

Because the Land Act of 1796 resulted in poor sales of tracts, Congress passed this act on 10 May 1800 to expedite purchases by allowing land "in the territory northwest of the Ohio, and above the mouth of the Kentucky River," to be bought on credit. This led to speculation, however, which increased the price of land and resulted in incomplete collections. It was repealed by the Land Act of 1820. The legislation established four land offices in Cincinnati, Chillicothe, Marietta, and Steubenville; mandated that the surveyor-general cause the "townships west of the Muskingum, which were directed [by the 1796 act] to be sold in quarter townships, to be subdivided into half sections of three hundred and twenty acres each"; and authorized that the sales of these properties be made for no less than $2 per acre, with the buyer paying either in cash or "in evidences of the public debt of the United States."

Land and Water Conservation Fund Act (78 Stat. 897)

In his message to Congress in 1962, President John F. Kennedy called for the establishment of a "Land Conservation Fund" to pay for the acquisition of lands, the establishment of facilities, and the creation of proper management programs for recreational areas. Kennedy's inspiration came from the Outdoor Recreation Resources Review Commission, which met from 1958 to 1962 and issued proposals on environmental protection and recreational resource preservation. On 18 February 1963, Rep. Wayne N. Aspinall of Colorado introduced H.R. 3846, the Land and Water Conservation Fund bill, and other conservation bills were added to it. It passed Congress and was signed into law by President Lyndon Johnson on 3 September 1965. The funds to be raised came from admission fees to federal areas, the sale of federal property specified in the act, and a federal tax on motorboat fuels. Sixty percent of the funds were to go to the states, and 40 percent to the federal government. The subsequent passage of the Endangered Species Conservation Act in 1969

siphoned off $15 million to purchase lands for the conservation and protection of endangered species.

See also Aspinall, Wayne Norviel; Emergency Wetlands Resources Act; Endangered Species Conservation Act.

Land Policy, Western

Federal land policy in the West has always been rooted in opposition to government dictates. In the last half of the nineteenth century, many western senators and congressmen—including Frank W. Mondell and Francis E. Warren of Wyoming; Henry Moore Teller, John Shafroth, and John Calhoun Bell of Colorado; William Andrews Clark of Montana; and William Morris Stewart of Nevada—advocated the opening of western lands to development and reclamation. It was their work that produced such legislation as the Lode Mining Law of 1866, the General Mining Law of 1872, and the Payson Act of 1891. Yet these men had varying agendas, and their overall plan was not completely effective.

It was not until Teddy Roosevelt called for a program of increased grazing fees and raised charges for coal and water power leases that organized protest from the West picked up steam. Congressional opposition to Roosevelt's program was especially fierce from Teller and Mondell. In the summer of 1907, with the backing of these two legislators, delegates met at the Denver Public Lands Convention to air their grievances. The conference's ultimate failure to take a firm stand against the Roosevelt administration's conservation policies came from a lack of cohesion and agreement among western delegates from states other than Colorado and Wyoming. In the end, the conference's final report merely criticized the Forest Service for its dictates on forests.

The Western Governors' Conference, held in Salt Lake City, Utah, in June 1913, called for giving western states the right to handle their own land matters, but the five governors who attended did not have enough support in Washington to accomplish their goal. Overall, the westerners' program never seemed to take hold with a majority of Congress.

See also Sagebrush Rebellion; Western Governors' Conference.

Lane, Franklin Knight (1864–1921)

Some people said that if Franklin K. Lane had been born in the United States instead of Canada, he might very well have been elected president. Instead, Lane served as secretary of the interior from 1913 to 1920 in the Woodrow Wilson administration. He was born on 15 July 1864 near the village of Charlottestown on Prince Edward Island, Canada, the eldest of four children of Christopher Smith Lane, a Presbyterian minister who later turned to dentistry, and Caroline (nee Burns) Lane, of Scottish descent. In 1871, the Lanes left Canada for Napa, California, where Franklin attended local schools and a private academy before the family settled in Oakland. From 1884 to 1886, Lane took a special course at the University of California prior to entering Hastings Law School in San Francisco (affiliated with the University of California). He earned a law degree and was admitted to the California bar in 1888.

For a year after his graduation, Lane was a special correspondent in New York for the *San Francisco Chronicle*. In 1891, after starting a law practice, he purchased a share of the *Tacoma* [Washington] *Daily Journal* and served as its editor until 1895, when it went out of business. Lane returned to the law but waited until 1898 to enter politics. He was appointed to a committee established to draft a San Francisco city charter; at the same time, he was nominated and elected to the first of three terms (1898–1902) as San Francisco city and county attorney. In 1902, he was a candidate for governor of California, running with the support of both the Democratic Party and the Non-Par-

tisan Party, but he was defeated. During the San Francisco earthquake of 1906, he was a leading member of the relief committee created to aid the disaster's victims.

In 1905, President Theodore Roosevelt solicited Lane's opinion of the Hetch-Hetchy Dam controversy and was so impressed by his knowledge that when a vacancy occurred on the Interstate Commerce Commission (ICC) in December 1905, Lane was named to fill it. Serving until 1913, Lane was perhaps one of the ICC's ablest members. James Hemphill, in the *North American Review* in 1917, wrote that Lane's commerce decisions were "probably the most important ... [in determining] the constitutional powers of the government in the regulation of common carriers." Every one of his decisions was upheld by the Supreme Court. On 1 January 1913, Lane was made chairman of the ICC. Less than two months later, on 6 March 1913, newly inaugurated President Woodrow Wilson chose Lane as his secretary of the interior. Lane was not happy about leaving the ICC, where he had great power and comfort, but he accepted the Interior post.

In his seven years at Interior, Lane's most controversial decision was to support his president in allowing the construction of the Hetch-Hetchy Dam. Otherwise, reported one source, "He revitalized the Interior Department, bringing to it new concepts of its functions and leading it into new fields of service." Said another, "A conservationist, [Lane] consistently maintained that the resources of the West should be used to develop the West."

In 1919, Lane suffered a stroke. He resigned his office on 29 February 1920 to enter private business in order to accumulate an estate for his family. He took a job with the Pan-American Petroleum Company. In May 1921, however, heart trouble drove him to seek medical treatment in Rochester, Minnesota. While on the operating table on 18 May, he suffered a heart attack and died. Lane was 56.

See also Department of the Interior.

League of Conservation Voters

This partisan (Democratic Party only) conservation group was founded by the environmental group Friends of the Earth (FOE) in 1970 because FOE's tax status did not allow it to raise funds for political purposes. The league's intended purpose is to affect public environmental policy while opposing those in conflict with its agenda. Secretary of the Interior Bruce Babbitt, who served as president of the league from 1988 to 1993, wrote, "we must identify our enemies and drive them into oblivion." Although a non-membership organization, it has an estimated 15,000 active supporters and about 60,000 backers in total; its budget in 1993 was estimated to be $1.3 million. Because it is a fund-raising group, it has had little direct influence on the enactment of conservation or environmental legislation, although with Bruce Babbitt as interior secretary, it has a friend in a high place.

As an advocacy group, the league has experienced only modest success. In 1978, it gave money to 34 candidates for office, 18 of whom were elected. In 1991, the Federal Election Commission declared the group to be bankrupt, limiting its contributions to a maximum of $1,000 for any single candidate.

Lenroot, Irvine Luther (1869–1949)

A progressive Republican who was a giant in the House and Senate, Irvine Lenroot was a leading congressional supporter of conservation measures. Born in Superior, Wisconsin, on 31 January 1869, Lenroot was the son of Lars Lenroot (original name, Larsson), a first-generation U.S. citizen, and Fredrika Regina (nee Larsdöttir; some sources say Larsdötter or Larson) Lenroot. Lars Lenroot, a blacksmith, emigrated from Sweden in 1855; he met his wife, also a Swedish immigrant, and married her that same year.

Irvine Lenroot attended school in a two-room schoolhouse. With the help of his education-minded parents and a caring teacher, he learned the virtues of hard work and enterprise—values that would make him one of the state's leading Republicans, although he never went beyond what would be considered a high school education. He worked at a series of odd jobs in the logging industry before learning shorthand at a business college in Duluth, Minnesota. With this experience, Lenroot worked first at a Superior, Wisconsin, law office and then as a court reporter in the Douglas County (Superior) Courthouse, a post he held until 1906. Meanwhile, in 1898, he was admitted to the state bar after studying law.

In 1900, Lenroot joined the progressive faction of the Republican Party, which was ruled in the state by Robert LaFollette, who later became governor of Wisconsin, U.S. Senator, and Progressive Party candidate for president in 1924. Because of LaFollette's influence, Lenroot ran for and was elected to a seat in the lower house of the Wisconsin legislature in 1900, where for two terms (1901–1907) he was part of a combine that advocated political reform and railroad taxation. From 1903 until the end of his second term, Lenroot was the assembly speaker. In 1908, he won election to the U.S. House of Representatives, where he teamed with other Republican "insurgents" such as George W. Norris to fight the iron power of House Speaker Joseph G. "Uncle Joe" Cannon. A *Washington Times* editorial said of the new congressman, "When ... Lenroot came to town as a new representative a few weeks ago, it was given out that the House organization was going to haze him, and make life a burden, because he had the temerity to run against and defeat Representative [John James] Jenkins, chairman of the Judiciary Committee, and one of the pets of the organization. ... Mr. Lenroot was to be shown how improper, how reprehensible, it is for an outsider thus to give demonstration that a member of the organization is but mortal."

Lenroot served in the House until 1918, when he ran for the U.S. Senate in a special election to fill the seat of the deceased Paul Husting. At this point, Lenroot parted company with LaFollette over the entry of the United States into World War I: Lenroot voted for war, LaFollette against it. From 1918 to 1925, Wisconsin was represented in the U.S. Senate by these two progressives who were at odds with each other. In 1920, Lenroot won a full term in the Senate, the only term he would serve. Wrote author Herbert Margulies, "A close friend of Gifford Pinchot and a longtime champion of regulated development of public resources, he contributed importantly to the Federal Water Power Act, which established a commission to authorize and license navigation improvements and hydroelectric plants on public land." The Republicans considered Lenroot for the vice presidency in 1920. In 1926, he was defeated for a second term in the Senate. After two years of law practice in Washington, during which time he was a member of the Anglo-American Conciliation Commission, Lenroot was appointed by President Herbert Hoover to serve as a judge on the federal court of customs and patent appeals in New York City. He remained in this post until his retirement in 1944. Lenroot died of cancer five years later on 26 January 1949, five days short of his eightieth birthday.

See also Federal Water Power Commission; Mineral Leasing Act.

Leopold, Aldo (1887–1948)

Aldo Leopold was one of the foremost game management experts and protectors of wildlife in the conservation movement. He was born Rand Aldo Leopold in Burlington, Iowa, on 11 January 1887, the eldest of the four children of Carl and Clara (nee Starker) Leopold, second-generation German immigrants. Leopold fell in love with nature at an early age, noting the diets of birds that lived along the Mississippi River environs near his

Aldo Leopold, 1946

home and jotting down the information in journals. He became skilled in ornithology and natural history and was an avid hunter. He received his education in the local schools of Burlington, at the Lawrenceville Prepatory School in New Jersey, and at Yale University, from which he earned a bachelor of science degree in forestry in 1908 and a master of forestry degree in 1909.

In 1909, Leopold joined the newly established U.S. Forest Service and was assigned to the Apache National Forest in the Arizona Territory. Later he was transferred to the Carson National Forest in the New Mexico Territory, where in 1912 he reached the rank of supervisor. The next year, his life was almost cut short by an attack of nephritis (an inflammation of the kidneys), and he returned home to recuperate. Eighteen months later, he was back at his old stomping grounds, establishing programs in the Park Service's

Southwestern District in fish and game management, setting up the administrative actions that led to the creation of Gila National Forest in New Mexico in 1924, and working to eradicate predators. In 1917, his attempts at predator control and his efforts to create refuges for animals and wildlife earned him the gold medal of conservationist William Temple Hornaday's Permanent Wildlife Protection Fund. He learned, however, that nature itself controls predators while maintaining a balance between the hunters and the hunted. Leopold applied this lesson to his teaching for the rest of his life.

In 1924, shortly before his plans for the Gila National Forest were completed, Leopold left the Forest Service for a post at the Forest Products Laboratory in Madison, Wisconsin. As assistant director, he wrote in 1926 of the goals of the laboratory: "1) Increase the value of forests so it will pay to grow them. 2) Increase the quantity of merchantable timber by utilizing waste. 3) Assist in the development of improved utilization of the National Forests." After two years as assistant director of the laboratory, with little chance that he would assume the directorship, he resigned and began to establish himself as an expert in game surveyorship, so that he would be able to formulate national policies on game management. His 1931 summary, *Report on a Game Survey of the North Central States*, and its 1933 follow-up, *Game Management*, are classics in the field. In 1933, Leopold served on President Franklin D. Roosevelt's Committee on Wildlife Restoration. That same year, he was made professor of game management at the University of Wisconsin, where he spent the rest of his life teaching new generations of students about game management and the wild. He was a founding member of the Wilderness Society in 1935. In 1949, his landmark work *A Sand County Almanac* appeared posthumously. A collection of essays, it combines graceful prose and a forceful defense of the environment, as in this example from "Back

From the Argentine": "There was a time in the early 1900s when Wisconsin farms nearly lost their immemorial timepiece, when May pastures greened in silence, and August nights brought no whistled reminder of impending fall. Universal gunpowder, plus the lure of plover-on-toast for post-Victorian banquets, had taken too great a toll. The belated protection of the federal migratory bird laws came just in time."

Aldo Leopold was fighting a grass fire near his home in Baraboo, Wisconsin, on 21 April 1948 when he had a heart attack and was caught in the fire. At his death, he was 61 years old. The Aldo Leopold Wilderness in Gila National Forest in New Mexico was named for him in 1980.

See also Forest Products Laboratory.

Live Oak Act of 1831
See United States v. Briggs.

Lode Mining Law (14 Stat. 233)
This legislation, enacted on 26 July 1866, was sponsored by Sen. William Morris Stewart of Nevada. A former miner and lawyer who dealt with mine cases, Stewart saw the need to legislate protection for lode and placer miners in the western United States. Lode, or vein, mines are distinguished by "lines or aggregations of metal imbedded in quartz or other rock in place" (*United States v. Iron Silver Mining Company*, 128 U.S. 673 [1888]). Stewart's legislation gave these miners, who had essentially been trespassing on federal lands, rights to their discoveries and the right to declare a patent for those lands. Six years after the 1866 act, Stewart helped enact the General Mining Law of 1872, which gave these same rights to placer miners.

See also General Mining Law; Stewart, William Morris.

Love Canal
A sleepy New York village near Niagara Falls, Love Canal became a symbol in the

In 1990 residents of Love Canal, New York, protested a proposal to reestablish residences near the site of a toxic waste dump.

1970s of the destruction of the earth. Between 1942 and 1953, the Hooker Chemical and Plastic Company stored toxic wastes in metal canisters buried in the ground—according to the *Toronto Star,* 21,800 tons of toxic wastes. After the company sold the canal and some surrounding land, the 16-acre area was divided into lots and houses were built. It wasn't until the late 1970s that samples of groundwater and soil showed massive amounts of toxic soup. Wrote Andrew Danzo in the *Washington Monthly,* "This wasn't a vague threat like smog or a dirty river or problems with the food chain. It was a chemical swamp 3,000 feet long. And it was in people's backyards—99 of them to be exact. It was drums popping from the ground where children played. It was ooze seeping through basement walls."

The government, both state and federal, was forced to step in. Eventually, the 237 houses in the immediate region of the canal were purchased and demolished. The Love Canal Revitalization Agency was established by the state to purchase an additional 495 houses in the surrounding area, designated the Emergency Declaration Area. The federal government, faced with one of the worst chemical spills ever, quickly enacted the Comprehensive Environmental Response, Compensation, and Liability Act of 1980 (CERCLA)—also known as the Superfund Act—to fund the cleanup of Love Canal and other such waste dumps. Today, although Love Canal was one of the first sites put on the Superfund's priority list, it remains virtually untouched except for containment of the problem; lawyers have been fighting for years over whose responsibility it is to pay for the cleanup. In 1988, the government reported:

The ROD [Record of Decision] of 10/26/87 altered an earlier decision at Love Canal to use onsite land disposal for dioxin-contaminated sewer and creek sediments. Now, a mobile thermal destruction unit will be used onsite to destroy and remove dioxin with an efficiency of 99.9999 percent. The cost for treatment will be twice that of land disposal, but the ROD selected thermal destruction on the basis of its ability to meet statutory requirements by eliminating toxicity and mobility. In addition, several site demonstrations elsewhere had successfully destroyed dioxin-contaminated soil with mobile thermal destruction units. EPA responded to extensive community comments against landfilling the contaminated material onsite and also decided not to attempt to separate material with less than 1 part per billion dioxin [EPA's cutoff for acceptable contamination] because of uncertain reliability in doing so.

In March 1994, a court ruled that the Occidental Chemical Corporation, which had bought the Hooker Chemical Company, was relieved of paying punitive damages in the case. In an out-of-court settlement, the company agreed to pay damages to New York State and to assume responsibility for cleanup work.

See also Superfund; Wastes, Toxic and Hazardous.

Lujan, Manuel, Jr. (1928–)

Manuel Lujan was the first Hispanic American to hold the post of secretary of the interior. He was born in the village of San Ildefonso on the Rio Grande, northeast of Santa Fe, New Mexico, on 12 May 1928, the eighth of eleven children. His father, Manuel Lujan, Sr., was a prosperous insurance salesman; he was mayor of Santa Fe and an unsuccessful Republican candidate for Congress in 1944 and for governor in 1948. His mother was Lorenzita (nee Romero) Lujan. Another of the Lujan children, Edward, was chairman of the New Mexico Republican Party. Manuel Lujan, Jr., attended local schools before entering St. Mary's College in San Francisco and the College of Santa Fe, from which he received a bachelor's degree in business administration in 1950.

After working in his father's insurance business for a number of years, Lujan ran for a seat in the New Mexico state senate in 1964. After his defeat, he moved to Albuquerque and extended the prosperity of his father's business. Four years later, in 1968, following the creation of a congressional district with a majority of Hispanic residents, Lujan was elected to the U.S. House of Representatives—the first and only Hispanic Republican ever elected to the House. During his 20 years in the House, Lujan was a member of the House Committee on Interior and Insular Affairs, where he was an outspoken advocate of the development of the West's natural resources. Wrote Martin Tolchin in the *New York Times*, "On the Interior committee, Lujan occasionally shunned the traditional role of the ranking Republican, who was expected to represent the [Reagan] Administration in committee deliberations. His close relationship with Morris K. Udall, an Arizona Democrat who was the committee chairman, resulted in more support for Democratic positions than the Administration wanted." Although he was poorly rated by environmental groups, he voted to overturn Ronald Reagan's veto of the Clean Water Act. During the Reagan years, Lujan was considered three times for the post of interior secretary.

After undergoing triple-bypass surgery in 1986, Lujan decided to retire from the House in 1989, claiming that he wanted to give someone else a chance to hold the seat. On 22 December 1988, George Bush named Lujan to be his secretary of the interior. Although environmental groups decried his selection, one Democrat, Rep. Bruce Vento of Minnesota, who had served with Lujan on the Interior Committee, said of his colleague, "He's a pragmatic guy who operates in the world of the possible. He is a known quantity to the members of Congress, and they will feel comfortable with him."

Lujan served from January 1989 until January 1993. During that time, he clashed with environmentalists over logging in the Pacific Northwest and water irrigation projects in California and was passed over for inside advice during the *Exxon Valdez* crisis. His tenure, although controversial, was considered an interregnum between that of his predecessor, Donald Hodel, who helped stabilize the department, and Bruce Babbitt, who was considered the first major environmentalist to head Interior.

See also Department of the Interior.

Lujan, Secretary of the Interior, et al. v. National Wildlife Federation et al.
(111 L Ed 2d 695, 110 S Ct 3177 [1990])

In 1985, an umbrella coalition of environmental groups sued Secretary of the Interior Donald P. Hodel and others in the department for reclassifying 180 million acres in 17 states in an "arbitrary and capricious manner." The environmental groups claimed that this reclassification would open the lands up for mining and thus would "destroy the natural beauty of the lands." The groups sued to stop the reclassification. A district court denied the Interior Department's motion to dismiss the case and found for the environmental groups. The U.S. Court of Appeals for the District of Columbia upheld the judgment of the lower court. By this time, Manuel Lujan, a former congressman from New Mexico, had been named secretary of the interior, and he became the plaintiff in the appeal to the U.S. Supreme Court. On 27 June 1990, the Court held 5–4 (Justices Harry Blackmun, William Brennan, Thurgood Marshall, and John Paul Stevens dissenting) that under Rule 56 of the Federal Rules of Civil Procedure, the environmental groups had no standing to sue a government agency for reclassifying public lands. The majority opinion was written by Justice Antonin Scalia.

McClelland, Robert (1807–1880)

Robert McClelland, one-time governor of Michigan, was secretary of the interior during the entire Franklin Pierce administration, from 1853 to 1857. McClelland was born in Greencastle, Pennsylvania, on 1 August 1807, the son of Dr. John McClelland and Eleanor Bell (nee McCulloh) McClelland. Robert McClelland attended local schools, graduated from Dickinson College in 1829, and was admitted to the Pennsylvania bar two years later. He began a law practice in Pittsburgh.

In 1833, McClelland moved to Monroe, Michigan, where he set up a law practice and dabbled in politics. He was an organizer of the Democratic Party in the Michigan Territory and served as a delegate to the Michigan constitutional convention in 1835. In 1839 and 1843, he was elected to single terms in the Michigan state house of representatives; during both terms, he served as speaker pro tempore. In 1841, he was elected mayor of Monroe.

During his second stint in the state legislature, McClelland was elected to the U.S. House of Representatives, where he sat until 1849. A friend of Lewis Cass and David Wilmot, McClelland supported the Wilmot Proviso (which allowed new western states into the Union as free and not slave states) and Cass for president in 1848. He was a delegate to the second Michigan constitutional convention in 1850 and was elected governor of Michigan for a short term (1851–1852). In 1852, he was reelected, defeating Zachariah Chandler, a future Michigan political powerhouse and, like McClelland, a future secretary of the interior.

Robert McClelland was still governor on 8 March 1853 when newly inaugurated President Franklin Pierce selected him as interior secretary. McClelland was chosen to appeal to those Democrats who supported the Wilmot Proviso, backers of Lewis Cass (who had lost the presidential nomination to Pierce), and Free-Soilers, who despised Pierce's proslavery stand. McClelland resigned the governorship immediately and went to Washington. He was the first interior secretary to serve a full four-year term. He tried to pick up where his predecessor, Alexander H. H. Stuart, had left off, advancing the department and overseeing the myriad offices dealing with land, Indians, patents, and pensions. McClelland set up his own system of hiring based on merit, strove to implement reforms, and tried to bring together the various offices of the department spread around Washington. He also attempted to set up a system of land purchases that would end fraud and abuse, but increased land and pension frauds and land speculators contributed to his failure in this area. According to historian Norman Forness's work on the early history of the Interior Department, McClelland "typified the weakness of the Pierce cabinet. As an individual he lacked prominence. He had been chosen to please many political factions. It was a blunder for Franklin Pierce, a man of small reputation, to have gathered about him such cabinet members equally lacking in status. Pierce's choice of advisers procluded the possibility of effective party leadership and hindered significant accomplishment."

Upon completing his term as secretary, McClelland returned to a private law firm in Detroit. His only other public service was as a delegate to the 1867 Michigan constitutional convention. His last 23 years were spent as a country attorney. McClelland died at his home in Detroit on 30 August 1880, less than a month after his seventy-third birthday.

See also Department of the Interior.

McGee, William John (1853–1912)

Geologist, hydrologist, archaeologist, anthropologist: Those are just a few of the accomplishments of William John McGee (known as "W J" almost all his life). He was one of the key players in the early conservation movement in the United States. Born in a log cabin near Farley, Iowa, on 17 April 1853, he was the fourth of nine children of James McGee, a lead miner, and Martha Ann (nee Anderson) McGee. James McGee was an Irish immigrant of Scottish descent who immigrated to the United States in 1831; Martha Anderson McGee was a native of Kentucky. In a privately published 1915 work, *The Life of WJ McGee*, his only sister, Emma, wrote, "WJ was a sickly infant who under healthy country life grew sturdy and filled with nervous energy. Although he performed his share of farm chores, his greater interest was the life of the mind." McGee was self-educated except for four years in rural schools.

It was his interest in the environment that led McGee to study the geological composition of his native Iowa. Wrote one source, "In 1875, he commenced researches concerning Indian mounds and other relics in Iowa and Wisconsin, and two years later began a geologic map of his own and neighboring counties." He reported his findings in the *American Journal of Science* from 1878 to 1882. His work brought him to the attention of U.S. Geological Survey (USGS) Director John Wesley Powell, who invited McGee to become a member of the survey in 1883. His immediate work centered on the extinct volcanic lakes of Nevada and California; soon after, Powell recalled him to Washington to head the survey's division of Atlantic

coastal plains geology. In this position, his inquiries resulted in works that were published as *The Geology of the Chesapeake Bay* (7th Annual Report of the USGS for 1885-1886, 1888) and "Three Formations of the Middle Atlantic Slope" (*American Journal of Science*, 1888), *The Pleistocene History of Northeastern Iowa* (11th Annual Report of the USGS for 1889-1890, part 1, 1891), *The Lafayette Formation* (12th Annual Report of the USGS for 1890-1891, part 1, 1891), and "The Gulf of Mexico as a Measure of Isostasy" (*American Journal of Science*, 1892). After Powell quit the survey, McGee joined him as ethnologist-at-large at the Bureau of Ethnology at the Smithsonian Institution. Wrote Forest Service historian Terry West, "Powell had lobbied Congress from 1873 until 1878 for a government-supported scientific bureau of anthropology. The Bureau of American Ethnology (BAE) was founded in 1879 to 'produce results that would be of practical value in the administration of Indian affairs,' according to Powell in the first Annual Report of the Bureau of American Ethnology (1881)." He conducted explorations of Tiburon Island—home of the Seri natives in Baja, Mexico—and studied the agricultural techniques of the Indian tribes of the Sonoran Desert of Mexico. In 1903, following the death of Powell, McGee clashed with Samuel Langley, secretary of the Smithsonian Institution, and resigned his position. During these years, McGee was an editor of *National Geographic Magazine* as well as president of the American Anthropological Association and the American Association for the Advancement of Science. He was also a founder of the Columbian Historical Association.

McGee was in charge of the anthropological, geological, and historical exhibit at the Louisiana Purchase Exposition in St. Louis. After the exposition ended in 1904, President Theodore Roosevelt named McGee to the Inland Waterways Commission. About his work on this

Anthropologist and geologist W J McGee stands with a Yuma man during a Nevada and California trip in the late 1800s.

committee, historian Whitney Cross related, "Indisputably, it was the Inland Waterways Commission which first put before a broad public the idea of multi-purpose river basin control, and it was out of this concept and out of the work of the Commission that the broader conservation movement developed. . . . It appears that the most important factor was one philosophically inclined individual who was consciously molding a political policy out of an idea developed earlier amidst the radical Darwinism of the 1880s and 1890s. This man was W J McGee."

In 1908, McGee was a major organizer (with Gifford Pinchot and Frederick Haynes Newell) of the Governors' Conference on the Environment, a meeting of the nation's governors and other administrators to discuss the environmental situation in the country. He was at the absolute forefront of the conservation movement. Gifford Pinchot wrote of him, "So far as such a thing can ever be said of any one man in a movement so extensive, W J McGee was the scientific brains of the conservation movement all through its early critical stages. Since from the first, the distinguishing fact about that movement was its joint consideration of all the natural resources as the working capital of humanity, McGee's wide and balanced knowledge of this continent and its resources gave him very special fitness to deal with this wide and weighty problem.... I have never met a man whose imaginative suggestiveness in scientific work, and in the application of scientific results to human problems, could equal his. It was always the *application* of knowledge that appealed to him. His mind passed easily across the details of scientific problems to their bearing on matters that would count for the welfare of the people." McGee died of cancer in Washington, D.C., on 4 September 1912. He was 59.

McKay, James Douglas (1893–1959)

The governor of Oregon, Douglas McKay (he did not go by his first name) was chosen by President Dwight D. Eisenhower as secretary of the interior. Born on 24 June 1893 in Portland, Oregon, he was the son of E. D. McKay, a carpenter, and Minnie (nee Musgrove) McKay and was descended from pioneers who had come to Oregon by wagon train in 1852 (seven years before Oregon became a state). While attending local schools, McKay earned money driving a butcher wagon, working as a meat cutter, and selling newspapers such as the *Portland Oregonian* and the *Daily News*. He never completed high school because of his family's financial situation, but he did attend the Oregon Agricultural College at Corvallis, which awarded him a bachelor of science degree in agricultural studies in 1917.

That same year, as the United States entered World War I, McKay volunteered for the U.S. Army. He was commissioned a first lieutenant in the 91st Division and saw action on the fields of France for more than a year. During the Argonne offensive in October 1918, he was shot in the shoulder and spent the next 13 months in hospitals recovering. Permanently disabled by the injury, McKay was limited to nonstrenuous jobs. Once out of the hospital, he entered the field of insurance and later sold cars. In 1927, he opened his own auto dealership, selling Chevrolets in Salem.

He surprised many by winning the mayoralty of Salem as a Republican in 1932—a year in which Republicans nationwide were swept aside in the revolution that brought Franklin Delano Roosevelt to power. In 1935, McKay was elected to the Oregon state senate and served until 1948. His terms in office were disrupted when, shortly after the start of U.S. involvement in World War II, he volunteered for the U.S. Army. He saw limited action because of his previous injury. In 1948, he was elected in a three-way race to fill the seat of Gov. Earl Snell, who had been killed in a plane crash. In 1950, McKay won a full four-year term.

In 1952, Governor McKay backed Gen. Dwight D. Eisenhower for president. Upon Eisenhower's election, the debt was repaid when he named McKay secretary of the interior. Eisenhower, speculated one author, wanted a "western man" who would back the president's resource-development policies. Just after he was confirmed, McKay announced that he "favored turning the offshore oil rights [belonging to the federal government] over to the states," according to the *New York Times*. Wrote historian Elmo Richardson, "Secretary McKay's accomplishments were undermined by the fact that he confronted several of the most complex and controversial resource problems of the post–World War II period. One of them was the proposal to develop the water and power potential of the Snake River, particularly the Hells Canyon site along the Idaho-Washington-Oregon border. The second was a costly multi-dam project stretching across the Upper Colorado Basin in the states of the central Rocky Mountain region." Further, McKay championed the passage of the Submerged Lands Act, which returned oil-rich tidelands to the states. However, his work with Horace M. Albright of the National Park Service on "Mission 66," which called for the development of programs for accommodating the estimated 80 million visitors to national parks by 1966, is often overlooked.

Throughout his tenure, McKay was perhaps the Eisenhower administration's most controversial figure. On 14 April 1956, he resigned to challenge Sen. Wayne Morse's reelection. Opposing Morse's left-wing views, McKay was defeated for the first and only time in his life. In 1957, soon after the loss, Eisenhower named him chairman of the International Joint Commission, which investigated water boundary problems between the United States and Canada. On 22 July 1959, McKay died at his home in Salem of a heart ailment and kidney complications. He was 66 years old.

See also Department of the Interior.

MacKaye, Benton (1879–1975)

A founding member of the Wilderness Society, Benton MacKaye is remembered chiefly for his work in establishing the Appalachian Trail, a nature path from Georgia to Maine. He was born on 6 March 1879 in Stamford, Connecticut.

He graduated from Harvard University in 1900 and received a master of forestry degree from that institution in 1905. That same year, he began work in the U.S. Forest Service under Gifford Pinchot. For the next 13 years, according to one source, "he investigated timber and water resources."

After a one-year assignment with the Department of Labor working on land planning (such as the concept of a highway system in his native Connecticut), MacKaye returned to Harvard to teach forestry. He wrote of his plans to create a nature trail along the eastern coast of the United States in the October 1921 issue of the *Journal of the American Institute of Architects* in an article titled "An Appalachian Trail: A Project in Regional Planning." He envisioned a series of communities along a nature trail that would stretch from Georgia to Maine. Although the communities never became a reality, the 2,050-mile Appalachian Trail was established.

MacKaye is also remembered for his work in the area of conservation. In 1923, he, Clarence Stein (then chairman of the American Institute of Architects [AIA]), and Charles Harris Whitaker, editor of the AIA's journal, formed the Regional Planning Association of America to outline the formation of communities and recreational areas with minimum use of natural resources. On 21 January 1935, MacKaye, Robert Marshall, Olaus Murie, Ernest Oberholtzer, and others formed the Wilderness Society. MacKaye served as president of the society from 1945 to 1950 and as honorary president from 1950 until his death. Even after his retirement in 1950, he continued to write articles on various conservation projects, including a proposed Missouri Valley Authority (along the lines of the Tennessee Valley Authority) and pioneering efforts such as urban renewal and the interstate highway system he envisioned.

Asked later in life what he felt was his most important work, he described a survey he had conducted for the Forest Service on the forest cover of the watershed in the White Mountains in New Hampshire, a plan that led to the creation of the White Mountain National Forest. He wrote over 80 books and articles from 1913 to 1973, including *Employment and Natural Resources* (1920), *The New Exploration: A Philosophy of Regional Planning* (1928), and *Expedition Nine: A Return to a Region* (1969), a collection of his essays published by the Wilderness Society. MacKaye died at his home in Shirley Center, Massachusetts, on 11 December 1975 at the age of 96. Wrote author Lewis Mumford, "As an explorer of natural resources, [MacKaye] was in the great pioneer tradition of Bartram and Audubon, Thoreau and George Perkins Marsh."

See also Marshall, Robert; Wilderness Society.

McKennan, Thomas McKean Thompson (1794–1852)
See Stuart, Alexander Hugh Holmes.

Maclure, William (1763–1840)
Biographer Charles Keyes called William Maclure "The Father of American Geology." It was his work in this field that led him to prepare the first in-depth geologic map of the United States. Apparently born with the name James McClure on 27 October 1763 in Ayr, Scotland, he was the son of David and Ann (nee Kennedy) McClure. At a later date he changed both his first name and the spelling of his last name, for unknown reasons. He was instructed by private tutors, leading one to believe that his family had some wealth. In 1782, he visited the United States to conduct business; upon his return, he joined the London mercantile house of Miller, Hart and Company. Between then and 1796, he prospered to such a degree that he was able to retire at age 33 with a sizable fortune.

In 1796, Maclure again visited the United States. His intense interest in the country led to the desire to establish U.S. citizenship, and one source speculates that he may have taken the first steps to do so during this time. Seven years later he was a full U.S. citizen. It was then, in 1803, that he was named to the French Spoliation Claims Commission, a committee set up to decide the merits of claims made by American merchants for shipping losses incurred at the hands of the French government between 1783 and 1800. This business took up Maclure's time for several years, and he published some of his opinions in "To the People of the United States: A Statement of the Transactions of the Board of Commissioners Appointed in 1803 for the Adjustment of Claims against the French Government" (1807). The U.S. government eventually settled the claims in 1865 after much wrangling.

While a member of the commission, Maclure spent some time traveling around the United States in search of its geologic history. He published his first conclusions in "Observations on the Geology of the United States, Explanatory of a Geological Map" in *Transactions of the American Philosophical Society* (no. 6, 1809). He issued a revised version in 1817 called "Observations on the Geology of the United States," which appeared the following year in the same journal (no. 1, new series). In 1815, Maclure visited France, where he met French naturalist Charles Alexandre Lesueur and hired him as a cartographer and naturalist to accompany him back in the United States. The two men spent the next two years investigating the wilderness of the Allegheny Mountains. In 1817, he embarked on a five-month voyage through Georgia and Florida with Lesueur and naturalists George Ord, Titian Ramsay Peale, and Thomas Say.

In the last years of his life, Maclure acquired an estate in Spain and lived there from 1820 to 1824, leaving only when revolution broke out and his home was confiscated. He returned to the United States and was involved in the New Harmony movement, centered at utopian thinker Robert Owen's Indiana commune, which Maclure visited soon after his return with his friends Lesueur, Say, and Dutch geologist Gerard Troost (1776–1850). In 1827, Maclure's health declined, and he traveled to Mexico with Say, where he found the climate more to his liking. He returned to the United States only once, in 1828, to preside over a meeting of the American Geological Society in New Haven. While in Mexico, he wrote on a number of economic issues, which he published as *Opinion on Various Subjects, Dedicated to the Industrious Producers* (2 volumes, 1831–1837). He attempted another trip to the States in 1839, but ill health forced him to stop at San Angel, Mexico, where he died on 23 March 1840 at the age of 76.

Biographer George W. White, the editor of Maclure's papers, wrote of the little-remembered geologist: "His best known work is the *Observations on the Geology of the United States*. Although proceeded by Johann D. Schöpf's work in German (1787) and Volney's in French (1803), Maclure's articles and book are the first connected account originally written in English on the geology of the United States.... In the text and map, Maclure divided the country into areas of 'primitive rocks,' 'transition rocks,' 'floetz and secondary rocks,' and 'alluvial rocks.'"

McNary, Charles Linza (1874–1944)

Perhaps the leading member of Congress during the twentieth century in the areas of conservation and forestry, Charles McNary represented Oregon in the U.S. Senate from 1918 to 1944. He was born on his father's farm just north of Salem, Oregon, on 12 June 1874, the son of Hugh Linza McNary, a farmer and teacher, and Mary Margaret (nee Claggett) McNary. Hugh McNary's father, James, had moved with a wagon train from Missouri to Oregon in 1845, and Mary Claggett McNary's family, of

English and Scottish ancestry, had made the same trip in 1852. Charles McNary was orphaned by his mother's death in 1878 and his father's passing in 1883. He spent his boyhood years with his three older sisters and an older brother in Salem, where he attended local schools. He attended Stanford University for a single year (1896–1897) before returning to Oregon to read the law; he was admitted to the state bar in 1898.

McNary then joined his older brother in a law practice in Salem. Up until this time, his interest in politics had been confined to a term as deputy county recorder of Marion County (1892–1896). McNary's brother John was elected district attorney of Marion County in 1904, and he named Charles as his deputy. McNary served in this position until 1911, when he was named special counsel for the state railroad commission. Two years later, he was appointed to a vacancy on the Oregon state supreme court. Unable to win the Republican nomination for that post, he served instead as chairman of the Republican State Central Committee for the next two years. In 1917, upon the death of Sen. Harry Lane, McNary was appointed to serve out the remainder of the term. Elected on his own in 1918, he was reelected four times, the last time just two years before his death.

A progressive Republican, McNary backed Woodrow Wilson's foreign policy during World War I. During the 1920s, he was instrumental in the passage of the Clarke-McNary Reforestation Act of 1924, which called for collaboration between the secretary of agriculture and the states to procure forest tree seeds and plants to regrow areas of forest that had been cut down or burned; the Mc-Sweeney-McNary Forest Research Act of 1928, which ordered the government to extend forestry research programs for ten years while using the experiment station system and mandated a national timber survey; and the Woodruff-McNary Act of 1928, which expanded federal jurisdiction over the acquisition of wood- and timberlands and authorized the government to establish new national forests. Wrote historian Earl Pomeroy, "As chairman of the Senate Committee on Irrigation and Reclamation [1919–1926], McNary supported development of the Tennessee, Colorado, and Columbia Rivers. In the Committee on Agriculture and Forestry [of which he was chairman, 1926–1933], he presided over the 1923 investigation of forest resources that led to the Clarke-McNary Reforestation Act, the McSweeney-McNary Act, and the Woodruff-McNary Act." As Republican minority leader (1933–1944), McNary was, in essence, the national leader of his party. His body of legislative work led him to be nominated for vice president in 1940 on the Wendell Willkie ticket, which was defeated by Franklin Roosevelt.

Charles McNary died on 25 February 1944 in Fort Lauderdale, Florida, where he was recuperating following a brain operation. He was 69.

See also Clarke-McNary Reforestation Act; McSweeney-McNary Forest Research Act; Woodruff-McNary Act.

McNary-Woodruff Act of 1928
See Woodruff-McNary Act.

McRae, Thomas Chipman (1851–1929)
A leading advocate of forest preservation in Congress, Thomas McRae was also governor of Arkansas. Born in Mount Holly, Arkansas, on 21 December 1851, McRae was the son of Duncan and Mary Ann (nee Chipman) McRae, both farmers. The McRae family was from Scotland, settling in North Carolina early in the eighteenth century. Thomas McRae attended private academies at Shady Grove, Mount Holly, and Falcon before enrolling at the Soulé Business School in New Orleans, Louisiana. After graduation, he spent three years at the law school of Washington and Lee University, earning a law degree in 1872.

Admitted to the Arkansas bar a year later, McRae practiced law in Rosston, Arkansas, until 1877 and later in Prescott, Arkansas; in 1877, he was elected to a single term in the Arkansas state house of representatives. In 1885, he was elected to the first of nine terms in the U.S. House of Representatives, where he was a member of the House Committee on Public Lands and its chairman for four years. One source said of him, "McRae has for a number of years been one of the most indefatigable advocates of a practical, business-like administration of the forest reserves of the United States."

In his 1893 message to Congress, incorporated in a letter to Secretary of the Interior Hoke Smith, President Grover Cleveland decried the fact that the Forest Reserve Act of 1891 had not protected timberlands included in forestry reserves. He asked Congress to pass legislation to strengthen the 1891 act and thereby protect the timber. McRae stepped forward and introduced the McRae reforestation bill. Although strengthening the right of the president to establish and preserve forest resources, its most important detail was that it allowed timber to be sold to the highest bidder, thereby cutting down on theft and waste. The bill had the backing of every major conservation group in the United States. Standing in its way, however, were interests that represented the West, including the "insurgent" legislators (the anticonservation members of Congress) from Colorado, namely, Reps. John Calhoun Bell and John Franklin Shafroth and Sens. Edward Oliver Wolcott and Henry Moore Teller, the latter a former secretary of the interior. Once debate on the McRae bill began, Bell stepped forward and condemned it as an infringement on the rights of farmers and ranchers. With the help of other western congressmen, he succeeded in shelving the legislation. A bill later introduced by McRae but watered down by the insurgents to include unrestricted mining on the reserves was passed by the House and Senate but died in conference. The bill was resurrected again in 1896, and Bell and Shafroth voted for it; Teller, however, helped kill it, this time permanently. Meanwhile, Cleveland acted anyway, creating 13 new reserves just before leaving office in 1897 and touching off a congressional whirlwind that culminated with the passage of the Pettigrew Act later that year, a piece of legislation that, ironically, included many of the provisions of McRae's bill.

After retiring from the House in 1904, McRae returned to his law practice in Prescott, Arkansas. For a time he was a banker and president of the Arkansas Bankers' Association. In 1920, he was elected governor of Arkansas and served from 1921 to 1925. As governor, he lobbied for a law that gave women the right to hold state office, as well as establishing a state tuberculosis sanitorium for Negroes, a state forestry commission, and a bureau to aid World War I veterans. After leaving office, he was named special chief justice of the Arkansas Supreme Court. Before his death, he donated two blocks of land that he owned in Prescott to be used as a park for blacks. Thomas McRae died in Prescott on 2 June 1929 at the age of 77.

See also Bell, John Calhoun; Pettigrew Act; Shafroth, John Franklin; Teller, Henry Moore; Wolcott, Edward Oliver.

McRae Reforestation Bill
See Pettigrew Act.

McSweeney-McNary Forest Research Act (45 Stat. 699–702)
This third federal legislation in four years cosponsored by Sen. Charles Linza McNary of Oregon authorized the government to expand forestry research programs for ten years using the experiment station system and called for the funding of a timber survey. Officially called "an Act to insure adequate supplies of timber and other forest products for the people of the United States, to promote the full use for timber growing and other purposes of forest lands in the United States, including farm wood lots and those abandoned areas not suitable for agricultural production, and to secure the correlation and the most economical conduct of forest research in the Department of Agriculture, through research in reforestation, timber growing, protection, utilization, forest economics, and related subjects, and for other purposes," it was cosponsored by Rep. John R. McSweeney (1890–1969) of Ohio and enacted by Congress on 22 May 1928.

See also Clapp, Earle Hart; Clarke-McNary Reforestation Act; Greeley, William Buckhout; Woodruff-McNary Act.

Magnuson Fishery Conservation and Management Act (16 U.S.C. 1801)
Enacted by Congress on 13 April 1976, this act was sponsored by Sen. Warren Magnuson of Washington. Under its provisions, Congress declared:

[T]he fish off the coasts of the United States, the highly migratory species of the high seas, the species which dwell on or in the Continental Shelf appertaining to the United States, and the anadromous species which spawn in United States rivers or estuaries, constitute valuable and renewable natural resources. These fishery resources contribute to the food supply, economy, and health of the Nation and provide recreational opportunities.... As a consequence of increased fishing pressure and because of the inadequacy of fishery conservation and management practices and controls, a) certain stocks of such fish have been overfished to the point where their survival is threatened, and b) other such fish stocks have been so substantially reduced in number that they could become similarly threatened.... [I]t is therefore declared to be the purposes of Congress in this [law] 1) to take immediate action to conserve and manage the fishery resources found off the coasts of the United States, and the anadromous species and Continental Shelf fishery resources of the United States, by establishing a) a fishery conservation zone within which the United States will assume exclusive fishery management authority over all fish, except highly migratory species, and b) exclusive fishery management authority beyond such zone over such anadromous species and Continental Shelf resources.

The act established Regional Fishery Management Councils (RFMCs) "to prepare, monitor, and revise" fishery management plans, which would "achieve and maintain, on a continuing basis, the optimum yield from each fishery." Seven regional RFMCs were created: the New England, Mid-Atlantic, South Atlantic, Caribbean, Gulf, Pacific, and North Pacific Councils.

In March 1994, the *New York Times* reported that even the Magnuson Act could not stop the depletion of fish. In 13 of the 17 fish zones worldwide, fish populations are in steep decline. "For several years [after the passage of the Magnuson Act], times were good," the *Times* narrated. "In the Pacific, fishermen took in $1.5 billion a year for their Alaskan pollock. In the Gulf of Mexico, the number of shrimpers increased significantly, although this led to a large decline in the species of fish, like groupers and red snappers, that are caught accidentally in shrimp nets. And in New England, many fishermen had record years as they went after the bounty that had been claimed by boats from Russia, Germany and Spain." The decline, most experts agree, seems to be the result of overfishing, the explosion of the world's population, and the massive dumping of pollutants into the waterways and oceans.

See also Magnuson, Warren Grant.

Magnuson, Warren Grant (1905–1989)

A congressional leader in the area of fishery protection, Warren Magnuson, as chairman of the Senate Appropriations Committee and the Commerce Committee, was once one of the most powerful politicians in Washington. Born in Moorhead, Minnesota, on 12 April 1905, he was orphaned at an early age and adopted by William and Emma Magnuson, a Swedish couple. Magnuson attended local schools before leaving Moorhead at age 17 and heading west to find his fortune.

Magnuson wandered across the Northwest and worked as a farm assistant to pay for classes at the University of North Dakota and North Dakota State University. He later moved to Washington State, where he enrolled in the University of Washington Law School at Seattle, which awarded him a law degree in 1929. During this time, he became involved in politics by working on the local campaign of Al Smith, the 1928 Democratic presidential candidate. After being admitted to the Washington bar, Magnuson worked for the law firm of Stern and Schermer from 1931 to 1932. In 1933, he was elected to the state legislature, where he served until 1934. During this time, he was also an assistant U.S. district attorney and an attorney for the Washington Emergency Relief Administration, a New Deal agency. In the state legislature, Magnuson sponsored a bill that became the first state unemployment compensation law in the United States. After leaving the legislature, he was elected King County (Seattle) prosecuting attorney, the first Democrat elected to the post since 1900. In 1936, he was elected to a seat in the U.S. House of Representatives, representing the First Washington District. Reelected three times, he served from 1937 until 1944. Among the bills he sponsored was the Bone-Magnuson Cancer Control Act of 1937 (which he co-wrote with Sen. Homer T. Bone [1883–1970] of Washington), which established the National Cancer Institute, the first federally financed cancer research center. During World War II, Magnuson volunteered for duty in the U.S. Naval Reserve as a lieutenant commander and spent about eight months on the aircraft carrier *Enterprise* in the Aleutian Islands and the South Pacific.

In 1944, Magnuson successfully ran against Republican Lt. Col. Harry M. Cain for the seat vacated by Sen. Homer Bone and held that seat for the next 36 years. As chairman of the Commerce Committee, he sponsored several pieces of legislation and became a champion of consumer and health protection. (He authored *The Dark Side of the Marketplace* in 1966 and was the impetus behind the creation of the Consumer Products Safety Commission in 1972.) He wrote the section of the 1964 Civil Rights Act that guarantees equal access to public accommodations. Finally, as chairman of the powerful Appropriations Committee, he helped finance the construction of shipyard projects in his home state. But it was in the area of wildlife conservation and the environment that Magnuson's impact was most felt. It is possible that he was the most successful legislator on conservation issues. The passage of such landmark laws as the Safe Drinking Water Act of 1972, the Toxic Substances Control Act of 1976, and the Magnuson Fishery Management and Conservation Act of 1976 was the result of his sponsorship and leadership. The latter act remains the leading legislation in the area of fishery conservation.

A quintessential liberal, Magnuson easily won reelection several times before the 1980 campaign. In that contest, however, his age was a factor—he was hard of hearing and walked with some difficulty. While he conducted an old-time canvass, his opponent, state Attorney General Slade Gorton, bicycled across the state to show off his youth. Magnuson was defeated for the first time in his career. He retired to his home in Seattle, where he died on 20 May 1989 at the age of 84.

See also Jackson, Henry Martin; Magnuson Fishery Conservation and Management Act.

Marine Game Fish Research Act (73 Stat. 642)

Enacted on 22 September 1959, this legislation authorized the secretary of the interior to "undertake continuing research on the biology fluctuations, status, and statistics of the migratory marine species of game fish of the United States and contiguous waters." The inquiry included studying migrations, identifying the personality of fish stocks, measuring growth rates, calculating mortality rates and variations in survival, examining all environmental influences both natural and artificial (including pollution), and probing the effects of fishing on each individual species.

Marine Mammal Protection Act (16 U.S.C. 1361)

Enacted on 21 October 1972, this law set out to establish policy regarding the declining species of marine mammals. "The Congress finds that certain species and population stocks of marine mammals are, or may be, in danger of extinction or depletion as a result of man's activities," the act concluded. "Such species and population stocks should not be permitted to diminish beyond the point at which they cease to be a significant functioning element in the ecosystem of which they are a part, and consistent with this major objective, they should not be permitted to diminish below their optimum sustainable population. Further measures should be immediately taken to replenish any species or population stock which has already diminished below that population. In particular, efforts should be made to protect the rookeries, mating grounds, and areas of similar significance for each species of marine mammal from the adverse effect of man's actions." The act mandated the establishment of a Marine Mammal Commission, called for consultation between the commission and the secretary of the interior, and decreed the issuance of reports on the condition of marine mammal habitats.

Marine Protection, Research, and Sanctuaries Act (16 U.S.C. 1431)

This legislation, enacted on 23 October 1972, was a groundbreaking effort in conserving marine wildlife sanctuaries. In the act, Congress declared that:

(1) this Nation historically has recognized the importance of protecting special areas of its public domain, but these efforts have been directed almost exclusively to land areas above the high-water mark; (2) certain areas of the marine environment possess conservation, recreational, ecological, historical, research, educational, or esthetic qualities which give them special national significance; (3) while the need to control the effects of particular activities has led to enactment of resource-specific legislation, these laws cannot in all cases provide a coordinated and comprehensive approach to the conservation and management of special areas of the marine environment; (4) a Federal program which identifies special areas of the marine environment will contribute positively to marine resources conservation and management; and (5) such a Federal program will also serve to enhance public awareness, understanding, appreciation, and wise use of the marine environment.

The legislation authorized the secretary of commerce to designate certain areas as marine sanctuaries. To be considered as factors were "the area's natural resource and ecological qualities, including its contribution to biological productivity, maintenance of ecosystem structure, maintenance of ecologically or commercially important or threatened species or species assemblages, and the biogeographic representation of the site," as well as "the socioeconomic factors of sanctuary designation." The act also mandated an environmental impact statement to assess the effects of sanctuary designation on the environment.

Marsh, George Perkins (1801–1882)

George Perkins Marsh's 1864 work, *Man and Nature; or, Physical Geography as Modified by Human Action*, was, for its time, comparable to the earth-shaking publication of Rachel Carson's *Silent Spring* almost a century later. The book spurred the conservation movement in the United States over the next several decades. Marsh was born on 15 March 1801 near the Quechee River in Woodstock, Vermont, the fifth of eight children of Charles Marsh, a local district attorney and U.S. congressman, and Susan (nee Perkins) Marsh. The family can be traced back to Joseph Marsh, the first lieutenant governor of Vermont. George Perkins Marsh supplemented his book education with one based on nature. He later wrote to a friend, "The bubbling brook, the trees, the flowers, the wild animals were to me persons, not things." What formal education he did receive was in the law, as per his father's instructions, but he also retained a deep interest in the history and languages of Europe. He attended local schools and nearby Dartmouth College.

Upon graduating at the head of his class, Marsh began to teach Greek and Latin at a local military academy, but recurring eye problems ended his tutoring career. He turned instead to the law, which he read at home. Upon his admission to the state bar, he opened a law office in Burlington, then the state's largest city. For the next 17 years, he was involved in the legal community and did some occasional writing, including such works as *Icelandic Grammar* in 1838. In *The Camel* (1856), he wrote, "The first command addressed to man by the

Creator ... predicted and prescribed the subjugation of the entire organic and inorganic world to human control and use."

Having served a single term in the Vermont Legislative Council in 1835, Marsh quit his law practice in 1842 and ran as a Whig for the U.S. House of Representatives. He won the election and served from 1843 to 1849 but left little evidence of his tenure, except for his support of the founding of the Smithsonian Institution. He was, however, a loyal Whig, and with the election of Zachary Taylor to the White House in 1848, Marsh was rewarded with an appointment as U.S. minister to Turkey. Marsh was hostile to the Turkish government during the Crimean War, and he was called home after five years.

Upon his return to Vermont in 1854, Marsh found himself near bankruptcy from the collapse of his woolen mill enterprises, and he was forced to teach again to keep himself solvent. He wrote a *Report on the Artificial Propagation of Fish* (1857) and authored two courses of *Lectures on the English Language and English Literature* at Columbia College in 1860. He also wrote *The Origin and History of the English Language, and of the Early Literature It Embodies* (1862), as well as an edition of the *Dictionary of English Etymology* (1862).

In 1861, President Abraham Lincoln named Marsh U.S. minister plenipotentiary to the Kingdom of Italy. Marsh spent the last 21 years of his life in this post, serving the cities of Turin, Florence, and Rome. It was here in 1864 that he wrote his greatest work, *Man and Nature*. According to one biographer, while in New England, "Marsh first noted important relationships between watersheds and their vegetation; forests were valuable because they retained moisture and prevented erosion that could cause flooding and drought. When sheep overgrazed, they exposed deep slopes to drainage from sun and rain." It was this interrelation between how man used his natural resources and how nature responded that inspired Marsh to write *Man and Nature*. It laid down with oratory and facts how man was abusing the environment far beyond what it could handle. Wrote forestry expert Harold K. Steen, "*Man and Nature* painted a sad picture of human excess in Europe, Asia and Africa, where land had been abused and misused for millennia. The evidence was erosion, floods, and abandoned or devalued agricultural land with its attendant hunger, disease, and dislocated populations. Marsh proposed the notion of responsible stewardship—use of the land and its resources but in such a way that it would retain its fecundity. Marsh also predicted the modern concepts—sustained yield, sustainable development, and land ethics."

Published in the United States by Charles Scribner, *Man and Nature* was simultaneously released in London by Sampson Low, Son, and Marston, as well as in Italy by a small publishing house. The Italian edition, however, was replete with errors in translation, and it was ultimately destroyed. A second printing came in 1872. In 1874, Marsh reworked his book into *The Earth as Modified by Human Action*, in which he wrote, "Almost all the processes of agriculture, and of mechanical and chemical industry, are fatally destructive to aquatic animals within reach of [human] influence." This work led the American Association for the Advancement of Science to call for congressional action on forestry protection; the resulting Payson Act, enacted in 1891, authorized the president of the United States to set aside forestry reserves.

George Perkins Marsh spent the rest of his life as the U.S. minister to Italy. He died there on 23 July 1882, aged 81. His biographer, David Lowenthal, said that following his passing, local foresters who knew Marsh came to his residence, wrapped his body in an American flag, covered this makeshift coffin with yellow daisies, and bore his body to a Protestant cemetery in Rome, where he was laid to rest.

Marsh, Othniel Charles (1831–1899)

Known as O. C., Othniel Charles Marsh was one of the leading geologists and paleontologists of the nineteenth century. He was born in Lockport, New York, on 29 October 1831, the eldest son of Caleb and Mary Gaines (nee Peabody) Marsh. The family was descended from one John Marsh who came to Salem, Massachusetts, in 1637. Mary Peabody Marsh was the sister of the philanthropist George Peabody; another relative was George Peabody Wetmore, U.S. senator from Rhode Island.

Othniel Marsh attended local schools in Lockport and the Wilson Collegiate Institute. He graduated from the prestigious Phillips Academy (now Phillips Exeter Academy) in Andover, Massachusetts, in 1856. He then enrolled at Yale University and graduated with honors in 1860. He took up the study of paleontology, geology, and mineralogy at Yale's Sheffield Scientific School for two years, then spent 1862 to 1865 at the Universities of Berlin, Heidelberg, and Breslau in Germany. In 1866, Yale established a chair of paleontology, and Marsh was asked to fill it. He served in this post for the rest of his life. That year, his uncle bequeathed to Yale $150,000, which was used to establish the Peabody Museum.

Building on trips to the western United States he had made as a youth, Marsh explored areas of fossil deposits in Nebraska, New Jersey, and Wyoming in 1868. From this trip came the 1869 reports "Notice of Some New Mosasauroid Reptiles from the Greensand of New Jersey," "Description of a New Gigantic Fossil Serpent (*Dinophis grandis*) from the Tertiary of New Jersey," and "Notice of Some Fossil Birds from the Cretaceous and Tertiary Formations of the United States." In 1870, Marsh founded the Yale Scientific Expedition, which explored Nebraska and Colorado. With this group, Marsh discovered in 1871 the first pterodactyl remains found in the United States. Marsh competed during this period with Edward Drinker Cope and

Joseph Leidy, two eminent scientists in mineralogy and paleontology. His 1880 work, *Extinct Toothed Birds of North America*, is a landmark.

With the death in 1878 of Joseph Henry, president of the National Academy of Sciences, the nation lost one of its pre-eminent scientific minds. According to Wallace Stegner, "Into Henry's empty shoes stepped . . . Othniel Charles Marsh of Yale, one of the greatest of American paleontologists, friend of Huxley and Darwin, contributor in real measure to the documentation of biological evolution, and nephew moreover of the philanthropic banker George Peabody. Marsh had single-handed[ly] run Columbus Delano out of his job as Secretary of the Interior in the 1875 scandal about the cheating of Red Cloud's Sioux. . . . For years he had been engaged in a bitter and rather disgraceful running fight with Professor Edward D. Cope of Pennsylvania in the collection and identification of vertebrate fossils—and Cope was a Hayden man, many of whose scientific papers had appeared in Hayden's reports and bulletins."

In 1880, Marsh authored volume 7 of Clarence King's *Professional Papers of the Engineer Department, U.S. Army*. Titled *Odontornithes: A Monograph on the Extinct Toothed Birds of North America*, it led to his being hired two years later as the vertebrate paleontologist of the U.S. Geological Survey. According to one source, as a member of the survey, his two most important discoveries were "the finding of a very extensive Cretaceous Mammalian fauna in the Laramie Beds of Wyoming" and the unearthing of "those curious horned dinosaurs, the Ceratopsia, in the same deposits." From these finds came his 1884 monograph, *Dinocerata: A Monograph of an Extinct Order of Gigantic Mammals*. His leading role in the formation of paleontological thought led to his election as the president of the National Academy of Sciences, where he served from 1883 to 1895.

Professor Othniel C. Marsh died at his home in New Haven, Connecticut, on 18 March 1899 at the age of 67. He had never married and was the last member of his family. *National Geographic* called Marsh "an eminent contributor to American science." The journal *Science* hailed him "the last of the famous trio of American vertebrate paleontologists" and spoke of the "rich legacy of discovery and advancement in biological knowledge which [he] has bequeathed to the world."

See also Cope, Edward Drinker.

Marsh, Secretary of the Army, et al. v. Oregon Natural Resources Council (490 U.S. 1989, 104 L Ed 2d 377)

In *Marsh*, the Supreme Court held that a federal agency's decision not to update an environmental impact statement with information that a project could damage the environment was not a violation of the National Environmental Policy Act (NEPA) of 1969. In the early 1970s, the Army Corps of Engineers set out to build the Elk Creek Dam, part of a three-dam project in southwestern Oregon, and, as required by NEPA, prepared an environmental impact statement (EIS). In 1980, the Corps of Engineers released its EIS, which stated that although some downwater turbidity might be affected by the dam, the changes were not overwhelming. In 1982, when work began on the dam, several Oregon state agencies alerted the Corps to new tests that showed a tremendous disruption in the fish population downstream. After careful consideration, the Corps decided that this information was not important and asked for appropriations for the project in 1985. At this time, four environmental groups under the umbrella of the Oregon Natural Resources Council filed suit to stop the dam's construction, contending that the Corps had failed to study a worst-case analysis of the dam's impact on the environment. A district court denied the plaintiff relief, claiming that under NEPA, the Corps had no duty to include a supplemental EIS in its deliberations. The Court of Appeals for the Ninth Circuit in San Francisco reversed, stating that the Corps' EIS was defective in not including such a worst-case analysis. John O. Marsh, secretary of the army and the chief of the Corps, took the case to the U.S. Supreme Court. On 1 May 1989, the Court unanimously struck down the court of appeals' judgment. In an opinion written by Justice John Paul Stevens, the Court held that the Corps of Engineers was not being "arbitrary and capricious" in not commissioning a supplemental environmental impact statement.

Marshall, Robert (1901–1939)

Socialist by ideology, called liberal by his staunchest defenders, Robert Marshall is considered one of the greatest forestry, wilderness, and recreational experts who ever lived. Born into a wealthy New York City family on 2 January 1901, he was the son of Louis Marshall—whom one source characterized as "a well-known constitutional lawyer, a Jewish leader and a conservationist noted for his pioneering efforts to defend the Adirondack wilderness"—and Florence (nee Lowenstein) Marshall. "As a boy," Marshall later recollected, "I spent many hours in the heart of New York City, dreaming of Lewis and Clark and their glorious exploration into an unbroken wilderness. Occasionally, my reveries ended in a terrible depression, and I would imagine that I had been born a century too late for genuine excitement." Louis Marshall's ownership of a summer residence on Lower Saranac Lake in the Adirondacks allowed his son to spend the first 21 summers of his life among the woods and wildlife of that region. When only 15, Robert climbed his first mountain in the Adirondacks, Ampersand; during the next several years, with his older brother George, he climbed all but 4 of the range's 46 peaks. Marshall wrote of this experience in *High Peaks of the Adirondacks* (1922).

Robert Marshall, 1935

Robert Marshall attended the Ethical Culture High School in New York before going to Columbia University and finally the New York State College of Forestry at Syracuse, where he received a bachelor's degree in forestry in 1924. He eventually received a master of forestry degree from the Harvard Forest School. In 1925, he joined the U.S. Forest Service and was assigned to the Northern Rocky Mountain Forest Experiment Station in Missoula, Montana, where he worked as a junior forester and assistant silviculturist. At the same time, he began his explorations of the West's wilderness areas. In 1928, he wrote an article for the *Forest Service Bulletin* that called for parts of the Montana wilderness to be set aside as protected territory. After earning his doctorate in plant physiology at Johns Hopkins University, he traveled through the northern Koyukuk region of Alaska. Marshall returned to Washington in 1931 and worked on the landmark report *A National Plan for American Forestry*, also known as the Copeland Report. He wrote a narration of his Alaska trip, which was contained in his *Arctic Village* (1933), and expressed his belief in the need to conserve wilderness areas in *The People's Forests* (1933).

Marshall's best work, however, was the 1930 article he wrote for *Scientific Monthly* magazine. Entitled "The Problem of the Wilderness," it called for the protection of wilderness areas by the federal government. Benton MacKaye, with Marshall one of the founders of the Wilderness Society, called the article the Magna Carta of the wilderness preservation movement. Discussing the reasons for such protection, Marshall wrote:

These steps of reasoning lead up to the conclusion that the preservation of a few samples of undeveloped territory is one of the most clamant issues before us today. Just a few years more of hesitation and the only trace of that wilderness which has exerted such a fundamental influence in molding American character will lie in the musty pages of pioneer books and the mumbled memories of tottering antiquarians. To avoid this catastrophe demands immediate action.... A step in the right direction has alrady been initiated by the National Conference on Outdoor Recreation, which has proposed twenty-one possible wilderness areas. Several of these have already been set aside in a tentative way by the Forest Service; others are undergoing more careful scrutiny. But this only represents the incipiency of what ought to be done.... A thorough study should forthwith be undertaken to determine the probable wilderness needs of the country. Of course, no precise reckoning could be attempted, but a radical calculation would be feasible. It ought to be radical for three reasons: because it is easy to convert a natural area to industrial or motor usage, impossible to do the reverse; because the population which covets wilderness recreation is rapidly enlarging and because the higher standard of living which may be anticipated should give millions the economic power to gratify what is today merely a pathetic yearning. Once the estimate is formulated, immediate steps should be taken to establish enough tracts to insure every one who hungers for it a generous opportunity of enjoying wilderness isolation.... To carry out this program it is exigent that all friends of the wilderness ideal should unite. If they do not present the urgency of their view-point the other side will certainly capture popular support. Then it will only be a few years until the last escape from society will be barricaded. If that day arrives there will be countless souls born to live in strangulation, countless human beings who will be crushed under the artificial edifice raised by man.

There is just one hope of repulsing the tyrannical ambition of civilization to conquer every niche on the whole earth. That hope is the organization of spirited people who will fight for the freedom of the wilderness.

While director of forestry at the Interior Department's Office of Indian Affairs, Marshall met with wilderness advocates Benton MacKaye, Olaus Murie, Ernest Oberholtzer, and others and formed the Wilderness Society in 1935. Wrote Roderick Nash in his *Wilderness and the American Mind*, "Marshall launched it [the Wilderness Society] financially with an anonymous contribution of a thousand dollars—the first of his many gifts climaxed by a bequest of close to $400,000." In 1937, Marshall was named chief of the new Forest Service Division of Recreation and Lands. That same year, Indian Affairs forestry chief John Collier approved Marshall's plan to establish 16 wilderness reserves on Indian reservations.

In his final years, Robert Marshall traveled back to Alaska twice, again visiting the Koyukuk River and Upper Anaktuvik River regions. His strenuous journeys, however, sapped the strength of his heart, which had been weak since birth. His final trip, through Washington's Cascade Range, has been called "punishing" by some. Two months later, while on a train between New York and Washington, Marshall's heart gave out in his sleep. He was 38 years old. In 1941, to honor the man who had single-handedly brought wilderness preservation to the forefront, the Montana wilderness he once worked in was renamed the Bob Marshall Wilderness Area.

See also Copeland Report; MacKaye, Benton; Murie, Olaus Johan; Wilderness Society.

Mason, David Townsend (1883–1973)

Wrote Carl A. Newport on David T. Mason, "[He] was a persistent advocate of sustained-yield management in American forestry and in doing so influenced or actually wrote much of the legislation that brought sustained-yield into widespread practice." Born in Newark, New Jersey, on 11 March 1883, Mason was the son of William Mason, publisher of the *Newark Chronicle*, and Rachel (nee Townsend) Mason. He grew up in Bound Brook, a small community in the northeastern corner of New Jersey. Mason attended local schools and worked on his father's paper. He received a bachelor's degree in civil engineering from Rutgers University in 1905 and a forestry degree from Yale's Forestry School in 1907.

After graduating from Yale, Marshall went to work for the U.S. Forest Service in Missoula, Montana. In 1912, according to author Elmo Richardson, "He succeeded Robert Y. Stuart as District One's assistant district forester." In 1915, he took a position as professor of forestry at the University of California School of Forestry. With the outbreak of World War I, Mason went to France as a member of the Tenth Engineers (Forestry) Regiment, attaining the rank of major.

After the war, Mason worked for the U.S. government as part of the Timber Valuation Section of the Bureau of Internal Revenue (now the Internal Revenue Service) before entering private business. He was a member of the Timber Conservation Board and the Lumber Code Authority, part of the New Deal's National Recovery Act (NRA). For the last 35 years of his life, according to one source, "he was chairman of the advisory committee to the Bureau of Land Management for Oregon and California Revested Lands, also a member of the Research Advisory Committee of the Pacific Northwest Forest and Range Experiment Station."

Wrote Robert Ficken about Mason: "He is an important figure in forest history because of his work on behalf of sustained yield. Of equal importance, he was one of the leading participants in the transformation of the forestry profession, as foresters moved from government and the universities into private industry." Mason was honored with the David T. Mason Professorship of Forest Land Use at the Yale University Forestry School. He died on 3 September 1973 at the age of 90.

Mather, Stephen Tyng (1867–1930)

Stephen T. Mather was the founder and first director of the National Park Service. Born in San Francisco on 4 July 1867, he was the son of Joseph Wakeman Mather and Bertha Jemima (nee Walker) Mather. He was a descendant of Rev. Richard Mather, a noted English Puritan clergyman who immigrated to the colonies in 1635. Stephen Mather attended local schools, then graduated from the University of California in 1887. He joined the *New York Sun* as a cub reporter for a five-year stint. In 1893, he took a job under his father in the New York office of the Pacific Coast Borax Company. In 1894, he went to Chicago and opened a Pacific Coast branch there. Desiring to own his own company, he founded the Thorkildsen-Mather Borax Company of Chicago in 1903, competing head-to-head with his former employer.

It was a love of the outdoors and the wilderness that attracted Mather to the Sierra Club as early as 1905, and he earned the nickname "the eternal freshman" because of his devotion to the many causes he later championed. He camped and hiked through many of the nation's parks and soon became disenchanted with the way they were being managed. He wrote a testy letter to Interior Secretary Franklin K. Lane in 1914, informing Lane that the parks were run down and mismanaged. Lane wrote back, "Steve, if you don't like the way the national parks are being run, come on down to Washington and run them yourself." Enticed by the offer, Mather gave up his lucrative job at Thorkildsen-Mather and took up Lane's challenge; as

soon as he arrived in Washington, he was named assistant secretary of the interior. He lobbied Congress to create an agency separate from the Interior Department bureaucracy that would oversee all the nation's parks. On 25 August 1916, mainly because of Mather's urging, Congress enacted the law that created the National Park Service (NPS). In 1917, when the service went into operation, Mather was named its first director; Horace M. Albright, a 27-year-old graduate of the University of California, became his assistant.

As NPS director, Mather became responsible for the 14 existing national parks, which were in varying states of disrepair. Using a three-step plan, he decreed that the parks were to be maintained in the most attractive setting possible, that they be used only for the enjoyment and entertainment of people seeking the pleasure of national parks, and that decisions involving national parks lands and their resources be made in the national interest, not for a single interest. To accomplish these aims, he had to restore the parks to their natural state. By lobbying Congress for increased appropriations, he was able to fund extensive cleanup projects. He fought some interests that wanted to turn Yellowstone National Park into an irrigation area and aided in the acquisition of new parks, such as Zion in Utah and Shenandoah in Virginia. He further lobbied state governments to preserve state parks. A conference called by him in Des Moines, Iowa, in 1921 led to the creation of the National Conference on State Parks.

Failing health forced Mather's retirement in 1929 after he had suffered a paralytic stroke the previous November. Although by June 1929 he had recovered sufficiently to leave the hospital and travel, he was rehospitalized in late 1929 for tests. On 22 January 1930, he suffered a second and fatal stroke. Mather was 62 years old.

See also Albright, Horace Marden; National Park Service.

Maxwell, George Hebard
(1860–1946)

Although born in California, George H. Maxwell is more closely associated with Arizona, where he was a leader in the drive to reclaim and irrigate the Colorado River. Hailed by one biographer as "Reclamation's Militant Evangelist," he is perhaps one of the United States' greatest conservationists. Maxwell was born on his family's ranch in Sonoma, California, on 3 June 1860, the son of John Morgan Maxwell, a land and gold speculator, and Clara (nee Hebard) Maxwell. John Morgan Maxwell, whom his son would write about in *The Argonauts of Golden California* (1934?), was, according to some sources, "a gold-miner and Forty-Niner" who "reached California via New Orleans and Panama." John Morgan Maxwell died when his son when 14, which left his widow and three children to subsist by working the family ranch.

George Hebard Maxwell

George Maxwell attended public schools in Sonoma and San Francisco and later Saint Matthew's Academy in San Mateo, California, where he studied law. At this time, he formulated a system of shorthand for use in courtrooms. At age 19, because of this talent, he was employed as a court stenographer for the circuit court and superior court in California. According to a contemporary, "That was the day before typewriters ... so Maxwell would read his notes to eight writers at a time, and the lawyers would get copies the next morning, something then unheard of."

Between 1880 and 1882, Maxwell had to travel to Tombstone, Arizona, to work on an irrigation case. There he met and became a good friend of Sen. William Morris Stewart of Nevada, who instilled in Maxwell an interest in land and water reclamation and irrigation. Upon Stewart's urging, Maxwell went back to law school to major in water and land law, graduating in 1882. That year, he opened a law firm, Messick and Maxwell, which was in business until 1899.

One of Maxwell's first cases was in defense of farmers' water rights under California's Wright Irrigation Act. After a defeat in the courts, Maxwell spent eight years traveling the state, educating people about irrigation law. After his tour, the case was reopened, and Maxwell won. He then quit the law, went to Washington, and became friends with men such as Sen. Henry Clay Hansbrough of North Dakota and Rep. Francis G. Newlands of Nevada. In 1896, he moved to Phoenix, Arizona, and attended the Fifth National Irrigation Congress as a delegate. According to one source, Maxwell "advocated systematic measures to promote public awareness of the organization's activities. He urged the Congress to establish a press committee, to enlist support from the business community, and also a committee on legislation. In debate of issues, he sided with advocates of federal projects against projects promoted by individual states or private organizations." From this conference, wrote one biographer, "Maxwell emerged into national prominence and became a central figure in the national irrigation movement."

In 1899, because the Irrigation Congress failed to heed his advice to lobby Congress, Maxwell established the National Reclamation Association, which set out to get Congress to pass a sweeping irrigation law. He became an intimate of such politicians as Stewart, Newlands, Frederick H. Newell, and Gifford Pinchot. As director of the National Reclamation Assocoiation, Maxwell had to contend with property rights in Arizona. Battling the forces of territorial Gov. Nathan O. Murphy, who advocated state and local, not federal, control of reclamation and irrigation projects, Maxwell worked hand in hand with such Arizonans as William "Bucky" O'Neill (who later went up San Juan Hill with Teddy Roosevelt as a Rough Rider). Maxwell convinced the Murphy backers that the anticipated Newlands bill, enacted in 1902 as the Newlands Reclamation Act, would preserve state and local rights. He won his argument, and by the time the act passed, he had secured Murphy's endorsement of the program.

The Newlands Act had made no provision for subsidizing lands in private hands. It was due to Maxwell's efforts that the law added the words "and lands in private ownership" to aid the farmers of the Salt River Valley in Arizona, for whom Maxwell had fought so long and hard.

For the rest of his life, Maxwell spoke out in favor of the reclamation and irrigation of the West. One source wrote, "At the first Conservation Congress held in Washington, D.C., in 1905, Maxwell launched a public attack against land grabbers and antiquated land laws that permitted them to despoil the public domain. His last great achievement was the starting in 1930 of the Muskingum Conservancy District in Ohio, a precursor of the Tennessee Valley Authority–type of

land and water administration." As an advocate of "homecrofting" (a term Maxwell coined to symbolize clean and efficient home life away from the crowded cities), he published *Maxwell's Talisman—The Homecroft Magazine*. He also authored *The First Book of Homecrofters; Our National Defense: The Patriotism of Peace* (1915), in which he stated his opposition to World War I; *Golden Rivers and Treasure Vallies: Wasted Wealth from Wasted Waters* (1929); and *The End of Employment: A Balance Wheel of Industry, the Nation's Greatest Asset* (1940). Forecasting in the 1890s that the United States would one day fight a ruinous Pacific war with Japan, Maxwell was truly a visionary. He died in Phoenix, Arizona, on 1 December 1946 at the age of 86. After cremation, his ashes were interred in the Sonoma Valley in California.

See also Hansbrough, Henry Clay; National Irrigation Association; National Irrigation Congress; Newlands Irrigation (or Reclamation) Act.

Merriam, Clinton Hart (1855–1942)

Zoologist and naturalist Clinton Hart Merriam was the first head of the Bureau of Biological Survey, the forerunner of the U.S. Fish and Wildlife Service. He was born in New York City on 5 December 1855, the son of Clinton Levi Merriam and Caroline (nee Hart) Merriam. Clinton Levi Merriam, a prosperous merchant and a Republican, served two terms in the U.S. House of Representatives. Clinton Hart Merriam's sister, Florence Augusta Merriam, became a renowned writer and ornithologist. The family moved to Locust Grove, the Merriam estate in Lewis County, New York, sometime after Clinton Merriam was born.

During his childhood at Locust Grove, Merriam learned about nature. Said one source, "His interest in natural history began early in life, and it broadened in scope and matured in character as time went on." Taught by tutors, he later attended the Pingry Military School in

Clinton Hart Merriam

Elizabeth, New Jersey, and the Williston Academy in Easthampton, Massachusetts. In 1872, at the age of 16, he and his talents in ornithology came to the attention of Spencer Fullerton Baird, assistant secretary of the Smithsonian Institution and head of the U.S. Commission on Fish and Fisheries. Baird was sufficiently impressed by Merriam that he attached the teenager to the Hayden survey of the western United States, where Merriam observed and collected many specimens of birds and other wildlife. He was the author of "Report on the Mammals and Birds of the Expedition," which appeared in the Sixth Annual Report of the Hayden survey. In 1874, he entered the Sheffield Scientific School at Yale University, from which he graduated three years later. During this time, he wrote *Review of the Birds of Connecticut* (1877), considered one of the best ornithological publications of its time. While at Sheffield, Merriam became interested in medi-

cine; he enrolled in the College of Physicians and Surgeons at Columbia University in New York and received a medical degree. After graduation, he practiced medicine at Locust Grove for several years while keeping active in the field of natural history. In 1878, he was a founding member and first president of the Linnaean Society in New York. In 1883, he served aboard the sealing ship *Proteus* while conducting seal studies in Greenland and Labrador. That same year, he was a founding member of the American Ornithologists' Union, later serving (1900–1902) as the group's president.

On 3 March 1885, Congress enacted the Agricultural Appropriations Act (23 Stat. 354), which provided funding for a section of "economic ornithology" in the Division of Entomology under the direction of the U.S. Department of Agriculture. Merriam was chosen to head this new department. In 1886, it became a separate entity when it was named the Division of Ornithology and Mammalogy. Said one source, "The agency's early emphasis was the scientific study of bird distribution and food habits in relation to agriculture." The agency's jurisdiction was expanded when Merriam undertook a number of surveys "to obtain data on life zones, distribution of animal and plant life, laws of temperature control, and [the] geographic distribution of life." While in Death Valley, California, in 1891, President Benjamin Harrison named Merriam to the U.S. Bering Sea Commission, which was established to study the plight of the fur seals in Alaska.

Under Merriam's leadership, the agency, which became the Division of Biological Survey in 1896 and the Bureau of Biological Survey in 1905, undertook the responsibility of enforcing the Lacey Act of 1900, which protected wild birds, as well as the Alaska Game Act of 1902, the 1908 Act for the Protection of Game in Alaska, the 1916 Migratory Bird Treaty Act, the Migratory Bird Conservation Act of 1929, the 1934 Migratory Bird

Hunting Stamp Act, and the 1937 Federal Aid in Wildlife Restoration Act.

Clinton Merriam retired from the Bureau of Biological Survey in 1910. Afterward, as a member of the Smithsonian Institution, he devoted much of his time to studying the linguistics of Indian tribes in California and Nevada. From 1917 until 1925, he was chairman of the U.S. Geographic Board. Among his numerous works were *The Mammals of the Adirondack Region* (1887), *Results of a Biological Survey of the San Francisco Mountain Region and Desert of the Little Colorado, Arizona* (1890), and *Review of the Grizzly and Big Brown Bears of North America* (1918). Following his wife's death in 1939, Merriam went to live with a daughter in California, where he died on 19 March 1942 at the age of 86.

Migratory Bird Conservation Act (45 Stat. 1222)

This legislation was enacted by Congress on 18 February 1929 to create the Migratory Bird Conservation Commission and establish a continuing refuge program for migratory birds. Before passage of this act, each separate wildlife refuge had to be created through an executive order by the president; this act allowed such refuges to be created through a refuge program. The original version of this legislation was the New-Anthony bill, named after Sen. Harry Stewart New (1858–1937) of Indiana and Rep. Daniel Read Anthony, Jr. (1870–1931), of Kansas. It was introduced in 1921 and backed by several conservation groups, including the Boone and Crockett Club. However, due to the machinations of western legislators led by Rep. Frank Mondell of Wyoming, the bill was defeated. The 1929 act, better known as the Norbeck-Andresen Act, was sponsored in the House of Representatives by August Herman Andresen (1890–1958) of Minnesota and in the Senate by Peter Norbeck (1870–1936) of South Dakota.

See also Mondell, Frank Wheeler.

Migratory Bird Hunting Stamp Act (48 Stat. 451)

This law, enacted by Congress on 16 March 1934, was the first to allow the use of a government fee—through the sale of "duck stamps" to hunters—to establish and maintain wildlife refuges. Officially called "an Act to supplement and support the Migratory Bird Conservation Act by providing funds for the acquisition of areas for use as migratory-bird sanctuaries, refuges, and breeding grounds, for developing and administering such areas, for the protection of certain migratory birds, for the enforcement of the Migratory Bird Treaty Act and regulations thereunder, and for other purposes," the act was supported most notably by cartoonist and conservationist Jay "Ding" Darling, who drew the first duck stamp.

See also Darling, Jay Norwood.

Migratory Bird Treaty Act (39 Stat. 2:1702)

This treaty between the United States and Great Britain in effect made international law of the Weeks-McLean Act of 1913, which sought governmental protection for wild migratory birds. The treaty was signed in Washington on 16 August 1916 by U.S. Secretary of State Robert Lansing and British Foreign Minister Cecil Spring Rice; it was ratified by the Senate on 29 August and signed by President Woodrow Wilson on 1 September. On 3 July 1918, amendments were added to toughen the act on poachers. This "enabling act" was the source of the controversy in the Supreme Court case *Missouri v. Holland*.

See also Missouri v. Holland; Weeks-McLean Act.

Mineral Leasing Act (41 Stat. 437)

Called "an Act to promote the mining of coal, phosphate, oil, oil shale, gas, and sodium on the public domain," this law was enacted by Congress on 25 February 1920 as a way to lease minerals found on public lands on a royalty basis; that is, as the minerals were found, the government would be paid for them. Sponsored by Sen. Irvine L. Lenroot of Wisconsin, this legislation amended the General Mining Law of 1872 by removing certain minerals from its dictates. These were important steps; previously, lands were leased at a flat rate, and anything found on them belonged to the lessee. The law, as one source noted, "removed oil, gas, sodium, sulphur, potassium, coal, phosphate, oil shale and certain other minerals from the 1872 act and set forth procedures for leasing—not selling or permanently disposing of—the rights to explore for and develop these minerals."

Mineral Patent Law of 1866
See Lode Mining Law.

Mining and Minerals Policy Act (Public Law 91-631)
Enacted by Congress on 31 December 1970, this law established a national mining and minerals policy. The act reads: "The Congress declares that it is the continuing policy of the Federal Government in the national interest to foster and ecourage private enterprise in (1) the development of economically sound and stable domestic mining, metal and minerals reclamation industries, (2) the orderly and economic development of domestic mineral resources, reserves, and reclamation of metals and minerals to help assure satisfaction of industrial, security, and environmental needs, (3) mining, mineral, and metallurgical research, including the use and recycling of scrap to promote the safe and efficient use of our natural and reclaimable mineral resources, and (4) the study and development of methods for the disposal, control, and reclamation of mineral waste products, and the reclamation on mined land, so as to lessen any adverse impact of mineral extraction and processing upon the physical environment that may result from mining or mineral activities."

Mississippi Flood Control Act (45 Stat. 534–539)
This legislation was enacted on 15 May 1928 to back up the Flood Control Act of 1917 in the aftermath of huge floods in 1927. Monies appropriated under the act were used to build a system of levees and reservoirs that seemed to tame the mighty Mississippi and prevent flooding of towns and agricultural areas. Although this system worked for a long time, massive floods along the Mississippi in 1993 led some to question whether the levees had helped save some areas from flooding or contributed to worse flooding in others.

Mississippi Valley Committee
This group of engineers and environmentalists banded together in 1934 to call for the development of "flood control, low-water control, navigation, power, water supply, sanitation, and erosion" programs in the Mississippi Valley. The committee, funded by a Public Works Administration grant from the Department of the Interior, submitted its two-volume report to Congress, but its recommendations were ignored.

Missouri v. Holland (252 U.S. 416 [1919])
This landmark Supreme Court case arose out of the question of the constitutionality of the Migratory Bird Treaty of 1916 and its enabling act of 1918. The state of Missouri claimed that the treaty infringed on its Tenth Amendment rights as well as the right to interstate commerce. The state brought suit against Ray P. Holland, a U.S. game warden, to enjoin him from enforcing the treaty. A lower court found for Holland and the regulations; the state of Missouri appealed to the U.S. Supreme Court. The Court held 7–2 (Justices Willis Van Devanter and Mahlon Pitney dissenting) that such regulations did not impinge on any of the states' Tenth Amendment

rights. With emotion and conviction, Justice Oliver Wendell Holmes wrote in the majority opinion:

It is said that a treaty cannot be valid if it infringes [upon] the Constitution; that there are limits, therefore, to the treaty-making power; and that one such limit is that what an act of Congress could not do unaided, in derogation of the powers reserved to the states, a treaty cannot do. An earlier act of Congress that attempted by itself, and not in pursuance of a treaty, to regulate the killing of migratory birds within the states, had been ruled out in the District Court (*United States v. Schauver*, 214 Fed. 154; *United States v. McCullagh*, 221 Fed. 288). Those decisions were supported by arguments that migratory birds were owned by the states in their sovereign capacity, for the benefit of the people, and under cases like *Geer v. Connecticut* . . . this control was one that Congress had no power to displace. . . . Here a national interest of very nearly the first magnitude is involved. It can be protected only by national action in concert with that of another power. The subject-matter is only transitorily within the state, and has no permanent habitat therein. But for the treaty and the statute, there soon might be no birds for any powers to deal with. We see nothing in the Constitution that compels the government to sit by while a food supply is cut off and the protectors of our forests and of our crops is destroyed. It is not sufficient to rely upon the states. The reliance is vain, and were it otherwise, the question is whether the United States is forbidden to act. We are of the opinion that the treaty and statute must be upheld.

See also Migratory Bird Treaty Act.

Mitigation Banking Policy

A relatively new policy being used in some states, mitigation banking requires that a developer desiring to build on an area deemed to be wetlands must restore damaged wetlands elsewhere as part of the cost of doing business. According to one source, "damaged wetlands to be made good would be designated as banks. Developers would buy credits in a bank—a large distressed wetland—to make up for the wetlands acreage they would develop. Those put in charge of the designated wetland would then use the developers' money for repairs."

See also Wetlands.

Mondell, Frank Wheeler (1860–1939)

An advocate for irrigation but a foe of conservation, Frank Mondell's name is one the lesser known among the congressmen who spoke out on the conservation issue. He was born in St. Louis, Missouri, on 6 November 1860, the son of Ephraim Mondell, a stable and hotel proprietor, and Nancy (nee Brown) Mondell. Frank Mondell was orphaned at age five and was brought up on the Iowa farm of a Congregationalist family named Upton. He received a limited rural education. Later he studied engineering and the law before moving to Chicago and then Denver, Colorado, in search of his fortune.

In 1887, Mondell was hired by the Chicago, Burlington & Quincy Railroad to travel to Wyoming and search out coal deposits along a proposed rail route. His exploration in Weston County led to the discovery of the Cambria coal mine, and he was told by the railroad to lay out a city that later became Newcastle, named after the English coal city. Mondell was the town's first mayor and established a water and sewage system there. In 1890, he was elected to the Wyoming state senate in its first session and served as senate president in the second session of 1892. Two years later, Mondell was elected to the U.S. House of Representatives for a single term; he was defeated for reelection because of his opposition to the free issuance of silver. He served two years (1897–1899) as assistant commissioner of the General Land Office. In 1898, the issue of free silver abated, and he was reelected to his old House seat, serving until 1923. Because of the state's small population, for those 25 years, Mondell, a Republican, was Wyoming's sole representative in the House.

Mondell used his experience in the General Land Office to advocate a national policy for the reclamation and irrigation of the western states, and he worked for the enactment of the Newlands Act in 1902. His zest for conservation stopped there, however. He was a bitter opponent of an act to establish national policy on the creation of wildlife refuges for birds. The New-Anthony bill, introduced in 1921 by Sen. Harry Stewart New (1858–1937) of Indiana and Rep. Daniel Read Anthony, Jr. (1870–1931), of Kansas, was endorsed by such groups as the Boone and Crockett Club, but Mondell and other western congressmen fought it until it was defeated. Such legislation was eventually passed in 1929 as the Migratory Bird Conservation Act. Wrote biographer John Garraty, "Mondell believed in the rapid expansion of the West even at the expense of the preservation of natural resources. Theodore Roosevelt's statement that Mondell 'took the lead in every measure to prevent the conservation of our natural resources' was perhaps an exaggeration, but Mondell was a consistent and vocal foe of the Forest Service, which he accused of withholding vast areas of nonforest land from the general public. He also opposed the reserving of coal lands on the grounds that this tended to cause a coal shortage in the West." In other areas, Mondell was a key ally of women's suffrage, and he introduced legislation in 1905 to achieve that aim. He was also responsible for helping to get funding for the construction of the Panama Canal. As Republican floor leader, he supported major post–World War I legislation, including reconstruction appropriations.

In 1922, Mondell ran for the U.S. Senate but was defeated by John B. Kendrick. Out of office for the first time in a quarter of a century, Mondell turned down offers of the ambassadorship to Japan and the governorship of Puerto Rico, but he accepted a position as director of the War Finance Corporation, a post he held for two years. In 1924, after being admitted to the bar, he entered into private law practice. His last political work was as permanent chairman of the 1924 Republican National Convention. The author of the autobiographical *My Story*, Mondell suffered from leukemia and died at his Washington, D.C., home on 6 August 1939 at the age of 78.

See also Migratory Bird Conservation Act.

Morton, Rogers Clark Ballard (1914–1979)

At one time chairman of the Republican National Committee, Rogers C. B. Morton also served as secretary of the interior during the Nixon administration. He was born on 19 September 1914 in Louisville, Kentucky, the son of Dr. David Cummins Morton and Mary Harris (nee Ballard) Morton, both members of socially prominent Republican families. In fact, Rogers Morton was a seventh-generation Kentuckian, descended from the explorer George Rogers Clark. His older brother, Thruston Ballard Morton, served in the U.S. Senate from 1957 until 1968. Rogers Morton, who grew up on his family's estate on the Wye River on Maryland's eastern shore, attended the prestigious Woodberry Forest preparatory school near Orange, Virginia, before enrolling at Yale University. He graduated with a bachelor's degree in 1937.

After attending the College of Physicians and Surgeons at Columbia University for a short time, Morton returned home to Kentucky to manage his mother's family's milling firm, Ballard & Ballard, maker of Ballard biscuits. He ran the company until 1951, when it merged with Pillsbury. When the United States entered World War II, he volunteered for duty in the U.S. Army as a private in the field artillery division. His service in Europe lasted until 1945, and he rose to the rank of major.

Rogers Morton entered politics in the early 1940s when he managed his brother's political campaigns. After he sold the family business to Pillsbury, he and his family moved permanently to Maryland, where he purchased a 1,000-acre farm. Even though Maryland was heavily Democratic, Morton decided to manage Republican Edward T. Miller's ultimately unsuccessful 1960 campaign for the U.S. House of Representatives. Two year later, however, Morton ran in his own district and was elected. He eventually served five terms (1963–1970). A member of the Ways and Means Committee and the Interior and Insular Affairs Committee, Morton was noted for his conservative votes, although in 1964 he voted against his party and for the Civil Rights Act. After Barry Goldwater's crushing defeat in the 1964 presidential election, Morton, in a speech called "Where the Votes Are," scolded Republicans for "writing off the Negro, writing off labor, writing off young people, writing off ethnic groups." In 1968, he served as both convention manager and campaign manager for former Vice President Richard Nixon in a winning cause.

On 26 February 1969, President Nixon appointed Morton to succeed Ray C. Bliss as chairman of the Republican National Committee. (His brother Thruston had held the same position during the late 1950s and early 1960s.) Less than two years later, on 25 November 1970, Secretary of the Interior Walter J. Hickel was forced to resign, and Morton was named in his place. "The Ouster was Abrupt," said the *New York Times*. Morton would serve nearly five years in

this post in both the Nixon and Ford administrations. Although at first there was great apprehension among environmentalists as to his record, Morton immediately put off a decision on the trans-Alaska pipeline, which was eventually built. Although his was not considered a noteworthy tenure, he did succeed in stabilizing the department after Hickel's reign. On 30 April 1975, he resigned from the Interior Department to take over the post of commerce secretary, replacing Frederick B. Dent. Morton held that position for only a short time before he was replaced by Elliot L. Richardson. When Gerald Ford left office in January 1977, Morton retired to his estate. At the time, he was dying of cancer. He succumbed two years later on 19 April 1979 at the age of 64.

See also Department of the Interior.

Muir, John (1838–1914)

Perhaps the greatest of the preservationists, John Muir was described by one biographical sketch as "the man who saved Yosemite, the California Redwoods, the Grand Canyon, and the Painted Desert." He was born the eldest son of Scottish parents in Dunbar, a lowland port on the Forth River east of Edinburgh, on 21 April 1838. His parents, Daniel Muir and Anne (nee Gilrye) Muir, were Scottish Highlanders who had married in 1833. Daniel Muir was a devoutly religious man who expected complete obedience from his family.

It was at an early age that John Muir began to explore his surroundings, and the Forth basin and Scottish Lowlands provided him ample opportunity to acquire an overwhelming passion for nature. In grammar school, he was instructed in Latin and French, but the lessons in natural history, particularly biographical treatments of John James Audubon and other naturalists, instilled in Muir a greater love of the natural world. In 1849, when Daniel Muir broke with the Presbyterian Church, he and his

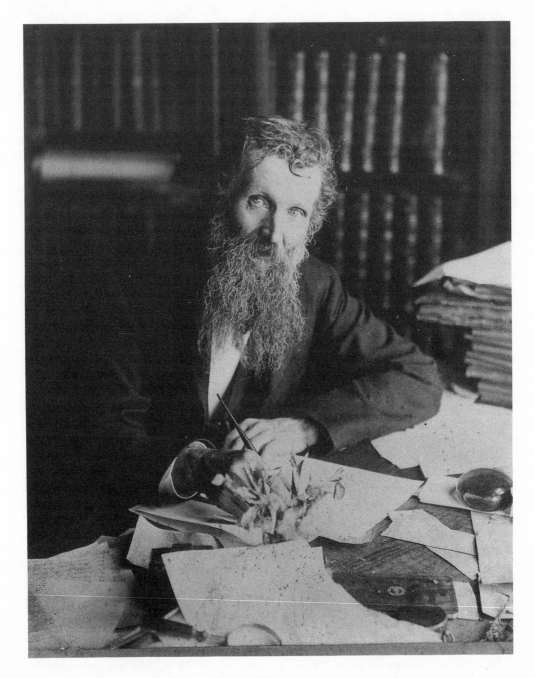

John Muir at his home in Martinez, California, about 1897

children David, John, and Sarah immigrated to the United States to find a home for the rest of the family.

On the family farm in Wisconsin, John Muir began to rebel against his father's religious "fanaticism," as one biographer called it, and with this rebellion came the dream to leave his family and pursue an education in nature. At 22, Muir left the farm and attended the University of Wisconsin. When his pacifism collided with the onset of the Civil War, he fled to Canada. Thus began his long journeys to find nature's bounty. In 1867, while employed at a wagon factory in Indianapolis, he was stabbed with a metal file and suffered temporary blindness. He spent the next three months fearing that he had lost the ability to see "God's beauty." When his eyesight returned, he wrote, "I bade adieu to all my mechanical inventions, determined to devote the rest of my life to the study of the inventions of God." He set out on "a grand sabbath day three years long." The journal he kept of his pilgrimage, in which he recorded observations of the places, flora, and animals he saw, was published posthumously as *A Thousand Mile Walk to the Gulf* (1916). After reaching the Texas side of the Gulf of Mexico, he headed west, arriving in California in 1868. Almost immediately he discovered the Yosemite Valley. He wrote: "June 6 1869 . . . We are now in the mountains, and they are in us, kindling enthusiasm, making every nerve quiver." After marrying the daughter of a Polish expatriate, Muir attended to his father-in-law's botanical garden and became one of the leading botanical experts in the nation.

Yet it was Yosemite that dominated his life. Almost from the time of his discovery of the valley, he lobbied Congress to protect it and give it national park status. Among the people to whom Muir gave personal tours of the Yosemite Valley was Robert Underwood Johnson, editor of *Century* magazine, who took up Muir's cause in his periodical. Muir wrote a series of articles in *Century* that gave his views on the subject. Their work culminated in the creation of Yosemite National Park in 1890.

Muir was a new type of explorer—a naturalist, preservationist, and self-appointed protector of the environment's limited resources. He called the forests "God's First Temples" and discussed the situation of the state's forests in a letter to the editors of the *Sacramento Record-Union*: "The forests of coniferous trees growing on our mountain ranges are by far the most destructible of the resources of California. Our gold, and silver, and

cinnabar are stored in the rocks, locked up in the safest of all banks, so that notwithstanding the world has been making a run upon them for the last twenty-five years, they still pay out steadily, and will continue to do so centuries hence, like rivers pouring from perennial mountain fountains." In 1892, he and several other California naturalists met and established the Sierra Club, which, author Holway R. Jones wrote, "has devoted itself to the study and protection of national scenic resources, particularly those of mountain regions."

A prolific author, Muir edited *Picturesque California and the Regions West of the Rocky Mountains from Alaska to Mexico* (1888) and wrote *The Mountains of California* (1894), *Our National Parks* (1901), *Stickeen* (1909), *My First Summer in the Sierras* (1911), and *The Yosemite* (1912). Later works that were published posthumously included *Travels in Alaska* (1915), *The Cruise of the Corwin* (1917), and *Steep Trails* (1918).

However respected and admired he was, Muir was not always successful in lobbying for the preservation of Yosemite. Even after inviting President Theodore Roosevelt for a tour of the park in 1903, Muir had to fight the forces of water power who wanted to dam the Hetch-Hetchy Valley. For the next several years, Muir battled California congressmen and the forces of water power, but President Woodrow Wilson signed the Hetch-Hetchy bill into law in 1913. This act killed the spirit of this Scottish immigrant who had spent his life in defense of nature. The shock of losing his beloved Hetch-Hetchy never subsided. Muir caught pneumonia and died on 24 December 1914 at the age of 76.

John Muir is considered one of the greatest naturalists to have ever lived. Botanist Charles Sprague Sargent dedicated the eleventh volume of his *Silva* to Muir, a volume on Muir's beloved conifers. Wrote author Robert L. Gale, "[Muir] adored nature in all of its wild and gentle aspects, observed it as God's living and eternally beautiful system, and wrote about it in fresh, delightful, and informative prose."

See also Hetch-Hetchy Valley Controversy; Johnson, Robert Underwood; Roosevelt, Theodore; Sierra Club.

Multiple Use

This action by government is defined as "the management of all renewable resources of the forests so that they are used in the combination that will best meet the needs of the American people. It provides for the judicious use of the several land resources with adjustments and coordinated management to conform with changing needs and conditions."

See also Multiple Use–Sustained Yield Act; Sustained Yield.

Multiple Use–Sustained Yield Act (74 Stat. 215)

This law, also known as the Multiple Use Act, was enacted by Congress on 12 June 1960 for the purpose of adding the recreational use of national forests to the list of allowable uses, which had previously been limited to such acts as timber collection and protection, watershed protection, grazing, and wildlife and fish protection. The Multiple Use Act recognized that national forests could be utilized by the public as outdoor recreation areas if such areas could be maintained as wildlife and nature refuges.

See also Classification and Multiple Use Act.

Murie, Olaus Johan (1889–1963)

The *Washington Post* called Olaus Murie "Mr. Wilderness." A longtime wilderness advocate, his posthumously published *Wapiti Wilderness* (1966) marked the end of a long life dedicated to the preservation of the natural world. Born in Moorhead, Nebraska, on 1 March 1889, Murie was the son of Norwegian immigrants. His father died when Olaus was a boy, leaving his mother to care for him and

his brother Adolph. Murie's boyhood in the natural surroundings of Minnesota (Moorhead lies on the Red River) led him to take an early interest in the environment and conservation. This interest led him to study biology at Fargo College in North Dakota and later at Pacific University in Oregon, which awarded him a bacheor's degree in that subject in 1912.

Within two years, Murie was able to fulfill his dream of studying the wild when he was hired as a field mammal curator for a Carnegie Museum of Natural History expedition to Canada. After a short time as a balloonist and observer in World War I, he joined the U.S. Biological Survey, which sent him to northern Canada and Alaska to study the wildlife there. Accompanied by his brother Adolph, Murie examined a multitude of wildlife, including waterfowl, bears, and elk, becoming an expert in the latter. A noted biologist in his own right, Adolph later wrote of his brother in *A Naturalist in Alaska:* "He was familiar with the north country and was a self-sufficient traveler. These were the blessed days before the advent of the airplane in the north. In the winter he journeyed alone by dog team, in summer he hiked cross-country, packing his dogs and living on blueberries, ptarmigan, and what other meat he could secure for himself and [his] dogs." While in Alaska in 1924, Olaus met Margaret ("Mardy") Thomas, the first female graduate of the University of Alaska. The two were married that year and became a team in their wilderness studies.

Murie's fame rests on his writings about the delicate balance of the wild and man's intrusion on it. Wrote Crandall Bay, "His early writings for scientific and popular journals pointed out the links between wildlife and wildland habitats and what he saw as growing threats to this way of life from a quickly modernizing world."

In 1927, the U.S. Biological Survey sent Murie to Jackson Hole, Wyoming, to discover the reason for the drop in the elk

Margaret and Olaus Murie at a 1953 Wilderness Society meeting in Montana

population there. He ascertained that because their natural predators had been killed off and unnatural animals had been introduced into the environment, the elk were forced to stray from their normal feeding patterns. The bushes they were foraging off were ripping their mouths (a disease called sore-mouth), causing fatal infections. By this time, however, Murie was in full disagreement with the survey's predator-control policy. Wrote historian Stephen Fox, "Taking its cue from farming and ranching interests, the Survey regarded wolves, coyotes, and even some rodents as harmful species to be eradicated by mass poisoning and bounty campaigns." Murie's anger was expressed sufficiently when he exclaimed, "I think we should go beyond proving the rights of animals to live in utilitarian terms. Why don't we just admit we like having them around? Isn't that answer enough?" An intimate of Benton Mac-Kaye and Robert Marshall, Murie joined with these two men and others in 1935 to found the Wilderness Society. As a member of the society's council from 1937 until his death, and a director from 1945, Murie was a leading member of the conservation movement's attempt to obtain congressional protection for the national wilderness areas. His and his wife's nearly 40 years of work were encapsulated in *The Elk of North America* (1951) and *Wapiti Wilderness* (1966), completed by his wife after his death. He was awarded the Audubon Medal of the National Audubon Society in 1959 and the John Muir Award of the Sierra Club shortly before his death. Murie died on 21 October 1963, before the passage of the landmark Wilderness Act of 1964, but it was his contributions that made it possible.

See also Wilderness Society.

National Air Pollution Control Administration

This federal agency was created as a section of the Air Quality Act of 1967. According to author Daniel Rohrer, the National Air Pollution Control Administration "manages extensive grant-in-aid, research, and development programs that deal with air pollution, [and] administers the air quality standards and enforcement provisions contained in the Air Quality Act of 1967."

See also Air Quality Act.

National Audubon Society

This well-known and highly profiled environmental activist group was originally founded in 1886 by writer George Bird Grinnell. Later a member of the hunting lobbyist organization the Boone and Crockett Club, Grinnell editorialized in the pages of his journal *Forest and Stream* about the need to form an organization that would, according to author James Trefethen, stop hunters from killing birds not used for food and end the killing of wild birds for decorative plumage on ladies' hats. Within three years, however, Grinnell shut down group operations when expenses became overwhelming.

In 1896, salesman William E. Dutcher began to advance the cause of establishing state Audubon clubs to advocate wildlife protection, principally bird conservation. These clubs grew in number and eventually became Audubon Societies; in 1905, when more than 40 such societies existed, Dutcher and ornithologist Frank Chapman established the National Association of Audubon Societies—now called the National Audubon Society. Dutcher served as its president from 1905 until his death in 1920. In 1935, Chapman gave his journal *Bird-Lore* to the society, which now publishes it as *Audubon* magazine.

As of 1991, the National Audubon Society had over 1.1 million card-carrying members in the United States, Canada, and Central and South America, with over 510 chapters. Led by activist Peter A. A. Berle, the society oversees 150,000 acres of habitat in 100 sanctuaries and six nature centers. In addition to *Audubon* magazine, it publishes *The Audubon Activist* (a lobbying news journal) and produces the award-winning "The World of Audubon" television specials.

See also Chapman, Frank Michler; Dutcher, William E.

National Biological Survey (NBS)

This federal division was created as part of the Interior Department's appropriations for fiscal year 1994. The survey's function is to "provide the scientific knowledge necessary to balance the compatible goals of ecosystem protection and economic progress." Its mission is to "perform research in support of biological resource management; inventory, monitor, and report on the status and trends in the Nation's biotic resources; and develop the ability and resources to transfer the information gained" in these surveys to those responsible for the upkeep and protection of the nation's wildlife resources. This new agency merged portions of the biological research and surveying activities of three government agencies: the Bureau of Land Management, the U.S. Fish and Wildlife Service, and the National Park Service. Secretary of the Interior Bruce Babbitt said that the survey "will take the field biology of the Department of the Interior—the best in the world—and redeploy it in a new way."

National Bison Range Act
(3 Stat. 267)

This legislation was passed by Congress on 23 May 1908 to establish a national refuge for the nearly extinct American bison, or buffalo. This was Congress's first appropriation of funds for the creation of a wildlife refuge. The act reads, "The President is hereby directed to reserve and exempt from the unallotted lands now embraced within the Flathead Indian Reservation, in the state of Montana, not to exceed twelve thousand eight hundred acres of said lands, near the confluence of the Pend d'Oreille and Jocko rivers, for a permanent national bison range for the herd of bison to be presented by the American Bison Society." After Congress appropriated $30,000 for the secretary of the interior to pay the Flathead, Kootenai, and Upper Pend d'Oreille Indian tribes for their land, 37 bison were released onto the reserve on 17 October 1909.

National Conservation Commission

This commission was set up on 8 June 1908 following the landmark Governors' Conference on the Environment. The commission was organized into several panels to discuss the conservation of natural resources—a waters commission (chaired by Rep. Theodore Elijah Burton of Ohio), a forests commission (chaired by Sen. Reed Smoot of Utah), a lands commission (chaired by Sen. Knute Nelson of Minnesota), and a minerals commission (chaired by Rep. John Dalzell of Pennsylvania). An executive committee was set up consisting of Gifford Pinchot, Senators Smoot and Nelson, Congressmen Burton and Dalzell, W J McGee, Overton W. Price, George W. Woodruff, and John A. Holmes.

National Environmental Policy Act of 1969 (NEPA) (83 Stat. 852)

This legislation, enacted on 1 January

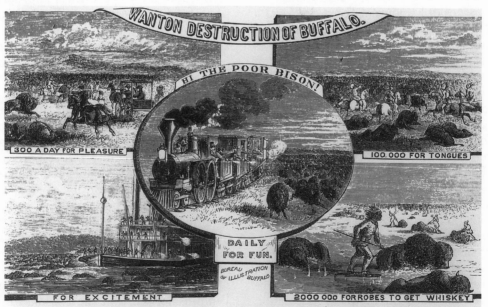

Five pictures for the consideration of Uncle Samuel, suggestive of a game law to protect his comb-horns, buttons, tallow, dried beef, tongues, robes, ivory-black, bone-dust, hair, hides, etc.

An 1872 illustration shows how bison were hunted to the brink of extinction. The 1908 National Bison Range Act established a refuge in Montana, where a herd of 37 bison contributed by the American Bison Society were released a year later.

1970, spelled out federal environmental policy and directed all pertinent federal agencies to carry out that policy. During the 1960s, there was much debate in Congress regarding a national environmental policy. The man most responsible for this debate, said author Patrick Parenteau, was Sen. Henry "Scoop" Jackson of Washington. The act states:

Congress, recognizing the profound impact of man's activity on the interrelations of all components of the natural environment, particularly the profound influences of population growth, high-density urbanization, industrial expansion, resource exploitation, and new and expanding technological advances and recognizing further the critical importance of restoring and maintaining environmental quality to the overall welfare and development of man, declares that it is the continuing policy of the Federal Government, in cooperation with State and local governments, and other concerned public and private organizations, to use all practicable means and measures, including financial and technical assistance, in a manner calculated to foster and promote the general welfare, to create and maintain conditions under which man and nature can exist in productive harmony, and fulfull the social, economic, and other requirements of present and future generations of Americans.

The act's stated goals were "to declare a national policy which will encourage productive and enjoyable harmony between man and his environment; to promote efforts which will prevent or eliminate damage to the environment and biosphere and stimulate the health and welfare of man; to enrich the understanding of the ecological systems and natural resources important to the Nation." Although as a Senate bill the act originally declared "that each person has a fundamental and unalienable right to a healthful environment," this language was stricken for substitute language that notes the "critical importance of restoring and maintaining environmental quality to the overall welfare and development of man." Section 101(b) spells out the goals of NEPA: (1) to fulfill the responsibilities of each generation as trustee of the environment for the succeeding generation; (2) to assure for all Americans safe, productive, and aesthetically and culturally pleasing surroundings; (3) to attain the widest range of beneficial uses of the environment without degradation, risk to health or safety, or other undesirable or unintended consequences; and (4) to preserve important historic, cultural, and natural aspects of our national heritage and to maintain, wherever possible, an environment that supports diversity.

NEPA also mandated the creation of a Council on Environmental Quality (CEQ) and required the preparation of environmental impact statements (EISs), summaries of how federally funded or licensed projects will affect the environment and what federal agencies have to do to mitigate or avoid environmental damage. Wrote Parenteau, "NEPA was not meant to be a quiet addition to the United States Code. It was meant to shake things up, to challenge conventional thinking, even to make people uncomfortable, most of all the bureaucrats concealed within the thick gray walls of the federal establishment in Washington."

See also Jackson, Henry Martin.

National Forest Commission

This commission was set up in the last years of the nineteenth century to lobby the government for a national policy on forest preservation. In 1896, Secretary of the Interior Hoke Smith asked the National Academy of Sciences to bring together prominent people in the area of forestry conservation to present

Congress with a draft of a national policy on forest management and preservation. It took a year before the commission was set up. Appointed commissioners were Charles Sprague Sargent, head of the Arnold Arboretum in Boston and author of *Forests of North America*, an addendum to the tenth census (1880); William Henry Brewer, a nationally known botanist from Yale University, who wrote the landmark work *The Woodlands and Forest Systems of the United States* for the ninth census (1870); Alexander Agassiz, one of the foremost U.S. scientists at the time; Wolcott Gibbs, president of the National Academy of Sciences; and Gifford Pinchot, soon to become chief forester of the United States. With a budget of $25,000, the commission spent three months investigating how forest policy could be changed for the better. Its final analysis, *Report of the National Forestry Committee of the National Academy of Sciences upon the Inauguration of a Forest Policy for the Forested Lands of the United States* (1897), called for a complete overhaul of the way the government treated its forests and recommended the formation of new preserves to save what forests were left. Based on this report, President Grover Cleveland created 13 new forest preserves in the waning months of his presidency. The anger over Cleveland's action led to the enactment of the Forest Management Act of 1897, also known as the Pettigrew Act.

See also Brewer, William Henry; Pettigrew Act; Pinchot, Gifford; Sargent, Charles Sprague.

National Forest Reservation Commission

This commission, established under the Weeks Act of 1911, was set up to choose lands to be purchased by the federal government that had forest reserves on them or that could aid in the free navigation of waters. The commission, whose members included the secretaries of war (now defense), interior, and agriculture and two members each from the U.S. House of Representatives and the Senate, was "authorized to consider and pass upon such lands as may be recommended for purchase as provided for in section six" of the Weeks Act, "and to fix the price or prices at which such lands may be purchased, and no purchases shall be made of any lands until such lands have been duly approved for purchase by said commission." Section six of the act directed the secretary of agriculture "to examine, locate, and recommend for purchase such [lands] as his judgment may [deem] necessary to the regulation of the flow of navigable streams, and to report to the National Forest Reservation Commission the results of such examinations." The commission was abolished under the Federal Land Policy and Management Act of 1976, and its jurisdictional functions were assigned to the Department of Agriculture.

See also Federal Land Policy and Management Act; Weeks Act.

National Forests Act (34 Stat. 1256)

This legislation was passed in 1907 to halt President Theodore Roosevelt's rapid creation of national forests, parks, and wildlife reserves. In his first six years as president, Roosevelt had set aside land for 5 national parks, 53 wildlife reserves, and 16 national monuments. In this act, Congress demanded that future set-asides or enlargements of parks, monuments, or wildlife refuges in six western states be done only through acts of Congress. As the bill sat on the president's desk awaiting his signature, he created or expanded 32 forest reserves totaling some 75 million acres, thus bypassing the law's intent.

See also Roosevelt, Theodore.

National Industrial Pollution Control Council

See Executive Order 11523 of 9 April 1970.

National Irrigation Association

This lobbying and advocacy group was formed in a hotel room in Wichita,

Kansas, on 2 June 1889 by George Hebard Maxwell (later called the father of irrigation) and other irrigation advocates, including Guy Mitchell, H. B. Maxson, and John Henry Smith. Author Andrew Hudanick wrote that the National Irrigation Association (NIA) "would be the vehicle through which Maxwell would 'educate' the nation by influencing the general public, businessmen, farmers and politicians, and aid in the promulgation of the National Reclamation Act." Although drafted to "supplement, not supplant" the National Irrigation Congress (NIC), the NIA overtook the NIC and became the leading promoter of a national irrigation and reclamation policy. With help from the association, Congress passed the Carey Irrigation Act in 1894 and the Newlands Reclamation Act in 1902.

Maxwell and the association came under fire from some congressional critics: According to one source, Sen. Francis E. Warren of Wyoming (at one time an ally) called Maxwell "an upstart, a mercenary, and an agitator, one who would foist upon the West rigid and exclusive National control of its waters." Maxwell was joined in his fight, however, by many noteworthy conservation leaders, including John Wesley Powell, William Ellsworth Smythe, and Richard J. Hinton. Even some forestry advocates came to Maxwell's defense; the American Forestry Association renamed its official journal *Forestry and Irrigation*.

Maxwell's message was heard when three of the major political parties (Republican, Democratic, and Silver Republican) made national irrigation part of their respective platforms in 1900. Although the association's influence waned after passage of the Newlands Act in 1902 (and it changed its name to the National Reclamation Association [NRA] to reflect a broader agenda), Maxwell remained one of the loudest voices in the area of irrigation. He worked for passage of a flood-control law that was embodied in the Newlands River Regulation Amendment to the Rivers and Harbors Bill of 1917. With his death in 1946, the NRA ceased to exist.

See also Maxwell, George Hebard; National Irrigation Congress; Newlands, Francis Griffith; Smythe, William Ellsworth.

National Irrigation Congress

The National Irrigation Congress (not to be confused with George H. Maxwell's National Irrigation Association) was influential in the passage of the Carey Irrigation Act of 1894 but remained in the shadows and was virtually powerless at the end of the 1890s. It began when Gov. Arthur L. Thomas of Utah called for a congress of irrigation interests to meet in Salt Lake City in 1891 to help create national policy on irrigation and reclamation in the arid western states. Leading the way in the field was journalist William Ellsworth Smythe, editor of the influential *Irrigation Age*. The congress, held from 15 to 17 September 1891, brought together some 350 delegates, including Smythe, Rep. Francis G. Newlands of Nevada, and Sens. Francis E. Warren of Wyoming and William Stewart of Nevada. This meeting called for the revocation of the Desert Land Act of 1877, which gave each settler a square mile of desert but required them to irrigate the land themselves, leaving many with poor farmland. Speakers at the first congress, including James J. Hill, president of the Great Northern Railroad, and William Mills, president of the Southern Pacific Railroad, called for the federal government to cede to the individual states all the arid lands in the public domain; the states would then take on large irrigation projects. The second congress met in Los Angeles in October 1893 and the third in Denver in October 1894. The national economic collapse in 1893 brought investment in private irrigation companies to a standstill and led to the passage in 1894 of the Carey Irrigation Act. The congress continued to meet, although Smythe did not take an active role after 1895. At that point, the congress was headed by Sen. (later Gov.) Joseph M. Carey of Wyoming, father of the Carey Act.

With the establishment of George H. Maxwell's National Irrigation Association in 1889 and its rise throughout the 1890s, the congress's days were numbered. Maxwell's fiery spirit and zest for lobbying made his organization the main advocacy group for irrigation legislation after 1895. Although congresses continued to assemble after 1900 (they met in Chicago in 1900, in Colorado Springs in 1902, and in Ogden, Utah, in 1903), their power waned. At one of the last assemblies in Portland, Oregon, in 1905, Frederick Haynes Newell, head of the Bureau of Reclamation, spoke about the need to make irrigation resources more available for small landowners. With the end of the National Irrigation Congress, George Maxwell's group, renamed the National Reclamation Association, was the last organization dedicated to irrigation and reclamation policy.

See also Carey, Joseph Maull; Irrigation Congresses; Maxwell, George Hebard; National Irrigation Association; Newell, Frederick Haynes; Smythe, William Ellsworth.

National Park Service

William Howard Taft's Secretary of the Interior Richard Achilles Ballinger recommended the creation of a National Park Service in his annual report in 1910. By 1916, national parks were run down and badly mismanaged. Stephen T. Mather, a businessman and nature lover, took notice of park conditions and complained to Secretary of the Interior Franklin K. Lane in 1914; Lane invited Mather to come to Washington, lobby Congress to pass a law setting up a government agency to protect the parks, and run the agency himself. Several conservationists spoke out in favor of a national park service law—among them Mather, Horace Marden Albright, Frederick Law

Olmsted, Robert B. Marshall, and Gilbert Grosvenor of *National Geographic* magazine. With their backing, the National Park Service Act became law on 25 August 1916. Stephen T. Mather became its first director, with Horace Albright of California serving as his assistant.

See also Albright, Horace Marden; Mather, Stephen Tyng.

National Park Service Act
(39 Stat. 535)

The National Park Service Act, which created the National Park Service in the Department of the Interior, was sponsored in the House of Representatives by William Kent (1864–1928) of California and in the Senate by Reed Smoot (1862–1941) of Utah. It was enacted on 25 August 1916 to "promote and regulate the use of the Federal areas known as national parks, monuments, and reservations ... to conserve the scenery and the natural and historic objects and the wild life therein and to provide for the enjoyment of the same in such manner and by such means as will leave them unimpaired for the enjoyment of future generations."

See also Albright, Horace Marden; Mather, Stephen Tyng; National Park Service.

National Parks and Recreation Act
(Omnibus Parks Act) (92 Stat. 3467)

This legislation was enacted on 10 November 1978 chiefly to add eight new wilderness areas under the jurisdiction of the National Park Service, essentially tripling the acreage under its control. Part of the nearly 2 million acres of new land was given to the wilderness areas; also included were additions to the Wild and Scenic Rivers System and the National Trails System (tripling the size of it). New preserves such as the Golden Gate, Santa Monica, and Gateway recreational parks were established. Further, funds were authorized for the creation of an Urban Park and Recreation Recovery Program.

The 1915 dedication of Rocky Mountain National Park, Colorado, attracted early Park Service and conservation luminaries including (left to right) then Assistant Secretary of the Interior Stephen T. Mather and first Park Service Director Robert Sterling Yard, a founder of the Wilderness Society in 1935, acting park supervisor Charles R. Trowbridge, photographer Herford Cowling, and Horace Albright, assistant to Stephen Mather and his successor as National Park Service director from 1929 to 1933.

National Trails System Act
(82 Stat. 919)

Enacted on 2 October 1968, this legislation was passed in response to the report *Trails for America*, prepared by the federal Bureau of Outdoor Recreation. The act's purpose was "to provide for the ever-increasing outdoor recreational needs of an expanding population and ... to promote public access to, travel within, and enjoyment and appreciation of the open-air, outdoor areas of the Nation." It called for the establishment of a trails system "primarily . . . near the urban areas of the North, and secondly ... within established scenic areas more remotely located."

National Water Commission

Established by the National Water Commission Act of 26 September 1968 (Public Law 90-515), this commission released its final report, *Water Policies for the Future*, on 14 June 1973. Chaired by Charles F. Luce, chief executive officer of Consolidated Edison of New York, the commission included Howell Appling, Jr., owner of a wholesale farm equipment firm in Portland, Oregon; James R. Ellis, a Seattle attorney; Roger C. Ernst, president of the Central Arizona Water Conservation District; Ray K. Linsley, professor of hydraulic engineering at Stanford University; James E. Murphy, a Montana attorney; Josiah Wheat, a Texas attorney and legal counsel for the Texas Water Quality Board; and Russell E. Train, chairman of the Council on Environmental Quality and a future administrator of the Environmental Protection Agency.

The commission's mission, according to the act that created it, was to "review present and anticipated national water resource problems, making such projections of water requirements as may be necessary, and identifying alternative ways of meeting these requirements— giving consideration, among other things, to conservation and the more efficient use of existing supplies, increased usability by reduction of pollution, innovations to encourage the highest economic use of water, interbasin transfers, and technological advances including, but not limited to, desalting, weather modification, and waste water purification and reuse; consider economic and social consequences of water resource development, including, for example, the impact of water resource development on regional economic growth, on institutional arrangements, and on esthetic values affecting the quality of life of the American people."

The commission's final report recommended, among other things, that (1) all agencies responsible for planning and carrying out water channelization projects evaluate channelization procedures to avoid flooding and other environmental risks; (2) the 1972 Clean Water Act be strengthened; (3) a national policy be established to define the role of the federal government in the conception of municipal water systems; and (4) the Water Resources Council, established under the Water Resources Planning Act of 1965, be improved.

See also Water Resources Planning Act.

National Water Quality Commission
See Federal Water Pollution Control Act Amendments of 1972.

National Wilderness Preservation System (16 U.S.C. 1131)
This complex of wilderness areas was established by Congress on 3 September 1964 as part of the Wilderness Act. In the declaration of policy, the act says that "in order to assure that an increasing population, accompanied by expanding settlement and growing mechanization, does not occupy and modify all areas within the United States and its possessions, leaving no lands designated for preservation and protection in their natural condition, it is hereby declared to be the policy of the Congress to secure for the American people of present and future generations the benefits of an enduring resource of wilderness. For this purpose there is hereby established a National Wilderness Preservation System to be composed of federally owned areas designated by Congress as 'wilderness areas,' and these shall be administered for the use and enjoyment of the American people in such manner as will leave them unimpaired for future use and enjoyment as wilderness, and so as to provide for the protection of these areas, the preservation of their wilderness character, and for the gathering and dissemination of information regarding their use and enjoyment as wilderness." The Eastern Wilderness Act of 2 January 1975 (88 Stat. 2096) mandated that all lands east of the 100th meridian be distinguished, investigated, and earmarked for incorporation into the National Wilderness Preservation System.

See also Wilderness Act.

National Wildlife Federation
One of the largest environmental groups in existence, the National Wildlife Federation was founded as a result of the North American Wildlife Conference held in Washington on 3–7 February 1935 under the auspices of cartoonist and conservationist Jay Norwood "Ding" Darling and Ferdinand A. Silcox, chief forester of the United States. At the conference, the General Wildlife Federation was established, with Darling as president. Its first goal was to lobby Congress for the passage of a national policy under which hunters would help pay for wildlife refuge management and future purchases through an ammunition tax. On 2 September 1937, the Pittman-

Robertson Act, also known as the Federal Aid in Wildlife Restoration Act, levied an 11 percent excise tax on the manufacturer's price of sporting equipment and ammunition sold to hunters, with the proceeds to fund state fish and game commissions nationwide. According to one source, this total amounted to some $30 million from 1938 to 1966.

The federation, with a 1991 budget of $77.2 million, has approximately 6.2 million members and supporters; it publishes *National Wildlife* and *International Wildlife* magazines, as well as the *Ranger Rick* journal for young children. Its statement of purpose is "to coordinate all agencies, societies, clubs and individuals which are or should be interested in the restoration, wise use, conservation and scientific management of wildlife and other natural resources into a permanent, unified, active agency for the purpose of securing adequate public recognition of the needs and values of wildlife resources and other natural resources." Its creed is: "I pledge myself, as a responsible human, to assume my share of the stewardship of our natural resources; I will use my share with gratitude, without greed, or waste; I will respect the rights of others and abide by the law; I will support sound management of the resources we use, the restoration of the resources we have despoiled, and the safekeeping of significant resources for posterity; I will never forget that life and beauty, wealth and progress, depend on how wisely we use these gifts ... the soil, the water, the air, the minerals, the plant life, and the wildlife."

See also Pittman, Key; Pittman-Robertson Act.

Nature Conservancy

One of the largest and wealthiest (1992 budget: $274 million) of the national environmental organizations, the Nature Conservancy's main purpose is to protect environmentally sensitive land though perpetual land-use agreements and outright purchase. Founded in 1951 from a group called the Ecologists Union (a subgroup of the Ecological Society of America), the Nature Conservancy has purchased more than 5 million acres of private land in the United States and more than 20 million acres in Canada. Although it aims to have these lands managed and maintained to prevent development and destruction, it has been accused of going too far. For example, the Carrizo Plains Natural Area in California became scrub when grazing cattle were removed. The group is headquartered in Arlington, Virginia, and has a membership of 680,000. It publishes the bimonthly *Nature Conservancy Magazine* and the internal newsletter *On the Land.*

Newell, Frederick Haynes (1862–1932)

Frederick H. Newell, an expert in irrigation and reclamation, was the first head of the U.S. Reclamation Service. The son of Augustus William Newell and Annie Maria (nee Haynes) Newell, he was born in Bradford, Pennsylvania, on 5 March 1862. After his mother's death when he was young, Frederick Newell was raised by maternal aunts in Newton, Massachusetts. He attended schools there before enrolling at the Massachusetts Institute of Technology (MIT) and earning a bachelor's degree in mining in 1885. For the next three years, he was involved in petroleum research in various states.

On 2 October 1888, he was hired by the U.S. Geological Survey as a hydraulic engineer to head the important hydrographic branch and soon became a protégé of survey director John Wesley Powell. Sent west to investigate various irrigation projects and their water resources, "Newell and his colleagues helped to fashion a multipurpose philosophy of water conservation and utilization—the coordination of all uses, including navigation, irrigation, power, and flood control," according to historian Michael Robinson. Historian Samuel

Hays reported that "during the 1890s, while ... Newell was undertaking hydrographic investigations and agitating for a national irrigation law, he also served as secretary of the American Forestry Association, which fought to extend the national forests and to adopt a sound national forest management program." Said another source, Newell "assisted Francis G. Newlands, later Senator from Nevada, and George H. Maxwell, president of the National Irrigation Association, in the preparation of various Congressional bills, including the Reclamation Act of 1902." After passage of the landmark Newlands Act in 1902, which established the U.S. Reclamation Service, Newell was appointed head of the new agency. Under the authority of the secretary of the interior, Newell began work on the Roosevelt Reservoir (now the Roosevelt Dam) on the Salt River in Arizona in 1903; in 1905, the Reclamation Service concluded its first venture, the Truckee-Carson ditch in Nevada, and five years later it was involved in 24 other projects. Newell headed the service from 1902 until 1914. In those years, he oversaw the building of "100 dams, 25 miles of tunnels, 13,000 miles of canals, and other facilities that served 20,000 farms," reported Robinson. In the 20 years after the passage of the Newlands Act, the Reclamation Service spent about $150 million on projects.

Complaints to Interior Secretary Franklin K. Lane over farm problems resulting from the irrigation projects led Lane to dismiss Newell from the service in 1914, although he was retained as a consulting engineer until 1915. After leaving government, he spent four years as head of the civil engineering department of the University of Illinois, then five years as president of the American Association of Engineers. In 1924, he helped found the Research Service, a private irrigation and reclamation consulting firm based in Washington, D.C. Newell was the author of many scientific works, including *Hydrography of the Arid Regions* (1891),

Report on Agriculture by Irrigation in the Western Part of the United States (1894; part of the eleventh census), *The Public Lands of the United States and Their Water Supply* (1895), *Irrigation in the United States* (1902), *Principles of Irrigation Engineering* (1913), *Irrigation Management* (1916), and *Water Resources, Present and Future Uses* (1920). Newell died suddenly in Washington, D.C., of heart failure on 5 July 1932 at the age of 70.

See also Bureau of Reclamation; Newlands Irrigation (or Reclamation) Act.

Newlands, Francis Griffith (1848–1917)

Francis Newlands of Nevada is best known as the sponsor in the House of Representatives of the nation's most important piece of irrigation legislation. He was born on 28 August 1848 in Natchez, Mississippi, to James Birney Newlands, a physician, and Jessie (nee Barland) Newlands, both Scottish immigrants. Sometime between 1848 and 1851, the Newlands family moved to Quincy, Illinois. James Newlands died when Francis was three, and Jessie Newlands raised her four sons and a daughter alone until she married Eben Moore. Moore's financial stability enabled Francis Newlands to attend a prestigious high school in Chicago and be tutored privately before entering Yale University. He attended Yale for two years until financial panic drove his stepfather into bankruptcy. Newlands obtained work with the federal government in Washington, D.C., and enrolled at the Columbian College (now George Washington University) Law School. He earned a law degree in 1869 and was admitted to the bar the same year

Newlands opened a law office in San Francisco and proceeded to work on court cases and a trusteeship. A Democrat, he supported Supreme Court Justice Stephen Johnson Field for president in 1884. Although Field—a Californian who was unpopular in the state for his decisions on land rights—never ran for the White House, Newlands's backing of

Francis Griffith Newlands

him aroused bitter feelings in the party. This, plus an interest in mining, drove Newlands east to Nevada, the state he would eventually represent in Congress. In 1888, Newlands took most of his money—an inheritance from his father-in-law—and opened the Truckee Irrigation Project. Wrote author Marc Reisner, "It was one of the most ambitious reclamation efforts of its day, and it failed—not because it was poorly conceived or executed (hydrologically and economically, it was a good project) but because squabbles among its beneficiaries and the pettiness of the Nevada legislature ruined its hopes. In the process, Francis Griffith Newlands lost half a million dollars and whatever faith he had in the ability of private enterprise to mount a successful reclamation program. 'Nevada,' he said bitterly as his project went bust in 1891, 'is a dying state.'"

In 1892, Newlands, now a member of the Republican-backed Silver Party, won a seat in the U.S. House of Representatives. In 1896, when Democrat William Jen-

nings Bryan ran for president on the platform of the free coinage of silver, Newlands returned to the Democratic fold. Although a leading member in the House Foreign Relations Committee, Newlands is remembered more for his work toward a national irrigation and reclamation policy. An intimate of George H. Maxwell, founder of the National Reclamation Association, Newlands fought to secure the passage of legislation that would help irrigate the western states, particularly Arizona and Nevada. He cosponsored, with Sen. Henry Clay Hansbrough of North Dakota, the Newlands Reclamation (or Irrigation) Act of 1902. In a 21 January 1902 speech, he said, "What is the demand of the West? Is it that the West should be enabled to reclaim itself, without taxing the Federal Treasury and without inflicting any burden upon the general taxpayer, by appropriate legislation dedicating the proceeds of sales of the public lands in the arid region to the reclamation of arid lands?"

Newlands was also a moving force in the area of water development programs, particularly in the western United States. Elected to the U.S. Senate in 1902, he spent the last 15 years of his life working for the enactment of a national water strategy. As he said in a House speech in 1901, "What improvements are required by our rivers? In the first place the navigable rivers are subject to floods, and we seek to prevent the overflow by constructing levees. Immense sums have been expended on the lower part of the Mississippi in an effort to confine the stream and to prevent overflow of the adjoining land ... the evils which attach to both navigation and irrigation are the same. The streams are overflowing at a time when the water is not needed and they are attenuated threads at a time when the water is most in demand. We of the arid regions contend that both navigation and irrigation can be promoted by the storing of these waters at the sources of these mountain streams which are tributaries to the great rivers." He sponsored a bill in the Senate to make Theodore Roosevelt's temporary Inland Waterways Commission a permanent government entity. He was also responsible for the creation of the Federal Trade Commission (1914) and for a 1913 act on the mediation of labor disputes, the forerunner of the National Labor Relations Board set up during the New Deal.

Newlands spent much of his time in the Senate attempting to get a waterways commission established by law. Stymied by the Taft administration, his proposal received only grudging support from Woodrow Wilson. After he made a deal to vote for another piece of legislation, Newlands's waterways commission was attached to the Rivers and Harbors Bill of 1917 as the River Regulation Amendment, but the outbreak of World War I delayed the appointment of commissioners. Newlands did not live to see the commission meet. He died on 2 December 1917 at the age of 69. With his death, support for the waterways commission dried up, and the temporary one was disbanded in 1920 by the passage of the Water Power Act.

See also Bureau of Reclamation; Federal Water Power Commission; Hansbrough, Henry Clay; Newlands Irrigation (or Reclamation) Act.

Newlands Irrigation (or Reclamation) Act (30 Stat. 388)

This legislation, known officially as the Newlands-Hansbrough Reclamation Act of 1902, was sponsored in the House of Representatives by Francis Griffith Newlands of Nevada and in the Senate by Henry Clay Hansbrough of North Dakota. Although Congress had passed an irrigation law in 1894 (known as the Carey Act), there had been little government backing of such irrigation projects, and many private attempts had failed. The Newlands Act set out to create a governmental bureaucracy to aid states and cities with reclamation and irrigation projects through a national Reclamation Service. Newlands drafted the legislation with the help of Frederick Haynes Newell, who would become the first chief of the U.S. Reclamation Service. Although he was a Democrat, Newlands's legislation was backed by Republican President Theodore Roosevelt and endorsed by such western senators as Francis E. Warren of Wyoming. In fact, Roosevelt had to implore fellow Republican and House Speaker Joseph Gurney Cannon not to block the bill. "I do not believe that I have ever before written to an individual legislator in favor of an individual bill, but I break through my rule to ask you as earnestly as I can not to oppose the Irrigation measure," Roosevelt wrote to Cannon. Although many eastern legislators opposed the bill, during the Senate debate Hansbrough read off the list of eastern newspapers supporting passage of the bill, including the *New York Times, Tribune, World,* and *Sun; Boston Transcript* and *Globe;* and the *Philadelphia Inquirer, Ledger,* and *Press.* With the backing of such organizations as the National Irrigation Congress and National Reclamation Association and the lobbying of irrigation guru George H. Maxwell, the Newlands Act ultimately passed Congress on 17 June 1902. It stated that the use of "all moneys received from the sale and disposal of public lands in Arizona, California, Colorado, Idaho, Kansas, Montana, Nebraska, Nevada, New Mexico, North Dakota, Oklahoma, Oregon, South Dakota, Utah, Washington and Wyoming ... shall be reserved, set aside, and appropriated as a special fund in the Treasury to be known as the 'reclamation fund,' to be used in the examination, and survey for and the construction and maintenance of irrigation works for the storage, diversion, and development of waters for the reclamation of arid and semiarid lands in the United States and Territories." Within four years, the act was used to irrigate some 3 million acres, mainly in the western United States.

See also Carey Irrigation Act; Hansbrough, Henry Clay; Irrigation and Reclamation Policy, National; Maxwell, George Hebard; Newlands, Francis Griffith.

Noble, John Willock (1831–1912)

John Willock Noble served as secretary of the interior from 1889 to 1893. He was born in Lancaster, Ohio, on 26 October

1831, the son of John and Catherine (nee McDill) Noble, two devout Presbyterians. John Noble attended local schools in Cincinnati and then Miami University in Ohio for three years before transferring to Yale University, where he earned a B.A. in 1851. In 1852, he received his law degree from the Cincinnati Law School. Noble got caught up in the westward movement and settled in St. Louis, Missouri, in 1855. A year later, he moved to Keokuk, Iowa.

In Iowa, Noble started a law practice with Samuel Freeman Miller, who would become an associate justice on the U.S. Supreme Court. During the Civil War, Noble served in the 3d Iowa Cavalry, rising to the rank of brevet brigadier general. Following the war, he returned to Missouri, where in 1867 he was named U.S. district attorney for the eastern district of the state. As district attorney, he prosecuted mostly whisky-revenue frauds. In 1870, poised to rise higher in the judiciary, he instead took a job as a corporate attorney representing southeastern railroads. For the next two decades, he turned his practice into a lucrative business—so much so that in 1872, he declined an offer from Ulysses S. Grant to be the higher-profile but lower-paid U.S. solicitor general.

In 1889, President Benjamin Harrison named Noble as interior secretary, a post the 57-year-old Noble was willing to take. Noble served for the entire four years of the Harrison administration, only the second man to serve a full term since the department's creation in 1849. As secretary, Noble oversaw congressional action that authorized the president to permit American Indians to cut and sell dead timber on Indian reservations, and he empowered the General Land Office to dispose of thousands of land claims by using a liberal interpretation of the land laws to find in favor of settlers. He is considered by historians to have been a prime lobbyist for the amendment to the General Revision Act of 1891 that allowed the president to set aside forestry reserves for national forests; however, new evidence disputes

this. In letters preserved in the National Archives, Arnold Hague, a U.S. Geological Survey worker, wrote that he merely showed Noble the amendment in question as it was being passed in Congress and explained its far-reaching implications. Hague later wrote a letter to *Forest and Stream* magazine crediting Noble with its insertion, apparently trying to protect any congressman who might be disadvantaged by its results. Noble, however, is properly credited with using the new law to the environment's advantage; his numerous recommendations to the president were accepted by Harrison and led to the creation of 15 reserves totaling some 13 million acres. Robert Underwood Johnson, the editor of *Century* magazine, said, "I had no idea that Noble was going to do so much." In 1892, President Harrison, upon Noble's recommendation, established the Afognak Forest and Fish Culture Reserve in Alaska, considered the first wildlife refuge.

With Harrison's defeat in the 1892 election, Noble left public office. He returned to St. Louis to pick up his old law practice, but after a four-year absence, it was difficult to reestablish himself; he nearly went bankrupt. He finally obtained a mining company as a client, which brought him financial stability. He spent his final years as a lecturer to veterans' groups. John Willock Noble died in St. Louis on 22 March 1912 at the age of 80.

See also Afognak Forest and Fish Culture Reserve; Department of the Interior; Hague, Arnold; Holman, William Steele; Payson Act.

Norbeck-Andresen Act (45 Stat. 1222)
See Migratory Bird Conservation Act.

Norris, George William (1861–1944)
Known by historians as the man responsible for the Tennessee Valley Authority, Sen. George W. Norris of Nebraska was a rogue politician who voted against U.S. involvement in World War I and was more

populist than Republican. He was born on his family's farm near the village of Clyde in Sandusky County, Ohio, on 11 July 1861, the second son and eleventh child of Chauncey Norris, a farmer, and Mary Magdalene (nee Mook) Norris. His father, of Scotch-Irish decent, and mother, of Pennsylvania Dutch stock, were married in New York in 1838 and made their way to Ohio by wagon train in about 1846. Their eldest child, John Norris, enlisted in the Union army and was killed at the Battle of Resaca in Georgia. This event shaped much of George Norris's philosophy about war. He later wrote, "I remember my brother John's letters and the faded ribbon with which they were tied up into a bundle. Mother used to read them to me when I was a small boy." When George was three, Chauncey Norris died of pneumonia, leaving his widow to care for the children.

Norris attended local schools and spent his summers working on the family farm and others in the area. He attended Baldwin University (now Baldwin-Wallace College) in Berea, Ohio, and, after a year of teaching to earn some money, enrolled in the Northern Indiana Normal School and Business Institute (now Valparaiso University), from which he was awarded a law degree in 1883. Although he was admitted to the Ohio bar that same year, Norris taught school in Ohio and in Washington Territory for two years before settling in Beatrice, Nebraska, in 1885 and starting a law office, which he moved to nearby Beaver City six months later.

Norris became a successful attorney and seemed destined for high political office. He began his upward climb by completing two unexpired terms and then one elected term as the prosecuting attorney for Furnas County (1890–1895). He subsequently became a state judge for the 14th Nebraska Judicial District (1896–1903), the office from which he resigned upon his election to the U.S. House of Representatives as a Republican from the Fifth Nebraska District. He served five terms in the House

(1903–1913). Although he was a liberal to moderate Republican, Norris was loyal to the conservatives who helped get him elected. According to author Richard Lowitt, Norris "displayed neither marked ability nor marked liberal tendencies. He sympathized with Theodore Roosevelt's domestic policies, but he was indebted to the party organization and to railroad officials for help in his campaign." Known as a member of the "insurgents"—the Republican freshmen who sought to break the absolute power wielded by House Speaker Joseph G. Cannon—Norris was denied patronage following the overthrow of "Cannonism" in 1910. Nevertheless, he ran for a seat in the U.S. Senate in 1912 and was elected. Norris held this Senate seat for the last 31 years of his life.

To list the numerous issues that Norris was involved in during his Senate tenure would fill a book. He was one of a handful of senators who voted against the war declaration and was bitterly criticized nationwide. When World War I ended, he likewise voted against U.S. participation in the League of Nations. He sponsored the bill that eventually became the 20th Amendment to the Constitution—the "lame duck" amendment—as well as the bill that set up Nebraska's unicameral legislature (the only body of its kind in the nation), the Norris-LaGuardia Labor Relations Act of 1932 (which allowed the use of collective bargaining and strikes by unions), the Norris-Doxey Farm Forestry Act of 1937, and the Rural Electrification Administration (REA) during the New Deal. Norris's pet project, however, during the 1920s and 1930s was the establishment of a water and electric power policy for the Tennessee Valley. Spurned by fellow Republicans Calvin Coolidge and Herbert Hoover, Norris found Franklin D. Roosevelt receptive. In 1933, with Roosevelt's backing, Norris's Tennessee Valley Authority became a reality.

In the last decade of his life, Norris's votes changed somewhat, and his record shows support for lend-lease (the policy that "loaned" ships and war material to

Great Britain before direct U.S. participation in World War II) and for U.S. involvement in World War II after the Japanese bombed Pearl Harbor. Challenged by his own party during his re-election run in 1942, he was defeated by Kenneth S. Wherry. Norris's biography, *Fighting Liberal,* was finished just weeks before he suffered a cerebral hemorrhage, which led to his death on 2 September 1944 at the age of 83.

See also Norris-Doxey Act; Tennessee Valley Authority.

Norris-Doxey Act (50 Stat. 188)

This legislation, passed on 18 May 1937, is officially known as the Cooperative Farm Forestry Act. Sponsored in the House by Rep. Wall Doxey (1892–1962) of Mississippi and in the Senate by George W. Norris of Nebraska, the act appropriated funds for the encouragement of farm forestry. It reads: "That in order to aid agriculture, increase farm-forest income, conserve water resources, increase employment, and in other ways advance the general welfare and improve living conditions on farms through reforestation and afforestation in the various States and Territories, the Secretary of Agriculture is authorized ... to produce or procure and distribute forest trees and shrub planting stock; to make necessary investigations; to advise farmers regarding the establishment, protection, and management of farm forests and forest and shrub plantations and the harvesting, utilization, and marketing of the products thereof; and to enter into cooperative agreements for the establishment, protection, and care of farm- or other forest-land tree and shrub plantings within the States and Territories." The act authorized an appropriation of $2.5 million.

See also Norris, George William.

North American Waterfowl Management Plan

This document was signed in 1986 by the U.S. secretary of the interior, the Canadian minister of the environment, and, in 1988, the Mexican director general for ecological conservation of natural resources. According to the North American Wetlands Conservation Act of 1989, this 15-year agreement "provides a framework for maintaining and restoring an adequate habitat base to ensure perpetuation of populations of North American waterfowl and other migratory bird species." It is "a statement of needs, objectives, and strategies designed to protect, restore, and enhance habitat for North American waterfowl. It establishes a goal, which is to be achieved by the year 2000, to restore the continent's waterfowl breeding populations to 1976 levels, approximately 62 million ducks, and the 'fall flight' population of 100 million ducks. To achieve this, the plan calls for the protection, restoration, and enhancement of 1.93 million acres of waterfowl habitat in the U.S. and 3.67 million acres in Canada.... [T]he plan goes beyond land acquisition programs to recommend that 'public works projects planning should include the prevention or mitigation of destruction or degradation of waterfowl habitats' and that 'public land management agencies should be encouraged to zone or otherwise regulate land uses to prevent the destruction or degradation of waterfowl habitats.' ... Principal habitat areas slated for protection include the Central Valley of California, the prairie pothole region of the Great Plains, the lower Mississippi River region, the Gulf Coast, the Atlantic Coast from Maine to South Carolina, the lower Great Lakes–St. Lawrence River basin, portions of five eastern Canadian provinces, and the prairie-parkland provinces of Canada."

North American Wetlands Conservation Act (103 Stat. 1968)

This legislation was "an Act to conserve North American wetland ecosystems and waterfowl and other migratory birds and fish and wildlife that depend upon such habitats." It was enacted by Congress on 13 December 1989 in response to the ever-increasing destruction of native wetlands. House Report 101–269, an investigation into wetlands loss in the continental United States, revealed disturbing trends as to the devastation of this unique habitat: "More than half the wetlands that existed in what is now the lower 48 states have been lost since European colonies were established on the continent more than three centuries ago. From the mid-1950s to the mid-1970s, 9 million acres of wetlands were drained, filled, and cleared in the 48 contiguous states, and the destruction of wetlands continues today at a rate of 450,000 acres annually."

The report continues, "Waterfowl populations, which have been carefully monitored by the U.S. Fish and Wildlife Service [FWS], are declining due to the destruction of wetlands. The average number of North American ducks in recent years has been lower than any comparable period on record. In 1985, there was an estimated breeding population of 30 million ducks in the U.S. and Canada, compared to 62 million ducks in the breeding population 10 years ago [1975]. The continued drought in the North American prairies has exacerbated the destruction and degradation of habitat. Mallard and pintail ducks are two species of waterfowl that have suffered the greatest decline in numbers." The report concludes, "Although less is known about the specific habitat needs of migratory non-game birds, it is clear that their populations also suffer from the loss of wetlands. Of the 30 species of migratory non-game birds that are of concern to the FWS because of their uncertain status, nearly one-half are dependent upon coastal and freshwater wetlands. Among the species whose populations have declined are white ibises (by an estimated 90%), sanderlings (80%), woodstorks (50%), and short-billed dowitchers (40%)."

In response to this report, Congress enacted the North American Wetlands Conservation Act of 1989. It includes the establishment of a North American Wet-lands Conservation Council, the recommendation and approval of wetlands-preserving projects, and the amending of the Emergency Wetlands Resources Act of 1986 to include a resource assessment that compares U.S. wetlands in 1780 with those in 1990.

See also Emergency Wetlands Resources Act; Wetlands.

North American Wildlife Conference

This conference—the idea of Jay Norwood "Ding" Darling, the Pulitzer Prize–winning cartoonist—was held 3–7 February 1935 in Washington, D.C. According to author James Trefethen, its purpose was to "bring together all North American agencies, organizations, and interests concerned with wildlife resources and the soils, waters, and forests that supported them." For the first time, nationally known conservationists met to discuss the problems facing the wildlife of the continent. Organized by Ferdinand A. Silcox, chief forester of the United States, the conference included speeches by Darling and Ira N. Gabrielson, Darling's successor at the Bureau of Biological Survey. Some 1,200 people attended, and the conference was eventually endorsed by the members of 36,000 wildlife groups. At the end of the meeting, the conference members formed the General Wild Life Federation and elected Jay Darling as its first president. Commented the *Sierra Club Bulletin*, "Probably nothing has happened in recent years to stimulate the conservation movement [more] than the Wild Life Conference." The annual conference is now called the North American Wildlife and Natural Resources Conference and has been sponsored by the Wildlife Management Institute since 1946.

See also Darling, Jay Norwood.

Nuclear Waste Policy Act (42 U.S.C. 10101)

This act became law on 7 January 1983 (it was passed by Congress in December

1982) and attempted to set national policy for nuclear waste handling and disposal. In it, Congress found that:

(1) radioactive waste creates potential risks and requires safe and environmentally acceptable methods of disposal; (2) a national problem has been created by the accumulation of (A) spent nuclear fuel from nuclear reactors; and (B) radioactive waste from (i) reprocessing of spent nuclear fuel; (ii) activities related to medical research, diagnosis, and treatment; and (iii) other sources; (3) Federal efforts during the past thirty years [1952–1982] to devise a permanent solution to the problems of civilian radioactive waste disposal have not been adequate; (4) while the Federal Government has the responsibility to provide for the permanent disposal of high-level radioactive waste and such spent nuclear fuel as may be disposed of in order to protect the public health and safety and the environment, the costs of such disposal should be the responsibility of the generators and owners of such waste and spent fuel; (5) the generators and owners of high-level radioactive waste and spent nuclear fuel have the primary responsibility to provide for, and the responsibility to pay the costs of, the interim storage of such waste and spent fuel until such waste and spent fuel is accepted by the Secretary of Energy in accordance with the provisions of this chapter; (6) State and public participation in the planning and development of repositories is essential in order to promote public confidence in the safety of disposal of such waste and spent fuel; and (7) high-level radioactive waste and spent nuclear fuel have become major subjects of public concern, and appropriate precautions must be taken to ensure that such waste and spent fuel do not adversely affect the public health and safety and the environment for this or future generations.

By passing this act, Congress intended to find an appropriate site for the first waste and fuel repository in a place that would not jeopardize human health and the environment, establish policy for the management of such a repository and repositories to be started in the future, and constitute a nuclear waste fund, which the act states would be "composed of payments made by the generators and owners of such waste and spent fuel, that will ensure that the costs of carrying out activities relating to the disposal of such waste and spent fuel will be borne out by the persons responsible for generating such waste and spent fuel." Ben Rusche of the Department of Energy (DOE), the government agency with jurisdiction in this area, wrote, "The passage of the Nuclear Waste Policy Act of 1982 was a major milestone.... [It] established a schedule and a step-by-step process by which the President, the Congress, the States, affected Indian tribes, DOE and other Federal agencies can work together in the siting, design, construction and operation of deep, geologic repositories for [the] disposal of spent nuclear fuel generated by civilian nuclear powerplants and high-level waste resulting from atomic energy defense activities."

See also Uranium Tail Millings Radiation Control Act; Wastes, Nuclear.

Nuttall, Thomas (1786–1859)

Although British by birth, Thomas Nuttall was one of the United States' earliest naturalists and botany experts. He was born in Yorkshire, England, on 5 January 1786, the son of Jonas Nuttall, a merchant of limited means. This may explain Thomas's apprenticeship at an early age to his uncle, a printer in Liverpool. Nuttall was employed by his uncle until 1807, when they had a falling out; Nuttall then moved to London, where within a year he became mired in poverty. In

1808, at the age of 22, he immigrated to the United States and landed in Philadelphia, where he came under the influence of Dr. Benjamin Smith Barton of the Medical College of the University of Pennsylvania. Having studied the writings of German botanist Alexander von Humboldt, Nuttall was directed by Barton to explore what one source called "the lower part of the peninsula between the Delaware and Chesapeake Bays, and later to the coasts of Virginia and North Carolina." After returning from these excursions, Nuttall met and befriended Scottish naturalist John Bradbury (1768–1823), who was setting out to retrace the path of Lewis and Clark along the Mississippi River valley and the Missouri River. Nuttall joined this expedition (1809–1811), which explored the Mandan Indian villages visited by Lewis and Clark and probed areas of Wisconsin with members of the John Jacob Astor fur-trading party. This work resulted in Nuttall's election as a fellow of the prestigious Linnaean Society of London in 1813.

From 1815 to 1820, Nuttall made several other expeditions, including to Savannah, Georgia, through the southern Appalachians, and into North and South Carolina. His records of these travels became *Genera of North American Plants with a Catalogue of the Species through 1817*, a leading work in its field. That year, Nuttall was made a member of the Academy of Natural Sciences in Philadelphia and of the American Philosophical Society.

Nuttall decided to take a great trip through the Arkansas country. He started out on 2 October 1818, reaching the mouth of the Arkansas River in January 1819 and passing forward to the Osage Indian lands, where he was attacked and robbed. Sickened to the point of death, he set out for New Orleans and arrived in February 1820. He returned to Philadelphia the next year and set about collating his plant discoveries and preparing to publish his journal of the trip. In a letter dated 2 August 1821, Nuttall wrote, "I wish now to finish as soon as possible my Arkansa Flora wh[ich] drags on so heavily." That year, he published his *Journal of Travels into the Arkansa Territory during the Year 1819*.

Nuttall became one of the most prolific botanists, explorers, and scientists in the field of North American flora ever. He wrote numerous articles for the *Journal of the Academy of Natural Sciences*, as well as authoring *An Introduction to Systematic and Physiological Botany* (1827) and volumes four through six of *The North American Sylva* (1842–1849). Never married, Nuttall subsisted on a small income for many years until the death of an uncle and the devise of an estate in England required him to return to his native land. He returned to Philadelphia in 1847 and worked for a time at the Academy of Natural Sciences, transcribing the collections brought back by Dr. William Gambel from the Rocky Mountain region and California. Among the specimens was *Callipepla gambelii*, named by Nuttall the Gambel's quail, a native bird of Arizona. After completing his work, Nuttall returned to England, where he died on 10 September 1859 at the age of 73.

Oberholtzer, Ernest Carl (1884–1977)

Although barely known outside the environmental community, Ernest C. Oberholtzer was the impetus behind the establishment of the Quetico-Superior Council, which "developed and advocated a regional plan for conservation of natural resources and preservation of wilderness in the United States–Canadian border lakeland of Minnesota and Ontario." Born in Davenport, Iowa, on 6 February 1884, Oberholtzer was the son of Henry Risced and Rosa (nee Carl) Oberholtzer. He attended Davenport public schools and Harvard University, where he was awarded a bachelor's degree in 1907. According to Oberholtzer, "Probably a severe bout with rheumatic fever at 17 made more difference than anything else, for it put a premium on health throughout college and for long afterwards." After a year of graduate studies in landscape architecture, he quit and devoted his time to studying and writing about geography, wildlife, and American Indians.

After a short stint as the editor of the *Rock Island* [Illinois] *Union* from 1908 to 1909, Oberholtzer traveled by canoe some 3,000 miles in the Rainy Lake watershed of Minnesota and Ontario, one of the dreams of his life. Wrote Oberholtzer, "This watershed with its thousands of connected lakes in both countries is larger than Massachusetts, Connecticut, and Rhode Island combined. It was then little known and almost entirely wild. Its Ojibway Indians were little changed from the time of Columbus. Here I traveled that summer with Indians some three thousand miles by canoe, improving steadily in health." In 1910, Oberholtzer went to England for three months to lecture on Canadian geography and even presented a paper, "Habits of Moose," to the London Zoological Society. While in England, he was named U.S. vice and deputy consul in Hanover, Germany, but resigned after only a few months to return to his beloved wilderness. According to his biography, a 1912 trip involved "one Canadian Indian named Billy Magee" and stretched "to the Barren Lands of Northwest Canada, including an exploration and sketch-mapping of Nueltin Lake and the Thlewiaza River." After 1913, Oberholtzer made his home on an island in Rainy Lake, Minnesota.

After fighting a dam proposal in 1926 that he believed would destroy Rainy Lake, Oberholtzer helped form the Quetico-Superior Council, a conservation group concerned about the lands and rivers on the Minnesota-Ontario border. He lobbied for passage of the Shipstead-Nolan Act of 1930, which withdrew from private use all the lands along that border. In 1935, he was with Benton MacKaye, Olaus Murie, Robert Marshall, and others when they founded the Wilderness Society. He was a member of its governing council from its inception until 1968. Oberholtzer was awarded the U.S. Department of the Interior's Conservation Award in 1967. Ernest Oberholtzer died at his home on Rainy Lake on 6 June 1977 at the age of 93.

See also Shipstead-Nolan Act; Wilderness Society.

Oil Pollution Act (43 Stat. 604)

Enacted on 7 June 1924, this legislation was one of the first congressional actions to control the pollution of water by oil. The act called on the secretary of war to punish those who discharged oil products (including oil sludge and oil refuse) with a fine of no more than $2,500 nor less than $500, or with imprisonment of no more than one year nor less than 30 days. This law was later supplemented by the Oil Pollution Act of 1990, which was enacted because of the *Exxon Valdez* oil spill in 1989. This amendment clarified certain terms and allowed for compensation due to oil spills not caused by "an act of God."

Olmsted, Frederick Law (1822–1903)

Frederick Law Olmsted, who designed and planned Central Park in New York City and over 80 other recreational areas, was a pioneer landscape architect and as such was also a supporter of the conservation movement in the late nineteenth century. Born in Hartford, Connecticut, on 26 April 1822, he was the son of John Olmsted, a prosperous merchant, and Charlotte Law (nee Hull) Olmsted, who died when Frederick was about four. His father subsequently married Mary Ann Bull, who raised Olmsted and his brother John. Frederick attended private rural schools and was supposed to follow his brother to Yale, but he suffered a near-fatal case of sumac poisoning. After recovering, he studied civil engineering for three years. He eventually found work at a dry-goods importer in New York. Two years later he moved on to study at Yale for a short time before heading off to the Far East by sea. When he returned to the United States, he was 22 years old.

A love of nature and agriculture was the impetus behind Olmsted's study of farming techniques under his uncle and the establishment of a farm on what is now Staten Island, New York. He worked on the farm unitl 1850, when he began literary pursuits. His adventures in Great Britain were recorded in *Walks and Talks of an American in England* (1852). This work led the *New York Times* to hire him as an observer of conditions in the antebellum South. Olmsted's commentaries in the *Times* were encapsulated in his *A Journey in the Seaboard Slave States* (1856). The success of this work led to a trip with his brother John through Texas, which resulted in *A Journey through Texas* (1857) and *A Journey in the Back Country* (1860).

It was during this time that men such as Asa Gray, Peter Cooper, and Washington Irving recommended that Olmsted be hired to create a park in New York City that would resemble those of the great European cities. Olmsted's "conservation ethic" was embraced by the city, which asked him to establish a massive park in its center. With his partner, British-born landscape artist Calvert Vaux, he submitted a grand design that was to become Central Park. Wrote author Bill Vogt, "It's easy to see why Central Park became the cornerstone of Olmsted's reputation during the 1800s, after he and Vaux submitted the winning entry in a contest for the park's design. The entire area was man-made, literally from the ground up. It sprang from an 843-acre eyesore of stinking quagmire, rubbish heaps, rocky outcroppings and squatters' shacks. Under Olmsted's direction as park administrator, more than 2,000 workers installed miles of ducting to create a lake, hauled hundreds of thousands of tons of topsoil to create the impression of pastoral meadows, and planted thousands of trees to screen out the burgeoning metropolis." Originally called Greensward, this became Olmsted's chief work. Yet he created much more. His love of the environment led him to spend $2,000 of his own money to map the Yosemite Valley and advocate for its federal protection. His new-found reputation led to an appointment during the Civil War as executive secretary of the U.S. Sanitary Commission (1861–1863), which was established to assist the wounded and provide medical assistance for the Union army. In 1865, he returned to New York, where Vaux had begun work on Prospect Park in Brooklyn. He later laid out Fairmont Park in Philadelphia (at 3,845 acres, it became the largest park in the United States); the Stanford University campus in 1888; the lands around George Vanderbilt's estate, called Biltmore, in Asheville, North Carolina; and the acreage around the Columbian Exposition in Chicago in 1893. This latter work is considered, aside from Central Park, his greatest legacy. In all, he planned over 80 public parks.

Olmsted believed in an integrated system of city parks rather than haphazardly designed individual parks. He discussed these ideas in an 1871 article, "Public Parks and the Enlargement of Towns," which appeared in the *Journal of Social Science:*

A park fairly well managed near a large town, will surely become a new centre of that town. With the determination of location, size, and boundaries should therefore be associated the duty of arranging new trunk routes of communications between it and the distant parts of the town existing and forecasted.... They should be so planned and constructed as never to be noisy and seldom crowded, and so also that the straightforward movement of pleasure-carriages need never be obstructed, unless as absolutely necessary crossings, by slow-going heavy vehicles used for commercial purposes. If possible, also, they should be branched and reticulated with other ways of a similar class, so that no part of the town should finally be many minutes walk from some one of them; and they should be made interesting by a process of planting and decoration, so that in necessarily passing through them, whether in going to or from the park, or to and from business, some substantial recreative advantage may be incidentally gained. It is a common error to regard a park as something to be produced in complete in itself, as a picture to be painted on a canvas. It should rather be planned as one to be done in fresco, with constant consideration of exterior objects, some of them at

quite a distance and even existing as yet only in the imagination of the painter.

Olmsted worked well into the late 1890s, even when his mind began to falter. By 1898 he was senile and was committed to McLean's Hospital outside Boston, where years earlier he had designed all the landscaping. Five years later, on 28 August 1903, he died there at the age of 81.

See also Yosemite National Park Act.

Olson, Sigurd Ferdinand (1899–1982)

A noted biologist, Sigurd Olson also wrote about the wilderness. Born on 4 April 1899 in Chicago, he was the son of Lawrence and Ida May (nee Cedarholm) Olson. Olson wrote in an autobiographical sketch years later, "When I first became interested in the wilderness, I do not know. It was a long time ago and lost in the nebulous background of childhood associations. As long as I can remember, I have always headed for the bush, always had a yearning to explore wild country." He spent much of his early life on a northern Wisconsin farm exploring his surroundings, which were mostly wilderness.

Olson attended Northland College and received a bachelor's degree from the University of Wisconsin in 1923. (In 1931, he earned a master's degree in biology from the University of Illinois.) After serving in the army in 1918, he moved on to a position in a high school. In 1922, he became a nature guide. According to Paul Hansen, one of Olson's first trips was "when he guided ... Will H. Dilg of the Izaak Walton League and others on a fishing trip through the northern Minnesota lakeland that would become the Boundary Waters Canoe Area."

Sigurd Olson's fame came from writing about nature and fighting with the federal government and developers in an attempt to halt the construction of dams and roads in the Boundary Waters area. His work and the work of others, such as Ernest Oberholtzer, resulted in a declaration by the Department of Agriculture that roads would not be built in those areas. Olson spent the next several decades as an environmental writer and advocate. In 1930, he helped lobby with Oberholtzer for passage of the Shipstead-Nolan Act, which prohibited the construction of public and private roads into and out of protected wilderness areas and the altering of shorelines. Olson served as a consultant (1947) to President Truman's Quentico-Superior Council, a conservation group concerned about protecting the lands and rivers on the Minnesota-Ontario border. He was president (1953–1958) of the National Parks Association as well. In 1956, he became a member of the governing council of the Wilderness Society and became president of that organization upon the death of Harvey Broome in 1968. In 1962, Secretary of the Interior Stewart L. Udall named Olson a consultant to the interior secretary and to the director of the National Park Service "on major problems in the field of wilderness preservation."

In "The Paddle," a selection from his 1958 work *Listening Point*, Olson wrote:

Paddles mean many things to those who know the hinterlands of the north. They are symbolic of a way of life and of the deep feeling of all voyageurs for the lake and river country they have known. Some time ago I received an envelope bordered in black, one of those old-fashioned conventional letters of mourning which today are no longer used. I glanced at the date and address, tried hard to remember from whom it might be. With hesitation and foreboding, I tore open the seal. Inside was a simple card edged in black and across the face of it the sketch of a broken paddle. In the lower corner was the name.

The significance of this death announcement struck me like a blow. The paddle was broken and my friend who had been with me down

the wilderness lakes of the border regions on many trips had cached his outfit forever. That broken blade meant more than a thousand words of eulogy, said far more than words could ever convey. It told of the years that had gone into all of his expeditions, of campsites and waterways. In its simple tribute were memories of the rushing thunder of rapids, the crash of waves against cliffs, of nights when the loons called madly and mornings when the wilds were sparkling with dew. It told of comradeship and meetings on the trail, of long talks in front of campfires and the smell of them, of pine and muskeg and the song of whitethroats and hermit thrushes at dusk.

Sigurd Olson lived in his beloved wilderness area near Ely, Minnesota, for the rest of his life. After snowshoeing near his home on 13 January 1982, he suffered a heart attack and died. He was 82.

Ord, George (1781–1866)

A naturalist and writer, George Ord is noted for his work on the seventh, eighth, and ninth volumes of Alexander Wilson's landmark series *American Ornithology*. Ord was born on 4 March 1781, probably in Philadelphia, the son of George Ord, a sea captain and rope maker, and Rebecca (nee Lindemeyer) Ord, the descendant of Swedish immigrants. George Ord received his education in the public schools of Philadelphia before going to work for the rope-making firm his father owned. After his father's death in 1806, Ord carried on the business.

From his earliest years, Ord was interested in science, especially ornithology. In about 1805, he became acquainted with Scottish naturalist Alexander Wilson, 15 years his senior, who was in the process of publishing what would became the first volume of *American Ornithology; or, the Natural History of the Birds of the United States*. Ord worked closely with Wilson on the preparation of the exhibits and illustrations for the volumes, more than once accompanying the Scot on expeditions for research material. Upon Wilson's death from pneumonia in 1813, Ord took it upon himself as Wilson's assistant to edit volume 8 and write a tribute to the late naturalist in volume 9. Because Ord did not break down which of the passages and research were the product of his own work, it is impossible to determine which parts of the books were his. From 1824 to 1825, he reissued Wilson's entire series with updated and additional information.

Ord is also noted for the famous Wilson-Audubon controversy, which occurred several years after the former's death. While preparing volume 9 of *American Ornithology*, Ord discovered that Audubon was releasing plates of his illustrations of American birds and tried to discredit the famed illustrator. The ensuing contention between the backers of Wilson and Ord and those of Audubon may have done more damage to Wilson's reputation than to Audubon's.

Ord's greatest work was on Wilson's volumes, but his compositions on the lives of naturalists Thomas Say and Charles Alexandre Lesueur, and the anonymous essay on American zoology he authored in William Guthrie's *New Geographical and Commercial Grammar* (1815), were also noteworthy undertakings. Because of the financial stability of his father's business, Ord was able to retire in 1829 to a life of research. Married in 1815, he had two children; his son, Joseph Benjamin Ord, became a noted artist. George Ord was a member of the Linnaean Society of London and served as president of the Philadelphia Academy of Sciences from 1851 to 1858. He died in Philadelphia on 24 January 1866 at the age of 84.

See also Say, Thomas; Wilson, Alexander.

Oregon Waste Systems, Inc. v. Department of Environmental Quality of the State of Oregon et al. (U.S. 1994)
This landmark Supreme Court case dealt with whether states could charge a higher fee for the disposal of garbage generated out of state than for in-state garbage. Oregon imposed a disposal fee of 85 cents per ton of garbage generated inside the state and $2.50 a ton for garbage sent from other states. The petitioner, Oregon Waste Systems, Inc., sued the Department of Environmental Quality, claiming that the higher fee was unconstitutional, based on the commerce clause of the U.S. Constitution. Another petitioner, Columbia Resource Company, joined the lawsuit to have the fee overturned. However, the Oregon Court of Appeals upheld the surcharge, and the Oregon Supreme Court upheld the judgment. The plaintiffs appealed to the U.S. Supreme Court. On 4 April 1994, the Court voted 7–2 (Chief Justice William Rehnquist and Justice Harry Blackmun dissenting) that the fee was invalid under the commerce clause. Writing for the Court, Justice Clarence Thomas said, "In making [a] geographic distinction, the surcharge patently discriminates against interstate commerce."

Organic Act of 3 March 1879 (20 Stat. 394)
See United States Geological Survey.

Outdoor Recreation Resources Review Act
See Outdoor Recreation Resources Review Commission.

Outdoor Recreation Resources Review Commission (ORRRC)
This council, composed of U.S. senators and representatives, strove from 1958 to 1962 to establish government policy on outdoor recreation. Sponsored in the Senate by Clinton P. Anderson of New Mexico, the Outdoor Recreation Resources Review Act (72 Stat. 238) was signed into law on 28 June 1958. The main section of the act decreed the establishment of the ORRRC, which was composed of Anderson; Sens. Henry

Dworshak of Indiana, Henry "Scoop" Jackson of Washington, Frank Barrett of Wyoming, Jack Miller of Iowa, and Arthur Watkins of Utah; and others, including seven presidential appointees and the commission chairman, Laurance S. Rockefeller.

The commission consulted with a panel of 15 government workers from various agencies and 25 people from the business and recreation communities. It probed a diversity of recreational pursuits held dear by the American people. The ORRRC's final report, *Outdoor Recreation for America*, which included 27 separate study reports, endorsed a national program of multiple uses for the nation's forestry resources while maintaining a preservationist attitude, advocated the enactment of a permanent recreation bureau in the federal government (which became the Bureau of Outdoor Recreation in 1962), and lobbied for the passage of the Land and Water Conservation Fund Act, which was accomplished in 1965. One source called the report and its 27 appendixes "the most exhaustive analysis available of recreation in the United States." The ORRRC's recommendations were reiterated by its future counterpart, the President's Council on Recreation and Natural Beauty (1966–1968), which also involved Rockefeller.

See also Anderson, Clinton Presba; Land and Water Conservation Fund Act.

Outer Continental Shelf Lands Act
(67 Stat. 462; 42 U.S.C. 1332(1)–(2))

Enacted by Congress on 7 August 1953, this legislation dealt with "all submerged lands lying seaward and outside of the area of lands beneath navigable waters ... and of which the subsoil and seabed appertain to the United States and are subject to its jurisdiction and control." These areas are commonly referred to as the outer continental shelf. In 1953, Congress sought to bring under federal authority those waters lying within the nation's three-mile boundary so as to es-

tablish a policy of leasing plots of these underwater lands for mining and oil recovery. It enacted this law, which characterized the waters above these submerged lands as "high seas" and stated that "the right of navigation and fishing therein shall not be affected." This law was amended in 1978 by the Outer Continental Shelf Lands Act Amendments.

See also Outer Continental Shelf Lands Act Amendments of 1978.

Outer Continental Shelf Lands Act Amendments of 1978
(92 Stat. 629; 42 U.S.C. 1332(3)–(6))

This amendment to the Outer Continental Shelf Lands Act of 1953 added sections 3–6 to the original act. It reads as follows:

It is hereby to be the policy of the United States that—(3) the Outer Continental Shelf is a vital national resource reserve held by the Federal Government for the public, which should be made available for expeditious and orderly development, subject to environmental safeguards, in a manner which is consistent with the maintenance of competition and other national needs; (4) since exploration, development, and production of the minerals will have significant impacts on coastal and non-coastal areas of the coastal States, and on other affected States, and, in recognition of the national interest in the effective management of the marine, coastal, and human environments, (a) such States and their affected local governments may require assistance in protecting their coastal zones and other affected areas from any temporary or permanent adverse effects of such impacts; and (b) such States, and through such States, affected local governments, are entitled to an opportunity to participate, to the extent consistent with the national interest, in the policy and planning decisions made by the Federal Government relating to exploration for, and [the] development and production of, minerals of the outer Continental Shelf; (5) the rights and responsibilities of all States and, where appropriate, local governments, to preserve and protect their marine, human, and coastal environments through such means as regulation of land, air, and water uses, of safety, and of related development and activity should be considered and recognized; and (6) operations in the outer Continental Shelf should be conducted in a safe manner by well-trained personnel using technology, precautions, and techniques sufficient to prevent or minimize the likelihood of blowouts, loss of well control, fires, spillages, physical obstruction to other users of the waters or subsoil and seabed, or other occurrences which may cause damage to the environment or to property, or endanger life or health.

Under the provisions of this act, the Department of the Interior submitted a Five-Year Offshore Oil and Gas Leasing Program for July 1987 to July 1992, which was approved by Congress. It allowed for the sale of 38 plots in 21 of 26 approved outer continental shelf planning areas.

Parker, Theodore A., III (1953–1993)

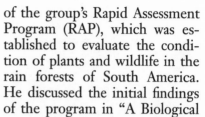

Perhaps one of the world's greatest field biologists, Theodore Parker III was also an expert in ornithology. Born 1 April 1953 in Lancaster, Pennsylvania, to Dorothy and Theodore Parker II, Parker attended local schools before enrolling at the University of Arizona in Tucson, where he was awarded a bachelor of arts degree in 1977.

In 1974, Parker began to do ornithological work at the Museum of Natural Science at Louisiana State University in Baton Rouge and joined a museum journey to Peru. After graduation in 1977, he became a research associate at the museum, where, according to Parker, he "coordinated zoological surveys for the ... museum ... primarily in Peru and Bolivia." His work in the field resulted in numerous publications on the systematics, behavior, and distribution of neotropical birds; several thousand pages of field notes; 300 specimens; and 12,000 tape recordings. Wrote Sara Engram in the *Baltimore Sun*, "Ted Parker had a good set of eyes. But it was his ears that inspired legends—his unique ability to identify 3,500 or so species of birds by their songs alone." Parker's amazing ability led him to discover the sounds of two previously unknown bird species: the orange-eyed flycatcher and a member of the *Herpsilochmus* ant wren family that he heard on a tape recording. He is also credited with codiscovering two new species: *Grallaricula ochraceifrons* and *Thryothorus eisenmanni*. A third species, *Herpsilochmus parkeri*, and a subspecies, *Metallura theresiae parkeri*, were named after him.

In 1989, Parker joined the exploration team of the environmental group Conservation International and was a creator of the group's Rapid Assessment Program (RAP), which was established to evaluate the condition of plants and wildlife in the rain forests of South America. He discussed the initial findings of the program in "A Biological Assessment of the Alto Madidi Region and Adjacent Areas of Northwest Bolivia," "Status of Forest Remnants in the Cordillera de la Costa and Adjacent Areas of Southwestern Ecuador," and "A Rapid Biological Assessment of the Columbia River Forest Reserve, Belize," all research papers produced for RAP. He and RAP team member Alwyn H. Gentry, a noted botanist, were flying over the coast of Ecuador, scouting out the dimensions of a new preserve for plants and wildlife, when their plane crashed on 3 August 1993, killing both men.

See also Gentry, Alwyn Howard.

Payne, John Barton (1855–1935)

Lawyer John B. Payne was noted during his lifetime for his position as head of the American Red Cross. Only a few historians have recognized his tenure as secretary of the interior from March 1920 to March 1921. He was born in Pruntytown, Virginia, on 26 January 1855, one of ten children of Amos Payne, a physician, and Elizabeth Barton (nee Smith) Payne. He was descended from another John Payne, who emigrated from England in 1620 to the Jamestown Colony, eventually settling in what is now Fauquier County, Virginia. His great-grandfather, Francis Payne, was an ensign during the American Revolution. When the Civil War began, Dr. Payne moved his large family to the family farm in Fauquier County. Here, following the war, John Payne attended the local rural school until he reached the age of 15. He

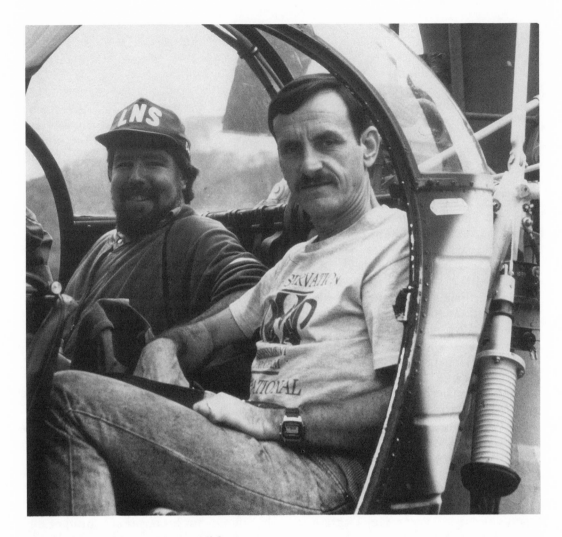

Theodore Parker (right), with botanist Al Gentry

later clerked and eventually returned to Pruntytown to read the law.

In 1876, Payne was admitted to the West Virginia bar, and the following year he began a practice in Kingwood, West Virginia. At the same time, he began to take an interest in politics. He published a small paper, the *West Virginia Argus*, and served as chairman of the Preston County Democratic Party. In 1880, he was named a special chancery court judge in nearby Tucker County. Two years later, he was elected mayor of Kingwood, a post he held until November 1882, when he picked up stakes and moved to Chicago. Some biographers speculated

that the move was intended to broaden his horizons, both physical and political. In Chicago, his expertise in the law was recognized immediately, and after several years of private practice, he was chosen to head the Chicago Law Institute. Four years later, in 1893, he was elected judge of the Cook County Superior Court. Biographer Oliver McKee, Jr., wrote that "as a judge, Payne was noted for his speed in dispatching the business of the court, his alertness on the bench, and the incisiveness and clarity of his decisions." Although elected to a six-year term, Payne resigned after only five years to return to private practice. He eventually formed a practice with Edwin Walker, a leading Chicago attorney, and then became a member of the firm of Winston, Payne, Strawn & Shaw, where he stayed until 1914. His most famous cases dealt with the defense of Chicago meat packers indicted under the Sherman Antitrust Act.

In 1913, President Woodrow Wilson offered Payne the post of solicitor general, but the overture was refused. Instead, following the outbreak of World War I, he became an arbitrator of shipbuilding strikes on the West Coast. That same year, Wilson appointed him general counsel of the U.S. Shipping Board of the Emergency Fleet Corporation, a federal agency handling shipping. He was also general counsel of the U.S. Railroad Administration. With these two jobs, Payne essentially had control over most of the transportation in the country.

On 29 February 1920, Secretary of the Interior Franklin K. Lane resigned. Two weeks later, on 15 March, Wilson named Payne as Lane's successor. There was little time for Payne to establish an imprint at Interior; he held the job for only a year. However, he immediately set out to sustain the policies of his predecessor; among these, according to Interior Department historian Eugene Trani, were "construction of the Alaskan Railroad and supervision of the terms on resource and hydroelectric leasing established during Lane's term." Added biographer Oliver McKee, "He devoted particular attention to the development of the national parks, and to the conservation of the navy's petroleum reserves." On 4 March 1921, with the inauguration of Republican Warren G. Harding as president, Payne was out of office.

Just seven months later, on 15 October 1921, Harding reached across party lines and named Payne chairman of the American Red Cross, a post he held for the rest of his life. During his tenure, he was the point man during the Japanese earthquake in 1923, the Florida hurricane of 1926, and the Mississippi Valley floods of 1927, among other disasters. Reappointed by Presidents Coolidge, Hoover, and Roosevelt, Payne earned many honors for his noble work; he was considered one of the most distinguished of political figures. In late 1934, upon returning from a Red Cross meeting in Japan, he suffered an attack of appendicitis. He underwent surgery but died on 24 January 1935. He was buried two days later, on what would have been his eightieth birthday.

See also Department of the Interior.

Payson Act (26 Stat. 1095–1103)

This legislation, sponsored in the House of Representatives by Lewis E. Payson of Illinois, was officially known as the General Revision Act of 1891, or the Creative Act of 1891. It became law on 3 March of that year to repeal the effects of the Preemption Act of 1841 and the Timber Culture Act of 1873, both of which had led to the destruction of many acres of timber and rampant land speculation in the West. More importantly, it overruled the Supreme Court case *Buford v. Houtz*, decided in 1890, which found that there was a fundamental right to graze and use the public lands.

Portions of the bill were originally drafted by forestry expert Edward A. Bowers, secretary of the American Forestry Association. Its major backers included Forestry Division chief Bernhard Eduard Fernow, Sens. Preston B. Plumb (1837–1891) and Richard Franklin Pettigrew, and Robert Underwood Johnson of *Century* magazine. By the accounts of most historians, the chief lobbyist in the government for an amendment allowing the president to set aside forestry reserves for use in national forests was Secretary of the Interior John Willock Noble. However, according to evidence found in the National Archives in the letters of Arnold Hague of the U.S. Geological Survey, Noble did not know about the amendment until its passage. In congressional conference, the important amendment later became section 24 of the act. Members of the conference included Senator Plumb of Kansas, chairman of the Senate Public Lands Committee; Senator Pettigrew of South Dakota; Sen. Edward Carey Walthall, Democrat of Mississippi; Rep. Lewis Edwin Payson, Republican of Illinois, chairman of the House Public Lands Committee, and author of the bill; Rep. John Alfred Pickler, Republican of South Dakota; and Rep. William Steele Holman, Democrat of Indiana. Author Ron Arnold has uncovered evidence that it was Holman who originally wrote section 24 and was its chief backer. Section 24 authorized the president of the United States, "from time to time, [to] set apart and reserve, in any State or Territory having public land bearing forests, in any part of the public lands wholly or in part covered with timber or undergrowth, whether commercial value or not, as public reservations, and the President shall, by public proclamation, declare the establishment of such reservations and the limits thereof." Another part of the law, section 14, was inserted at the request of Commissioner W. J. McDonald of the U.S. Commission on Fish and Fisheries to allow that agency to investigate fish resources on Afognak and Kodiak Islands in Alaska.

President Benjamin Harrison signed the bill into law and, using Secretary Noble's numerous recommendations, within a month had authorized the first preserve to be established in an area next to Yellowstone National Park; before the end of his term in 1893, he had created 15 reserves totaling 13 million acres of forest and timberland. Harrison also used section 24 to create, on 24 December 1892, the first wildlife "refuge" in the form of the Afognak Forest and Fish Culture Reserve. Gifford Pinchot called section 24 "the most important legislation in the history of Forestry in America." Representative Payson remarked later, "Ten years of continuous research has resulted in placing upon the statute books of this nation, to endure, I trust, as long as the nation shall endure, these two propositions: that not an acre of the public domain shall be taken by anybody except as a home of a poor man or for reclamation, where now there is only desert."

See also Afognak Forest and Fish Culture Reserve; Bowers, Edward Augustus; Buford v. Houtz; Hague, Arnold; Holman, William Steele; Noble, John Willock.

Peale, Titian Ramsay (1799–1885)

A noted naturalist and nature artist of the nineteenth century, Titian Peale has been all but forgotten by most historians of conservation. Born in Philadelphia on 2 November 1799, he was the son of Charles Willson Peale, a museum director and naturalist, and his second wife Elizabeth (nee De Peyster) Peale. Titian Peale had an older half brother of the same name who died at age 18 during a yellow fever epidemic.

At age 13, because of his adeptness with machinery, Peale's tutoring was suspended and he was apprenticed to a spinning machine manufacturer, with an eye toward going into the cotton-spinning business with his brother Franklin. Instead, Peale followed in his father's footsteps and began to take an interest in natural history. When he was 17, he worked with his older brother Rubens at their father's Philadelphia Museum of Natural History. Peale's education was furthered when he attended lectures in anatomy at the University of Pennsylvania.

When he turned 18, Peale traveled to Georgia and Florida on a specimen-collecting expedition with well-known naturalists William Maclure, George Ord, and Thomas Say. In 1818, Peale was selected as an assistant naturalist and expedition painter for a journey to the Rocky Mountains being conducted under the leadership of Maj. Stephen H. Long. Peale's illustrations were later used to document the expedition's reports. In 1824, Peale was sent by Charles Lucien Bonaparte to assemble flora and fauna specimens for Bonaparte's *American Ornithology* (four volumes, 1825–1833); Peale's work appeared in volumes 1 and 4. He also traveled with an 1831 expedition to South America that explored the Magdalena River (now the Rio Magdalena in Colombia). Peale's illustrations also appeared in Thomas Say's *American Entomology* (three volumes, 1824–1828). Peale's own *Lepidoptera Americana: Prospectus* (1833) was never completed.

In 1838, Peale was chosen along with Charles Pickering to serve on Charles Wilkes's U.S. Exploring Expedition. Peale's research during this trip, which lasted until 1842, was documented by his illustrations and text for what became volume 8 of the expedition's report. But a dispute with Wilkes led to his being dropped as the author of the manuscript and his replacement by ornithologist John Cassin (1813–1869), who received all the credit. Peale's original specimens from the expedition are preserved in the Academy of Natural Sciences in Philadelphia.

Because of financial difficulties, Peale left the Philadelphia Museum in 1849 to take a post as an examiner in the U.S. Patent Office in Washington, D.C. He retired from this position in 1872. His final years were spent painting, dabbling in the new field of photography, and working on a book on the butterflies of North America, which was never published. Peale died in Philadelphia on 13 March 1885. He was 85.

See also United States Exploring Expedition.

Pearson, Thomas Gilbert (1873–1943)

A noted ornithologist who headed the National Audubon Society, Thomas Gilbert Pearson was, as the *New York Times* said, "one of the best known conservationists in the country." He was the younger son and one of five children of Thomas Barnard Pearson and Mary (nee Eliott) Pearson, both farmers. He was born in Tuscola, Illinois, on 10 No-

vember 1873. The nomadic family moved to Indiana and, in 1882, to Florida, where they became citrus farmers. During these numerous travels, Thomas Pearson became interested in birds; he collected eggs and specimens to research his first article, which was published in the magazine *Oologist* in January 1888. The proceeds from the article paid the tuition for his first two years at Guilford College in Greensboro, North Carolina, where he earned a bachelor of arts degree in 1897. While attending school, Pearson served in the state geologist's office, which allowed him to continue to study nature. He eventually attended the University of North Carolina, from which he earned a bachelor of science degree in 1899.

After graduation, Pearson returned to Guilford, where he taught biology for two years and then biology and geology at the State Normal and Industrial College for Women in Greensboro from 1901 to 1904. It was during this time that he became intensely interested in the preservation of endangered bird species, particularly those whose feathers were used to decorate ladies' hats. Wrote the *New York Times*, "His interest in wildlife conservation was his life-long passion. Dr. Pearson decided to do something practical about stopping the slaughter of wild life when he was a young biology professor." His first published book, *Stories of Bird Life* (1901), inspired William E. Dutcher, president of the National Association of Audubon Societies, to suggest that Pearson form a state Audubon society, which he did in 1902; the state bestowed on it wide-ranging policing powers over wildlife protection. In 1905, Pearson helped incorporate all the state Audubon societies into one organization, the National Audubon Society, and became its secretary, serving from 1905 to 1920. In 1910, Dutcher suffered a paralytic stroke, and although he continued as Audubon's president until his death in 1920, it was essentially Pearson who ran the organization during that time. In those years, Pearson was a leading advocate of the protection of animal sanctuaries and endangered species. He was responsible for advocating passage of the all-important New York State Plumage Law in 1910 and the Weeks-McLean Migratory Bird Act in 1913. In 1922, he helped found the International Committee for Bird Preservation and headed the organization until 1938.

However, as the years went on, Pearson's effectiveness became limited by his lack of opposition to hunting. It left him vulnerable to intense criticism and led, ultimately, to his downfall. As a member of the Boone and Crockett Club, a leading hunting organization at the time, Pearson considered it "a healthful sport." Conservation historian Stephen Fox wrote, "Pearson was the last survivor from the old wildlife establishment. For almost two decades he had dominated the Audubon Association. As president he was the chief administrator; as virtual head of the board of trustees he oversaw his own presidential activities. His control was complete, almost beyond criticism." Yet starting in 1926 and continuing for eight years, critics challenged his leadership. Among them was Mabel Edge, a rich socialite from New York, who accused Pearson of being cozy with gun manufacturers. Working with activist William G. Van Name, whose book *A Crisis in Conservation* documented the National Audubon Society's receipt of funds from firearms manufacturers, she formed the Emergency Conservation Committee (ECC), which set out to remove Pearson and reform the society. Four years after its formation, the ECC helped drive Pearson from the Audubon presidency.

In his last years, Pearson worked as national director of the Izaak Walton League, served on President Herbert Hoover's Yellowstone Park Boundary Commission, and acted as an advisor to many environmental and conservation organizations. He wrote *The Bird Study Book* (1917), *Tales from Bird Land* (1918), and his autobiography, *Adventure in Bird Protection* (1937); edited *Birds of America* (three volumes, 1917) and *Portraits and Habits of Our Birds* (1920); coedited *The Book of Birds* (1937); and coauthored *Birds of North Carolina* (1919). He died in New York on 3 September 1943 at the age of 69. Frank M. Chapman, editor of the Audubon journal *Bird-Lore*, called him "the leading bird conserver of his generation."

See also Edge, Mabel Rosalie Barrow.

Pelican Island Wildlife Refuge
See Executive Order 1014 of 14 March 1903.

Pelly Amendment (Public Law 95-376)
This amendment to the Fisherman's Protective Act of 1967 was enacted on 18 September 1978 to protect sea wildlife from overharvesting. Its major provision reads, "When the Secretary of Commerce or the Secretary of the Interior finds that nationals of a foreign country, directly or indirectly, are engaging in trade or taking which diminishes the effectiveness of any international program for endangered or threatened species, the Secretary making such finding shall certify such fact to the President.... Upon receipt of any certification ... the President may direct the Secretary of the Treasury to prohibit the bringing or the importation into the United States of fish products (or wildlife products) from the offending country for such duration as the President determines appropriate and to the extent that such prohibition is sanctioned by the General Agreement of Tariffs and Trades." As defined in the body of the act, "wildlife products" means "fish and wild animals, and parts (including eggs) thereof, taken within an offending country and all products of any fish and wild animals, or parts thereof, whether or not such products are packed, processed, or otherwise prepared for export in such country or within the jurisdiction thereof."

Pennsylvania v. Union Gas Company (491 U.S. 1989)

In this Supreme Court decision, it was held that citizens were allowed to sue states under the Superfund Amendments and Reauthorization Act of 1986 (SARA), a legislative addition to the Comprehensive Environmental Response, Compensation, and Liability Act of 1980 (CERCLA). In an attempt to control flooding, the commonwealth of Pennsylvania purchased an easement to a piece of property along a creek and began the construction of a levee. While excavating, state workers struck a deposit of coal tar, which began to leak into the creek. The Environmental Protection Agency ordered Pennsylvania to clean up the spill. The state paid for some of the costs of the cleanup (the federal government reimbursed it for other costs) but sued the original owner of the property, the Union Gas Company, under the auspices of CERCLA. The company countersued, claiming that it was not responsible for the release of the coal tar into the creek and that, under CERCLA's definition of the liable parties ("the owner and operator of a vessel or a facility"; 42 U.S.C. 9607(a)(1)), it was not obligated to pay for the cleanup. A district court found that the gas company could not sue the state, and the Third Circuit Court of Appeals affirmed.

While the parties were waiting for the Supreme Court to hear the case, Congress enacted SARA, which allowed for "citizen suits" against states. With this new amendment in effect, the Supreme Court remanded the case to the court of appeals for a rehearing. On remand, the court of appeals reversed itself, finding that Congress intended SARA to allow citizens to sue states. On appeal to the Supreme Court, the case was upheld by a vote of 5–4. Speaking for the Court, Justice William Brennan found that CERCLA, as amended, clearly allowed for the filing of lawsuits by citizens, and such lawsuits did not violate the Eleventh Amendment powers given to the states. Dissenting in part were Justices Sandra Day O'Connor, Byron White, and Anthony Kennedy and Chief Justice William Rehnquist. On several other points, White joined the majority and Justice Antonin Scalia joined the minority.

Pettigrew Act (30 Stat. 34–36)

This legislation, known officially as the Forest Management Act of 1897, laid out how forest reserves would be established and provided for the management, protection, and administration of these reserves. The act, which became law on 4 June 1897, was sponsored in the Senate by Richard Franklin Pettigrew (1848–1926) of South Dakota. It was part of an attempt by western senators to curb the president's right to create forest reserves that were off-limits to commercial production and use. This type of legislation had been advocated in Congress for several years, originally by Rep. Thomas Chipman McRae (1851–1929), a member of the House Public Lands Committee. The idea for such a law, however, was that of forestry expert Edward A. Bowers, who had been working for its passage since the early 1890s. The McRae bill attempted to provide for the reforestation of the upper Mississippi and Missouri River basins to provide flood control to the lower Mississippi delta. Opposition to the McRae bill came from western congressmen who opposed giving the government control over the creation of new forest reserves. After several unsuccessful attempts in the Senate, Pettigrew was able to attach his revised version of the McRae bill onto a civil appropriations act that passed Congress. The final act authorized the secretary of the interior to sell "the dead, matured, or large growth trees after such trees . . . had been marked and designated" by a forester.

The Pettigrew Act was passed in response to two events: the passage of the General Revision Act of 1891, which gave the president almost exclusive authority in creating forest and wilderness reserves; and the 1894 decision by Interior Secretary Hoke Smith to stop sheep grazing on federal lands. Just before he left office in 1897, President Grover Cleveland created 13 new land preserves totaling 21.4 million acres. Outraged by that action, Congress passed the Pettigrew Act and denied Cleveland's authority to set aside all but two of the reserves. Later in 1897, Interior Secretary Cornelius Bliss decided that sheep ranchers could graze their sheep on the lands and even found that some lumbering could be done as well. This decision outraged the conservation community, even though Gifford Pinchot called the law "the most important Federal forest legislation ever enacted."

See also McRae, Thomas Chipman; National Forest Commission.

Pinchot, Gifford (1865–1946)

Chief Forester of the United States Gifford Pinchot was perhaps the leading figure in the conservation movement. He was born in Simsbury, Connecticut, on 11 August 1865, the son of James W. Pinchot, an affluent New York businessman (he ran a wallpaper enterprise), and Mary (nee Eno) Pinchot. James Pinchot's father, Cyril Constantine Désiré Pinchot, was a French Huguenot who came to the United States in 1815 to find political and religious freedom. Mary Eno Pinchot was an immediate descendant of William Phelps, one of the founding fathers of Windsor, Connecticut. Gifford Pinchot attended the prestigious Phillips Exeter Academy in New Hampshire before enrolling at Yale University in 1885; he graduated four years later.

Interested in forestry even before entering Yale, Pinchot went to Washington during his senior year to confer with government officials about becoming a forester. Disheartened by their negativism, Pinchot nevertheless went to Europe to study forestry in England, France, Germany, and Switzerland. His work at the prestigious École Nationale du Génie Rural d'Eaux et de Forêt (the National School of Forestry) in Nancy, France, was most impressive. He returned to the United States in 1890 and began a systematic program to encourage national scientific forestry management policy. He put his ideas into action while working from 1892 to 1897 in the forest preserves of Biltmore, George Vanderbilt's Asheville, North Carolina, estate. In the latter year, he was chosen as a member of the National Forest Commission, designed to establish national forestry policy. The commission's final report urged Congress to pass sweeping legislation designed to determine such policy and a plan of management for forest reserves. Because of his work on this commission, Pinchot was named chief forester of the United States in 1898 by President William McKinley.

During his tenure, Pinchot was responsible for carrying out the federal government's policies on forests. His other work during this time included being a key organizer of the Society of American Foresters (1900), the Public Lands Commission of 1903, the Inland Waterways Commission of 1907, and the Governors' Conference on the Environment (1908). As head of the Division of Forestry (renamed the U.S. Forest Service in 1905), Pinchot became perhaps the most recognized leader in the conservation movement. Wrote one source, "During his tenure, the number of national forests increased from 32 to 149, with 193 million acres. Speaking of men in his administration, President Theodore Roosevelt praised Pinchot as the man who, 'on the whole, stood first.'" His close relationship with the president gave him virtually a free hand in establishing policy during the Roosevelt years. However, he came into conflict with Roosevelt's successor, William Howard Taft, and butted heads with Secretary of the Interior Richard A. Ballinger over the issue of land frauds in Alaska. Pinchot was fired for insubordination in 1910.

Secretary of the Interior Gifford Pinchot (right) talks with President Theodore Roosevelt during a meeting of the Inland Waterways Commission on the Mississippi River in 1907.

A member of Roosevelt's Progressive Party, he ran against Republican Sen. Boies Penrose in 1914 but was defeated. In 1922, he ran for governor of Pennsylvania on the Progressive ticket and was elected. In 1926, he lost reelection in a bitter race. (He was later elected to a second term, 1931–1935.) During his life, he was a prolific writer, authoring *Government Forestry Abroad* (1891), *Biltmore Forest* (1893), *Timber Trees and Forests of North Carolina* (1896), *A Study of Forest Fires and Wood Production in New Jersey* (1899), *The Profession of Forestry* (1901), and his posthumously published autobiography, *Breaking Ground* (1947). He died of leukemia in New York City on 4 October 1946 at the age of 81.

In 1909, in one of his last reports as chief forester, Pinchot wrote:

> For the sake of the Forests themselves as well as in furtherance of the principle of the best use of all kinds of land, the settlement of such areas within National Forests as can with advantage to the public be given over to agriculture is encouraged. Settlers on or near a Forest help, under a proper administrative policy, both its protection and development. Decision as to whether or not it is to public advantage that particular tracts should be opened to settlement presents, however, a complex problem. In deciding whether the land is chiefly valuable for agriculture, the future needs of the community for timber and the expectation value of immature timber on the land must be considered, as well as the value of the agricultural crops which the land will produce if cleared. The National Forests are primarily a provision for the future. In many parts of the East, the desire for new land in earlier days led to much clearing of tracts which have since reverted to forest, and usually inferior forest, because the land was not in reality adapted to permanent agriculture. The mistake must not be repeated in the West, where the consequences would be far more serious, because in dry climates the forest is much more easily destroyed. Even in the case of land which will permanently grow good crops, but which is covered with timber certain to be in great demand later, or with young timber just nearing market size, clearing at the present time may mean a loss like

that caused by drawing money from a savings bank a few days before interest falls due.

See also Ballinger-Pinchot Affair; Forest Homestead Act; Governors' Conference on the Environment; Graves, Henry Solon; National Forest Commission; Public Land Commission of 1903.

Pittman, Key (1872–1940)

Key Pittman was the U.S. senator who sponsored the Federal Aid in Wildlife Restoration Act of 1937, also known as the Pittman-Robertson Act. Born in Vicksburg, Mississippi, on 19 September 1872, Pittman was the first of four sons of William Buckner Pittman, an attorney and Confederate veteran of the Civil War, and Catherine (nee Key) Pittman, a relative of Francis Scott Key, author of the *Star-Spangled Banner*. Key Pittman was educated by private tutors at an early age.

After studying law at Southwestern Presbyterian University in Clarksville, Tennessee, Pittman wandered to the northwestern corner of the country, where in 1892 he settled in Seattle, Washington, and was admitted to that state's bar. Within five years, he had tired of the law and cast his lot to the wind, which blew him to the Klondike for the gold rush in 1897. Two years later, he moved on to Nome, Alaska, where he worked as a miner and prospector, as well as serving as an attorney against claim jumpers. He married Mimosa June Gates in 1900, and they moved to Tonopah, Nevada. Pittman soon began to build a small fortune in mines futures.

After a decade in the mining business, Pittman decided to enter politics. In 1910, he easily won the Democratic nomination for the U.S. Senate seat up for grabs that year and faced incumbent Republican George S. Nixon in the general election. Although the Eighteenth Amendment to the Constitution, allowing for the popular election of senators, had not yet been enacted, Nixon and Pittman mutually de-

Key Pittman

cided to let the people of Nevada vote in the election instead of having the state legislature decide. It was a close race, but Nixon won by 1,105 votes. Two years later, however, Nixon died, and Pittman was named to serve the four remaining years of his term. Pittman was ultimately elected to this seat a total of five times—the last time just days before his death. In total, he spent the last 28 years of his life in the U.S. Senate.

In the Senate, Key Pittman was best known for having supported suffrage for women as early as 1916, for being a leading member and later chairman of the Senate Foreign Relations Committee, and for being a spokesman for the West's mining interests. In fact, in mid-1940, the magazine *American Mercury* called him a "Frontier Statesman." Because of his support of silver and mining, most conservationists did not consider him a friend of the environment. In fact, he was always at odds with Secretary of the Interior Harold L. Ickes over land conservation policy. However, a closer look at Pittman's Senate record shows that he was responsible for the passage of several pieces of conservation legislation, including the Pittman-Robertson Act of 1937, and for the introduction of the first bills to construct Boulder Dam. Further, he was a close friend of Sen. Francis G. Newlands, father of the Newlands Reclamation Act. After Newlands's death, Pittman wrote, "It was due to his remarkable vision, patient energy, and unanswerable arguments, during a fight over 10 years, that national reclamation is today a fact."

Pittman had just been elected to his fifth full term in the Senate when he died in Reno, Nevada, of a coronary thrombosis on 10 November 1940. He was 68 years old.

See also Pittman-Robertson Act.

Pittman-Robertson Act (50 Stat. 917)

This act, known officially as the Federal Aid in Wildlife Restoration Act of 1937, was enacted by Congress on 2 September of that year to raise monies to fund state fish and game departments. It was sponsored in the House of Representatives by A. Willis Robertson of Virginia (1887–1971), chairman of the House Conservation Committee, and in the Senate by Key Pittman of Nevada. The act placed an 11 percent excise tax on the manufacturer's price of sporting equipment and ammunition sold to hunters, with the proceeds to fund state fish and game commissions nationwide. According to one source, this total amounted to some $30 million from 1938 to 1966.

See also Pittman, Key.

Pollution Prevention Act (42 U.S.C. 13101)

This legislation, enacted on 5 November 1990, sought to establish a national policy on pollution emissions. Congress found that

the United States annually produces millions of tons of pollution and spends tens of billions of dollars per year controlling this pollution; that there are significant opportunities for industry to reduce or prevent pollution at the source through cost-effective changes in production, operation, and raw materials use. Such changes offer industry substantial savings in reduced raw material, pollution control, and liability costs as well as help protect the environment and reduce risks to worker health and safety; [and] that the opportunities for source reduc-

tion are often not realized because existing regulations, and the industrial resources they require for compliance, focus upon treatment and disposal, rather than source reduction; existing regulations do not emphasize multimedia management of pollution; and businesses need information and technical assistance to overcome institutional barriers to the adoption of source reduction practices.

Because of such findings, Congress authorized the administrator of the Environmental Protection Agency to "develop and implement a strategy to promote [pollution] source reduction." As part of this arrangement, basic procedures for the measurement of source reduction would be established; better techniques for the coordination, streamlining, and assurance of public access to the data collected as part of this act's functions would be created; and goals would be set as to when this policy would be met.

Powell, John Wesley (1834–1902)

According to one contemporary journal, explorer and geologist John Wesley Powell was "the pioneer in the scientific study of the possibilities of the reclamation of the arid region. There have been other explorers of the West who, before his time, have called attention to the opportunities for hunting, mining, and the cattle industry, but Major Powell was the first to demonstrate, from a carefully detailed consideration of the various conditions of climate and soil, that irrigation would furnish the greatest opportunities for intensive agriculture on this continent."

John Wesley Powell was born on 24 March 1834 in Mt. Morris in the Genesee Valley of western New York, the fourth of nine children of Joseph and Mary (nee Dean) Powell. He was named after John Wesley, theologian and founder of Methodism, and grew up under the stern influence of his parents,

John Wesley Powell talks with a Paiute man in Arizona in 1869.

who were English immigrants and abolitionists. Joseph Powell was, one source noted, "a licensed exhorter in the Methodist Episcopal Church, a man of strong will, deep earnestness, and indomitable courage." The family moved several times after John's birth, eventually settling in Walworth County, Wisconsin. John Wesley Powell grew up working on his father's farm while attending rural schools. He spent some time at a Methodist preparatory school and, when he was 16, began to teach in rural schools to earn money. Eventually, he enrolled at Illinois College in Jacksonville. Interested in botany and geology, Powell changed his life's vocation from the ministry to science. He took long excursions for the college to collect plant material.

With the outbreak of the Civil War, Powell enlisted in Company H of the 20th Illinois Infantry as a private, and he was later promoted to captain. (Although later in life he was addressed as Major Powell, he never attained that rank.) He saw action at the Battle of Shiloh (Pittsburg Landing, Tennessee), where he lost his right arm. Although he wanted to remain in the army, Powell was honorably discharged on 14 January 1865.

Powell was soon named professor of geology at Illinois Wesleyan College at Bloomington, and he continued to make trips west. On one of these trips, he saw waters that turned out to be the Colorado River, and he conceived the idea to explore that great waterway. Powell raised the money to finance the first Colorado River expedition from the

Smithsonian Institution and a congressional appropriation. He used four small boats—the *Emma Dean* (named after his wife), the *Maid of the Canyon*, the *No Name*, and *Kitty Clyde's Sister*—commencing the journey from Green River City, Wyoming, on 24 May 1869 and ending near what is today Las Vegas, Nevada. The trip lasted three months, and the places that Powell and his men gave names to—Desolation, Flaming Gorge, Cataract—remain so named to this day. At one point, the expedition was joined by English explorer Frank Goodman, who eventually struck out on his own. When the *Emma Dean's* three-man crew beached for a time in one of the canyons, a group of Shivwits Indians attacked the men and killed them. The boat was abandoned. In early August, according to author Joseph Judge, "Powell's party arrived where Lees Ferry, Arizona, is today. Burned by desert sun, drenched by almost daily thunder squalls, their energy sapped by constant portaging of their heavy boats, the explorers faced murderous Marble Canyon and beyond it the awesome void of the Grand Canyon." Powell wrote on 13 August, "We are now ready to start our way down great unknowns. The rushing waters break in great waves on the rocks and lash themselves in a mass of white foam. We must run the rapids or abandon the river. There is no hesitation." At the end of the trip, Powell was met by three Mormon scouts sent by Brigham Young. Four men in the pilgrimage decided to take the river to its end in California.

The findings of this expedition, as well as similar inspections in 1871, 1874, and 1875, led to the establishment of the Powell survey. At the same time, the government was subsidizing the surveys of Clarence King, George Montegue Wheeler, and Ferdinand Vandeveer Hayden. Powell's 1877 survey, officially called the Survey of the Rocky Mountain Region, led to the consolidation of all the western surveys into the U.S. Geological Survey in 1879, led by geologist Clarence King. In 1880, Powell succeeded King as director of the Geological Survey, serving until 1894.

His 1878 summary, *Report on the Lands of the Arid Region of the United States*, is a landmark. Wrote author Paul Gates, "Powell's report is important because for the first time a man of considerable scientific attainment whose major geological and topographical work had been centered in the Interior Basin, particularly Utah and the watershed of the Colorado River, attempted to draft plans for future government policy toward the remaining public lands." In 1888, Congress asked Powell to determine to what extent irrigation and reclamation policies could be used to make the arid lands of the West more livable. The U.S. Geological Survey then established the Powell Irrigation Survey under the direction of irrigation expert Frederick Haynes Newell. Powell's work laid the foundation for the creation of the Reclamation Bureau in the first decade of the twentieth century. His view that government should pay to reclaim and irrigate the West put him in direct conflict with western interests that wanted to develop the lands privately. After Congress became convinced that Powell was aiming for government control over irrigation projects, Sen. William Morris Stewart of Nevada led the fight in the Senate to block Powell's program. Under congressional authority, Stewart set up the Special Committee on Irrigation and Arid Lands, which chastised Powell. In 1894, Powell—chagrined and blocked by the government, as well as suffering from ill health—resigned as head of the Geological Survey. His death from a brain tumor on 23 September 1902 at his summer home in Haven, Maine, left the nation bereft of a scientific giant.

See also Hayden, Ferdinand Vandeveer; King, Clarence Rivers; Newell, Frederick Haynes; Powell Irrigation Survey; Stewart, William Morris; United States Geological Survey; Wheeler, George Montegue.

Powell Irrigation Survey

This survey, carried out under John Wesley Powell, director (1881–1894) of the U.S. Geological Survey, was part of a joint congressional resolution passed on 20 March 1888 (25 Stat. 526). It instructed the secretary of the interior to "make an examination of that portion of the United States where agriculture is carried on by means of irrigation, as to the natural advantages for the storage of water for irrigating purposes with the practicability of constructing reservoirs, together with the capacity of streams, and the cost of construction and capacity of reservoirs, and such other facts as bear on the question." Congress appropriated $100,000 for the study, later adding $250,000.

Wrote Everett Sterling, "Major Powell began work at once in Montana, Nevada, Colorado, and New Mexico." Added Wallace Stegner, "His [Powell's] work was threefold. . . . He had to complete the topographical mapping, make a survey of reservoir sites, catchment basins, stream flow, canal lines, and the lands in which water could most economically and efficiently be brought, and conduct an exploratory engineering survey to determine the practicability of headworks and canals." Within a year, one source noted, "the survey had determined 127 reservoir sites totalling some 2,500 acres, and 30 million acres of irrigable land in five separate river basins." In 1975, the Senate Committee on Energy and Natural Resources, in a history of the survey, called the inquiry "of inestimable value . . . because of the stream measurements that were made and the topographical maps that resulted."

On 14 February 1889, angered by what he saw as shortcomings in Powell's program, Sen. William Morris Stewart of Nevada helped establish the Senate Select Committee on Irrigation and Reclamation of Arid Lands (with himself as committee chairman) to investigate Powell's governmental irrigation program. As Stewart related, "The ambition of Major Powell to manage the whole subject of irrigation, without regard for the views of others, led him to induce the Interior Department to withdraw vast regions of the public lands preparatory to the selection of the necessary sites [for irrigation projects]. This withdrawal of public land from settlement practically closed many of the land offices in the West, and created much complaint. It became necessary to secure legislation to restore the public domain to settlement . . . [and] the result was that Major Powell was removed from his powerful position as Director of the Geological Survey."

Powell had kept the committee up to date on his program, but Stewart and the other westerners on the Irrigation Committee wanted quick results. When most of the original survey appropriation went for topographical mapping of the region, Stewart accused Powell of using the funds illegally. This, added to his other concerns about Powell's agenda, pushed the Nevada senator to act. Eventually, he succeeded in having the survey moved from the jurisdiction of the Geological Survey to the Department of Agriculture. Faced with this controversy, as well as declining health, Powell resigned in 1894. The irrigation surveys were collated under the authority of the Reclamation Service in 1902.

See also Powell, John Wesley; Senate Committee on Irrigation and Reclamation of Arid Lands; Stewart, William Morris.

Preemption Act (5 Stat. 453)

This law, enacted on 4 September 1841, allowed settlers to locate a piece of unsurveyed land and stake a claim to it at $1.25 an acre, thus opening up much of the West to legal settlement. The act was eventually repealed by the General Revision Act of 1891, also known as the Payson Act.

See also Payson Act.

President's Council on Recreation and Natural Beauty

This committee was established by Executive Order 11278, signed by President

Lyndon B. Johnson on 4 May 1966. The council, headed by Vice President Hubert H. Humphrey, worked hand in hand with the Citizens' Advisory Committee on Recreation and Natural Beauty, a nongovernmental group headed by Laurance S. Rockefeller of the Rockefeller Brothers Fund of New York. Other members of the President's Council on Recreation and Natural Beauty included Secretary of Defense Clark M. Clifford; Secretary of the Interior Stewart L. Udall; Secretary of Agriculture Orville L. Freeman; Secretary of Commerce John T. Connor; Secretary of Health, Education, and Welfare Wilbur J. Cohen; Secretary of Housing and Urban Development Robert C. Weaver; Secretary of Transportation Alan S. Boyd; Federal Power Commission chairman Lee C. White; Tennessee Valley Authority chairman Aubrey J. Wagner; and administrator of the General Services Administration Lawson B. Knott. The council released its final report, *From Sea to Shining Sea: A Report on the American Environment—Our Natural Heritage*, in 1968. In discussing the report's findings, the council wrote:

> Since the beginning of the Industrial Revolution, the people of the United States have achieved progressively higher standards of living through the increasing division of labor and the continuing development of a mechanical technology... [and] the overall result is a system of large-scale industries organized for the production of specialized goods, and government agencies organized to perform specialized services or construct specialized public works. The consequence is a cornucopia of goods and services raising the standard of living for most Americans to hitherto undreamed levels of material abundance. Yet, the sum of these specialized activities of the specialized industries and public agencies fails to satisfy certain basic human needs—

particularly the need for an orderly, balanced, attractive environment. As the specialist—technician, industrialist, or administrator—increasingly narrows the focus of his work to achieve greater efficiency, he tends to a certain narrowness of vision. In concentrating on the immediate purposes of his own work, he gives insufficient attention to the fact that certain byproducts of his activities are having a harmful effect on the common environment. The industrialist, for example, is so intent on turning out increasing quantities of goods that he neglects the fact that his factory also is fouling the public air and waters. The government official charged with building roads is so concerned with developing safe, efficient means of moving traffic that he fails to calculate the damaging impact of highways on neighborhoods or farms or scenic lands.

Wrote Humphrey, "The report can lead to more and better use of existing tools for environmental improvement. It can lead to wider application of tested techniques. It can mobilize public support for new measures. It is the Council's hope that the report will help stimulate further activities by increasing numbers of Americans."

Price, Overton Westfeldt (1873–1914)

Overton Price was one of the founders of the Society of American Foresters and served as Chief Forester Gifford Pinchot's assistant. Born in Liverpool, England, on 27 January 1873 to American parents who had moved to England following the Civil War, Price returned to the United States as a young man and attended high school in Virginia. After a year at the University of Virginia, he began a course of study in forestry that took him to the Biltmore Forest School at the estate of George W. Vanderbilt in North Carolina (where he became an acquaintance of Gifford Pinchot). He also spent two years studying at the University of Munich and a year under the tutelage of European forestry expert Sir Dietrich Brandis in European forests. Upon his return to the United States, he again spent time at Biltmore.

In 1899, Price began working as Pinchot's assistant at the Division of Forestry of the U.S. Department of Agriculture. He started with the title of agent; in 1900, he was promoted to superintendent of working plans. When the agency's name was changed to the Bureau of Forestry in 1901, Price was again promoted, this time to associate forester. As Pinchot's right-hand man, Price was integral in making agency policy. In 1905, when the bureau was renamed the U.S. Forest Service, authority over federal forest reserves was moved from the General Land Office to the Forest Service, bringing Price's leadership qualities into focus.

Price became interested in the formation of a private organization devoted to forestry matters. In 1900, he was one of seven founding members of the Society of American Foresters. He served as the society's first chairman and was a member of the editorial board that published the association's proceedings.

In 1910, Pinchot and Price were fired from the Forest Service over the Ballinger-Pinchot affair. Although this was a crushing blow both mentally and physically (Price's health soon began to fail), he went on to serve as vice president of the National Conservation Association, which had been founded by Pinchot. However, his ouster from federal forestry policy was more than Price could take. Overcome by depression, he ended his life in 1914 by shooting himself in the head. He was 41 years old.

See also Ballinger-Pinchot Affair.

Proctor, Redfield (1831–1908)

One early-twentieth-century source on forestry called Sen. Redfield Proctor of Vermont "one of the most effective friends of forest conservation in the U.S. Senate." He was born in Proctorsville, Vermont, on 1 June 1831, the youngest of four children of Jabez Proctor, a merchant, and Betsey (nee Parker) Proctor. Jabez Proctor's father, Leonard, a Revolutionary War veteran who fought at the Battles of Monmouth and Trenton, founded the small village of Proctorsville on the Vermont frontier. The early death of her husband left the care of the children to Betsey Proctor. Whatever early education her son Redfield received is not known (although one source noted that he obtained a good preparatory education); what is known is that he graduated from Dartmouth College in 1851 and Albany Law School in 1859. He was admitted to both the New York and Vermont bars that same year, but he practiced in Boston with his cousin, Judge Isaac F. Redfield.

With the outbreak of the Civil War, Proctor enlisted as a lieutenant with the 3d Vermont Regiment as a quartermaster. In July 1861, he was named to the staff of Gen. W. F. "Baldy" Smith and was commissioned a major in the 5th Vermont Volunteer Militia. In 1862, he was promoted to colonel, overseeing the 15th Vermont Volunteers. It was later that year, during the peninsular campaign, that he contracted tuberculosis and was sent home. He disobeyed orders and returned to the front, seeing action at Gettysburg before being mustered out of the service. He returned to his law practice with a new partner but quit the law in 1869.

After the war, Proctor was elected a selectman for the town of Rutland; he was later elected to the Vermont state senate, where he served as president pro tempore. In 1876, he was elected to a single term as lieutenant governor; two years later, he was elected governor over Democrat W. H. H. Bingham. In 1889, he was appointed secretary of war in the cabinet of President Benjamin Harrison and served until 2 November 1891, when he was named to the U.S. Senate to succeed George F. Edmunds, who had resigned.

In the Senate, Proctor was a member, and later chairman, of the Committee on Agriculture and Forestry. He sponsored bills on forestry and was supportive of the conservationist program. He was re-elected in 1892, 1898, and 1904 and served in the Senate until his death on 4 March 1908. Barely remembered today as one of the congressional friends of the environment, he was buried in his hometown of Proctorsville (now Proctor), Vermont.

Public Land Commission of 1879

This first-of-its-kind group was formed to delve into the issue of public land law and policy. It was established by Congress on 3 March 1879. Between the beginning of the nation and 1879, Congress had passed some 3,500 land laws—many of them overlapping or conflicting with other laws. The land commission's main job was to codify these laws, recommend changes, and examine the survey system of the West.

Appointed by President Rutherford B. Hayes to this commission were James A. Williamson, commissioner of the General Land Office (GLO); Clarence King, the new chief of the U.S. Geological Survey; Alexander T. Britton, a former GLO clerk and a member of the law firm of Britton and Gray, one of the largest firms appearing on behalf of land claimants before the GLO; John Wesley Powell, explorer of the West; and Thomas Donaldson, a loyal Republican, former registrar of the land office in Boise, Idaho, and an expert in land law. The commission spent three and a half months touring the West, conducting extensive interviews with people across the political spectrum, and recording their observations. Although the members traveled to every state, much of their final report was slanted toward California.

The final report that the commission delivered to Congress in 1880 ran to five

volumes: Volume 1 was sometimes referred to as the *Preliminary Report*. It consisted of 96 pages detailing the commission's conclusions and recommendations. There were another 673 pages of testimony and documents. Volume 2 contained a codification of all existing land laws considered permanent, and volumes 3 and 4 listed laws thought to be temporary. Volume 5, written entirely by Thomas Donaldson, was entitled *The Public Domain, Its History, with Statistics, with References to the National Domain, Colonization, Acquirement of Territory, the Survey, Administration and Several Methods of Sale and Disposition of the Public Domain of the United States, with Sketch of Legislative History of the Land, States and Territories, and References to the Land System of the Colonies, and also that of Several Foreign Governments*. This last volume contained 1,343 pages.

The commission recommended that (1) separate classifications of laws be created for those lands regarded as forestry lands, those regarded as mineral lands, and those regarded as agricultural lands; (2) all public forestlands be withdrawn from sale to individuals and companies and sold only at the discretion of the secretary of the interior; and (3) the government and railroads exchange lands to consolidate ownership. Most of the commission's report was ignored, although one historian believes that passage of the Forest Reserve Act in 1891 was made easier by the report's existence. Few if any people are interested in the commission report now, although it was a landmark in land policy discussion in its time.

Public Land Commission of 1903

Author Wayne Hage called this council "an unpaid commission composed of men in the executive departments to investigate the public lands and to give recommendations for their management." Different from the Public Lands Commission of 1879, this committee was appointed by President Theodore Roosevelt on 2 October 1903. It was chaired by General Land Office commissioner and former Wyoming governor William A. Richards; the other members were Frederick Haynes Newell of the Reclamation Service and Gifford Pinchot as the commission secretary. Roosevelt authorized the three-man panel "to report upon the condition, operation, and effect of the present land laws and to recommend such changes as are needed to effect the largest practicable disposition of the public lands to actual settlers who will build permanent homes upon them, and to secure the fullest and most effective use of the resources of the public lands."

The Public Land Commission held a series of hearings in Washington with politicians and policy makers to fulfill the president's directive. Newell and Pinchot personally toured the nation's public land areas to investigate the effect of previously passed legislation. Two reports from the commission were issued—the first on 7 March 1904 and the second, final report on 13 February 1905. The first advocated the repeal of the Timber and Stone Act of 1878, as well as the opening of agricultural lands in forest reserves for homestead entry. The final report, which was transmitted by Roosevelt to Congress, summarized: "The present laws are not suited to meet the conditions of the remaining public domain ... the number of patents issued is increasing out of all proportion to the number of homes." This final report again called for the refutation of the Timber and Stone Act; it also advocated broad presidential authority to establish grazing districts in the West. Papers inserted in the commission's final report included two on grazing by Albert F. Potter of the U.S. Forest Service and Frederick V. Coville of the Department of Agriculture. The results of this commission paved the way for the landmark Governors' Conference on the Environment in 1908.

See also Governors' Conference on the Environment; Newell, Frederick Haynes; Pinchot, Gifford.

Public Land Commission of 1929
See Garfield, James Rudolph.

Public Land Law Review Commission (1965–1969)
This committee was formed in response to the battle between environmentalists and mining interests. Pushing for enactment of what would become the Wilderness Act of 1964, environmentalists traded mining in wilderness areas and the formation of this commission in 1965 for having the chairman of the House Interior and Insular Affairs Committee, Wayne N. Aspinall of Colorado, let the bill go through his committee and come to the House floor for a vote.

The commission was composed of 19 members—including six U.S. senators and representatives and six presidential appointees—with a consulting council of people from environmental groups and representatives of all 50 state governors. It met from 1966 to 1969 for a total of ten hearings held in different areas of the nation. Some 900 witnesses delivered testimony. The commission's mandate was to review the nation's laws, rules, and codes as they pertained to public lands, wildlife refuges, forest reserves, national parks, and ranges. While hearings were ongoing, some 33 studies of the commission's work were produced, including *Digest of Public Land Laws* (1968) and *History of Public Land Law Development* by committee members Paul W. Gates and Robert W. Swenson.

In 1970, the commission's final report, *One Third of a Nation's Land*, was issued. Recommending that the public lands be opened for use by private interests, it called into question the right to set them off as part of the public domain, as demanded by environmentalists. It said that "the highest and best use of particular areas" was "dominant over other authorized uses." It recommended the enactment of laws that reflected this new national policy. The commission's suggestions were enacted in such laws as the

Federal Land Policy and Management Act of 1976, which permitted the multiple-use management of public lands under the care of the Bureau of Land Management.

See also Federal Land Policy and Management Act.

Public Rangelands Improvement Act (92 Stat. 1803)

Enacted on 5 October 1978, this legislation, according to one source, "established a long-term program to improve the condition of public rangelands" by imposing "a grazing fee to be used until 1985 which incorporated the cost of production and forage values" of livestock on federal lands into the determination of the fees to be charged for grazing.

Pumpelly, Raphael (1837–1923)

Raphael Pumpelly was a noted nineteenth-century geologist and explorer. He was born in Owego, in Tioga County, New York, on 8 September 1837, the son of William Pumpelly and his second wife Mary (nee Welles) Pumpelly. He was descended from Jean Pompilie, a Huguenot who fled religious persecution in France and ended up in Canada sometime in the seventeenth century. Mary Pumpelly was, according to one source, an accomplished artist and musician.

After attending local schools and Owego Academy, Raphael Pumpelly attended Gen. William Russell's Collegiate and Commercial Institute at New Haven, Connecticut. Two years later, he went on a tour of Europe with his mother. In 1855, while traveling through the Rhine Valley, he wrote in his *Reminiscences*, "Truly I was entering, though gropingly, into geology through the gate of romance." In 1856, he disappeared and was given up for dead by his mother. Four months later he returned, having run off to see Corsica and Elba, where he encountered mining and geological ex-

plorations. Because of his new interest in geology, he entered the Royal School of Mines in Freiberg, Saxony, Germany, the *Bergakademie*. Among his classmates were James Hague, who would later serve as a geologist with the Clarence King survey of the American West. In 1859, Pumpelly returned to the United States.

Soon after his return, Pumpelly was hired to manage the development of the Santa Rita silver mine in what is now Arizona. However, according to Pumpelly biographer Andrew Wallace, "Pumpelly's chosen field of endeavor was not the state of Arizona as Americans know it today, but rather it included the greater part of Dona Ana County, New Mexico Territory, lying west of the Rio Grande." Pumpelly administered a geological survey of the area and later told of his discoveries before a meeting of the California Academy of Sciences in 1861. While in California, he was hired by the Japanese government to conduct a study of that country. He later did further analyses in China's coal fields and Mongolia and Siberia. He summed up these investigations in the monographs *Geological Researches in China, Mongolia, and Japan* (1867) and *Across America and Asia* (1870).

In 1865, geologist Josiah Dwight Whitney asked Pumpelly to join the faculty of Harvard University as the school's first professor of mining and geology. Although he worked for Harvard in various capacities, Pumpelly never officially joined the faculty. Instead, he worked as the state geologist for Michigan and investigated the copper and iron deposits of the Lake Superior region from 1866 to 1877. He reported on these studies in "The Paragenesis and Derivation of Copper and Its Associates on Lake Superior" (*American Journal of Science*, 1874), "Copper District" (*Geological Survey of Michigan: Upper Peninsula*, 1869–1873), and "Metasomatic Development of the Copper-Bearing Rocks of Lake Superior" (*Proceedings of the American Academy of Arts and Sciences*, 1878). In 1925, two years after his death, a mineral found in Lake Superior was named pumpellyite in his honor.

After settling down, marrying, and having a family, Pumpelly served on several more geological explorations. In 1884, he was put in charge of the New England division of the U.S. Geological Survey and carried out analyses of the Precambrian remains in the area. His report, *Geology of the Green Mountains of Massachusetts*, was published as volume 23 of the U.S. Geological Survey monographs (1894). In 1903, at age 66, he set out with his son Raphael Welles Pumpelly (who later became a noted ge-

ologist in his own right) on a trip backed by the Carnegie Institution in Washington to explore the central Asian steppes of Turkestan. Their work there led to the discovery of the ancient civilization of Anau. Pumpelly's treatises on this trip, *Explorations in Turkestan: Expedition of 1903* (1905) and *Ancient Anau and the Oasis World* (1908), are perhaps his greatest works. Following this trip, he retired to his homes in Newport, Rhode Island, and Dublin, New Hampshire. He died in Newport on 10 August 1923 at the age of 85. His friend Charles Poston, in his 1878 work *Apache Land*, called Pumpelly "the Prince of Mining men."

Rafinesque, Constantine Samuel (1783–1840)

The techniques and writings of Constantine Rafinesque, considered one of the best early American naturalists, were later used by Charles Darwin. Rafinesque was born in Galata, near Constantinople (now Istanbul), Turkey, on 22 October 1783, the son of Georges Rafinesque, a French merchant from Marseilles, and Madeleine (nee Schmaltz) Rafinesque, who was born in Greece of German parentage. Until 1814, Rafinesque was known by the name Constantine Rafinesque-Schmaltz. The family returned to Marseilles before Constantine was a year old. In 1793, while on a voyage to the United States, Georges Rafinesque died. Constantine was tutored privately in Leghorn, Italy, and Marseilles, and he devoured books on travel and the natural sciences.

In 1802, Rafinesque traveled to the United States with his brother and settled in Philadelphia. Although he worked for the Clifford Brothers countinghouse, he spent much of his time exploring the general area in search of unique botanical specimens. He met with such dignitaries as Dr. Benjamin Rush and President Thomas Jefferson and visited the Great Dismal Swamp in Virginia. In 1805, Rafinesque returned to Leghorn, Italy, where he was employed as the secretary to the U.S. consul to Italy until 1815. While on Sicily, he discovered the medicinal squill—an herb in the lily family—which later became a leading export of the island. He published the results of his discoveries on Sicily in 1810.

Rafinesque sailed for the United States in 1815 and was almost killed when his ship wrecked. All his botanical specimens and notes were lost in the disaster. He eventually made it back to Philadelphia, where he began work on his *Antikon Botanikon*, which was not published until 1840, the year of his death. Wrote one source, "In the United States, [he] traveled more extensively than any other naturalist." His subsequent journeys and investigations, including treks through New York's Hudson Valley, Long Island, the Alleghenies, and along the Ohio River into Kentucky, led to over 900 manuscripts (including his work in Europe) and countless pages of notes and examinations of American botanical and animal specimens. From 1818 until 1826, he was a professor of botany and natural history at Kentucky's Transylvania University. In 1820, his work *Ichthyologia Ohioensis* appeared; four years later, his *Ancient History, or Annals of Kentucky* was released. His other works include *Medical Flora, etc. of the United States* (two volumes, 1828–1830), *American Manual of the Grape-Vines* (1830), *American Florist* (1832), *A Life of Travels and Researches in North America and South Europe* (1836), *New Flora and Botany of America* (1836), *Flora Telluriana* (four volumes, 1836–1838), and *The American Monuments of North and South America* (1838). Other publications dealt with economics, the Bible, and even some poetry. Most if not all of these works are rare and nearly impossible to find.

Rafinesque's extensive travels depleted what wealth he had. In his final years in Philadelphia, he lived in obscurity and poverty. He died on 18 September 1840 and was buried in an unmarked (and now lost) pauper's grave. Today, his research is barely known outside the scientific community, even though he pioneered many of the naturalist doctrines that paved the way for later naturalists of the nineteenth and twentieth centuries. Wrote biographer Joseph Ewan, "Rafinesque missed greatness by embracing too many fields of knowledge, yet the rule of priority in systematic biology, which requires the earliest validly published description to be honored, has forced the recognition of this rejected naturalist. His contemporaries gave scant heed to his voluminous, erratic writings. . . . Yet in Rafinesque's espousal of the emerging schemes of natural plant classification over the artificial 'sexual system' of Linnæus, he was ahead of his time."

Ransdell-Humphreys Flood Control Act
See Flood Control Act, First.

Ray, Dixy Lee (1914–1994)
In *Trashing the Planet*, Dixy Lee Ray wrote, "Reading the headlines and lead stories in newspapers or listening to television and radio news, one could conclude that we Americans are the most gullible of people and certainly the most easily frightened. From simple scare sto-

Dixy Lee Ray, August 1957

ries about carcinogens lurking in everything we eat, breathe, and touch to truly stupendous claims of earth-destroying holes in the sky, global changes in climate, and doom for Western society, we have been panicked into spending billions of dollars to cure problems without knowing whether they are real." The first woman to be governor of Washington State, as well as the first woman to chair the Atomic Energy Commission, Ray was a lifelong skeptic of the claims made by environmental groups. She was born Margaret Ray in Tacoma, Washington, on 3 September 1914, the second daughter and one of five children of Alvis Marion Ray, a printer, and Frances (nee Adams) Ray. She grew up on Fox Island in Puget Sound (near Seattle), where her father owned 65 acres of beachfront property. Here she developed a love of nature and the outdoors. In fact, at age 12, she climbed Mt. Rainier, Washington's tallest peak, and was the youngest woman to do so. After attending local schools, she studied zoology at Mills College in Oakland, California, and graduated with a bachelor's degree in 1937.

For five years after graduation, Ray taught in the Oakland public schools before receiving a fellowship for graduate study at Stanford, where she earned a Ph.D. in biological science in 1945. That year, she was made an associate professor of zoology at the University of Washington at Seattle, where she stayed until 1976. She became a national figure for her outspoken views against environmental alarmism. Her writings, which included *Trashing the Planet* (1990) and *Environmental Overkill* (1993), made her a frequent target of the environmental movement. On 17 July 1972, President Richard Nixon named Ray to succeed Wilfred Johnson as a member of the Atomic Energy Commission (AEC). On 6 February, Nixon appointed her chairwoman of the AEC to succeed James Schlesinger. Schlesinger praised Ray upon her initial appointment to the commission, claiming that she would be "a particular asset in handling the commission's delicate dilemma of balancing the demands of energy and environment," reported the *New York Times*. On the commission, Ray was a friend of the nuclear power industry. Although appointed to a five-year term due to end in 1977, Ray left the commission in 1975.

A Democrat, Ray was elected governor of Washington in 1976—the first female governor of that state and the second female governor in U.S. history (Ella Grasso of Connecticut was the first) to be elected without succeeding her husband. Her administration was marked by fiscal and social conservatism. But the division within her party on environmental matters and their impact on jobs and the economy cost her the governorship in 1980 in the Republican sweep by Ronald Reagan. Retiring to her home, she wrote of environmental alarmism. She died on Fox Island on 2 January 1994 at the age of 79.

Reclamation Act
See Newlands Irrigation (or Reclamation) Act.

Recycling
The trend of reusing the aluminum cans, bottles, and other garbage we dispose of has been accelerating. Newspaper, according to *Audubon* magazine, "is the leading disposable product in this country." Americans generate some 160 million tons of garbage each year, packing landfills to their capacity. Many cities are finding that recycling—once a luxury—is becoming a necessity and require that trash be separated at the curbside for recycling. Others try to promote voluntary recycling, asking residents to take their recyclables to central collection locations.

Of all recyclable materials, plastics require the least energy to recycle. Because each kind of plastic has a unique characteristic, the plastics industry developed a coding system to identify the seven types of plastic: (1) PETE or PET (polyethylene terephthalate, made into beverage containers and the most widely recycled plastic), (2) HDPE (high-density polyethylene, used in milk containers), (3) V (vinyl/polyvinyl chloride [PVC]), (4) LDPE (low-density polyethylene), (5) PP (polypropylene), (6) PS (polystyrene), and (7) others (including other resins and layered multimaterials). Because of confusion by the public, the industry announced in March 1994 that the code would be abandoned.

According to author Stephen J. Bennett, "aluminum is one of the most expensive and polluting metals to produce." However, it is by far the easiest of all reusable materials to recycle. Amazingly, Americans dispose of enough aluminum cans in three months to rebuild the entire U.S. airline fleet. Presently, the nation is recycling about 61 percent of the 81.25 million cans sold each year.

Glass is also a significant recycling problem. Americans throw away 28 billion glass bottles and jars every year, amounting to about 8 percent of landfill space. Yet only 10 percent of glass is recycled. One expert notes that glass is among the easiest materials to recycle and reuse.

What is the future of recycling? As environmental awareness increases, more people will participate more often in recycling programs. Under the Resource Conservation and Recovery Act of 1976, the federal government was mandated to procure "items composed of the highest percentage of recovered materials [those made out of recycled products] practicable, consistent with maintaining a satisfactory level of competition." In 1989, the Office of Technology Assessment, a congressional watchdog agency, said that "a clear national policy on MSW [municipal solid waste] that addresses the use of materials is essential for providing a broader context in which specific MSW programs can be developed and implemented." Thus, further congressional action may be

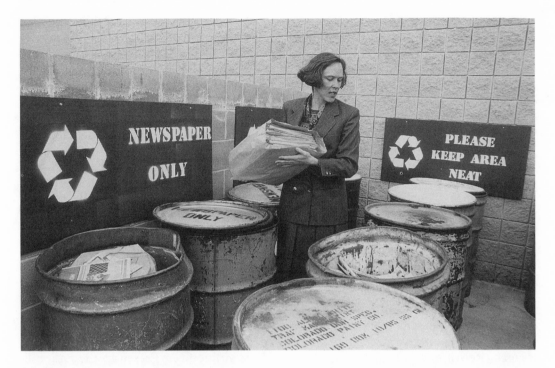

This Colorado company's recycling center, set up by employees, handles not only paper from the company but glass, newspaper, and aluminum the workers bring from home.

needed before the amount of unrecycled waste is substantially reduced.

See also Wastes, Solid.

Refuse Act (33 U.S.C. 407)

This legislation of 3 March 1899 was the first federal law dealing with the emission of refuse into navigable waters. Enacted with the Rivers and Harbors Act of 1899, this act reads:

It shall not be lawful to throw, discharge, or deposit, or cause, suffer, or procure to be thrown, discharged, or deposited either from or out of any ship, barge, or other floating craft of any kind, or from the shore, wharf, manufacturing establishment, or mill of any kind, any refuse matter of any kind or description whatever other than that flowing from streets and sewers and passing therefrom in a liquid state, into any navigable water of the United States, or into any tributary of any navigable water from which the same shall float or be washed into such navigable water; and it shall not be lawful to deposit, or cause, suffer, or procure to be deposited material of any kind in any place on the bank of any navigable water, or on the bank of any tributary of any navigable water, where the same shall be liable to be washed into such navigable water, either by ordinary or high tides, or by storms or floods, or otherwise, whereby navigation shall be impeded or obstructed; *Provided,* That nothing herein contained shall extend to, apply to, or prohibit the operations in connection with the improvement of navigable waters or construction of public works, considered necessary and proper by the United States officers supervising such improvement or public work; *And provided further,* That the Secretary of the

Army, whenever in the judgment of the Chief of Engineers anchorage and navigation will not be injured thereby, may permit the deposit of any material above mentioned in navigable waters, within limits to be defined and under conditions to be prescribed by him, provided application is made to him prior to depositing such material; and whenever any permit is so granted the conditions thereof shall be strictly complied with, and any violation thereof shall be unlawful.

See also Executive Order 11574 of 23 December 1970; Rivers and Harbors Act of 1899.

Refuse and Discharge Permits
See Executive Order 11574 of 23 December 1970.

Regional Clean Air Incentives Market (RECLAIM)
The world's first "smog market," the Regional Clean Air Incentives Market in Los Angeles, California, fixes limits for industrial pollution emissions and provides "trading credits" to 390 participating companies. The program is designed to allow smog producers to buy "credits" from other companies—those that have cut their pollution emissions—so that the total levels of emissions do not break federal clean air laws.

Reilly, William Kane (1940–)
William Reilly became the first professional environmentalist to head a major government bureau when he was chosen as the administrator of the Environmental Protection Agency (EPA) in 1989. Born in Decatur, Illinois, on 28 January 1940, Reilly was the son of George and Margaret Reilly. Raised on his family's farm until age 14, he then moved to Fall River, Massachusetts. He received a bachelor's degree in history from Yale University in 1962. Three years later, he was awarded a law degree from Yale Law School. After practicing law for a short time, he entered the U.S. Army, where he served from 1966 to 1967, rising to the rank of major. In 1968, he was named associate director of the Urban Policy Center in Washington, D.C., and in 1969 became a staff member of the Council on Environmental Quality (CEQ).

A member of the Rockefeller Task Force on Land Use and Urban Growth, Reilly was named president of the environmental group the Conservation Foundation (CF) in 1973. Over the next decade and a half, he transformed the group from a little-known think tank to a worldwide advocate for the protection of wildlife and its habitat. In 1985, Reilly was named president of the World Wildlife Fund (WWF) when CF merged with it. Under his leadership, the WWF had 600,000 members worldwide by 1989, with a budget of $35 million.

On 22 December 1988, President-elect George Bush named Reilly administrator of the EPA. Environmentalists, long skeptical of Republican administrations, cheered Reilly's appointment. "Bill Reilly is a world class environmental leader," said Jay Hair of the National Wildlife Federation. In his first official interview, Reilly stated that strengthening the Clean Air Act was the first thing on his agenda. Other issues of importance included stressing pollution prevention through waste reduction and recycling, supporting automobile fuel-efficiency standards, and endorsing stricter policing of hazardous-waste cleanup. Reilly's main job during his tenure (1989–1993) was as the government point man during the cleanup of the *Exxon Valdez* oil spill in Alaska. When he left office in 1993, he was considered one of the best administrators the EPA ever had.

See also Environmental Protection Agency.

Reorganization Plan No. 3 of 1970
See Environmental Protection Agency.

Resource Conservation and Recovery Act (RCRA)

This legislation was enacted in 1976 to supplement the Solid Waste Disposal Act of 1965. RCRA was intended to deal with the disposal of public solid waste. In 1985, the amount of such waste generated reached 275 million metric tons—more than 1 metric ton for every man, woman, and child in the nation. RCRA was passed specifically to protect human health and the environment, reduce waste and conserve energy and natural resources, and reduce or eliminate the generation of hazardous waste as quickly, safely, and cheaply as possible. In November 1984, Congress amended the act with the Hazardous and Solid Waste Amendments, which dealt with the problem of underground storage tanks that contain oil, fuel, or other hazardous wastes that may be leaking into the environment. In 1988, Congress again supplemented the act with the Subtitle J amendment, which included provisions for the identification, tracking, and proper disposal of medical waste, such as used syringes and tubing, in government-approved generators and disposal facilities.

See also Hazardous and Solid Waste Amendments; Solid Waste Disposal Act.

Right-of-Way Act (31 Stat. 790)

This law, enacted on 15 February 1901, authorized the secretary of the interior to "permit the use of rights of way through the public lands, forest and other reservations of the United States ... for electrical plants, poles and lines for the generation and distribution of electrical power, and for telephone and telegraph purposes, and for canals, ditches, pipes and pipe lines, flumes, tunnels, or other water conduits, and for water plants, dams, and reservoirs used to promote irrigation or mining or quarrying." A 1911 amendment gave the secretaries of interior and agriculture broad powers in complying with the act.

River Regulation Amendment to the Rivers and Harbors Bill of 1917

See Newlands, Francis Griffith.

Rivers and Harbors Act of 1890

See Rivers and Harbors Act of 1899.

Rivers and Harbors Act of 1899 (30 Stat. 1151)

This act of 3 March 1899 outlawed the emission or deposit into the nation's waterways of refuse that could inhibit navigation. Passed in conjunction with the Refuse Act, this law was a reincarnation of a similar one passed earlier. That act, the Rivers and Harbors Act of 1890 (26 Stat. 426), also prohibited dumping or disposing of refuse in waterways, but it was challenged in the Supreme Court in *United States v. Rio Grande Dam and Irrigation Company* (174 U.S. 690 [1899]). The 1899 legislation addressed the concerns of the Court raised in that case.

Under section 10 of the 1899 act, the U.S. Army Corps of Engineers was allowed to issue permits to dredge and fill waterways. (This act was later superseded by Executive Order 11574 of 23 December 1970, which authorized the granting of such permits to the secretary of the army, and by section 402 of the Federal Water Pollution Control Act Amendments of 1972, which gave that authority to the administrator of the Environmental Protection Agency.) Section 10 was, according to one source, "the primary statutory basis for the Corps of Engineers' program to regulate water quality through discharge permits."

See also Executive Order 11574 of 23 December 1970; Federal Water Pollution Control Act Amendments of 1972; Refuse Act; United States v. Rio Grande Dam and Irrigation Company.

Roadless Area Review and Evaluation (RARE)

Following the issuance of President Richard Nixon's Executive Order 11644 of 8 February 1972, all federal agencies concerned with natural resource conservation set out to develop policies that would balance people's right to use off-road vehicles on public lands with the conservation of those lands. The U.S. Forest Service asked its foresters to submit inventories of those forestry reserves affected under the plan. According to one source, a total of 1,449 roadless tracts totaling some 56.2 million acres were inventoried. Comments from 25,000 people at 300 hearings were solicited. Environmentalists, however, claimed that many areas were left unprotected. This led to the Eastern Wilderness Areas Act of 1974 (88 Stat. 2096), which added 16 new areas east of the 100th meridian, totaling 207,000 acres, to the National Wilderness Preservation System (NWPS) and assigned 17 other areas, totaling 125,000 acres, as wilderness study areas. Because RARE was received so poorly, the Carter administration began RARE II, which broke down wilderness areas into three categories: those slated for inclusion into the NWPS, those that could be opened for off-road vehicle use, and those under study. Because of its controversial subject, the RARE II study is still under review.

Robertson, Chief of the Forest Service, et al. v. Methow Valley Citizens Council, et al.
(490 U.S. 1989, 104 L Ed 2d 35)

In *Robertson*, the Supreme Court held that the Forest Service's environmental impact statement did not require a "fully developed plan" to alleviate environmental damage to a particular area. The U.S. Forest Service established a three-point plan to issue special-use permits for the operation of ski areas on federal lands. Because the National Environmental Policy Act of 1969 mandated that it prepare an environmental impact statement, the Forest Service decreed certain regulations on how to issue special-use permits. Under section 251.54(e)(4) of Title 36 of the Code of Federal Regula-

tions, such an applicant's "proposed measures and plans for the protection and rehabilitation of the environment during construction, operation, maintenance, and termination of the project shall be required."

Romans, Bernard (c. 1720–1784?)

An assistant to naturalist and surveyor William Gerard De Brahm, Bernard Romans later became a noted naturalist, civil engineer, and cartographer. He was born about 1720, probably in the Netherlands. No information is available on his parents. The only reports on his education suggest that he studied civil engineering in England.

About 1757, Romans was sent by the British government to North America to work as an engineer. Several years later, he was named deputy surveyor of Georgia. As a surveyor and part-time botanist, he was able to traverse what is now northern Florida and scrutinize the St. John's River. He later purchased tracts in northern Florida and southern Georgia and thus was able to expand his scientific observations. In 1769, Romans was appointed deputy surveyor of the southern district of North America, working under William Gerard De Brahm. Wrote biographer Clark Elliott, "At [his] personal expense, Romans completed an exploration of Florida and the Bahama banks and of the western coast to Pensacola, and there was engaged by the Florida Governor [Peter Chester] and the superintendent of Indian Affairs [John Stuart] to take part in a survey of West Florida." Chester asked the king of England for the right to use Romans's botanical skills to map the flora and fauna of Florida, and on his recommendation, Romans was named the king's botanist in Florida. In 1774, Romans wrote articles in the *Royal American Magazine* about indigo and madder (a southwest Asian plant from which a red dye is obtained).

It was at this time that Romans became involved in the American Revolution. At the outbreak of the war, George Washington approached him for his engineering skills and recruited him to serve on the New York Committee of Safety and, later, as a supervisor on fortifications for the Highlands, near West Point. Romans later served in a Pennsylvania artillery regiment and participated in the American invasion of Canada. In 1779, he was captured and sent back to England in chains. After the war, in 1784, he attempted to make his way back to America. It is alleged that he was murdered at sea for money, but no source can confirm this. His body was never found, and there is no account of his actual fate.

Few historians credit Bernard Romans as one of the first men to actively explore and survey the southern regions of the colonies. His works, however, are numerous, although not all are scientific treatises. His *Concise Natural History of East and West Florida* (1775; a proposed second volume was never completed) was followed by *The Complete Pilot for Gulf Passage* (1789), some of which was written by De Brahm. It was reprinted five years later under the title *A New and Enlarged Book of Sailing Directions*. His two-volume series, *Annals of the Troubles in the Netherlands, from the Accession of Charles V, Emperor of Germany; a Proper and Seasonable Mirror for the Present Americans, Collected and Translated from the Most Approved Historians in the Native Tongue*, appeared in 1778 and 1782.

Roosevelt, Theodore (1858–1919)

More than any other political or social figure, Theodore Roosevelt embodied the leadership of the conservation movement of the early twentieth century. Roosevelt was born in New York City on 27 October 1858, the son of Theodore Roosevelt, Sr., a businessman and banker, and Martha (nee Bulloch) Roosevelt. Paternally, the family can be traced back to Klaes Martensen van Roosevelt, a Dutch merchant who landed in New Amsterdam (New York) in 1644.

Because of ill health throughout his childhood, Theodore Roosevelt's early education was obtained through private tutors. He later attended Harvard College, from which he graduated in 1880. There, he began work on a historical monograph, *The Naval War of 1812*, which was published in 1881 to critical acclaim. In 1882, because of his family's connection to New York trade, he was elected at age 23 to the New York state assembly, where he was noted as an independent Republican (he was not shy about taking on Republican Party bosses he differed with). In particular, he advocated political reform, especially in the area of civil service. Roosevelt was also an ally of the poor, touring sweatshops with labor leader Samuel Gompers and calling for improved working conditions for laborers.

In 1884, following the deaths of both his wife and mother on the same day, Roosevelt traveled to the western United States, where he received an education in the ways of the frontier that shaped and molded his conservation ethic. The carnage he saw in the unceasing slaughter of wildlife earned his enmity. At the end of 1887, he returned to New York, where he was an unsuccessful candidate for mayor of New York City. At that time, he met with *Forest and Stream* editor George Bird Grinnell and other sportsmen and called for the formation of an organization that would "work for the preservation of the large game of this country, further legislation for that purpose, and assist in enforcing existing laws." It was Roosevelt's suggestion that the group be named after the frontiersmen Daniel Boone and Davy Crockett. The following month, January 1888, the Boone and Crockett Club was established.

After some time in Europe, where Roosevelt met his second wife, Roosevelt's friend Sen. Henry Cabot Lodge of Massachusetts, an old-guard Republican, asked President Benjamin Harrison to appoint Roosevelt a member of the National Civil Service Commission. The

President Theodore Roosevelt at Yosemite Valley, California, in 1903

New Yorker soon used this "bully pulpit" to attack corruption, even in his own party. Roosevelt served as president of the commission until May 1895, when, with the election of a reform mayor in New York, he became president of the city's Board of Police Commissioners, where law enforcement, graft, and corruption came under his watchful eye.

With the election of William McKinley to the presidency in 1896, Lodge intervened again for his old friend Roosevelt and secured him an appointment as assistant secretary of the navy. As secretary, he made a controversial decision to take charge during the Spanish-American naval confrontation in Manila Bay and order Commodore George Dewey to attack the Spanish warships. Not satisfied to sit at a desk in Washington during the conflict, Roosevelt resigned and formed the First Volunteer Cavalry Regiment, better known as the Rough Riders. Roosevelt saw action during the Spanish-American War, most notably at the Battle of San Juan Hill. Nominated for the Congressional Medal of Honor for bravery, he returned home a national hero. Just two months after his daring charge, he was nominated by the Republicans for governor of New York and won by a plurality. Once in office, however, he surprised his Republican backers again and sought to demolish corruption wherever he saw it. Just a year and a half into his term, party bosses resolved to get him out of New York and stick him in a post that had little power. They decided to put him on the national ticket with William McKinley in 1900. His election as vice president seemed to end what little political career he had had. All that changed, however, on 6 September 1901; McKinley was speaking at the Pan American Exposition in Buffalo, New York, when an anarchist assassin shot him. Eight days later, McKinley died of his wounds, and Roosevelt became the twenty-sixth president of the United States.

It would be impossible to relate all the successes and failures of Roosevelt's nearly eight years as president. His major success, however, was in the area of conservation. He used the presidency as a "bully pulpit" to call for the protection of the environment. As one source noted, "He was our first conservationist President. He awakened in Americans a new concern for our soil and water, and he stressed the need to preserve the remaining wilderness." When he became president, 41 forest reserves had been created, totaling 46,410,209 acres; in his first year in the White House, Roosevelt established 13 new reserves amounting to some 15,500,000 acres. After six years in the White House, he had created 5 new national parks, 53 wildlife reserves, and 16 national monuments. This led angry western congressmen to enact the National Forests Act of 1907, which directed that any future set-asides or enlargements of parks, monuments, or wildlife refuges in six western states be done only through an act of Congress. While the bill sat on the president's desk awaiting his signature, Roosevelt created or expanded 32 new forest reserves totaling some 75 million acres, thus bypassing the law's intent. Known as the conservation president throughout his later life, Roosevelt sermonized, "Conservation of our natural resources is the most weighty question now before the people of the United States." On 14 March 1903, he signed Executive Order 1014, which created the Pelican Island National Refuge, the first official government-sanctioned sanctuary in the United States for marine wildlife. In 1908, he was the impetus behind the Governors' Conference on the Environment, the first gathering of national political and social leaders to discuss ways to protect the nation's natural resources.

Retirement in 1909 left the 50-year-old Roosevelt with few outlets for his energy. A lifelong hunter, he traveled to Africa in search of big game and traversed the Brazilian wilderness. In 1912, he ran a dramatic but unsuccessful campaign (which included an attempt on his life) for a third term for president, beating his former friend William Howard Taft but losing to Democrat Woodrow Wilson. His establishment of the Progressive Party (known as the Bull Moose Party), which backed his campaign, was based on his stature alone; without him, the movement soon collapsed. With the start of World War I, Roosevelt volunteered to fight, but his offer was refused. The death of his son

Quentin during the war seemed to sap the former president of life. In the midst of planning a possible run for the White House in 1920, Roosevelt's health failed, and he died in the morning hours of 6 January 1919, apparently of a blood clot in the brain. He was 60 years old.

See also Afognak Forest and Fish Culture Reserve; Boone and Crockett Club; Governors' Conference on the Environment; National Forests Act.

Roth, Filibert (1858–1925)

Filibert Roth was a forestry expert who had great influence on U.S. forestry policy during the early twentieth century. Born in the village of Wilhelmsdorf in Wurttemberg, Germany, on 20 April 1858, he was the son of Paul Raphael Roth, a merchant, and his Swiss wife Amalie (nee Volz) Roth. After attending local schools, Filibert Roth was sent to his mother's homeland as an exchange student. In 1870, he and his parents immigrated to the United States, settling on a farm near Saul City, Wisconsin. After a brief period of sheep and cattle ranching in Texas and Montana, Roth entered the University of Michigan in 1885, where, according to one source, he specialized in the natural sciences. His microscopic examinations of wood led to his being named curator of the university's museum in 1887. In 1890, he graduated with a bachelor of science degree in timber studies.

Following his graduation, Roth was employed by the Division of Forestry of the Department of Agriculture as a special agent and specialist in timber physics. After 1893, he lived in Washington. Five years later, he left the division to work as an assistant professor of forestry at Cornell University in New York, under the direction of Professor Bernhard E. Fernow. He served for a short time (1901–1903) as the first chief of the Forestry Division of the General Land Office. While in this position, he authored the paper "Administration of U.S. Forest Reserves." In it, he wrote:

Though the first federal forest reserves were created as early as March 1891, there were no laws or appropriations made for their care until June 1897. At this time, twenty-nine reserves were in existence, with a total of nearly 26 million acres. The act of June 4, 1897, provided for an administration, but failed to supply the funds, and only $18,000 were [*sic*] available for the care of reserves for 1897–1898.... During this first year nothing more than a few field agents could therefore be employed, and the care of the reserves was practically the same as before. Nevertheless, the administration of the reserves, entrusted to the General Land Office in its special-service division, was organized, and a set of rules and regulations was established June 30, 1897, which are in force today and have proven quite satisfactory.

In 1903, Roth left the government a second time to found the School of Forestry at his alma mater, the University of Michigan. He taught his craft at the school until his retirement in 1923. An advocate for the establishment of state forestry reserves and for their proper management, Roth was the author of *Timber* (1894), *Uses of Wood* (1896), *Forest Conditions in Wisconsin* (1898), *Grazing in Forest Reserves* (1902), *The First Book of Forestry* (1902), *Forest Regulation* (1914), and *Forest Valuation* (1916). A president of the Society of American Foresters from 1917 to 1918, he was also a U.S. representative to the International Forestry Congress in Brussels in 1910 and 1912. Roth died at his home in Ann Arbor, Michigan, on 4 December 1925 at the age of 67.

Rothrock, Joseph Trimble (1839–1922)

A noted forester and botanist, Joseph Rothrock is famous in his native Pennsylvania. Born on 9 April 1839 in

269

McVeytown, Pennsylvania, he was the son of Dr. Abraham Rothrock, whose grandfather had immigrated to America from Prussia early in the eighteenth century, and Phoebe Brinton (nee Trimble) Rothrock. As a young child, Joseph Rothrock was educated in the village school; he then attended the Tuscarora Academy of Academia in Juniata County and the Freeland Academy (now Ursinus College) in Montgomery County, both in Pennsylvania. At these two institutions, he became aware of the environment around him. Later, he purchased thousands of acres of forests for the state, a portion of which is today called the Rothrock State Forest.

In 1860, Rothrock enrolled in the Lawrence Scientific School at Harvard University, where he came under the influence of Asa Gray, one of the most noted botanists in the United States. Within a short time, Rothrock became Gray's chief assistant. At this point, the first battles of the Civil War were being fought, and Rothrock broke away from his studies to enlist in Company D of the 131st Pennsylvania Infantry; later he served with the 20th Regiment of the Pennsylvania Cavalry. He saw action at Antietam and was wounded at Fredericksburg. In June 1864, he was honorably discharged and returned to Harvard, where a month later he earned his bachelor of science degree. He received his M.D. degree from the University of Pennsylvania in 1867.

For the next two years, Rothrock taught at the Pennsylvania State Agricultural College, where he was a professor of botany. He became an authority on forest mycology (a branch of biology dealing with fungi) and entomology (the study of insects). In 1864, Rothrock founded and later (in 1886) became the president of the Pennsylvania Forestry Association. In 1866, he was chosen by the Smithsonian Institution to accompany the Alaska survey expedition of Robert Kennicott and Maj. Frank Pope as a botany expert. After a short period of private medical practice, he joined George Montegue Wheeler's survey of New Mexico, California, and Colorado from 1873 to 1875. His resulting work, *United States Geographical Surveys West of the 100th Meridian*, included, according to one source, descriptions of "1,168 species belonging to 637 genera, representing 104 natural orders of plants." His name lives on in the plant genus *Rothrockia*.

Rothrock was a professor of botany at the University of Pennsylvania from 1877 to 1904. In 1893, his Rothrock Commission, appointed by the governor of Pennsylvania, investigated the physical condition of state forests. After the state established a Division of Forestry, Rothrock was named its first commissioner, serving from 1895 until 1904. In 1903, he was instrumental in organizing the Pennsylvania State Forest Academy at Mont Alto, where young forest students could be trained. The school is now an arm of Pennsylvania State University. A delegate to the American Forest Congress in Washington in 1905, Rothrock spent his final years as a member of the State Forest Commission. He died at his home in West Chester, Pennsylvania, on 2 June 1922, at the age of 83. Since his death, several memorials to him have been erected, including one at his birthplace in McVeytown, one at the Mont Alto Sanitarium he established for tuberculosis patients, and another at the state capitol in Harrisburg.

See also American Forest Congress.

Ruckelshaus, William Doyle (1932–)

William Ruckelshaus, longtime lawyer and Republican politician, was the first administrator of the Environmental Protection Agency (EPA). Born in Indianapolis, Indiana, on 24 July 1932, he was the son of John and Marion (nee Doyle) Ruckelshaus. John Ruckelshaus, son of a prominent Indiana Republican and county prosecutor, was himself a politician and considered running for the

First Environmental Protection Agency administrator William Ruckelshaus (right) introduces his successor, Russell E. Train.

U.S. Senate in the 1940s, but his son persuaded him that it was not the right time for a Roman Catholic to run for high office in Indiana. William Ruckelshaus went east to obtain an education; he attended the Portsmouth Priory School in Rhode Island and, later, Princeton University. His studies at Princeton were interrupted by a short time spent as an instructor in the Signal Corps, but he completed his education and received his bachelor of arts degree cum laude in 1957. Three years later, he was awarded a law degree from Harvard University Law School.

After being admitted to the Indiana bar in 1960, Ruckelshaus joined the family's Indianapolis law firm of Ruckelshaus, Bobbitt, and O'Connor, where he worked until 1968. During that time, he served as state deputy attorney general, chief counsel of the state attorney

general's office, and, from 1966 to 1968, in the Indiana state legislature, where he was the youngest man to be majority leader. In 1968, he ran an unsuccessful U.S. Senate campaign against incumbent Birch Bayh. After his defeat, Ruckelshaus worked in the Justice Department's civil division, where, reported the *New York Times*, he "emerged as a voice of conciliation and pragmatism." Following the invasion of Cambodia by U.S. troops and the resulting student demonstrations, Ruckelshaus was named as the Nixon administration's point man to handle the protests.

In 1970, after passage of the National Environmental Policy Act of 1969, President Nixon sent Congress Reorganization Plan no. 3, which consolidated all federal agencies with jurisdiction over environmental control matters into one office: the Environmental Protection

Agency. On 6 November, he named Ruckelshaus the first administrator of the new bureau. On 2 December, Ruckelshaus was confirmed by the Senate. He eventually served two terms, 1970–1973 and 1983–1985. In his first term, reported editor Donald Whitnah, "Ruckelshaus tried to carry out Nixon's [environmental] plan by developing a 'functional' organization for the agency. He called for amalgamating EPA's programs into functional administrative offices, such as planning and management, standards and compliance, and research and monitoring." Congress enacted several laws during Ruckelshaus's first term that expanded the EPA's authority, including in the areas of noise, water, and toxic pollution. On 26 July 1973, Ruckelshaus was named by President Nixon as deputy attorney general, but he was caught less than three months later in the "Saturday Night Massacre," when Ruckelshaus refused to fire Watergate special prosecutor Archibald Cox and resigned. His second term at the EPA came on the heels of the resignation of controversial administrator Anne Burford; Ruckelshaus's job involved cleaning up the bad reputation the agency had received. After leaving the EPA a second time, Ruckelshaus entered private industry. In the late 1980s, he became the head of Browning-Ferris Industries, Inc., a Houston-based waste-management firm.

See also Environmental Protection Agency.

Safe Drinking Water Act
(42 U.S.C. 300f)

This landmark legislation, enacted by Congress on 16 December 1974, mandated "primary drinking water regulation[s]" that "appl[ied] to public water systems" and that specified "contaminants which, in the judgment of the Administrator [of the Environmental Protection Agency] may have any adverse effect on the health of persons." A public water system was defined as "a system for the provision to the public of piped water for human consumption, if such system has at least fifteen service connections or regularly serves at least twenty-five individuals." If the administrator of the Environmental Protection Agency determined that it was "not economically or technologically feasible to so ascertain the level of . . . contaminant, each treatment technique known to the Administrator which leads to a reduction in the level of such contaminant [would be] sufficient to satisfy the requirements of section 300g–1 of this title." This act was supplemented in 1986 by the Safe Drinking Water Act Amendment.

See also Safe Drinking Water Act Amendment.

Safe Drinking Water Act Amendment
(42 U.S.C. 300g–l)

Enacted by Congress on 19 June 1986, this revision of the Safe Drinking Water Act of 1974 dictated the setting of more stringent standards than those in the earlier act. "If the Administrator [of the EPA] identifies a drinking water contaminant the regulation of which, in the judgment of the Administrator, is more likely to be protective of the public health, (1) the Administrator may publish a maximum contaminant level goal and promulgate a national primary drinking water regula-

tion for . . . the identified contaminant," the act stated.

Sagebrush Rebellion

This "revolt" by state governments and private interests took place in the late 1970s and early 1980s. It was an attempt by western states to get the federal government to return its public lands to the states. It was not the first such rebellion: Author Michael McCarthy documented a similar revolution against the federal ownership of western lands in the late 1890s, following passage of the General Revision Act of 1891, President Grover Cleveland's establishment of the "Midnight Reserves," and the enactment of the Organic Act of 1891. Participating in this early insurrection were Reps. John Calhoun Bell and John Franklin Shafroth and Sens. Henry Moore Teller and Edward Oliver Wolcott, all of Colorado. From 1891 until the end of the Teddy Roosevelt administration in 1909, they battled unsuccessfully to get the lands out from under federal government control. The second rebellion, of much shorter duration, occurred at the Western Governors' Conference held in Salt Lake City from 5 to 7 June 1913. Governors from five western states called for the dissolution of the federal government's landholdings in the West. Again, nothing came of this call.

In 1979, the movement arose again, over the same issue. Leaders of this third movement called for the disposal and selling of the lands as well as their use by private individuals. On 2 June of that year, Gov. Robert List of Nevada signed a law that effectively returned 49 million acres of federal land to state control. The federal government blocked the law from going into effect, and the state of Nevada sued, asserting that since the federal government was only holding the lands in "temporary trust," it was obligated to release them. In 1981, a district court found for the federal government, essentially ending the third rebellion. Many adherents of the rebellion are now part of the wise-use movement.

See also Bell, John Calhoun; Shafroth, John Franklin; Teller, Henry Moore; Western Governors' Conference; Wise-Use Movement; Wolcott, Edward Oliver.

Sargent, Charles Sprague (1841–1927)

Charles S. Sargent was an expert in forest conservation and arboriculture. Born on 24 April 1841 in Boston, he was the son of Ignatius Sargent, a wealthy merchant, and Henrietta (nee Gray) Sargent. Charles Sargent attended Harvard University, but he was a disinterested student and graduated at the bottom of his class in 1862. The following year, he enlisted in the U.S. Army, rising to the rank of major before the end of the Civil War. For the next three years, he traveled throughout Europe, where he developed an interest in gardening and arboriculture. After returning to the United States, he was appointed in 1872 to the Botanic Garden at Harvard University and in 1873 to the newly established Arnold Arboretum at the same institution. Work on the arboretum started from scratch; Sargent collaborated with renowned landscape artist Frederick Law Olmsted to create one of the best scientific gardens in the United States.

George Perkins Marsh's landmark 1864 work *Man and Nature* extended Sargent's interests to include the conservation of natural resources and forest preservation. For the tenth census in 1880, the Department of the Interior hired him to do a study of the conditions of forests. His work, *Report on the Forests of North America (Exclusive of Mexico)* (1884), volume 9 of the census, is considered one of the most extensive investigations into the subject. Wrote author Phillip Drennon Thomas, "Sargent's careful presentation of the dis-

tribution, habits, and taxonomy of 412 species of trees quickly made the *Report* a standard reference work. It warned that if timber management policies were not altered, the nation would experience a substantial loss in its forest resources." Sargent's other accomplishments included editing *Garden and Forest* magazine (1887–1897), supervising the 14-volume encyclopedia *Silva of North America*, editing the three-volume *Plantae Wilsonianae* (1913–1917), and writing *Manual of the Trees of North America* (1905). In 1883, he traveled with Raphael Pumpelly to northern Montana and investigated what is now Glacier National Park. On his return to the East, Sargent began to write articles and lobby for the creation of the park, which occurred in 1910. In 1897, the National Academy of Sciences formed a committee to lobby Congress for legislation to protect forest preserves in the nation. Sargent was chosen to head this group, known as the Sargent Commission. Its report, *Senate Report of the Committee Appointed by the National Academy of Sciences upon the Inauguration of a Forest Policy for the Forested Lands of the United States* (1897), led to the enactment of the General Revision Act of 1897, or the Payson Act.

Noted one source, "Until his death, the development and enlargement of the living collections of the Arboretum was Sargent's chief consideration, although with this was coordinated research work on ligneous plants, particularly those composing a part of the North American flora." In 1920, he was awarded the Garden Club of America's first medal of honor; in 1923, Sargent received the Frank N. Meyer horticulture medal from the American Genetics Association. Charles Sargent died on 22 March 1927, a month shy of his eighty-sixth birthday.

See also Payson Act.

Say, Thomas (1787–1834)

Entomologist and naturalist Thomas Say has been called the father of descriptive entomology in America. He was born on 27 June 1787 in Philadelphia, the son of Benjamin Say, a physician and apothecary, and Ann (nee Bonsall) Say, granddaughter of naturalist John Bartram. The family's ancestors were Huguenots in France. The first to come to America was one William Say, who arrived at the close of the seventeenth century; his son (Benjamin's father) was also an apothecary. Benjamin and Ann Say (who died when her son was six) were Quakers, and Thomas was educated at a Quaker boarding school until he was 15. Because of his mother's connection to the Bartrams, Say had a chance to study their collections and become interested in the flora and fauna of North America.

Although he was pushed by his father to continue the family tradition and become an apothecary, Thomas Say plunged deeper into scientific study. He and his circle of friends were responsible for founding the Academy of Natural Sciences in Philadelphia in 1812. It was here that Say became an intimate of William Maclure, the Scottish-born geologist. Say was at this time a rising star in the field of naturalist studies and was engaged to write a major article on conchology, which appeared in the American edition of *Nicholson's British Encyclopedia* (1817–1818). In 1818, Say joined Maclure and naturalists George Ord and Titian Ramsay Peale on a journey to the Sea Islands and northern Florida. The following year he was named chief zoologist on Maj. Stephen Long's expedition to the Rocky Mountains. He served in a similar capacity on Long's 1823 trip to find the headwaters of the Minnesota River. After his return to Philadelphia, Say was curator of the American Philosophical Society's collections and was named professor of natural history at the University of Pennsylvania, where he stayed until 1828.

Say became interested in the utopian New Harmony movement in Indiana, a commune run by idealist Robert Owen. On Maclure's urging, Say, along with French naturalist Charles Alexandre Lesueur and Dutch geologist Gerard Troost, traveled to New Harmony and lived there for a while. Wrote Say's biographer Elizabeth N. Shor, "This idealistic community had been established as an escape from the harshness of clamoring cities and as proof that beauty, culture, and science could flourish where all worked willingly together. It failed. Say was among its victims, for, although hopeless at financial matters, he stayed as Maclure's agent after the latter's departure; and the malarial climate on the Wabash River contributed to Say's early death." It was at New Harmony, however, that Say finished his most important and monumental work, *American Conchology*, which appeared in six volumes from 1830 to 1834; completed the third and final volume of *American Entomology; or Descriptions of the Insects of North America* (1817–1828); and edited the first volume of Charles Bonaparte's *American Ornithology; or the Natural History of Birds Inhabiting the United States* (1825). At the end of his stay at New Harmony, however, Say was seriously ill with malaria. He traveled to Mexico to recuperate but never fully recovered. He returned to New Harmony, where he died on 10 October 1834 at the age of 47. It was up to Dr. John Lawrence LeConte to gather and prepare Say's notes into *The Complete Writings of Thomas Say on the Entomology of North America* (two volumes, 1859). Even though Say's specimens had been allowed to degenerate, LeConte wrote in the preface to his compilation, "His [Say's] descriptors are so clear as to leave scarcely a doubt as to the objects designated." George Ord wrote a biography of Say, as did Harry B. Weiss and Grace M. Zeigler; theirs was titled *Thomas Say, Early American Naturalist* (1931).

Schurz, Carl (1829–1906)

Possibly the greatest secretary of the interior, Carl Schurz was a German-born

Secretary of the Interior Carl Schurz, appointed by President Rutherford B. Hayes, 1877–1881

Republican who served during the administration of Rutherford B. Hayes. Born in the town of Liblar on the Rhine River near Cologne on 2 March 1829, he was the son of Christian Schurz, a village teacher, and Marianne (nee Jüssen) Schurz, the daughter of a farmer. Carl Schurz attended a small school in the nearby town of Bruhl and was then enrolled at the gymnasium (a European secondary school that prepares students for the university) in Cologne. Although forced to leave the gymnasium to obtain his father's release from debtor's prison, he was allowed to enter the University of Bonn. There he came under the influence of Professor Gottfried Kinkel, a liberal revolutionary whose cause was German unification. Schurz participated in the abortive coup that took place on 11 May 1849, taking his place in the revolutionary army as a lieutenant. Schurz was nearly captured at the fortress of Rastatt before escaping to Switzerland. He returned to Germany to rescue Kinkel in a famous escape, but Schurz then moved on to France and then England (working as a correspondent for underground German newspapers) before leaving for the United States in August 1852. As he later wrote, "I turned my eyes across the Atlantic Ocean: America and Americans, as I fancied them, appeared to me as the last repositories of the hopes of all true friends of humanity."

Schurz spent four years traveling the United States before he bought a small farm in Watertown, Wisconsin, and settled down. During the election of 1856, he campaigned in German for the Republican presidential candidate, John Charles Frémont. The following year, the Republican state convention nominated him for lieutenant governor, a race he lost by a slim margin. In 1858, he was in Illinois campaigning for Abraham Lincoln in his unsuccessful U.S. Senate race. Said the *New York Times*, "In the interval before the Presidential campaign of 1860, Schurz performed one of the most remarkable achievements in his life. The so-called Native American element in the Republican party was important in strength and in activity, especially in New England.... During 1859 Schurz delivered a series of addresses ... in support of rights for naturalized citizens." His "True Americanism" speech helped put a halt to the nativist attempt to end suffrage for naturalized citizens in Massachusetts. There was an attempt to nominate him for governor of Wisconsin, but the effort failed. Instead, Schurz headed up the Wisconsin delegation to the 1860 Republican National Convention, where his friend Abraham Lincoln was nominated for president. Schurz spent the campaign speaking before German voters nationwide, traveling some 21,000 miles.

For his support, Schurz was offered the post of minister to Spain. Shortly after arriving, however, he returned home because the European view of the disunion favored the Confederacy. When the Civil War broke out in earnest,

Schurz volunteered for duty in the Union army, and he was given his own division and the rank of lieutenant general. Schurz saw action at Bull Run, Gettysburg, Chancellorsville, and Chattanooga. After the war ended, President Andrew Johnson sent him to the South to report on conditions. When his work there was completed, Schurz was made editor of the *Detroit Daily Post* and bought half of the German-language *Westliche Post* in St. Louis.

As he became more popular in Missouri, Schurz's political fortunes began to rise. In 1868, he was nominated for the U.S. Senate, and a combination of votes from a number of forces succeeded in electing him. He took his seat on 4 March 1869 and began to agitate for civil service reform in the government. Before long, he became an enemy of President Ulysses S. Grant, a fellow Republican, and was instrumental in the formation of the Liberal Republican Party. He was a prominent leader at the 1872 convention that nominated newspaper editor Horace Greeley for president. Greeley had been Schurz's last choice to sit in the White House, as evidenced by his unenthusiastic canvass for Greeley. As author Eugene Trani noted, "His [Schurz's] speeches were against Grant [rather] than for Greeley." Greeley's landslide defeat marked the end of Schurz's drive for reform within the Republican Party. He was soon an independent, but with the election of Rutherford B. Hayes in 1876, he was forgiven his political independence and given the plum post of secretary of the interior.

Schurz's tenure as secretary of the interior (1877–1881) is regarded as the most productive administration of that department. He attempted radical reforms in land and railroad laws, oversaw the report by Franklin B. Hough on the condition of the nation's forests, guided control over the Indian reservations, called for the assimilation of the Indians into American society, and lobbied unsuccessfully against enactment of the Timber and Stone Act. Wrote author Dian Zaslowsky, "Schurz . . . took timber depredation very seriously. He had been born and raised in Germany, where the science of forestry was commonly practiced and plundering had long since fallen out of favor. He fired land agents who had gained their positions through patronage or who had developed a knack for looking in the other direction when corruption approached. Schurz replaced them with tougher, independent men. He lobbied for the novel idea of selling timber-cutting privileges while retaining the land to prevent its mistreatment." Stymied by Congress's refusal to implement his reforms (as well as its disavowal of the recommendations of the Public Land Commission of 1879), Schurz resigned on 7 March 1881. In 1889, he commented about the frustration of governmental and national stonewalling on conservation matters. On the destruction of the nation's forests, he spoke of "a public opinion, looking with indifference on this wanton, barbarous, disgraceful vandalism; a spendthrift people recklessly wasting its heritage; a Government careless of the future and unmindful of a pressing duty." Before the American Forestry Association that same year, he said, "Let me say to you that the laws of nature are the same everywhere. Whoever violates them anywhere, must always pay the penalty. No country ever so great and rich, no nation ever so powerful, inventive and enterprising can violate them with impunity. We most grievously delude ourselves if we think we can form an exception to the rule."

The last two decades of Schurz's life were spent as a journalist for the *New York Evening Post* and *Harper's Weekly*. Moving to the left on the political spectrum, he came full circle in his ideology when he campaigned for Democrat William Jennings Bryan for president in 1900. Six years later, on 14 May 1906, Schurz died after a series of illnesses struck him in a single week. He was 77.

See also Department of the Interior.

Seaton, Frederick Andrew
(1909–1974)

Fred Seaton was a newspaper publisher, a U.S. senator from Nebraska, assistant secretary of defense in the Eisenhower administration, and secretary of the interior from 1956 to 1961. The son of Fay Noble Seaton, a newspaper publisher in the Midwest, Seaton was born on 11 December 1909 in Washington, D.C., where his father was an aide to Sen. Joseph Bristow of Kansas at the time. In 1915, the Seaton family moved to Manhattan, Kansas, where Fay Seaton owned and operated the *Mercury and Chronicle*. Fred Seaton attended local schools before enrolling at the Kansas State Agricultural College to study journalism. In 1931, he was awarded a bachelor's degree.

For several years, Seaton worked for his father's newspaper chain, first as wire editor for the *Mercury* and later as city editor. He eventually became associate editor of the Seaton Publishing Company. His first interest in politics surfaced during his college days (he was the head of a Hoover for President Club in 1928). Seaton served as chairman of the Young Republicans of Kansas and as chairman of the Kansas Republican State Committee from 1934 to 1937. In 1936, he became the secretary to Republican presidential candidate Alfred M. Landon, who was defeated by Franklin D. Roosevelt that year.

In 1937, Seaton moved to Hastings, Nebraska, where he began publishing the *Daily Tribune*. In 1945, he was elected to the first of two terms in the Nebraska state legislature (1945–1949). In 1948, he managed Harold Stassen's unsuccessful bid for the Republican presidential nomination. Three years later, upon the death of Sen. Kenneth S. Wheery, Gov. Val Peterson appointed Seaton to serve out the one year remaining in Wherry's term. During the 1952 election, Seaton was an aide to presidential candidate Dwight D. Eisenhower. With Eisenhower's election, Seaton was named assistant secretary of defense for legislative and public affairs. Later, Eisenhower moved him to the White House, where he served as administrative assistant and later deputy assistant to the president.

On 15 April 1956, Secretary of the Interior Douglas McKay resigned to run for the U.S. Senate in his native Oregon. Seaton was put in charge of finding a successor to McKay, who had clashed frequently with environmental groups. Eisenhower was looking for a man who could work with these groups. Ironically, it was Seaton, a member of the Republican Party's liberal wing, who fit the bill. He took control of the department on 6 June 1956. In his four and a half years as secretary, he advanced the cause of wildlife conservation, championed increased educational assistance for American Indians, backed improvements in national parks, and was the administration's greatest supporter of statehood for Alaska and Hawaii in 1959. He also supported increasing the fees for duck stamps (from two to three dollars) to acquire new areas to be designated wildlife refuges. When the Eisenhower administration ended in January 1961, Seaton left government. In 1962, he was the unsuccessful Republican candidate for governor of Nebraska. Having been considered for the Republican vice-presidential nomination in 1960, Seaton was an advisor to Nixon in the latter's 1968 presidential campaign. After Nixon's election, he gave the president frequent counsel on environmental matters. Nixon named him chairman of the Committee on Timber and Environment. Seaton died on 17 January 1974 in Minneapolis after a long illness. He was 64.

See also Department of the Interior.

Senate Committee on
Agriculture and Forestry

This Senate committee began as the Senate Committee on Agriculture in 1825; its jurisdiction was expanded to include forestry matters on 5 February 1884. It lasted until 11 February 1977, when its name was changed to the Committee on Agriculture, Nutrition, and Forestry. The present committee, as did its forebears, oversees soil conservation, flood-control, watershed, agricultural conservation, and resource conservation and development initiatives. In 1956, the committee reported out a bill creating the Great Plains Conservation Program to be administered by the Soil Conservation Service. Among those who have served on the committee are Redfield Proctor of Vermont, Charles L. McNary of Oregon, Asle J. Gronna (1858–1922) of North Dakota, George W. Norris of Nebraska, Joseph Ransdell (1858–1954) of Louisiana, Henrik Shipstead (1881–1960) of Minnesota, Henry C. Hansbrough of North Dakota, Francis E. Warren of Wyoming, Arthur Capper (1865–1961) of Kansas, and Jonathan P. Dolliver (1858–1910) of Iowa.

See also Hansbrough, Henry Clay; McNary, Charles Linza; Norris, George William; Proctor, Redfield; Warren, Francis Emroy.

Senate Committee on Environment and Public Works

An amalgamation of the Senate Committee on the Environment and the Committee on Public Works, this committee has jurisdiction over laws involving the environment. According to the Senate rules, these matters include air pollution, environmental aspects of outer continental shelf lands, the environmental effects of toxic substances other than pesticides, environmental policy, environmental research and development, fisheries and wildlife, flood control and improvements of rivers and harbors (including environmental aspects of deepwater ports), noise pollution, nonmilitary environmental regulation and control of nuclear energy, ocean dumping, solid waste disposal and recycling, water pollution, and water resources. Among the laws that originated in the committee are the Clean Water Act, the Water Pollution Control Act of 1948, and the Refuse Act of 1899. Former committee members in-

clude Zachariah Chandler of Michigan; Andrew Johnson of Tennessee, later the seventeenth president of the United States; Harry S Truman of Missouri, later the thirty-third president; Jefferson Davis, later president of the Confederacy (1861–1865); Albert B. Fall of New Mexico; and George Sutherland of Utah, who later served as an associate justice on the U.S. Supreme Court.

See also Chandler, Zachariah; Fall, Albert Bacon.

Senate Committee on Fisheries

Overseeing the fish resources of the United States was the narrow focus of this committee's oversight responsibilities. Established on 11 January 1884, it was disbanded on 18 April 1921 as part of Congress's consolidation of committees. Members included Thomas R. Bard (1841–1915) of California, Wesley L. Jones (1863–1932) of Washington, and Harry Lane (1855–1917) of Oregon.

Senate Committee on Forest Reservations and the Protection of Game

See Senate Select Committee on Forest Reservations.

Senate Committee on Interior and Insular Affairs

Originally the Committee on Public Lands, this committee has gone through many changes. Established on 10 December 1816, it included such members as Thomas Ewing of Ohio, who served as the first secretary of the interior; James Harlan, the eighth secretary of the interior; Richard Pettigrew (1842–1926) of South Dakota; and Henry Clay Hansbrough of North Dakota. This committee is best known for the 1862 legislation that became the landmark Homestead Act. On 18 April 1921, the committee became the Committee on Public Land and Surveys. In its nearly 26 years in this form, it had such members as Reed Smoot (1862–1941) of Utah, Gerald P. Nye (1892–1971) of North Dakota, and Robert F. Wagner (1877–1953) of New York. In 1923, the committee under Chairman Smoot and member Thomas J. Walsh (1859–1933) of Montana conducted hearings into the Teapot Dome affair. On 2 January 1947, the committee was renamed the Committee on Public Lands. This change lasted a year. On 28 January 1948, it became the Committee on Interior and Insular Affairs. The committee was renamed on 11 February 1977 the Senate Energy and Natural Resources Committee, its present designation. With the new name, authority over some environmental laws was transferred to the Environment and Public Works Committee.

The chairman of the committee from 1963 to 1981 was Sen. Henry M. Jackson of Washington. Jackson, an avowed environmentalist, helped pass bills that protected wilderness areas in Alaska and set limitations on strip mining. Under James A. McClure of Idaho, who served as chairman from 1981 to 1987, the committee looked toward the free market and tried to relax controls over natural gas extraction and strip mining, but these proposals were defeated. Wrote *Congressional Quarterly* on the slant of this committee, "The committee traditionally has been controlled by westerners whose states have vast amounts of federal parks and forests, and it is attractive to Senators whose states depend on oil and gas . . . but the committee also includes advocates of wilderness and proponents of energy conservation." Other members of this committee included Clinton P. Anderson of New Mexico and Hugh A. Butler (1878–1974) of Nebraska.

See also Anderson, Clinton Presba; Ewing, Thomas; Hansbrough, Henry Clay; Harlan, James; Jackson, Henry Martin.

Senate Committee on Irrigation and Reclamation of Arid Lands

The Senate Committee on Irrigation and Reclamation of Arid Lands was created on 16 December 1891 from a special committee that had been established on 14 February 1889 by Sen. William Morris Stewart of Nevada to investigate and later block John Wesley Powell's governmental irrigation program. The committee's name was shortened to the Senate Committee on Irrigation and Reclamation in 1921. Under the Legislative Reorganization Act of 1946, the committee was dissolved into the Senate Committee on Public Lands. Members of this committee included Stewart, William Boyd Allison (1829–1908) of Iowa, Francis E. Warren of Wyoming, Frank Hiscock (1834–1914) of New York, Preston B. Plumb (1837–1891) of Kansas, and James K. Jones (1839–1908) of Arkansas.

See also Powell, John Wesley; Powell Irrigation Survey; Stewart, William Morris; Warren, Francis Emroy.

Senate Committee on Public Lands
See Senate Committee on Interior and Insular Affairs.

Senate Committee on the Conservation of National Resources
The Senate Committee on the Conservation of National Resources was created on 21 March 1909 and dissolved on 18 April 1921 as part of Congress's work to consolidate and eliminate committees. This committee oversaw conservation matters involving the nation's natural resources, including minerals and forests. Members of this committee included Joseph M. Dixon (1867–1934) of Montana, James K. Vardaman (1861–1930) of Mississippi, and Marcus Aurelius Smith (1851–1924) of Arizona.

Senate Committee on the Geological Survey
The Senate Committee on the Geological Survey was originally a select committee that investigated the workings of

the survey. On 28 July 1892, the Select Committee to Investigate the Operations of the Geological Survey was established under the chairmanship of Edward O. Wolcott of Colorado. On 15 December 1899, Congress created the Committee of the Geological Survey as a standing member of that body. It was terminated on 18 April 1921 as part of Congress's reduction of committees. In its nearly 22 years, the committee oversaw all operations of the Geological Survey and its many projects. Members of this committee included Albert B. Fall of New Mexico, the secretary of the interior best known for his involvement in the Teapot Dome affair; Marcus Aurelius Smith of Arizona; Robert L. Taylor (1850–1912) of Tennessee; and Stephen B. Elkins (1841–1911) of West Virginia.

See also Wolcott, Edward Oliver.

Senate Select Committee on Forest Reservations

This committee was formed from the short-lived Select Committee on Forest Reservations in California, which existed from 28 July 1892 to 15 March 1893 with Charles N. Felton (1828–1914) of California as the only committee chairman. Responsibilities of this early committee included the oversight of forestry resources in California that were being targeted for protection and designation as reserves by the government. The Select Committee on Forest Reservations, which encompassed a wider authority nationwide, was headed from its creation on 15 March 1893 until its reformation on 19 March 1896 as the Committee on Forest Reservations and the Protection of Game by William V. Allen (1847–1924) of Nebraska. The committee was disbanded on 18 April 1921 as part of Congress's attempt to decrease the number of committees. Members of the committee in its three forms included Allen, Gilbert M. Hitchcock (1859–1934) of Nebraska, Albert J. Beveridge (1862–1927) of Indiana, and Joseph R. Burton (1850–1923) of Kansas.

Senate Select Committee to Investigate the Operations of the Geological Survey

See Senate Committee on the Geological Survey.

Shafroth, John Franklin (1854–1922)

U.S. representative, senator, and governor of Colorado, John Franklin Shafroth was a leader in the movement against the public ownership of land in the American West. Born on 9 June 1854 in Fayette, Missouri, he was one of six children of John Shafroth, a merchant, and Anna (nee Aull) Shafroth. John Shafroth was born in Berne, Switzerland, as John Gotlieb Schoffroth in 1810 and immigrated to the United States sometime before 1840. Anna Aull was a native of Frankfurt, Germany. John Franklin Shafroth apparently attended the local schools of Fayette and Central College before enrolling at the University of Michigan in 1872. Three years later, he graduated with a bachelor's degree.

After returning to Fayette, Shafroth began to study the law under local attorney Samuel C. Major. In 1876, after being admitted to the Missouri bar, Shafroth formed a law partnership with his teacher. The firm lasted only three years, after which Shafroth moved to Denver, Colorado, where he began a law partnership with Judge Andrew W. Brazee. In 1887, Shafroth was elected city attorney of Denver and served until 1891. Three years later, he was elected as a Silver Republican (a politican who supported the silver standard instead of the gold standard for currency) to the U.S. House of Representatives, where he worked more against conservation legislation than for the imposition of silver currency. In his five terms (1895–1904), Shafroth demonstrated an interest in many issues—he worked for a reclamation act and called for women's suffrage two decades before it became a reality. When rumors surfaced about fraud in his last election, he investigated and, finding them to be true, resigned his seat on 15 February 1904. A subsequent campaign that year to regain the seat legitimately ended in failure.

His reformist philosophy won him the governorship of Colorado, which he held from 1909 to 1913. As governor, he helped push for state constitutional amendments providing for the initiative, the referendum, and the direct primary. When his second term ended on 14 January 1913, the state senate elected him to the U.S. Senate. His two terms there were marked by a clash with President Woodrow Wilson over conservation matters in the West. Denied a second term in 1918, he later served as chairman of the War Minerals Relief Commission from 1919 to 1921. Shafroth died in Denver on 20 February 1922 at the age of 67.

See also Bell, John Calhoun; Teller, Henry Moore.

Shipstead-Nolan Act (46 Stat. 1020)

Enacted on 1 July 1930, this important act in the area of the recreational use of public lands withdrew from entry (that is, denied public access to for recreation, hunting, fishing, or grazing) all public lands north of Township 60 North in Cook and Lake Counties, Minnesota. It also directed the Forest Service to conserve for recreational use the lakes and streams of the Superior National Forest in Minnesota and prohibit the construction of public and private roads into and out of these protected areas. The act was sponsored in the House of Representatives by William Ignatius Nolan (1874–1943) of Minnesota and in the Senate by Henrik Shipstead (1881–1960) of Minnesota, a member of the Senate Committee on Agriculture and Forestry.

See also Oberholtzer, Ernest Carl; Olson, Sigurd Ferdinand.

Sierra Club

This internationally known environmental group was founded on 28 May 1892 by John Muir and a number of Cal-

The Sierra Club logo

ifornia conservationists, including Professor Joachim Henry Senger of the University of California; William Dallam Armes, an English lecturer at the University of California and the club's first director; Robert Martin Price, a college student; Professor Cornelius Beach Bradley of the University of California; and Joseph LeConte, a noted geologist who was a protégé of Alexander Agassiz. Muir, known for his advocacy of the preservation of the Yosemite Valley and the Tuolumne River area, was lobbied by Robert Underwood Johnson, editor of *Century* magazine, to establish a Yellowstone and Yosemite defense association because the Boone and Crockett Club, the hunters' support group, had refused to help preserve the two parks.

In 1893, the club began publication of the *Sierra Club Bulletin* (now known as *Sierra* magazine). The first issue described the first meeting (16 September 1892) as follows:

About two hundred and fifty members and friends of the Club met at the hall of the California Academy of Sciences, 809 Market Street, San Francisco. In the absence of the President [Muir] and the Vice-Presidents, Professor J. H. Senger occupied the chair.... The Secretary, Mr. Wm. D. Armes, gave a brief account of the organization of the Club, the objects that it hoped to attain, and the methods to be followed in attaining them.... Mr. R. M. Price read a paper narrating a trip that he had recently made through the Grand Cañon of the Tuolumne from Soda Springs to the Hetch-Hetchy.... Mr. W. W. Price described a hitherto unreported grove of Sequoias, north of those heretofore generally known.... Mr. Mark B. Kerr gave an account of his attempt to reach the summit of Mt. St. Elias, illustrating his remarks with very interesting lantern-slides.

In its 100-plus years of work, the club has been responsible for efforts to establish Mt. Rainier National Park (1899), Glacier National Park (1910), and Kings Canyon National Park (1940). Among its leaders was David R. Brower, dubbed the "archdruid" of the environmental movement. Wrote club historian Holway R. Jones, "The Sierra Club ... has devoted itself to the study and protection of national scenic resources, particularly those of mountain regions." The club's own statement of purpose is: "To explore, enjoy, and protect the wild places of the earth; to practice and promote the responsible use of the earth's ecosystems and resources; to educate and enlist humanity to protect and restore the quality of the natural and human environment; and to use all lawful means to carry out these objectives." The club claims 600,000 members in 58 chapters, which are divided into 409 different groups. In addition to *Sierra* magazine, the group publishes the *National News Report*, a synopsis of national environmental news, as well as local newsletters detailing local environmental initiatives and information.

See also Brower, David Ross; Johnson, Robert Underwood; Muir, John.

Smith, Caleb Blood (1808–1864)

Attorney and Republican Party loyalist Caleb Blood Smith served as the fifth secretary of the interior. Born in Boston on 16 April 1808, he moved with his par-

ents to Cincinnati when he was six. What public education he received is not known, but he entered Cincinnati College when he was only 15. He later enrolled at Miami University in Ohio but never earned a degree. Instead, he moved to Connorsville, Indiana, studied law under a local attorney, and was admitted to the bar in 1828.

Caleb Smith entered politics almost immediately. After failing to win a seat in the Indiana state house of representatives in 1831, he entered into a partnership and bought a local paper, the *Political Clarion*, and turned it into a Whig journal, the *Indiana Sentinel*. Within a year, this helped him win the seat he had lost in 1831. He served in the state legislature from 1833 to 1837, and again from 1840 to 1841. During the former period he served as speaker of the house, and in the latter term he was chairman of the important Canals Committee, where he took a leading role in internal improvements in the state. In 1842, after losing a similar race two years earlier, Smith was elected to the U.S. House of Representatives, where he served three terms (1843–1849). He refused renomination and was considered for the job of postmaster general in Zachary Taylor's administration, but instead he was appointed to the commission that listened to claims against Mexico. In 1851, he returned to Cincinnati to practice law. From 1854 until 1859, he served as president of the Cincinnati & Chicago Railroad.

During the 1850s, Smith joined the long list of antislavery Whigs who abandoned the dying party and moved to the Republican Party. In 1860, Smith was a delegate to the second party convention; when he spoke for his delegation in favor of Abraham Lincoln, his name became famous. For this small favor that swung the convention his way, Lincoln named Smith to his cabinet as secretary of the interior. Wrote Eugene Trani, "From his assumption of the office on 5 March 1861, Smith contemplated retirement. His health was failing and he found himself unsuited to the bureaucratic regime."

Although abdicating much of his authority to Assistant Secretary John Palmer Usher, Smith oversaw passage of the landmark Homestead Act in 1862. Unable to keep up with his duties, Smith looked for a way to leave government. He resigned on 31 December 1862 after being appointed to fill a vacancy on the Indiana Supreme Court. Ill with heart trouble, he held this final post a little more than a year. On 7 January 1864, he collapsed at the courthouse and died later that afternoon. Smith was 55.

See also Department of the Interior.

Smith, George Otis (1871–1944)

A noted geologist, George Otis Smith was the fifth director of the U.S. Geological Survey. He was born in Hodgdon, Maine, on 22 February 1871, the son of Joseph Otis Smith, publisher of the *Somerset Reporter* (Skowhegan, Maine), and his second wife, Emma (nee Mayo) Smith. George Smith had an older half sister from his father's first marriage. The family could be traced back on his father's side to one Joseph Smith, who came to the colonies in 1630 and settled in Barnstable, Massachusetts. In addition to being a newspaper publisher, George's father was a Civil War veteran, Maine secretary of state, and state insurance commissioner.

Although inclined toward a career in journalism, George Smith took up the study of geology before earning a bachelor's degree in English from Colby College in Waterville, Maine, in 1893. He moved on to Johns Hopkins University, where he was awarded a Ph.D. in geology in 1896. In that same year, he became a member of the U.S. Geological Survey. Wrote survey historian Mary Rabbitt, "He immediately joined the Survey as an assistant geologist in a field assignment in the state of Washington . . . [and] for the next six years he worked on assignments in various parts of the West and participated in the preparation of five geologic folios. In 1903, he was as-

signed to diverse investigations throughout New England." From 1896 to 1906, he published many scientific papers in the survey's bulletin. In that latter year, he was named head of the survey's petrographic division (dealing with the description and classification of rocks, especially by microscopic examination). After chairing a committee to analyze survey operations, Smith won high praise from many in the government.

In 1907, survey director Charles D. Walcott resigned to become head of the Smithsonian Institution. On 1 May, Secretary of the Interior James R. Garfield named Smith as the new director. He would serve until 1930, longer than any previous survey director. According to James Penick, Jr., in those 23 years, the survey's work in cataloging and organizing the nation's national resources, such as oil, gas, coal, potash, phosphates, and water power locations, "was continued as the Survey's chief concern, and developed into its principal contribution to the coordinated policies known as the conservation movement." Smith's tenure was marked by his support of embattled Secretary of the Interior Richard A. Ballinger; publication of *The World Atlas of Commercial Geology*, a two-volume compendium of the nation's mineral resources requested by President Woodrow Wilson; the correlation of the duties of the Federal Oil Conservation Board, appointed by President Calvin Coolidge in 1924; and support of the Federal Water Power Commission, which was established in 1920. The landmark "Superpower Survey," an inquiry into the power supplies for the Boston-to-Washington industrial region, established Smith as the leading power expert in the nation. He resigned from the Geological Survey when President Herbert Hoover named him to the Federal Power Commission in 1930. Because of Smith's conflicts with other power experts, President Franklin Roosevelt asked him to resign in 1933. Smith retired to Skowhegan, Maine, and died of a heart attack while in

Augusta for a conference on 10 January 1944.

See also United States Geological Survey.

Smith, Michael Hoke (1855–1931)

Secretary of the Interior Hoke Smith, a southern Democrat, was also a U.S. senator and governor of Georgia. Born Michael Hoke Smith on 2 September 1855 in Newton, North Carolina, he was the son of Hosea Hildreth Smith, president of Catawba College in Newton, and Mary Brent (nee Hoke) Smith, daughter of a prominent North Carolina attorney. In 1857, Hosea Smith—a native of New Hampshire and a graduate of Dartmouth College—became a professor of Greek and Latin at the University of North Carolina, and his son grew up in the college town of Chapel Hill. Following the Civil War, Hosea Smith had to leave the university because of his proslavery views. He started his own private school in nearby Lincolnton, where young Michael Hoke Smith received a limited education under his father's supervision. In 1872, the family moved to Atlanta, where Smith studied law and was admitted to the state bar in 1873. To make ends meet, he practiced law and taught school at the same time. In 1883, his brother Burton joined his law practice.

An avid Democrat, Smith was chairman of the Fulton County (Atlanta) Democratic Committee by 1876. He purchased the *Atlanta Journal* in 1887 as a political sheet, which he edited and wrote articles for. In 1888, he was chairman of the Georgia state Democratic convention. Smith's support of New York Gov. Grover Cleveland, and Cleveland's subsequent election to the White House, helped Smith become one of the leading Democrats in the nation and the party's spokesman in Georgia. In 1892, Smith was rewarded for his party loyalty by being named secretary of the interior. He served in this post from 6 March 1893 to 1 September 1896. His tenure was marked by great activity in

the pursuit of the preservation of natural resources and the creation of forestry reserves. In 1893, John Muir and Robert Underwood Johnson, editor of *Century* magazine, approached Smith with a flaw in the 1891 General Revision Act—that sheep were allowed to graze on the public lands, thus destroying them. Smith took it upon himself to ban such grazing and earned the wrath of sheep ranchers in the West. Smith was an ally of Gifford Pinchot and worked with the forester to shuttle important forestry legislation through Congress. He asked the National Academy of Sciences to convene a committee to draft legislation that would protect the forestry reserves created by the president. Although Smith was not in office when the committee report was delivered, he can be credited with its conception.

Although Smith did not support William Jennings Bryan for president in 1896, he believed that once Bryan won the nomination, it was up to him, as a good Democrat, to back his party's nominee. When he received signals that President Cleveland was against Bryan, Smith resigned. For the next ten years, he used the pages of the *Journal* to advocate progressive state measures in the areas of civil service reform, new highways, and railroads. In 1906, he was elected governor of Georgia. During his single two-year term (1907–1909), according to one source, "Smith advocated many progressive reforms which were enacted. The Railroad Commission was strengthened; the convict lease system was abolished; juvenile courts and a parole system were established; a primary election law was passed; and appropriations to public schools were increased." Smith, however, was a white supremacist, and the disenfranchisement of African Americans increased during his administration. Defeated in 1909, he was reelected in 1911 but served only a few months before being elected by the state legislature to fill the U.S. Senate seat of Alexander S. Clay, who had died. In the Senate, Smith was at odds with fellow Democrat Woodrow Wilson over U.S. entry into World War I. Although elected in 1914 to a full term, he was defeated in 1920 by another Georgia progressive, Thomas E. Watson. He worked as a lobbyist in Washington until 1925, when he returned home to his law practice. Hoke Smith died after a long illness at his home in Atlanta on 27 November 1931 at the age of 76. Said the *Atlanta Constitution*, "In going Senator Smith leaves an indelible imprint upon the history of the State which he served long and well."

See also Department of the Interior; Johnson, Robert Underwood; Muir, John.

Smog
See Regional Clean Air Incentives Market.

Smokey Bear
This important symbol of the environmental movement was created by U.S. Park Service artist Harry Rossoll in the 1930s to foster awareness of the dangers of accidental forest fires. Smokey's loud voice is used to proclaim to listeners, "Only you can prevent forest fires." He has his own ZIP code, and 5 million children have written letters to him and become junior rangers.

Smythe, William Ellsworth (1861–1922)
William Ellsworth Smythe stands out as the true father of irrigation in the United States. Born in Worcester, Massachusetts, on 24 December 1861, he was the son of William Augustus Somerset Smythe—who, according to writer Lawrence B. Lee, was "a wealthy shoe manufacturer [who] traced his ancestry to Plymouth Plantation"—and Abigail (nee Bailey) Smythe. Although pushed by his father to enter college, William Smythe had read of journalist Horace Greeley and was determined to enter the

printing and writing trade instead. He was an apprentice at the *Medford Mercury*, the *Brockton Gazette*, and the *Boston Herald*. In 1881, Smythe tried to start his own book publishing enterprise, but failure drove him back to printing.

Heading west in 1888, he established a daily paper in Kearney, Nebraska, called the *Kearney Expositor*. Two years later he was named editorial assistant of the *Omaha Bee*. While working at the *Bee*, Smythe was able to view the situation regarding irrigation of the western states. Based on his observations, he proposed to his boss, Edward Rosewater, a series of articles on the need for a national irrigation policy. Smythe biographer Martin E. Carlson noted that the articles, which appeared in the *Bee* in January and February 1891, "attracted widespread attention and led to the organization of a popular movement to obtain irrigation laws and to interest farmers and capitalists in ditch building. Smythe stated . . . in his autobiography *The Conquest of Arid America* (1905) . . . that within three months [of the publication of the articles] he had succeeded in making irrigation the foremost issue in Nebraska." Lawrence Lee, another Smythe biographer, claimed that "Smythe founded the national irrigation congress and the journal, *Irrigation Age*." Historian Donald J. Pisani called Smythe "the foremost publicist of the reclamation movement in the years from 1891 to 1895." Whatever his title or claim to fame, Smythe was responsible for Utah Gov. Arthur L. Thomas's call for a "congress" of irrigation advocates to meet in Salt Lake City to lobby Congress to establish a national policy on irrigation and reclamation in the arid western states. The congress, held from 15 to 17 September 1891, attracted about 350 delegates, including Smythe, Rep. Francis G. Newlands of Nevada, and Sens. Francis E. Warren of Wyoming and William Stewart of Nevada—the last three powerful members of Congress who were at the forefront in advocating congressional action on irri-

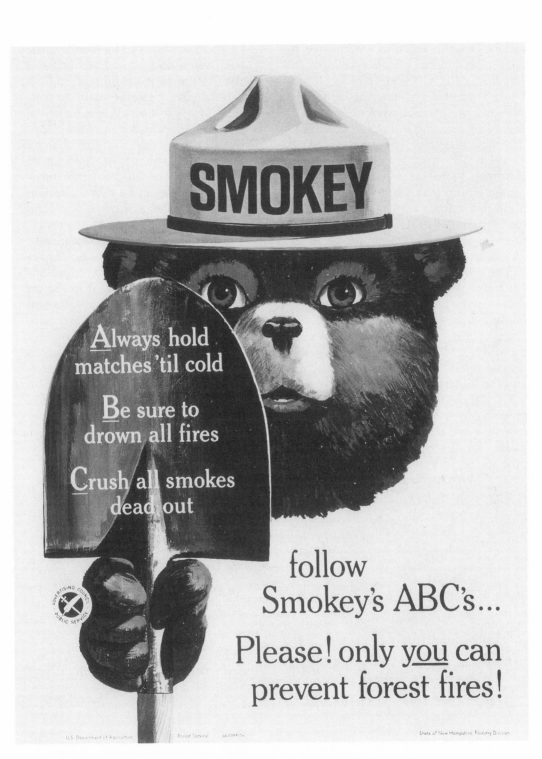

Smokey Bear, one of the most successful advertising icons, is a steward of the U.S. Forest Service.

gation. Smythe was also responsible for congresses held in October 1893 in Los Angeles and in October 1894 in Denver. During these years, Smythe edited the advocacy publication *Irrigation Age*. In addition to writing numerous articles for this newsmagazine, he later aired his views in *The Conquest of Arid America*. Historian Pisani wrote that in the book, "irrigation became a coherent, optimistic philosophy that bordered on religion. And settlement of the arid West became providential, the culmination of American history."

The collapse of the national economy in 1893 and Smythe's almost zealous ambition to settle and irrigate the West based on the colony system at Greeley, Colorado (an unsuccessful utopian vision of communelike settlements that culminated in the colony of New Plymouth, Idaho), cost him the leadership of the irrigation movement. Although he attended the 1895 congress in Albuquerque, he was edged out as chairman and relieved of his duties at *Irrigation Age*. Instead, Smythe turned full time to the idea of colonization. He also wrote *Constructive Democracy* (1906) and *A History of San Diego* (1906). In his final years, he was a writer for the Los Angeles–based magazine *Out West*. William Ellsworth Smythe died in New York City on 6 October 1922 at the age of 60. Wallace Stegner, in *Beyond the Hundredth Meridian*, wrote that Smythe's "persistent publicizing of irrigation problems, and his organization of arid-belt farmers into a politically coherent group, made him the single most influential figure, with the exception of Major [John Wesley] Powell, in the early years of reclamation."

See also Maxwell, George Hebard; National Irrigation Association; National Irrigation Congress.

Soil and Water Resources Conservation Act (16 U.S.C. 2001)

Enacted by Congress on 18 November 1977, this legislation called for an appraisal of soil and water resources in the nation. Congress found:

[T]here is a growing demand on the soil, water, and related resources of the Nation to meet present and future needs.... Congress, in its concern for sustained use of the resource base, created the Soil Conservation Service of the United States Department of Agriculture which possesses information, technical expertise, and a delivery system for providing assistance to land users with respect to conservation and use of soils; plants; woodlands; watershed protection and flood prevention; the conservation, development, utilization, and disposal of water; animal husbandry; fish and wildlife management; recreation; community development; and related uses.... Resource appraisal is basic to effective soil and water conservation.

The act mandated an appraisal of these resources, which included, but was not limited to:

data on the quality and quantity of soil, water, and related resources, including fish and wildlife habitats; data on the capability and limitations of those resources for meeting current and projected demands on the resource base; data on the changes that have occurred in the status and condition of those resources resulting from past uses, including the impact of farming technologies, techniques and practices; data on current Federal and State laws, policies, programs, rights, regulations, ownerships, and their trends and other considerations relating to the use, development, and conservation of soil, water, and related resources; data on the costs and benefits of alternative soil and water conservation policies; and data on alternative irrigation techniques regarding their costs, benefits, and impact on soil and water conservation, crop production, and environmental factors.

The first appraisal was to be completed by 31 December 1979, with further assessments to be made every five years.

Soil Conservation Act (49 Stat. 163)
This law came after a five-year period of soil erosion studies by the Forest Service. In 1928, Hugh H. Bennett, a soil scientist at the U.S. Department of Agriculture, wrote a paper on soil erosion, calling it "a national menace." On 16 February 1929, Rep. James P. Buchanan attached an amendment, called the Buchanan Amendment, to the Department of Agriculture's appropriations for 1930 that called for soil erosion studies to be done by Bennett. On 27 April 1935, following the release of these studies, Congress enacted the Soil Conservation Act of 1935, which authorized the creation of the Soil Conservation Service (SCS) as part of the Department of Agriculture, with Hugh Bennett as the agency's chief. The SCS was one of the better-run agencies of the New Deal. By using Civilian Conservation Corps (CCC) laborers as soil conservation workers, the SCS was able to spread its program nationwide at a minimal cost. In 1938, Secretary of Agriculture Henry Wallace added to the SCS's mission by ordering it to exploit nonessential lands under the direction of the Bankhead-Jones Farm Tenancy Act of 1937, create flood-control programs as directed in the Flood Control Act of 1936, build water-storage facilities as part of the Water Facilities Act of 1937, and follow the mandates of the Norris-Doxey Farm Forestry Act of 1937 regarding work on farm forestry programs.
See also Bennett, Hugh Hammond; Flood Control Act of 1936; Norris-Doxey Act; Soil Conservation Service.

Soil Conservation Service (SCS)
This government agency, a division of the Department of Agriculture, was established by the Soil Conservation Act of 1935. Its creation can be directly attributed to the soil erosion studies done by soil scientist Hugh H. Bennett in the late 1920s and early 1930s. According to the *United States Government Manual, 1993–1994*, the SCS "has the responsibility for developing and carrying out a national soil and water conservation program in cooperation with landowners and operators and other land users and developers, with community planning agencies and regional resource groups, and with other Federal, State, and local government agencies. The Service also assists in agricultural pollution control, environmental improvement, and rural community development." The SCS's programs include conservation, river basin surveys and investigations, watershed planning, watershed and flood-prevention operations, the Great Plains Conservation Program, and programs dealing with rural abandoned mines.
See also Soil and Water Resources Conservation Act; Soil Conservation Act.

Solid Waste Disposal Act (79 Stat. 997)
This law, enacted on 20 October 1965, was passed to discover methods of recycling materials being wasted in garbage dumps. Although the legislation did not specify which materials these might be, reporter Paul Hodge of the *Washington Post* found in 1968 that in garbage dumps nationwide, about $7 million of gold and silver was being lost. If this was recovered, industries that use such materials could increase their annual supply by 10 percent.
See also Wastes, Solid.

Sparks, William Andrew Jackson (1828–1904)
William Sparks of Illinois was considered one of the most effective commissioners of the General Land Office; he served

from 1885 to 1887, during the first Grover Cleveland administration. Born near New Albany in Harrison County, Indiana, on 19 November 1828, Sparks was the youngest of ten children. He moved with his family to Macoupin County, Illinois, in about 1835. When William was 12, his father died; the death of his mother three years later forced him to seek work. He found employment on a local farm and continued to educate himself; he later worked as a teacher. From 1847 to 1850, he attended Mc-Kendree College in Lebanon, Illinois, from which he graduated. He studied law and was admitted to the Illinois bar in 1851.

After a number of years in law practice, Sparks served a single term in the state legislature and one in the state senate. A Democrat, he was elected to the U.S. House of Representatives in 1874 and served until 1882, when he declined to run for reelection. He returned to Illinois and again took up the practice of law. On 26 March 1885, President Grover Cleveland named him commissioner of the General Land Office (GLO). Sparks had had no previous dealings with land matters, and the appointment could be seen as a political payback for his staunch support of the party. Sparks, however, set out to fix what he saw wrong at the GLO.

Sparks was called by one environmental study "the most aggressive reformer to serve [as GLO commissioner] in the post–Civil War period." He fought bitterly against those opposed to land reform and curbed the power of so-called land lawyers. His struggle to remove land grants from railroad companies that had profited illegally from generous land claims led many to call for his firing, but he resigned instead on 15 November 1887. Having seen his political demise on the horizon, Sparks wrote in the commissioner's report of 1887: "All efforts to secure a reform in the land laws by a repeal or amendments of particular acts and provisions have failed through

the opposition of interests.... What is needed in my opinion is an entire reforestation of existing laws, retaining an absolute homestead law and obsoleting all other forms of disposal of agricultural lands."

In 1900, Sparks moved to St. Louis, where he died four years later on 7 May 1904 at the age of 75.

See also General Land Office.

State Implementation Plan
See General Motors Corporation v. United States.

Stewart, William Morris (1827–1909)
Known to historians as the chief proponent of the 1872 mining law, William Morris Stewart was the son of Frederick Augustus Stewart and Miranda Dodd (nee Morris) Stewart. He was born in Galen, New York, on 9 August 1827 (although Stewart sometimes gave the year of his birth as 1825). As Stewart later wrote in his *Reminiscences*, "My father's family were of Scotch origin, and were among the early settlers of Massachusetts. My grandfather was a soldier in the Revolution, and shortly after peace was declared he moved to Vermont and settled upon a tract of land where he with his family resided many years." When William was six, the family moved to a farm in Trumbull County, Ohio. After spending several years at the Farmington Academy in Ohio, William Stewart returned to New York, where he became an instructor in mathematics at a high school in Lyons. When he enrolled at Yale College (now Yale University) in 1848, his interest was in the law, but according to one source, he was soon "lured by the discovery of gold in California." He traveled to the West Coast by ship through Panama, arriving in San Francisco on 7 May 1850. He spent the next two years mining for gold in Nevada City, California. When he amassed some $8,000, he left the mines and took up the

study of law in the offices of John R. Mc-Connell; in 1852, he was admitted to the state bar and began to practice.

Elected district attorney of Nevada County in 1853, Stewart was named acting state attorney general the following year. He formed a law partnership with Henry Foote, a former governor of Mississippi, and married Foote's daughter Elizabeth. In 1859, the Comstock lode was discovered in Nevada, and Stewart moved there in search of riches. But in Virginia City, Nevada, instead of becoming a miner, Stewart was employed by other miners to defend their claims. Through four years of litigation he received about $500,000. One source said that he was a "domineering" figure in the courtroom, and once he allegedly drew a gun while interrogating a witness. His wit and intelligence served him well. C. Elizabeth Raymond, the preparer of his papers, wrote, "Stewart was instrumental in shaping the state of Nevada from its territorial beginnings in 1861." That year, he was elected to the territorial council, and in 1863, he served as a member of the constitutional convention to codify state laws. When Nevada entered the Union on 31 October 1864 as the thirty-sixth state, the new state legislature elected William Morris Stewart as Nevada's first U.S. senator.

Stewart served in the Senate from 1864 to 1876 and again from 1888 to 1905. In the first interval, writes one biographer, Stewart was "an active supporter of the [Civil W]ar legislation before the Fourteenth Amendment was offered. Mr. Stewart proposed a plan of reconstruction which provided for universal amnesty and universal suffrage.... His plan was not adopted.... When President Grant was elected [1868] Mr. Stewart, as a member of the Senate Judiciary Committee wrote, reported and secured [the passage of] the Fifteenth Amendment.... He was the author of our national mining laws, recognizing and continuing all local mining restric-

tions then in existence." Historian William Condit has traced Stewart's active role, one that has been forgotten, in the passage of the General Mining Law of 1872. "Stewart proved [to be] an eloquent 'apologist' for lode and placer mines occupying the public lands in technical trespass," Condit reports. "Since no federal statute authorized the settlement and mining of mineral lands, he championed their system of self-governance. He believed 'free mining' by U.S. citizens should be encouraged by enactment of a law granting patents to the discoverers of mineral wealth who diligently worked their deposits under the rule of their mining districts."

A placer mine is "ground with defined boundaries containing loose deposits of mineral in the earth, sand, or gravel on or near the surface." A placer mine differs from a lode mine "in the amount of land which may be taken, the price per acre to be paid, the rights conferred by the respective patents, and the conditions upon which they are held" (*United States v. Iron Silver Mining Company*, 128 U.S. 673 [1888]). Stewart's first bill, the Lode Mining Law of 1866 (which included the Mineral Patent Law), was enacted by Congress on 28 July 1866. One source noted that under this legislation, "mineral lands of the public domain, surveyed and unsurveyed, [were] declared free and open to exploration and occupation." Four years later, in 1870, Stewart attempted to convince Congress that the 1866 laws needed to be amended to include placer miners under the protection of the law along with lode miners. On 10 May 1872, he got Congress to enact the General Mining Law of 1872, one of the landmarks in public land legislation. Along with these distinctions, Stewart also claimed to have been one of three people in attendance when Vice President Andrew Johnson took the oath of office upon the assassination of Abraham Lincoln. He supported Johnson's program until the president deviated on Reconstruction policy; Stewart later voted in the affirmative during Johnson's impeachment trial.

Faced with a well-financed opponent in the 1874 election, Stewart chose to bow out and left the Senate. He returned to Nevada and picked up his old law practice. In 1887, he was reelected by the legislature to his old seat, and he spent much of this final term (1888–1906) calling for the remonetization of silver. His zeal on this issue triggered him to leave the Republican Party for Nevada's short-lived Silver Party. While in Washington, he published the *Silver Knight* (later called the *Silver Knight–Watchman*), a paper devoted to silver issues. With the end of the Silver Party's influence in 1900, Stewart rejoined the Republicans. That year, he endorsed William McKinley, the pro-gold candidate, and attacked Democrat William Jennings Bryan's silver position.

Concerned about issues relating to the irrigation and reclamation of the western states, Stewart followed the irrigation surveys of John Wesley Powell, director of the U.S. Geological Survey (USGS). In February 1889, furious at what he saw as inadequacies in Powell's program, Stewart got Congress to form the Senate Select Committee on Irrigation and Reclamation of Arid Lands as an oversight mechanism for the surveys and got himself installed as chairman. Powell did his best to keep the committee informed as to the workings of the surveys, which were going slower than expected. But Stewart and the other westerners on the committee wanted irrigation projects to begin quickly; when most of the survey's appropriations went for topographical mapping of possible irrigation sites, Stewart charged that Powell was using the funds for his own benefit. Ultimately, Stewart helped enact legislation that transferred the Powell survey from the jurisdiction of the USGS to the Department of Agriculture. This led Powell to resign as head of the USGS in 1894.

Returning to private life in 1905, Stewart spent his last few years at his home in Bullfrog, Nevada. He was counsel for the Roman Catholic Church in a controversy with the Mexican government, a case that he argued before a court at The Hague in the Netherlands. In 1908, his *Reminiscences of Senator William Morris Stewart* were published by the Neale Publishing Company in New York under the editorship of George Rothwell Brown. On 23 April 1909, while in a Washington, D.C., hospital following an operation, Stewart died at the age of 81.

See also General Mining Law; Lode Mining Law; Powell, John Wesley; Powell Irrigation Survey; Senate Committee on Irrigation and Reclamation of Arid Lands.

Stuart, Alexander Hugh Holmes (1807–1891)

Alexander H. H. Stuart served as the third secretary of the interior from 1850 to 1853 during the Millard Fillmore administration. He was born on 2 April 1807 in Staunton, Virginia. Of Scotch-Irish descent, he was named after his grandfather Alexander, who commanded a regiment at the Revolutionary War battle of Guilford Courthouse in 1781. Stuart was also named after a family friend, Hugh Holmes.

Alexander Stuart attended the noted educational academy at Staunton as well as the College of William and Mary before earning a law degree at the University of Virginia in 1828. Admitted to the state bar that same year, he began a law practice in Staunton. Active in politics at a young age, he was elected to the state house of delegates in 1836 and served three years. In 1840, he was elected to the U.S. House of Representatives, where he sat for a single term. As a congressman, he was noted for his fight against the "gag rule," the attempt by southern congressmen to end the debate on slavery. For the next decade after leaving office, he was involved chiefly in the law but supported the Whig Party in politics.

Following the resignation of Interior Secretary Thomas Ewing in July 1850, President Millard Fillmore nominated Sen. James A. Pearce of Maryland to replace Ewing, but Pearce wanted to remain in the Senate. Instead, Fillmore turned to Thomas McKean Thompson McKennan of Pennsylvania, an intimate of Pearce. Unfortunately for Fillmore, McKennan served only a week, 16 to 23 August 1850, before resigning due to what he called "his peculiar nervous temperament." Into the clouded picture stepped Alexander Hugh Holmes Stuart. Stuart received a letter from Daniel Webster announcing that the president wished to nominate him for secretary of the interior. Stuart accepted and, on 12 September, Fillmore sent his name to Congress. That same day, Stuart was confirmed and began his tenure immediately.

Alexander H. H. Stuart served as interior secretary until 7 March 1853. In that span of two and a half years, Stuart was, according to Eugene Trani, the "first Secretary to establish significant policies. He introduced a civil service system for judging subordinates, standardized the procedures and attempted to clarify the responsibilities for each position in the Department, appealed for a commission to insure issuance of clear and free land titles, advocated outright land sale rather than leasing of mineral lands, renewed the plea for the creation of an agricultural bureau and the appointment of a solicitor, begun by Ewing, as well as for a building to house the entire Department. Stuart also asked for an index for all patented inventions and discoveries, which would include explanations and drawings.... He championed the cause of building a transcontinental railroad." When Fillmore left office, Stuart returned to his native Virginia and the law.

Stuart lived for nearly 40 years after leaving the government, during which time he was deeply involved in politics. With the collapse of the Whig Party, he joined the nationalistic and antiforeign American Party, which he supported in the 1856 campaign by writing a series of messages called the "Madison letters." Although his letters received wide attention, the party's candidate for president, Stuart's old boss Millard Fillmore, lost. Stuart declined a nomination to the Virginia state senate in 1853, but four years later he accepted and was elected for a single term (1857–1861). He was one of the lone voices in his state to decry secession, and he fought its effects with "silence" during the war. Elected to Congress in 1865, he was refused his seat because of Confederate sympathies. He later served in the Virginia house of delegates but resigned because of ill health. During his last years, he implored the nation's sections to reunite. He served as trustee of two educational institutions in Virginia, resigning from the last one just two years before his death on 13 February 1891 at the age of 83.

See also Department of the Interior.

Submerged Lands Act (67 Stat. 29)

Enacted on 22 May 1953, the Submerged Lands Act defined such areas as "all lands within the boundaries of each of the respective States which are covered by nontidal waters that are navigable under the laws of the United States at the time such State became a member of the Union, or acquired sovereignty over such lands and waters thereafter, up to the ordinary high water mark as heretofore or hereafter modified by accretion, erosion, [or] reliction [the gradual withdrawal of water, leaving land permanently exposed]; all lands permanently or periodically covered by tidal waters up to but not above the line of mean high tide and seaward to a line three geographical miles distant from the coast line of each State and to the boundary line of each such State where in any case such boundary as it existed at the time such State became a member of the Union, or as heretofore approved by Congress, extends seaward [or into the Gulf of Mexico] beyond three geographical miles; and all filled in, made, or reclaimed lands which formerly were lands beneath navigable waters." This law granted the states rights and authority over navigation, flood control, and the production of power but did not conflict with conservation laws, such as those protecting fish and wildlife sources.

Superfund

Known officially as the Comprehensive Environmental Response, Compensation, and Liability Act of 1980 (CERCLA), this law was enacted in response to the environmental horror known as Love Canal. In 1986, the Superfund Amendments and Reauthorization Act was enacted to supplement the original law. Originally given an appropriation of $1.6 billion to be spent over five years, Superfund was reauthorized in 1986 with an additional $9 billion. Wrote Betsey Carpenter in *U.S. News & World Report*, "Superfund was created as a short-term effort to rid the nation of hazardous waste dumps." It authorized the administrator of the Environmental Protection Agency (EPA) to designate, state by state, pollution sites to be cleaned up. These sites are then placed on the National Priorities List, which designates the order for the cleanups. Fines can be levied against the polluters.

Superfund has proved to be a far more expensive and difficult project than anticipated. Of nearly 1,200 sites immediately targeted for cleanup, by the end of 1993 only 150 had been completed, with new sites being added to the Superfund list every year. Another side of CERCLA's failure is in the area of lawyers' fees. To combat the growing expense of pollution cleanup, chemical and oil companies began hiring attorneys to fight the EPA for as long as possible to minimize their liability for site cleanup. *Time* magazine dubbed Superfund's creation of a new layer of "attorney bureaucracy" the "Lawyers' Money Pit." A Rand Corporation study in November 1993 found that one out of every three dollars spent by Superfund goes either to lawyers or to shift the responsibility for cleanups. Said Sen. Daniel Patrick Moynihan of New York, "In 1980, we thought that $1.6 billion spent over five years would bring us well along the way to solving this problem. If EPA estimates are right, the final cleanup for only the most hazardous dumps could reach $22.7 billion and their numbers are very optimistic. Probably unrealistic."

See also Love Canal.

Surface Mining Control and Reclamation Act (30 U.S.C. 1292)

This law, enacted on 3 August 1977, was passed to set national policy on the control of surface mining and its impact on the environment. The law established the Office of Surface Mining Reclamation and Enforcement, an agency in the Interior Department. Its major mission is to aid the states in administering national policy on protecting the environment from the effects of coal mines.

Sustained Yield

This action in the maintenance of forestry resources is defined as "continuous achievement and maintenance of a high-level input of forest resources without impairing the productivity of the land."

See also Multiple Use; Multiple Use–Sustained Yield Act.

Swamp Land Acts of 1849, 1850, and 1860 (43 U.S.C. 982–988)

These three acts were passed to allow states with excessive swamplands to use the acreage for drainage. In effect, they were the first congressional actions in the area of reclamation. The first act was signed into law on 2 March 1849. Its purpose was "to aid the State of Louisiana in constructing the necessary levees and drains to reclaim the swamp and overflowed lands therein." It granted to the state "the whole of those swamp and overflowed lands which may be or are found unfit for cultivation." The second, enacted 28 September 1850, mandated that "it shall be [the] duty of the Secretary of the Interior to make accurate lists and plats of all such lands, and transmit the same to the governors of the several States in which such lands may lie, and at the request of the governor of any State in which said swamp and overflowed lands may be, to cause patents to be issued to said State therefor, conveying to said State the fee-simple of said land...the proceeds of said lands, whether from sale or by direct appropriation in kind, shall be applied exclusively, as far as necessary, to the reclaiming [of] said lands, by means of levees and drains." The final law, enacted 12 March 1860, extended the terms of the previous two acts to the states of Minnesota and Oregon, which had entered the Union in 1858 and 1859, respectively. It read:

The provisions are extended to Minnesota and Oregon ... provided the grant shall not include any lands under which the Government of the United States may have sold or disposed of under any law, enacted prior to March 12, 1860, prior to the confirmation of title to be made under the authority of said sections—and the selections to be made from lands already surveyed in each of the States last named, under the authority of said sections, shall have been made within two years of the adjournment of the legislature of each State, at its next session after the 12th day of March, A.D. 1860—and as to all lands surveyed or to be surveyed, within two years from such adjournment, at the next session after notice by the Secretary of the Interior to the governor of the State, that the surveys have been completed and confirmed.

Taft-Barkley Water Pollution Control Act
See Water Pollution Control Act.

Tawney, James Albertus (1855–1919)

Rep. James A. Tawney was responsible for the passage of the Tawney Amendment, the 1909 addition to the Civil Sundry Appropriations Bill that forbade the president of the United States to form any investigative committee unless Congress gave its approval in advance. Tawney was born in Mount Pleasant Township near Gettysburg, Pennsylvania, on 3 January 1855, the son of John and Sarah (nee Boblitz) Tawney, both farmers. James Tawney learned blacksmithing and the machinist's trade at an early age and dropped out of school. In 1877, he moved west to Winona, Minnesota, where he worked as a machinist while studying law. After attending law school at the University of Wisconsin, he was admitted to the bar in 1882.

From 1890, when he won a term in the Minnesota state senate, until 1910, Tawney was a member of the nation's body politic. In 1892, he was elected to the first of eight terms in the U.S. House of Representatives. Known as a conservative Republican, Tawney was against the conservationist policies of Presidents Grover Cleveland and Theodore Roosevelt. In 1909, after Roosevelt presented Congress with the final report of the Commission on Country Life, Tawney wrote the amendment that would preclude the president from convening any future commission without the authority of Congress. Roosevelt wrote later, "As almost my last official act, I replied to Congress that if I did not believe the Tawney Amendment to be unconstitu- tional, I would veto the Sundry Civil Bill which contained it, and that if I remained in office I would refuse to obey it." Wrote author Roger Wyman, "Roosevelt's pioneering work in the field of conservation was regarded by conservatives like Tawney as another example of executive usurpation." President William Howard Taft later ended the authority of the National Conservation Commission because of the Tawney Amendment.

Because of his actions against the conservation movement, Tawney was targeted for defeat in his 1910 reelection run. The Democrats chose Harry L. Buck, a judge, and progressive Republicans picked Syndey Anderson, an attorney and veteran of the Spanish-American War. Gifford Pinchot and Henry Wallace (father of Franklin Roosevelt's secretary of agriculture) came to Minnesota to speak against Tawney. Anderson beat Tawney in the Republican primary and went on to defeat Buck in the general election. Tawney was appointed by President Taft to the United States–Canada Joint Boundary Commission, a post he held until his death on 12 June 1919 in Excelsior Springs, Missouri.

See also Commission on Country Life.

Taylor, Edward Thomas (1858–1941)

Rep. Edward Taylor of Colorado was one of the leading forces in the House of Representatives in the area of federal control over grazing on federal lands. He was born on a farm near Metamora in Woodward County, Illinois, on 19 June 1858, the son of farmers Henry and Anna (nee Evans) Taylor. Later, he moved to a cattle ranch in western Kansas. He was educated in the public schools in Illinois and Kansas, apparently receiving only a high

Ranchers round up Herefords on range near Gunnison, Colorado, in 1949.

school education. He moved to Colorado in 1881 and was the principal of a high school in Leadville. He attended law school at the University of Michigan and was awarded a law degree in 1884.

Over the next 20 years, Taylor held several jobs, including superintendent of schools in Lake County (Leadville), deputy district attorney of Lake County, district attorney of northwestern Colorado, county attorney, and city attorney of Glenwood Springs, Colorado. In 1896, he was elected to the state senate, where he served until 1908. In that year, he was elected representative-at-large in the U.S. House of Representatives for three terms (1909–1915). In 1914, he was elected to the House from the Fourth Colorado District and held that seat until his death, one of the longest tenures on

record. Among the more than 100 bills he got passed was the Taylor Grazing Act of 1934, which opened public lands in the West for grazing. He was lauded as the "worthy son of western America" for his lifetime of work. Taylor died on 3 September 1941 at the age of 83.

See also Taylor Grazing Act.

Taylor Grazing Act
(48 Stat. 1269–1275)

This act, sponsored by Rep. Edward T. Taylor of Colorado, was drafted to minimize the impact of rampant grazing on federal lands and set up a fee system for the use of those lands. The measure, known officially as "an Act to stop injury to the public grazing lands by preventing over-grazing and soil deterioration, to provide for their orderly use, improvement, and development, to stabilize the livestock industry dependent upon the public range, and for other purposes," was enacted by Congress on 28 June 1934. It contained 15 sections, most of which authorized the secretary of the interior to protect grazing areas by implementing programs to cut down on grazing and instituting grazing fees.

See also Taylor, Edward Thomas.

Teale, Edwin Way (1899–1980)

Naturalist Edwin Way Teale was awarded the Pulitzer Prize for general nonfiction in 1966 for his work *Wandering through Winter.* He was born Edwin Alfred Teale on 2 June 1899 in Joliet, Illinois, the son of British expatriate Oliver Cromwell Teale, a railroad worker and mechanic, and Clara Louis (nee Way) Teale, a teacher. At the age of 12, Teale, who liked to imitate naturalist and writer Henry David Thoreau—Thoreau had rearranged his name—changed his middle name to the more distinguished "Way." He grew up in Illinois but spent his summers on his maternal grandfather's farm in northern Indiana. His exploits and education in nature on the farm are highlighted in his *Dune Boy: The Early Years of a Naturalist* (1943).

Teale's early education in nature was an integral part of his later writing. Commenting on the part that nature plays in a person's upbringing, he wrote in *Days without Time* (1948), "For here there is something more, something magical, something that fills a deep need of the human heart." He was awarded a bachelor's degree from Earlham College in Richmond, Indiana, in 1922, and afterward worked as a tutor at Friends University in Wichita, Kansas. He met Nellie Donovan at Earlham and married her the year after he graduated. Their only son, David Allen Teale, was killed during World War II.

Desiring to write about the nature he loved, Teale moved to New York to attend Columbia University, from which he received a master's degree in English in 1926. For 13 years starting in 1928, he worked as a science writer for *Popular Science Monthly.* At the same time, he worked on different angles of nature studies, perfecting nature photography. His first work on the subject, *Grassroot Jungles,* (1937) was an analysis of insect photographs. Further entomological investigations resulted in *The Junior Book of Insects* and *The Boys' Book of Photography* (both 1939). These were followed by an examination of bees in *The Golden Throng* (1949) and *Near Horizons,* detailing the work of an "insect garden" he had nurtured.

Following the publication of his *Dune Boy* in 1943, Teale was awarded the John Burroughs Medal by the American Museum of Natural History in New York for distinguished nature writing. *The Lost Woods* (1945) followed. Starting in 1951, his four-volume series on the seasons and nature appeared. *North with the Spring* was followed by *Autumn across America* (1956), *Journey into Summer* (1960), and *Wandering through Winter* (1965). The latter earned him the Pulitzer Prize in 1966 for general nonfiction. He traveled to Ravensthorpe, England, where his father had left in 1884, and commented on his father's homeland in *Springtime in Britain* (1970).

Shortly before his death, Teale's *A Walk through the Year* (1978), a diary of the year 1977, appeared. About Teale, writer Orville Prescott wrote in the *New York Times,* "No other American naturalist of his generation has found so large a public in his books as Mr. Teale and none deserves such success more. Scientifically learned in botany, zoology, geology and diverse other ologies, Mr. Teale somehow manages to remain as sensitively aware of beauty as a poet." Teale died in Norwich, Connecticut, on 18 October 1980 at the age of 81. Two years after his death, *A Conscious in Stillness,* cowritten with Ann Zwinger, was published.

Teller, Henry Moore (1830–1914)

U.S. senator and influential Colorado politician Henry Moore Teller served as the fourteenth secretary of the interior. He was born on his family's farm in Allegany County, New York, on 23 May 1830, the eldest son of John Teller, a farmer, and Charlotte (nee Moore) Teller. In between doing farming chores, Henry attended school and received a limited education. He later attended academies at Rushford and Alfred, New York. After completing his studies and working for a short period as a teacher, Teller moved to Angelica, New York, where he read the law and was admitted to the bar in 1858.

Looking for opportunity, Teller headed west, first settling in Morrison, Illinois, for three years and finally in Central City, Colorado, in 1861. He learned the complexities of mining law and had a stint as a major general in charge of the Colorado militia, defending Denver from potential Indian attacks. He became one of the leading politicians in Colorado after the Civil War. He was a key player in the fight for Colorado statehood following the war, and he served as president of the Colorado Central Railroad (1872–1876).

When Colorado was admitted into the Union in 1876, Teller was elected one of the state's first two U.S. senators. A Republican, he served until 1882, speaking on behalf of western interests and advocating the free coinage of silver. On 18 April 1882, President Chester A. Arthur named Teller secretary of the interior, replacing the outgoing Samuel Jordan Kirkwood. While secretary, Teller established a Court of Indian Offenses to have Indian magistrates judge other Indians, urged repeal of the ineffective Timber Culture Act and other related laws, and became embroiled in a scandal involving a railroad land claim that he attempted to grant. His tenure was considered a successful one, and he left the post in 1885 with the change in administrations.

The Colorado legislature soon returned Teller to the U.S. Senate, and he served until his retirement in 1909. He teamed with fellow Colorado legislators Rep. John Calhoun Bell, Rep. John Franklin Shafroth, and Sen. Edward Oliver Wolcott to stymie President Grover Cleveland's attempt to create forestry reserves in 1893. Teller was a firm supporter of the Pettigrew Act that arose out of the controversy.

Although a conservative Republican, Teller bolted the party in 1896 over the issue of the free coinage of silver. He was a founding member of the short-lived Silver Republican Party; following that group's demise, he became a Democrat. During his tenure in the Senate, he served on such committees as Mining and Mines, Public Lands, and Patents. Recognized by the sobriquet "defender of the West," he is remembered chiefly for the landmark legislation known as the Teller Amendment, which called for Cuba's independence and its ultimate inclusion as a U.S. protectorate. His last years in the Senate were spent on the problems of land distribution, reclamation, and Indian affairs. Teller left the Senate in 1909, served until 1912 as a member of the National Monetary Commission, and then retired to Colorado. He died in Denver at the home of his daughter on 23 February 1914 at the age of 83.

See also Bell, John Calhoun; Department of the Interior; Pettigrew Act; Shafroth, John Franklin; Wolcott, Edward Oliver.

Tennessee Valley Authority (TVA)

The Tennessee Valley Authority was the federal flood-control program covering the entire Tennessee Valley from Paducah, Kentucky, south along the Tennessee River to Guntersville, Alabama, and then north through the Blue Ridge Mountains into West Virginia. Dams were to be built along the Mississippi at points such as Muscle Shoals, near the town of Florence in northern Alabama, and Hale's Bar.

The first bill passed by Congress dealing with the development of hydro-

Dams constructed by the Tennessee Valley Authority beginning in the 1930s controlled flooding and brought electricity to large areas of Tennessee and Kentucky.

electric power at Muscle Shoals was introduced by Rep. Joseph Wheeler of Alabama and Sen. Edmund W. Pettus of Alabama on 21 March 1898. It passed both houses of Congress and was signed into law by President William McKinley on 2 March 1899. Called "General Wheeler's bill," it honored this military hero of the Civil War and Spanish-American War with a dam bearing his name along what is now the TVA's path. The bill gave the Muscle Shoals Power Corporation the right to construct power stations and canals at Muscle Shoals. The company had a time limitation for the start and end of construction, however, and when it could not meet these requirements, Congress granted this power to another company, N. F. Thompson and Associates, in 1903. Theodore Roosevelt vetoed the "Thompson bill," writing in his veto message, "Wherever the Government constructs a dam and lock for the purpose of navigation there

is a waterfall of great value. It does not seem right or just that this element of local value should be given away to private individuals of the vicinage, and at the same time the people of the whole community should be taxed for the local improvement." This battle between private and public development was key in the failure to use Muscle Shoals for the next 30 years.

By the 1930s, people were looking for the government not only to supply cheap power in the Tennessee Valley but also to implement flood-control measures. Such a flood-control program was the dream of Republican Sen. George Norris of Nebraska. Long frustrated by Republican presidents in his fight to control the flooding of the Tennessee River, Norris finally found a friend in Franklin Delano Roosevelt. In his message to Congress on 10 April 1933, Roosevelt said, "It is clear that the Muscle Shoals development is but a small part of the potential public usefulness of the entire Tennessee River. Such use, if envisioned in its entirety, transcends mere power development: it enters the wide fields of flood control, soil erosion, afforestation, elimination from agricultural use of marginal lands, and distribution and diversification of industry." A month later, on 18 May, Congress enacted the Tennessee Valley Authority Act (48 Stat. 58–72). Among the TVA's other purposes were to use the dams to provide the poverty-stricken residents of the Tennessee Valley with cheap power and fertilizer, initiate soil and forest conservation techniques, and form waterways for commercial use.

See also Norris, George William.

Thompson, Jacob (1810–1885)

Jacob Thompson was the fiery southerner who served all but two months of his term as James Buchanan's interior secretary, then worked for the Confederacy during the Civil War. He was born on 15 May 1810 in the village of Leasburg, North Carolina, the third son of Nicholas Thompson, an industrious tanner from Virginia, and Lucretia (nee Van Hook) Thompson. Jacob Thompson attended the Bingham Academy in nearby Orange County, then entered the University of North Carolina and proceeded to graduate with honors in 1831. Although his father wanted him to become a minister, Jacob chose to study law at the university, remaining there as a tutor for 18 months. He then read the law in an office in Greensboro and won admittance to the bar in 1835. To escape from the influence of their stern and religious father, Thompson and his older brother James Young Thompson, a physician, moved first to Natchez, Mississippi, and then to nearby Pontotoc, where Jacob opened a law office.

In Pontotoc, and later in nearby Oxford, Thompson got caught up in the politics surrounding the ceding of Chickasaw Indian lands to the government. He demanded to represent the new counties

carved out of the new territory but was refused. Undaunted, he penned a brutal attack on Gov. Charles Lynch. Thompson's fiery attitude earned him a nomination for state attorney general in 1837, and although he was unsuccessful in that election, one year later he was elected to the U.S. House of Representatives. In his six terms in the House (1839–1851), Thompson served on the Committee on Public Lands as well as on the Committee on Indian Affairs, which he chaired during his final term. Speaking out for slave owners' rights and the annexation of territory from Mexico, Thompson was defeated in 1838 by a coalition of Unionists and Whigs.

With the election of Franklin Pierce in 1852, Thompson was offered the post of ambassador to Cuba, but he refused. In 1855, he ran a losing campaign for the U.S. Senate. Thompson backed James Buchanan in the 1856 presidential race, and upon winning, Buchanan chose Thompson as secretary of the interior. Politics played a major role in Thompson's selection. The previous interior secretary, Robert McClelland, had been staunchly antislavery. Having narrowly defeated John Charles Frémont, the Republican antislavery candidate, Buchanan sought to fill his cabinet with proslavery politicians.

Thompson served all but two months of his term. He concentrated much of the department's duties in his own office, personally overseeing the Indian and pension bureaus and increasing the department's overall efficiency. During his tenure, the number of bureaus of the department expanded, with the Texas Boundary Survey, the Copyright Office, and the California Boundary Survey being added. Although he is little remembered in this capacity, Thompson may have been one of the best nineteenth-century interior secretaries.

Thompson's reputation has been tarnished, however, by his conduct during the Civil War. With tensions building following the election of Abraham Lincoln in 1860, South Carolina seceded from the Union and demanded that the federal government turn over Fort Sumter in Charleston Harbor. Instead, President Buchanan sent the ship *Star of the West* to resupply the fort. Thompson resigned over this breach of southern sensibilities on 8 January 1861. He wrote, "I went to Mississippi as a young man and everything I am she made me." Thompson was apparently involved in secessionist intrigue before his resignation. In a diary, fellow Confederate Thomas Bragg wrote for 6 January 1861: "At night went with Wheeler to see Secretary Thompson. He read me his letter to the Gov. of Miss'i giving an acc't of his visit as a Comm'r to No. Ca. Expresses the opinion in it that the State will be prepared to secede by 4th March, and advises that the ordinance of secession to be passed by Miss. Convention take effect on third March."

When the war started, Thompson enlisted in the Confederate army as a lieutenant colonel, serving under Gen. P. G. T. Beauregard. In 1862, Gen. John L. Pemberton appointed Thompson inspector general of the Confederate army. Captured at Vicksburg in 1863, he was soon released and subsequently elected to a seat in the Mississippi state legislature. A year later, Confederate President Jefferson Davis named Thompson commissioner to Canada; his mission was to cause havoc in the North by giving monetary aid to the anti-Union "Sons of Liberty" in Ohio and Indiana and helping Confederate prisoners escape. In a letter from Confederate blockade runner W. H. Brawley to southerner Thomas Muldrup Logan, Brawley wrote about his ship outrunning a Union cruiser from North Carolina to St. George, Bermuda. "My fellow passengers, Hon. Jacob Thompson of Miss. & Ex-Senator C. C. Clay of Alabama & Mr. Cleary of Kentucky who go on a mission of the Gov't at Canada, left us yesterday on the Halifax ship [*sic*]," Brawley related. Working with the shadowy group the Knights of the Golden Circle, Thompson attempted to arm 25,000 Confederate troops to raid the North. When Lincoln was assassinated, Thompson's name was mentioned in a possible conspiracy. The former interior secretary and his wife fled to Europe and lived in France for the next several years.

Thompson returned to Mississippi to find all his property gone. In 1876, he was caught up in a burgeoning scandal surrounding funds stolen from the Indian bureau while he was interior secretary. There were also questions concerning the whereabouts of monies he was given for his mission in Canada. Although he managed to clear his name before his death, many suspect Thompson's guilt. One historian believes that he "may very well rank as the greatest scoundrel of the Civil War." Thompson died a rich man, with an estate estimated at more than $500,000, on 24 March 1885.

See also Department of the Interior.

Thoreau, Henry David (1817–1862)

Perhaps best known as the independent and fiery author of *Walden*, Henry David Thoreau was a member of the group of transcendentalist thinkers (which included Ralph Waldo Emerson) that dominated the conservationist movement in the mid-nineteenth century. Thoreau was born David Henry Thoreau on 12 July 1817 in the home of his materal grandmother in Concord, Massachusetts, the third of four children of John Thoreau, a pencil maker, and Cynthia (nee Dunbar) Thoreau. His early childhood, at least until his fifth birthday, was spent in Boston and Chelmsford, Massachusetts; for the rest of his life, however, he would call Concord home. Named after a deceased uncle, Thoreau changed the order of his first and middle names in 1837.

After attending Concord Academy, Thoreau enrolled at Harvard University and graduated in 1837. It was at this time that he became acquainted with author

Ralph Waldo Emerson (1803–1882), who was 15 years Thoreau's senior. Known as the father of transcendentalism, Emerson was an advocate of civil disobedience and nonconformism. The tenets of his teachings were that human beings could "transcend" to a higher spiritual plane, and that this movement occurred through intuition, not reason. Works such as his "Essay on Self-Reliance" and numerous public speeches held that the individual, and not God, had much to do with the traditional beliefs of man. Thoreau was an early and enthusiastic follower of this doctrine.

Thoreau taught for a while in Concord, Massachusetts, and then worked at his father's pencil shop. In 1838, he started a small academy with his brother John, and they kept the enterprise alive until 1841. In 1839, he and John traveled for 13 days down the Concord River to its convergence with the Merrimack River and to Concord, New Hampshire. After John's untimely death in 1842,

Henry David Thoreau, 1856

Thoreau wrote about the trip in his *A Week on the Concord and Merrimack Rivers* (1847). Thoreau had begun writing even earlier. In 1834, he started to keep a journal that he continued for the remainder of his life. *The Journal of Henry Thoreau* was published in 14 volumes by Houghton Mifflin in 1906.

In 1845, at the age of 27, Thoreau made his noted move to Walden Pond on the eastern edge of Concord. There he constructed a cabin on land owned by Emerson. It was during his two years there that he wrote parts of the only two books published during his lifetime: *A Week on the Concord and Merrimack Rivers* and *Walden*. The latter work remains one of the most important books written by an American author. Published just eight years before his death, this composition—subtitled *A Life in the Woods*—sets forth Thoreau's ideas on what author Kurt Kehr called "his timeless message: mankind and nature form an ecological unity; the materialism of the industrial age robs man of the real basis of his life; skepticism is appropriate whenever the advance of scientific knowledge is lauded." Wrote author William Howarth, "Some readers of *Walden*, Thoreau's most famous book, think of him as a hermit, a solitary crank who hated society and never strayed from his backyard. Others, remembering his essay 'Civil Disobedience,' see him as a radical dissenter, the man who went to jail (though for only a night) for refusing to pay his poll tax, as a protest against slavery and the Mexican War." Thoreau was not either of these men, Howarth argued. "One of America's first backpacking tourists, he traveled widely and wrote several books, all expressing a firm social ethic: 'In Wildness is the preservation of the World.'" That attitude was expressed in this passage from *Walden*: "If a man walks in the woods for love of them half of each day, he is in danger of being regarded as a loafer, but if he spends his whole day as a speculator shearing off those woods and making the earth bald before her time, he is esteemed as an industrious and enterprising citizen."

The two years spent at Walden Pond provided Thoreau with continuing material for his observations on man and nature. Authors Gordon Whitney and William Davis wrote, "Thoreau's journals and manuscripts provide a wealth of information on the response of ground vegetation to coppice management in the nineteenth century. The earliest phases of the coppice cycle, recently cut oak woods or land dominated by sprouting oaks, frequently supported a rather transient community of weedy, sun-loving plants."

But nature was not Thoreau's only interest. Arrested in 1846 for not paying his poll tax as a protest against slavery and the ongoing Mexican-American War, he refused to pay the bail and spent the night in jail. He was angered when a member of his family paid the bail for him. Thoughts on the defiance of governmental power resulted in his 1848 lecture "Resistance to Civil Government," which was published posthumously as "Civil Disobedience" (1866). His 1854 speech on "Slavery in Massachusetts" was in response to the worsening conditions in the Kansas and Nebraska territories. His interest in travel produced *Excursions* (1863), *The Maine Woods* (1864), *Cape Cod* (1865), and *A Yankee in Canada* (1866), all published posthumously. At the end of 1860, he developed tuberculosis and traveled to Minnesota. The disease worsened and finally claimed him on 6 May 1862. Thoreau was 44. His humble life is reflected in the epitaph on his tombstone—it simply reads "Henry."

Three Mile Island

The near-meltdown at Three Mile Island, a nuclear power facility near Harrisburg, Pennsylvania, in 1979 caused the American public to rethink the benefits and dangers of nuclear power. The accident began at 3:53 A.M. on 28 March 1979 when, according to the *New York Times*, "Because of a complex series of human and mechanical errors, signaled by a harmless blast of steam, a reactor at the Metropolitan Edison Company plant on Three Mile Island in the middle of the Susquehanna River began to tear itself apart, loosing small whiffs of radioactivity into the predawn chill of central Pennsylvania." Wrote Mark Stephens, who was a member of the President's Commission on the Accident at Three Mile Island, "Following a dozen less serious accidents, a hundred near-misses and a thousand malfunctions, the nuclear industry had that day to face up to conditions it had claimed were 'impossible.' And the people of Pennsylvania and the world found themselves having to stand by, accepting the impossible and hoping that the men who had been so wrong in claiming that a serious accident could not happen would be able to deal successfully with the accident now that it had come."

Because a valve that was supposed to cool the reactor failed, fuel rods were exposed and began to crack open. Water from the stuck valve backed up into a container and overflowed onto the floor. Gases from the water rose to the air-conditioning system, which pumped it out into the environment. The reactor fuel rods had come close to a full meltdown. Over the next several days, as pregnant women and children were evacuated, workers inside the plant dumped thousands of gallons of radioactive water into the Susquehanna River. President Jimmy Carter visited the plant, the fuel rods were cooled, and the crisis essentially ended. Unfortunately, following the release of radioactivity into the air and water, there were reports of retardation in children and unusual numbers of cancers in the surrounding area. Stories detailing the illnesses included the *Harrisburg Patriot*'s "31 Blame TMI for Health Problems." As of April 1994, there were over 2,100 outstanding lawsuits against the company that owns

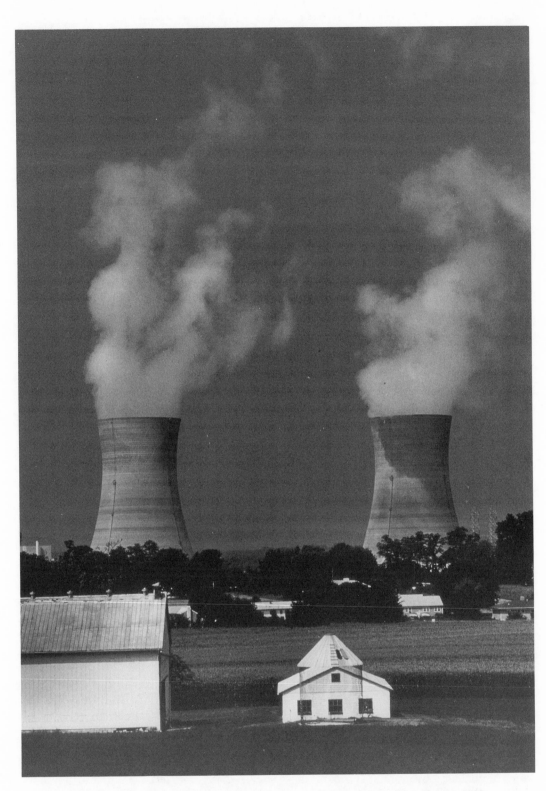

Cooling stacks at the Three Mile Island nuclear power facility, south of Harrisburg, Pennsylvania, 1992

Three Mile Island. It was not until August 1993 that the last radioactive water was removed from the damaged reactor. Inside the reactor, a ton of spent fuel rods damaged in the initial accident remain, too dangerous to be touched until at least the year 2014.

Tiffin, Edward (1766–1829)

British-born Edward Tiffin, a friend of James Madison, served as the first commissioner of the General Land Office. He was born the second of five children to Henry and Mary Parker Tiffin on 19 June 1766 in Carlisle, England. He began his education at a Latin School in Carlisle, then served a five-year apprenticeship in the study of medicine beginning in 1778. In 1783, along with the rest of his family, Tiffin immigrated to the newly independent United States. They settled near Charles Town, Virginia, now in Jefferson County, West Virginia. Although accounts vary, Tiffin allegedly returned to England to study and then attended Jefferson Medical College in Philadelphia, from which he graduated with a medical degree. In either 1787 or 1789, he married Mary Worthington, the daughter of a local landowner; Tiffin's brother-in-law, Thomas Worthington, later served as the second governor of Ohio.

After 1790, Tiffin and his wife became devout followers of the Methodist movement in the United States; two years later, Tiffin was ordained as a lay preacher in the Methodist Church. In 1797 (or, according to other biographies, 1798), Tiffin, his wife, and Thomas Worthington's family set out with their possessions and slaves to Chillicothe, Ohio, located in the Scioto valley, to start a new life in an unsettled land. In 1798, George Washington, in a letter to Ohio's territorial governor Arthur St. Clair, gave a glowing reference to Tiffin for some sort of governmental post, although there is no evidence that Tiffin ever met Washington. St. Clair responded by appointing Tiffin prothonotary (chief notary public) of the Territorial Court of Common Pleas. From 1799 to 1801, Tiffin served as speaker in the territorial legislature, representing the Scioto valley.

Following the elections of 1802, Tiffin and St. Clair became political enemies as the nation split between Federalist and Anti-Federalist. On 1 March 1803, Ohio became the seventeenth state to enter the Union, and Tiffin was elected the state's first governor. Much of his two terms was spent setting up laws for the new state. On 1 January 1807, Tiffin, having refused a third term, was elected by the state legislature to fill the U.S. Senate seat of his brother-in-law Thomas Worthington, who had resigned after being elected Ohio's second governor. Tiffin served only until 1809, but he was a leader in the discussion over land problems in what is now the middle section of the United States. After returning to Ohio, Tiffin served two terms as speaker in the state legislature from 1809 to 1811.

On 25 April 1812, Congress created the General Land Office as part of the Treasury Department to deal with problems in the newly settled lands of the growing United States. President James Madison named Tiffin as the first land commissioner. In his two years on the job, Tiffin, with a chief clerk and a staff of only eight people, was able to refine and coalesce record keeping on land acquisition, set up surveys of unknown lands, and report to Congress on his progress. When the British invaded Washington, D.C., during the War of 1812, Tiffin was one of only a handful of department heads to move his records out of the capital to safety. Following his return to Washington, Tiffin requested to switch jobs with Josiah Meigs, surveyor general of the Northwest Territory. With the president's approval, Tiffin assumed the surveyorship in Ohio. He served in this post until 1829, when politics caused President Andrew Jackson to remove him from the job. Tiffin lived less than a year after his removal, dying on 9 August 1829. The town of Tiffin, Ohio, is named after him.

See also General Land Office.

Timber and Stone Act (20 Stat. 89–91)

This action of 3 June 1878, passed in accordance with the Timber Cutting Act, allowed the sale of 160 acres of land of a nonmineral nature to persons in Washington State, Oregon, California, and Nevada. Such land was basically to be used as areas of timber and stone cultivation on lands classified as unfit for agricultural use. This act was soon circumvented by large forestry companies that grabbed up thousands of acres of forest reserves.

Timber Culture Act (17 Stat. 605–606)

This legislation, enacted on 3 March 1873, was passed because vast areas of the western United States lacked proper timber cover. The law provided 160 acres of land to any person who planted 40 acres of trees and kept them growing and healthy for ten years. Unfortunately, soon after passage of the act, more settlers moved west, destroyed acres of land and timber reserves to make their 160 acres, and planted 40 acres of trees—in essence, creating a net loss of timber reserves. In 1878, the acreage requirement was dropped from 40 acres to 10, but this did not solve the problem. In addition, rampant speculation among land dealers caused certain lands to be used for sale rather than for the protection of timber. The Timber Culture Act was repealed by Congress through the General Revision Act of 1891, also known as the Payson Act.

See also Payson Act.

Timber Cutting Act (20 Stat. 88–89)

This act was passed in response to the wholesale and illegal cutting down of the nation's forests in the states of Colorado and Nevada and in the western territories by miners and farmers for domestic and industrial use and for the creation of agricultural lands. The goal of this legislation was to slow the progress of clear-cutting. Congress passed the act on 3 June 1878, in essence making legal what was once illegal in hopes of managing the cutting. But the unchecked cutting of whole forests continued, leaving the act toothless.

Torrey, Bradford (1843–1912)

The *National Cyclopaedia of American Biography* wrote of Bradford Torrey before his death, "He has devoted much time to the study of birds, their habits, peculiarities, and domestic traits." Born on 9 October 1843 in Weymouth, Massachusetts, Torrey was the son of Samuel Torrey, a shoemaker, and Sophronia (nee Dyer) Torrey. A descendant of Lt. James Torrey, who emigrated from England in about 1640, Bradford was also related to the eminent botanist John Torrey (1796–1873). After graduating from high school in Weymouth, Torrey worked in a local shoe factory; later, he taught school for several years.

After moving to Boston (date unknown), Torrey began the study of birds. His first published article, "With the Birds on Boston Common," appeared in the February 1883 issue of the *Atlantic Monthly*. Two years later his first book, *Birds in the Bush*, emerged. From 1886 to 1901, he was on the editorial staff of the magazine *Youth's Companion*. For a time, Torrey was involved in various excursions through the Northeast, particularly New England. He wrote *A Rambler's Lease* (1889), *The Foot-path Way* (1892), and *A Florida Sketch Book* (1894). After 1894, he traveled to the West through Arizona and California, eventually settling in Santa Barbara, California, where he built a log cabin like another New England writer and naturalist, Henry David Thoreau. In fact, Torrey edited Thoreau's journals for publication in 1906. Torrey's

Nineteenth-century mining required lumber for buildings and timbers for shoring up tunnels. The Gold King mill processed timber from forests near Telluride, Colorado, at the end of the century.

later works included *Spring Notes from Tennessee* (1896), *A World of Green Hills* (1898), *Everyday Birds* (1901), *Footing It in Franconia* (1901), *The Clerk of the Woods* (1903), *Nature's Invitation* (1904), and *Field Days in California* (1913).

Several sources on Torrey's life concluded that although he was not a professional ornithologist, he was a keen observer and included concise detail in his works. "He was first of all an essayist," Henry S. Chapman wrote in *The Dictionary of American Biography*. It was this lack of academic standing that made Torrey one of the lesser-known ornithologists of the nineteenth century, however important his contributions. He died in a hospital in Santa Barbara on 7 October 1912, two days short of his sixth-ninth birthday.

Townsend, John Kirk (1809–1851)

An influential force in ornithology during the first half of the nineteenth

century, John Kirk Townsend contributed substantial work and research to John James Audubon's *Birds of America*. Townsend was born on 10 August 1809 in Philadelphia, the son of Charles Townsend, a watchmaker, and Pricilla (nee Kirk) Townsend, both Quakers. Townsend was educated at a Quaker boarding school in Westtown, Pennsylvania, and later studied dentistry in Philadelphia. While at boarding school, he was influenced by Ezra Michener (1794–1887), who later became a noted ornithologist and botanist, to take up bird studies. On one field trip, Townsend discovered a bird that Audubon later named Townsend's Bunting in his honor.

In 1834, explorer Nathaniel Jarvis Wyeth (1802–1856) began an overland expedition to Oregon. Townsend and naturalist Thomas Nuttall joined the journey, which originated at Independence, Missouri, and ended at Fort Vancouver, in the

Oregon Territory. Townsend later spent a year in the Hawaiian Islands doing research before returning to Fort Vancouver as its surgeon. Townsend's collections from these trips and stops were discussed in an article he wrote for the *Journal of the Academy of Natural Sciences of Philadelphia*, which appeared in two parts, one in 1837 and the second in 1839. In the latter year, Townsend undertook to document his discoveries in a grandiose undertaking that he called *Ornithology of the United States of North America*. Although he intended it to comprise several volumes, it consisted of only one, which was printed in 1839. It is now considered one of the rarest ornithological books ever written. (Biographer Witmer Stone reported that Townsend abandoned the project because of John James Audubon's more extensive published work in the field.) The bird specimens he collected appeared as part of the last volume of Audubon's massive *Bird of America* (1844), and his mammalian collections were used by Audubon and John Bachman in their *Viviparous Quadrupeds of North America* (three volumes, 1845–1849). Townsend himself published *Narrative of a Journey across the Rocky Mountains to the Columbia River* (1839), an account of his trip west.

After 1842, Townsend was curator of the Academy of Natural Sciences in Philadelphia and then had the same position in the National Institute, based in the U.S. Patent Office in Washington, D.C. He returned to Philadelphia in 1845 because of declining health and became a practicing dentist. In 1851, he planned to sail aboard a U.S. Navy ship on its way to the Cape of Good Hope, but his health took a turn for the worse. John Kirk Townsend died in Washington, D.C., on 6 February 1851 at the age of 41.

Toxic Substances Control Act
(7 U.S.C. 136)

This law, enacted on 11 October 1976, established national policy on toxic elements in the environment. Section 2 of the act read, "Congress finds that (1) human beings and the environment are being exposed each year to a large number of chemical substances and mixtures; (2) among the many chemical substances and mixtures which are constantly being developed and produced, there are some whose manufacture, processing, distribution in commerce, use, or disposal may present an unreasonable risk of injury to health or the environment; and (3) the effective regulation of interstate commerce in such chemical substances and mixtures also necessitates the regulation of intrastate commerce in such chemical substances and mixtures." The act set several policy measures to deal with this problem, including that "adequate data should be developed with respect to the effect of chemical substances and mixtures on health and the environment and that the development of such data should be the responsibility of those who manufacture and those who process such chemical substances"; that "adequate authority should exist to regulate chemical substances and mixtures which present an unreasonable risk of injury to health or the environment, and to take action with respect to chemical substances and mixtures which are imminent hazards"; and that "authority over chemical substances and mixtures should be exercised in such a manner as to not impede unduly or create unnecessary economic barriers to technological innovation while fulfilling the primary purpose of this Act to assure that such innovation and commerce in such chemical substances and mixtures do not present an unreasonable risk of injury to health or the environment." The act authorized the administrator of the Environmental Protection Agency to set policy that would regulate the emission and production of chemical wastes "in a reasonable and prudent manner."

Turner, Frederick Jackson (1861–1932)

Turner was the author of an essay on the American West titled "The Significance of the Frontier in American History," a groundbreaking work giving Turner's interpretation of the importance of the West in American history, culture, and habits. Born the first child of Andrew Jackson Turner, a printer, and Mary (nee Hanford) Turner in the small village of Friendship, Wisconsin, on 14 November 1861, Frederick Jackson Turner's American roots ran deep. On his father's side, patriarch Humphrey Turner came to the New World from England in 1634 and settled in Scituate, Massachusetts; Andrew Jackson Turner's father, Abel, was a veteran of the War of 1812. Mary Hanford's ancestor, Rev. Thomas Hanford, immigrated to the colonies in 1642, settling in Norwalk, Connecticut. Andrew Jackson Turner's wanderlust and desire to succeed as a yeoman printer led him to travel from rural New York to the untamed wilderness of Wisconsin in 1858. Mary Hanford Turner was likewise from New York, although she had been in Wisconsin since 1838. It was this background that gave impetus to their son's outlook on the American psyche.

Frederick Turner's initial schooling was local, but he later attended the University of Wisconsin, where he earned bachelor's and master's degrees. Johns Hopkins University (where he was a student with future President Woodrow Wilson) awarded him a Ph.D. in history; his dissertation was called "The Character and Influence of the Indian Trade in Wisconsin." At the University of Wisconsin, he was an assistant professor of history (1889–1891), professor of history (1891–1892), and professor of American history (1892–1910). It was his paper "The Significance of the Frontier in American History," delivered at the Columbian Exposition in Chicago in July 1893, however, that made him famous.

The Columbian Exposition was an event to remember—a combination world's fair, circus, and scientific showcase of the future. Turner was a member of the American Historical Association, which had come to the exposition to de-

liver papers on historical subjects. "The Significance of the Frontier in American History" has been called the quintessential essay on the foundations of American mannerisms and style. Turner wrote, "Stand at Cumberland Gap and watch the procession of civilization, marching single file—the buffalo following the trail to the salt springs, the Indian, the fur-trader and the hunter, the cattle-raiser, the pioneer farmer—and the frontier has passed by. Stand at South Pass in the Rockies a century later and see the same procession." He spoke of the psyche created by the West: "We note that the frontier promoted the formation of a composite nationality for the American people. The coast was preponderantly English, but the later tides of continental immigration flowed across to the free lands. This was the case from the early colonial days. The Scotch-Irish and the Palatine Germans, or 'Pennsylvania Dutch,' furnished dominant elements in the stock of the colonial frontier. With these peoples were also the freed indent[ur]ed servants, or redemptioners, who at the expiration of their time of service passed to the frontier. Gov. Alexander Spotswood of Virginia wrote in 1717, 'The Inhabitants of our frontiers are composed generally of such as have been transported hither as Servants, and being out of their time, and settle themselves where Land is to be taken up and that will produce the necessarys of Life with little Labour.'" In closing, Turner wrote, "What the Mediterranean was to the Greeks, breaking the bond of custom, offering new experiences, calling out new institutions and activities, that, and more, the ever retreating frontier has been to the United States directly, and to the nations of Europe more remotely. And now, four centuries after the discovery of America, at the end of a hundred years of life under the Constitution, the frontier is gone, and with its going has closed the first period of American history."

Wrote one commentator on Turner's essay, "In it, he viewed the American past from the standpoint of his native ground, the Middle West, whereas most historians before him had looked at the subject from the perspective of the Atlantic Coast. He was interested in basic questions, such as the origin of the national character and of American democracy. Older historians had traced the 'germs' of democratic institutions as far as the forests of ancient Germany. Turner found the roots of American democracy as well as other American ways in the forests of his own country."

Turner was a professor at Harvard from 1910 to 1924. He wrote little else during his lengthy academic career. He combined his thesis on the frontier with other essays in *The Frontier in American History* (1920) and *The Significance of Sections in American History* (1932), for which he won the Pulitzer Prize posthumously. Turner was working on a manuscript entitled *The United States, 1830–1850: The Nation and Its Sections* when he died in Pasadena, California, on 14 March 1932 at the age of 70.

Udall, Morris King (1922–)
Wrote James M. Perry, "'Mo' Udall may not be the best friend the environmental movement has in high places, but it would be difficult to find anyone more important." The scion of a well-known Arizona family, Udall served as chairman of the prestigious House Committee on Interior and Insular Affairs. He was born on 15 June 1922 in the Arizona desert town of St. Johns, the second son and fourth of six children of Levi Stewart Udall, a lawyer and judge on the Arizona Superior and Supreme Courts, and Louise (nee Lee) Udall, a writer. The first Udall to come to the United States was David Udall, who was born in Kent, England, in 1829 and immigrated in 1851. His son, Bishop David King Udall (Morris's grandfather), was a Mormon who led a daring band of Mormon pioneers from Salt Lake City to the small town of St. Johns in 1880. His neighbors looked down on his practice of bigamy, however; they arrested him and sent him to jail in Detroit on a charge of perjury. It took an official pardon from President Grover Cleveland to get him released. Bishop Udall was an intimate of men such as Miles Romney, a pioneer from Michigan, and Morris Goldwater, a Jewish merchant from Arizona. Ironically, all three men had grandchildren who ran for president of the United States and lost: Morris Udall, George Romney, and Barry Goldwater.

Morris Udall lost his right eye when he was six. He attended local schools, then entered the University of Arizona to study law. His studies were interrupted by World War II, and he joined the U.S. Army Air Corps, serving in a noncombat unit in Utah from 1942 until 1945. Discharged in 1946 with the rank of captain, he went back to the University of Ari-

zona and then to the University of Denver, where he earned an LL.D. degree.

After being admitted to the Arizona bar in 1949, Udall joined his older brother Stewart to form the Tucson law firm of Udall & Udall. In 1950, Morris Udall was elected Pima County (Tucson) chief deputy attorney, and county attorney three years later. The following year, he ran unsuccessfully for the Pima County Superior Court. Meanwhile, Stewart Udall was elected in 1954 to the U.S. House of Representatives. In 1961, President John F. Kennedy named Stewart Udall secretary of the interior, and Morris was elected to the vacant House seat, which he held from 1961 to 1991. In his 30 years in the House, Udall was a giant in the support and passage of conservation and environmental legislation. The *Arizona Republic* reported that "as chairman of the Interior and Insular Affairs Committee, Udall recorded several major environmental achievements, including the regulation of reclamation on strip-mined land and the Alaska Lands Act [1980] that doubled the size of the national park system and tripled the amount of wilderness area. Truly, Mr. Udall was ahead of his time on environmental concerns."

With his brother being a key figure in the Lyndon Johnson administration, Udall supported the president's Vietnam policy until 1967. In a speech in Tucson on 23 October of that year, he came out against the war and called for a full U.S. withdrawal, becoming one of the first Democrats in Congress to demand an end to the war. The move drew instant headlines. Udall's independence was a keystone of his failed 1976 presidential run, in which he placed second in all seven primaries he entered but was ultimately defeated by Jimmy Carter.

In the late 1980s, Udall was diagnosed with Parkinson's disease. On 4 May 1991, after being hospitalized for several months, he retired from Congress and returned to his home in Arizona.

Udall, Stewart Lee (1920–)

Stewart Udall was the youngest man (41 at the time) to hold the post of secretary of the interior and was one of the longest serving. Stewart Udall, said one biographer, "finds conservation good politics." Born on 31 January 1920 in St. Johns, Arizona, he was the first son and one of six children of Levi Stewart Udall, an Arizona attorney and judge, and Louise (nee Lee) Udall, a writer. Wrote author James M. Perry, "St. Johns was a town of about 1,325 church-going Mormons and Roman Catholics." Stewart Udall attended local schools before entering Eastern Arizona College in Thatcher, where he stayed for a year. He later went to the University of Arizona in Tucson but left to travel through the East as a Mormon missionary. When World War II broke out, he volunteered for service in the U.S. Air Force and flew 56 missions over Europe. On his return to the United States, he reentered the University of Arizona, which awarded him a law degree in 1948.

Stewart Udall started a law practice with his younger brother Morris and entered the political arena when he was elected to the U.S. House of Representatives in 1954. He served until 1961. Wrote Interior Department historian Eugene Trani, "His career in Congress showed love of the West. He sought appointments on the Committee on Interior and Insular Affairs, as well as Education and Labor. He became a member of the Joint Committee on Navajo-Hopi Administration. On these three committees he did his best work. In his first term, he unsuccessfully fought for [the] return of 100,000 acres to the Coconino and Sitgreaves National Forests. He favored the Colorado River Project, and in 1960 made his position on national parks clear by warning that 'the one overriding principle of the conservation movement is that no work of man (save the bare minimum of roads, trails, and necessary public facilities in access areas) should intrude into the wonder places of the National Park System.'"

After stumping the country on behalf of Democratic presidential candidate John F. Kennedy in 1960, Udall was rewarded with the secretaryship of the Interior Department. Serving through two administrations, Udall was on the job until 1969, one of the longest tenures at Interior. As secretary, he stressed water development, emphasized the protection of water resources and the improvement of water quality, advocated the expansion of national park and recreation areas, and called for a moratorium on the selling of public lands. He formed the Bureau of Outdoor Recreation to report to Congress on ways to improve that area of American life. Udall oversaw the administration of the Wilderness Act (1964), the Highway Beautification Act (1965), the Endangered Species Preservation Act (1966), the National Trails System Act (1968), and the Wild and Scenic Rivers Act (1968). When the Johnson administration left office in 1969, Udall left the Interior Department. Since then, he has written several books, including his *Quiet Crisis: The Next Generation* (1988) and *The Myths of August: A Personal Experience of Our Tragic Cold War Affair with the Atom* (1994), which supplemented his *Quiet Crisis* (1963) and *1976: Agenda for Tomorrow* (1968).

See also Udall, Morris King.

United States Biological Survey

See Merriam, Clinton Hart.

United States Exploring Expedition

Known as the "Ex Ex," this journey headed by Lt. Charles Wilkes of the U.S.

Navy was the first major government-sponsored nature research mission. By an act of Congress on 14 May 1836, the president of the United States was authorized "to send out a surveying and exploring expedition to the Pacific Ocean and South Seas"; the sum of $150,000 was appropriated for the mission. Originally, Capt. Thomas ap Catesby Jones was named by President Andrew Jackson as the expedition's commander (Jones had been one of his aides at the Battle of New Orleans), but after several years of delay and arguments over the size of the ships he would have to use, he resigned and was replaced by his aide, Wilkes. Charles Wilkes (1798–1877) was a lifelong navy man who took a strong interest in geophysics and astronomy. Born in New York, he was related to the English politician John Wilkes (1725–1797), who fought the British government for colonists' rights in America and for civil liberties at home. (The city of Wilkes-Barre, Pennsylvania, is named for him.) Because of his family's wealth, Charles Wilkes was able to attend a prestigious preparatory school before joining the navy as a midshipman in 1818. Eight years later, he was commissioned a lieutenant. In 1830, he was named head of the newly established Depot of Charts and Instruments, which later became the Naval Observatory and the U.S. Hydrographic Office. In the years before he was named commander of the Ex Ex, Wilkes demonstrated a commitment to research voyages; in 1836, he was sent to Europe to obtain scientific devices for another exploratory expedition, and he conducted important surveys of the St. George's Bank and the Savannah River in 1837 and 1838.

Expedition biographer Daniel C. Haskell wrote of Wilkes, "Impetuous and dominating, he was a resolute, determined man of great driving power. A strict disciplinarian, he was often in conflict with both his superiors and his subordinates, and was known as the 'stormy petrel' of the Navy. His scientific attain-

ments, which greatly exceeded those of the average naval officer of his day, had much to do with his final selection for the command." The expedition comprised the U.S.S. *Vincennes* and five other ships—the *Peacock*, the *Porpoise*, the *Relief*, the *Sea Gull*, and the *Flying Fish*. The scientific staff consisted of horticulturist William D. Brackenridge, conchologist J. P. Couthey (or Couthouy), mineralogist James D. Dana, botanist William Rich, philologist Horatio Hale, naturalists Charles Pickering and Titian Ramsay Peale, and draughtsmen (artists) Alfred T. Agate and Joseph Drayton. They set sail from Hampton Roads, Virginia, on 18 August 1838 and did not return until July 1842. Several of the ships were lost during the trip, and a vessel bought and renamed the *Oregon* was added. The expedition traveled from the Cape Verde Islands across the sea to Brazil, wrapped around Cape Horn, and ended at Peru. Thereafter, some of the ships visited the Pacific Islands, Australia, New Zealand, Hawaii, the northwestern coast of the United States, the Philippines, Singapore, and St. Helena. At one point, Wilkes's ship explored the coastal region of Antarctica. In honor of his apparently being the first to sight it, the area was later named Wilkes Land. Wilkes reported on his scientists' accomplishments: "They made frequent and long excursions into the interior of several countries," he wrote. "Mr. Drayton travelled along the Columbia River as far as the Blue Mountains; another party—among whom were Messrs. Peale, Dana, Rich, Agate, and Brackenridge—went from Fort Vancouver, southward, across the land to San Francisco; a third party left one of the ships at Fort Nesqually [now near Tacoma, Washington], travelling east to Fort Colville, then southeast to the head waters of the Kookooske River.... Mr. Hale made considerable excursions in New Holland, New Zealand, and Oregon, and returned from California overland to the United States. Dr. Pickering was a bold explorer, and made numerous excursions in Brazil and elsewhere."

After the expedition returned to the United States, arguments over credit for reports and other controversies held up publication for many years. Wilkes's narrative of the trip, appearing in five volumes with an atlas of maps, was published in 1844. Volume 6, *Ethnography and Philology*, was written by Horatio Hale in 1846. James Dana's *Zoophytes* (volume 7) appeared in 1846, with an atlas coming out three years later. Titian Peale's *Mammalia and Ornithology* was taken away from him in a dispute and given to ornithologist John Cassin (1813–1869); it was released in 1858. Other volumes, including William Brackenridge's *Botany, Cryptogamia, Filices* and Edward Tuckerman's *Licenes*, were released later; they are now collector's items. Wilkes himself also edited some 200 volumes of reports on the exploration's scientific collections.

Charles Wilkes became better known in his later life. At the outbreak of the Civil War, he was named commander of the Union ship *San Jacinto*, which in 1861 captured the British mail steamer *Trent* and forcibly removed James M. Mason and John Slidell, the Confederate commissioners to Great Britain and France, respectively, triggering the so-called Trent affair. Wilkes was congratulated by the U.S. government and sent to the Bahamas to capture Confederate ships. Incompetence caused his mission to fail, and he was court-martialed in 1864. He retired, before being convicted, with full rear admiral status and spent his last years in retirement in Washington, D.C. Historian Ralph Andrist wrote:

Wilkes was dead long before Antarctic exploration was resumed, and then the new generation of explorers did little to rescue him from oblivion. In several cases they, too, found only ocean where Wilkes had charted land—although none could deny that the polar continent Wilkes had dis-

covered was there, even though it might be fifty or a hundred miles from where he had charted it.... [T]he cause of these discrepancies was revealed only as polar phenomena came to be better understood. Under certain conditions, atmospheric refraction over polar ice is so pronounced that a mountain peak more than a hundred miles away may appear to be only a few miles distant.

But no cold and scientific explanation is going to bring Wilkes any recognition at this late date. Moreover, it has been chiefly Australians who have explored along the shores of Wilkes Land and have taken the lead in re-establishing Wilkes' reputation. In his own country he is still a footnote in most histories, or at best a brief paragraph. Eccentric, prickly, occasionally wrongheaded, he was nevertheless one of his country's great explorers. He deserves a better place in her pantheon of heroes.

See also Peale, Titian Ramsay.

United States Fish and Wildlife Service (USFWS)

Created under the Fish and Wildlife Act of 1956 (16 U.S.C. 742a), the U.S. Fish and Wildlife Service actually existed for more than 85 years as separate agencies within the government. By a joint resolution passed on 9 February 1871, Congress created the U.S. Fish Commission to oversee fish and water animal reserves in the nation. Spencer F. Baird, assistant secretary of the Smithsonian Institution, was named the first commissioner of fisheries, holding that post until his death in 1887. During his administration, the Atlantic and Pacific Oceans' rich diversity of fish species was explored and scrutinized; the nation's first marine laboratory at Woods Hole, Massachusetts, was constituted; and preservation of the environment of fish and other marine animals as well as the guardianship of the animals themselves was advocated.

Under an act of Congress on 14 February 1903, the Fish Commission was renamed the Bureau of Fisheries and placed under the authority of the Department of Commerce and Labor (now the Department of Commerce). According to *Government Agencies*, "the primary function of the Commission, specified in the 1871 act creating the agency, concerned the scientific investigation of the causes for the decrease of commercial food fish and aquatic animals, such as mussels, oysters, and sponges, in the coastal and inland waters of the United States. Later related activities included studying the effects of river impoundment projects on fish migration patterns, determining the feasibility of using fishways to direct fish around obstructions, and investigating the effects of red tides and water pollution in coastal and inland waters on commercial fish and shellfish." Between its establishment in 1871 and its becoming the Fish and Wildlife Service in 1940, laws that were under its jurisdiction included the Alaska Fisheries Act of 1906, the Federal Black Bass Law of 1930, and the Whaling Treaty Act of 1936.

In 1885, Congress established the Division of Economic Ornithology and Mammalogy in the Agriculture Department. Renamed the Division of Biological Survey in 1896 and the Bureau of Biological Survey in 1905, it was merged in 1940 into the Fish and Wildlife Service. Naturalist and zoologist Clinton Hart Merriam was in charge of the bureau from 1886 to 1910. The bureau's responsibilities covered the collection of statistics on state game laws as ordered by the Lacey Act of 1900, the preservation of endangered species in Alaska according to the Alaska Game Law of 1902, the inspection of fur seals and other such animals in the Pribilof Islands, and the management and protection of bird species under the Federal Migratory Bird Law of 1913. After 1903, the bureau was in charge of the numerous wildlife refuges created by presidential authority.

On 1 July 1940, under Reorganization Plan III, Congress combined the Bureau of Fisheries and the Bureau of Biological Survey into the Fish and Wildlife Service. Under the Fish and Wildlife Act of 1956, the service was renamed the U.S. Fish and Wildlife Service, with the responsibility of two new departments: the Bureau of Commercial Fisheries and the Bureau of Sport Fisheries and Wildlife. In 1970, the Bureau of Commercial Fisheries was transferred to the Commerce Department.

According to the *U.S. Government Manual 1993–1994*, the service "is responsible for migratory birds, endangered species, certain marine mammals, inland sport fisheries, and specific fishery and wildlife research activities. Its mission is to conserve, protect, and enhance fish and wildlife and their habitats for the continuing benefit of the American people."

See also Baird, Spencer Fullerton; Carson, Rachel Louise; Fish and Wildlife Act; Merriam, Clinton Hart.

United States Forest Service (USFS)

This federal agency, now a part of the Agriculture Department, was originally created by an appropriations act passed by Congress on 15 August 1876. Prior to its passage, the federal government had a hands-off policy as to the forestry reserves of the nation. It was up to the machinations of private groups, such as the American Academy for the Advancement of Science, to carry out forest management policies state by state. In 1873, the academy assembled a committee chaired by forestry expert Franklin Benjamin Hough to lobby the government for the creation of a federal entity to oversee forestry matters. In 1874, the Hough committee fashioned a statement that was given to both President Ulysses S. Grant and Congress. On 15 August 1876, Congress requested that the commissioner of agriculture (later the secretary of agriculture) appoint a specialist in forestry matters who could conduct nationwide research and report back to Congress; it appropriated $2,000 to be used to hire the expert and pay his salary. Franklin Hough was hired, and in 1881, the Division of Forestry was officially designated, with Hough as its chief. Five years later, in 1886, the division became a permanent agency of the government. Hough was replaced in 1883 by Nathaniel Hillyer Egleston, who also served as secretary of agriculture during the first Grover Cleveland administration. In 1886, Egleston was replaced by Bernhard Eduard Fernow. Along with Hough, Fernow was one of the leading forestry experts in the nation. During his tenure, Fernow oversaw passage of the General Revision Act (the Payson Act) of 1891, which authorized the president to preserve forest and timberland reserves in national parks and forests, and the Pettigrew Act of 1897, which set out the proper administration and use of these preserves. In 1898, Fernow was succeeded by Gifford Pinchot, who would become the best known and perhaps most controversial forest chief. During his reign, Pinchot sought to consolidate the power of the Forestry Division, which was then under the authority of the Department of Agriculture, as well as the administration of forestry reserves, which was handled by the Department of the Interior. His work was rewarded when the Forest Service was transferred from the Interior Department to the Agriculture Department under the Forest Transfer Act of 1 February 1905 (16 U.S.C. 472), which was sponsored by Rep. John Fletcher Lacey of Iowa, the supporter of the Lacey Game and Wild Birds Preservation Act of 1900 and the Antiquities Act of 1906.

An important sidebar to the formation of the Forest Service involves the American Indians. Since the signing of a number of peace treaties with the Indians during the nineteenth century, the federal government had placed lands al-

lotted to the Indians in trust. Surplus or dead timber found on these lands was sold by the Indians for profit. With the 1878 Supreme Court decision in *United States v. Cook* and the subsequent Down and Dead Timber Act of 1889, the forestry management of Indian lands became the Forest Service's responsibility. Much of this work was done separately through the Bureau of Indian Affairs (BIA). In 1908, Pinchot, Commissioner of Indian Affairs Francis Leupp, and Leupp's secretary, Robert G. Valentine, worked on a plan to consolidate all forestry programs under Pinchot's authority. On 22 January 1908, Secretary of Agriculture James Wilson and Secretary of the Interior James R. Garfield signed the agreement, which consolidated all the agencies. In 1910, Jay P Kinney, a graduate of the Cornell University forestry program, became the head of the new Bureau of Indian Affairs Forestry Service and served until 1933.

The U.S. Forest Service, according to the *U.S. Government Manual, 1993–1994,* "has the Federal responsibility for national leadership in forestry. Its mission is to provide a continuing flow of natural resource goods and services to help meet the needs of the Nation."

See also Egleston, Nathaniel Hillyer; Fernow, Bernhard Eduard; Graves, Henry Solon; Greeley, William Buckhout; Hough, Franklin Benjamin; Kinney, Jay P; Payson Act; Pettigrew Act; Pinchot, Gifford; United States v. Cook.

United States Geological Survey (USGS)

The U.S. Geological Survey is the federal agency established in 1879 to coordinate all the governmental land surveys of the United States and to explore and map the nation's metal, water, and mineral resources. During the nineteenth century, many federally sponsored groups of scientists and geologists explored various regions of the nation in an attempt to geographically map these mostly unexplored areas. Many of these surveys overlapped and were led by men who zealously guarded their conclusions, lest others be credited. Among these were the explorations of Frenchman François André Michaux in the area of the Missouri River (1792), Lewis and Clark in the territory of the Louisiana Purchase (1803–1807), Zebulon Pike in the Rocky Mountains (1805–1807), and Englishman George Featherstonhaugh in the western United States (1834–1835). These, however, were expeditions and not mapping or surveying parties.

In the decade following the Civil War, Congress appropriated funds for surveys across the western United States. These became Ferdinand V. Hayden's Geological and Geographical Survey of the Territories, George Montegue Wheeler's Survey of the Territory West of the One Hundredth Meridian, John Wesley Powell's Survey of the Colorado River, and Clarence King's Survey of the Fortieth Parallel. The act of 2 March 1867 (14 Stat. 471) designated "a geologic survey of Nebraska, said survey to be prosecuted under the direction of the Commissioner of the General Land Office." This became the Hayden survey, which was expanded by Congress in 1869 to include Colorado, New Mexico, Wyoming, Idaho, and Montana; its results were published between 1867 and 1883. The same act authorized the "geological and topographical exploration of the territory between the Rocky Mountains and the Sierra Nevada Mountains, including the route or routes of the Pacific Railroad." This excursion was conducted by Clarence King, a geologist and mining engineer from Newport, Rhode Island. One source called it "a survey entirely across the Cordilleran ranges from eastern Colorado to the California boundary." King's survey had the benefit of such talented geologists as Arnold Hague and Samuel Franklin Emmons. The third survey was begun in 1867 by the Smithsonian Institution and was headed by John Wesley Powell of Illinois Wesleyan University.

The act of 11 July 1868 (15 Stat. 253) empowered the secretary of war "to issue rations for twenty-five men of the expedition engaged in the exploration of the river Colorado under [the] direction of Professor Powell, while engaged in that work." Under the act of 23 June 1874, Powell's authority was extended to what is now Utah. One of the members of the Powell team was Clarence E. Dutton, pioneer geologist. The fourth survey was sanctioned by the act of 10 June 1872 (17 Stat. 367). Headed by Lt. George Montegue Wheeler of the U.S. Corps of Engineers, this exploration was aimed, according to Wallace Stegner, "toward a master atlas of the western states and territories." Wrote author Mary Rabbitt, "This survey ... covered all the territory west of the one hundredth meridian, which includes the western parts of the Dakotas, Nebraska, Kansas and Texas, the Rocky Mountain states, and the Pacific Coast states." According to the official report of the chief of engineers to the secretary of war, the survey's chief aim was to collect "all the information necessary before the settlement of the country, concerning the branches of mineralogy and mining, geology, paleontology, zoology, botany, archaeology, ethnology, philology, and ruins." Among the geologists on this excursion was Grove Karl Gilbert.

Between 1867 and 1877, reports from these four surveys flowed back to Washington. Much of their work overlapped, however, and the National Academy of Sciences asked Congress to consolidate them. In 1877, Clarence King himself recommended to President Rutherford B. Hayes that all federal geological surveys be conducted by a single federal agency; this would concentrate all power in the hands of an agency director, avoid overlapping or unnecessary expeditions, and allow dissemination of the acquired information. Two years later, under the Sundry Civil Appropriations Act of 3 March 1879, Congress created the U.S. Geological Survey and made King its first director. It directed the survey to complete "the classification of the public lands and the examination of the geological structure, mineral resources, and products of the national domain." In another part of the act, known as the Organic Act, Congress directed the Geological Survey to categorize the lands it examined as "arable, irrigable, timber, pasturage, swamp, coal, mineral lands and other classes as deemed proper, having due regard to humidity of climate, supply of water for irrigation, and other physical characteristics." King accepted the position as director only to set up offices and staff; he resigned two years later, handing the administrative reins to explorer John Wesley Powell. Thomas G. Manning wrote that, as director, "King ratified the intention of Congress at the founding [of the Geological Survey] and organized scientific parties for work in centers of gold and silver mining—in Virginia City and Eureka, Nevada, and in Leadville, Colorado. . . . [He] also initiated Annual Reports on the statistics and technology of minerals."

Powell served as director from 1881 to 1894. Wrote Lindsey Morris, "Under Powell's personal direction and guidance, the survey was built into an efficient organization, operating in a highly satisfactory manner. He selected geologists versed in special branches of each field and placed them in charge, while he acted in an advisory capacity, always free to assist and instruct the members of his project, which included many college professors and students." Powell's most important directive may have been his establishment of the irrigation surveys (called the Powell Irrigation Surveys) of the arid West under the direction of Frederick Haynes Newell, an expert in irrigation who later served as the first head of the Bureau of Reclamation. Under Powell's directorship, William John McGee headed the survey's division of the Atlantic Coastal Plains Geology from 1883 to 1894. Ill health and conflicts with William Morris Stewart of the Senate

Committee on Irrigation and Arid Lands led to Powell's resignation in 1894.

His successors, Charles Doolittle Walcott, George Otis Smith, and Walter Curran Mendenhall, led the survey through what have been called its "middle years." Responding to the rising conservation movement, Walcott in particular secured surveys of the forest reserves set up by Presidents Cleveland and Harrison, as well as overseeing for a short period the establishment of irrigation projects and surveys under the 1902 Newlands Reclamation Act. Under directors William Embry Wrather and Thomas B. Nolan, appropriations for the survey passed $100 million a year. Wrather's tenure was marked by an acceleration in the mapping of the United States using a process known as photogrammetric mapping (photographic mapping from the air). Under the act of 5 September 1962 (43 U.S.C. 31(b)), the survey's authority was expanded to include inquiries outside the "national domain." The *U.S. Government Manual 1993–1994* reported, "The Survey's primary responsibilities are: investigating and assessing the Nation's land, water, energy, and mineral resources; conducting research on global change; investigating natural hazards such as earthquakes, volcanoes, landslides, floods, and droughts; and conducting the National Mapping Program."

See also Dutton, Clarence Edward; Emmons, Samuel Franklin; Gilbert, Grove Karl; Hague, Arnold; Hayden, Ferdinand Vandeveer; King, Clarence Rivers; McGee, William John; Powell, John Wesley; Powell Irrigation Survey; Smith, George Otis; Walcott, Charles Doolittle; Wheeler, George Montegue; Wrather, William Embry.

United States Reclamation Service
See Bureau of Reclamation.

United States v. Briggs
(9 Howard [50 U.S.] 351 [1850])
One of the first cases heard by the U.S. Supreme Court regarding federal rights on timber reserves, *United States v. Briggs*

set the Court's policy on how such reserves would be managed. In 1831, Congress passed the Live Oak Act, which was officially called "an Act to provide for the punishment of offenses committed in cutting, destroying, or removing live oak and other timber or trees reserved for naval purposes." In 1845, Ephraim Briggs cut down 20 white oak and 20 hickory trees on public land that even the government admitted was not reserved for naval purposes. Briggs claimed that he could be charged only with simple trespass; the government disagreed and charged him with violating the Live Oak Act. The circuit court judges could not agree on Briggs's guilt, so the case was sent to the U.S. Supreme Court. Originally heard in 1847 and sent back to the circuit court, *Briggs* was reargued in 1850. Justice John Catron of Tennessee presented the Court's unanimous opinion, holding that Briggs was guilty of violating the Live Oak Act. He wrote that "cutting, removing, or using, for any other than naval purposes, any trees or timber standing, growing, or being on any land belonging to the United States, whether reserved for naval purposes or not, was a criminal offense." *Briggs* was a landmark case, setting timber policy for the next century.

United States v. Cook
(91 U.S. 389 [1878])

This Supreme Court case decided whether American Indians, having been allotted land parcels by the United States, could cut down and sell timber on those lands. Following an 1831 treaty, the Menominee Indians of Wisconsin ceded a portion of their lands to the United States at a cost of $20,000 to provide a settlement for certain parts of the Oneida Indian tribe of New York. In 1838, the Christian and Orchard sections of the Oneida sold their land to the United States, except for a small portion to be held in trust. This land was used by some of the Oneidas for farming. Others

cut timber from the land and sold the lumber to one George Cook, a lumber dealer. The United States sued Cook, claiming that under the trust agreement, such logs could not be legally purchased without the federal government's permission. Judges in the original trial court, the Circuit Court for the Eastern District of Wisconsin, heard the case but were deadlocked. The case was sent to the U.S. Supreme Court for a decision. The Court ruled unanimously that under the trust agreement such timber sales belonged solely to the United States and that the Indians' rights consisted only of residency, nothing more. Writing for the Court, Chief Justice Salmon P. Chase held:

The right of the Indians in the land from which the logs were taken was that of occupancy alone. . . . The fee [for the sale of the logs] was in the United States, subject only to this right of occupancy. . . . The right of use and occupancy by the Indians is unlimited. They may exercise it at their discretion. If the lands in a state of nature are not in a condition for profitable use, they may be made so. If desired for the purpose of agriculture, they may be cleared of their timber to such an extent as may be reasonable under the circumstances. . . . The timber while standing is a part of this realty, and it can only be sold as the land can be. The land cannot be sold by the Indians, and consequently the timber, until rightfully severed, cannot be.

This was a landmark case in that it set government policy on Indian land rights for much of the rest of the nineteenth century.

Following the decision, Secretary of the Interior Columbus Delano canceled all Indian timber sales in the allotted lands under federal control. This decision hampered Indian sales of dead timber for a number of years until Congress decided to act. Under the Dead and Down Timber Act (25 Stat. 673) of 16 February 1889, Congress authorized the president to allow for the sale of "dead timber, standing or fallen."

See also Dead and Down Timber Act.

United States v. Grimaud
(220 U.S. 506 [1910])

This Supreme Court case hinged on whether Congress can delegate the right to make administrative rules, and not the right of legislative power, to a member of the government. Under the Payson Act, also known as the General Revision Act of 3 March 1891, Congress authorized the president to set apart and reserve certain areas to be used as forest reservations. Another act, passed in 1906, gave the secretary of agriculture the right to set laws dealing with the use and management of these reservations. One such regulation said that any person wanting to use such lands for grazing must obtain a permit. Defendants Pierre Grimaud and J. P. Carajous allowed their sheep to graze on land in the Sierra Forest Reserve in California in 1907 without a permit. Grimaud and Carajous argued that their offense was not criminal, that the rule had been made not by Congress but by an officer of the government, and that such an action was unconstitutional. A lower court agreed with Grimaud and Carajous, and the District Court for the Southern District of California upheld the judgment. The United States appealed the case to the U.S. Supreme Court. The Court ruled unanimously on 3 May 1911 that although only Congress could set the limits of a penalty for a crime, it was constitutional to confer upon the secretary of agriculture the right to establish rules for forest reserves. Justice Lucius Quintus Cincinnatus Lamar, a former secretary of the interior, wrote the Court's opinion. "The Secretary of Agriculture could not make the rules and regulations for any and every purpose. As to those here involved, they all relate to matters clearly indicated and authorized by Congress. The subjects as to which the Secretary can regulate are defined. The lands are set apart as a forest reserve. He is required to make provision to protect them from depredations and harmful uses. He is authorized 'to regulate the occupancy and use and to preserve the forests from destruction,'" wrote Lamar. "A violation of reasonable rules regulating the use and occupancy of the property is made a crime, not by the Secretary, but by Congress. The statute, not the Secretary, fixes the penalty."

United States v. Midwest Oil Company
(236 U.S. 459 [1915])

This little-remembered Supreme Court case established that the president, as agent of public lands, has the right to withdraw certain lands for government use or for the protection of public causes, such as to save oil reserves or animal habitats. By an act of 11 February 1897, "all public lands containing petroleum or other mineral oils and chiefly valuable therefor, have been declared by Congress to be free and open to occupation, exploration and purchase by citizens of the United States ... under regulations prescribed by law." Because of this act, some pockets of oil were extracted so fast that supplies were threatened. The director of the U.S. Geological Survey, George Otis Smith, recommended to Secretary of the Interior Richard A. Ballinger that all oil claims in the state of California be suspended. Ballinger relayed the message to President William Howard Taft, who on 27 September 1909 issued a proclamation that withdrew all public lands in California and Wyoming from entry for purposes of oil extraction. On 27 March 1910, William T. Henshaw and other men from the Midwest Oil Company entered onto public land in Wyoming, drilled a well, and struck oil. They then filed a claim for the land. The government filed a claim to recover the property and reimbursement for the 50,000 barrels of oil Midwest had pumped from the site. A district court dismissed the suit; the Court of Appeals for the Eighth Circuit asked the Supreme Court to decide certain questions in the case. Justice Joseph Rucker Lamar answered for the Court (Justice James McReynolds did not participate). "We need not consider whether, as an original question, the President could have withdrawn from private acquisition what Congress had made free and open to occupation and purchase," Lamar wrote. "The case can be determined on other grounds and in the light of the legal consequences flowing from a long continued practice to make orders like the one here involved. For the President's proclamation of September 27, 1909, is by no means the first instance in which the Executive, by a special order, has withdrawn land which Congress, by general statute, had thrown open to acquisition by citizens. And while it is not known when the first of these orders was made, it is certain that 'the practice dates from an early period in the history of the government.' *Grisar v. McDowell*, 6 Wall. 381 (1867). ... [T]he case is therefore remanded to the District Court with directions that the decree dismissing the Bill be *reversed*."

United States v. Rio Grande Dam and Irrigation Company
(174 U.S. 690 [1899])
This Supreme Court case involved whether the River and Harbors Act of 1890, which prohibited the blockage of navigable waters, included dams. The Rio Grande Dam and Irrigation Company sought to construct a dam across the Rio Grande in what was then the territory of New Mexico. The U.S. government sought to block construction of the dam. The district court in the Third Judicial District of New Mexico ruled that the Rio Grande was not a navigable waterway as defined by the Rivers and Harbors Act of 1890; therefore, the gov-

ernment had no standing to block the company's plans. The New Mexico Territorial Supreme Court upheld the judgment, and the United States sued for relief to the U.S. Supreme Court. On 22 May 1899, the Court ruled 7–0 (Justices Horace Gray and Joseph McKenna did not participate) that "the prohibition by the act of Congress of 19 September 1890 against the creation of any obstruction to the navigable capacity of any waters, includes not only an obstruction in the navigable portion of the stream, but also anything, wherever or however done, to destroy the navigable capacity of one of the navigable waters of the United States." Justice David Josiah Brewer delivered the Court's opinion. "Without pursuing this inquiry further we are of the opinion that there was error in the conclusions of the lower courts; that the *decree must be reversed*, and the case remanded, with instructions to set aside the decree of dismissal, and to order an inquiry into the question whether the intended acts of the defendants in the contruction of a dam and in appropriating the waters of the Rio Grande will substantially diminish the navigability of that stream within the limits of present navigability, and if so, to enter a decree restraining those acts to the extent that they will so diminish," Brewer wrote.
See also Rivers and Harbors Act of 1899.

United States–Canada Joint Fish Commission of 1892
This council sat for four years to study the problems of fish on the U.S. and Canadian borders. Overfishing by commercial fishermen in the Great Lakes led to a plunging fish population. Establishment of the U.S. Fish Commission in 1871 was the first step in addressing this issue. The Fish Commission's first director, Spencer Fullerton Baird, appointed James W. Milner, a scientist from Kenosha, Wisconsin, to study the problems in the lakes and determine what could be done. Milner found that fishermen themselves

had noticed a dramatic reduction in fish stocks in the lakes. In his summary to Congress, *Report on the Fisheries of the Great Lakes: The Result of Inquiries Persecuted in 1871 and 1872* (1873), Milner stated that restocking was the only solution. Unfortunately, the Canadian government had rules that prohibited a large-scale program without both countries cooperating.

For the next 20 years, as fish stocks dwindled, the two countries refused to budge toward cooperation. Finally, in 1892, both nations set up the United States–Canada Joint Fish Commission. Sitting as U.S. commissioner was Richard Rathbun (1852–1918), a member of the U.S. Commission on Fisheries. Rathbun, a 40-year-old naturalist originally from Buffalo, New York, had already had a long and distinguished scientific career; prior to his appointment to this commission, he had been in charge of fishery inquiries for the tenth census in 1880 and had been in charge of preparing material for the Paris fur seal tribunal in 1893. Great Britain, with authority over Canada, chose as its commissioner Dr. William Wakeham, commander of *La Canadienne*, a ship situated in the Gulf of St. Lawrence to protect fish stocks.

The final report of the commission was issued on 30 December 1896. Wrote historian Margaret Beattie Bogue, "The report portrayed a once-bounteous natural resource seriously eroded and urgently in need of constructive regulation. In every lake and river examined, fieldwork checked against official records obtained in other ways verified the decline of one or more of the commercially preferred species. The commissioners attributed the sorry state of affairs to overfishing and to a vastly improved fishing technology used by ever larger numbers of fishermen in pursuit of the wealth of the Great Lake waters." The commissioners suggested the licensing of all U.S. fishermen (Canadian fishermen had been licensed since 1869), the establishment of regulations as to the number

and size of fish caught, and programs to repopulate fish stocks through artificial propagation techniques and sanctuary protection. Officials and fishermen in the affected states fought the regulations for several years as an infringement of their rights. It took the signing of the 1909 Boundary Waters Treaty between the United States and Canada, which created the International Joint Commission, an investigative and regulatory council; the 1972 Great Lakes Water Quality Agreement; and the 1978 Great Lakes Quality Agreement before stringent regulations were set down as to water quality and fishing. In 1985, authors Jane Elder and Marge Wetzel reported that although the number of fish has increased in the Great Lakes as a whole, pollution by PCBs and DDT residues may threaten their future growth.

Uranium Tail Millings Radiation Control Act (42 U.S.C. 7901)

This legislation, enacted on 8 November 1978, was the first congressional action in the area of nuclear waste. Uranium tailings, or tail millings, are the residues of uranium brought to the surface for use. This act attempted to deal with the disposal of such wastes. Congress found that "uranium mill tailings located at active and inactive mill operations may pose a potential and significant radiation health hazard to the public, and that the protection of the public health, safety, and welfare and the regulation of interstate commerce require that every reasonable effort be made to provide for the stabilization, disposal, and control in a safe and environmentally sound manner of such tailings in order to prevent or minimize radon diffusion into the environment and to prevent or minimize other environmental hazards from such tailings." The act mandated the secretary of energy to designate no later than 8 November 1978 "processing sites" where uranium tailings would be disposed. These sites included the areas of

Salt Lake City, Utah; Green River, Utah; Mexican Hat, Utah; Durango, Colorado; Grand Junction, Colorado; two sites at Rifle, Colorado; Gunnison, Colorado; Naturita, Colorado; Maybell, Colorado; two sites at Slick Rock, Colorado; Shiprock, New Mexico; Ambrosia Lake, New Mexico; Riverton, Wyoming; Converse County, Wyoming; Lakeview, Oregon; Falls City, Texas; Tuba City, Arizona; Monument Valley, Arizona; Lowman, Idaho; and Cannonsburg, Pennsylvania.

Usher, John Palmer (1816–1889)

John Palmer Usher was the sixth secretary of the interior, serving from 1863 to 1865 in the Lincoln administration. Born in Brookfield, New York, on 9 January 1816, he was the son of Nathaniel Usher, a physician, and Lucy (nee Palmer) Usher. The family was descended from Hezekiah Usher, who emigrated from England and settled in Boston sometime in the mid-seventeenth century. John Palmer Usher attended the local schools of Madison County, read the law in the offices of Henry Bennett and John Hyde in New Berlin, New York, and was admitted to the bar at the age of 23. For a time he was Hyde's partner, but a lack of business drove him west. He settled in Terre Haute, Indiana, where he started a law practice with several lawyers. Among his acquaintances was another young attorney named Abraham Lincoln.

Elected as a Whig to the state legislature from Vigo County in 1850, Usher declined to run for reelection. He was active in the founding of the state Republican Party in 1854 and ran an unsuccessful campaign for the U.S. House two years later. During the campaign of 1860, Usher was busy speaking on behalf of Lincoln throughout the state. Although in November 1861 he became attorney general of Indiana, he resigned four months later when Lincoln asked him to take the post of assistant secretary of the interior under Caleb B. Smith. Within a year, Smith's poor health forced him to resign, and Usher became the interior secretary. During his tenure, he oversaw enactment of the Homestead Act, altered Indian policy to force the Indians onto reservations, and wrote in his last annual report as secretary about the range of public lands in the nation. Usher was, like most of the cabinet, at Lincoln's side on the night of his assassination. One source noted that he was the only person who was able to get some sleep that long night—Secretary of the Navy Gideon Welles found Usher asleep on a small bed in the back parlor of the Peterson House sometime after 6:00 A.M. on the morning of Lincoln's death.

Before Lincoln's assassination, radical Republicans had been looking to cleanse the cabinet of all who did not follow their line. Usher was counted among this group, and Lincoln had been pressured to fire him. Instead, Usher resigned on 9 March 1865, to take effect on 15 May. He settled in Lawrence, Kansas, where he worked as chief counsel and solicitor of the Kansas Pacific Railroad, a post he held until his retirement in 1880. He also served for two years (1879–1881) as mayor of Lawrence. He died in Lawrence on 13 April 1889. The *Lawrence Daily Journal* wrote of him, "It has long been the practice here to close closely contested campaigns with a rally, at which Judge Usher invariably presided. His speeches on such occasions, though audible to but a few, were as decisive in effect as a volley of musketry in the faces of a wavering foe. They swept away disguises, steadied the line of his comrades, and were the key-notes of assured victory."

See also Department of the Interior.

Vermont Yankee Nuclear Power v. Natural Resources Defense Council (435 U.S. 519 [1978])

In this Supreme Court case, it was held that the National Environmental Policy Act of 1969 (NEPA) did not require federal agencies to call for safeguards in the establishment of nuclear power plants when reviewing applications for the construction of such plants. The Vermont Yankee Nuclear Power Corporation applied to the Atomic Energy Commission (AEC) (which in 1974 became the Nuclear Regulatory Commission) for an operating license for its unopened nuclear power plant. At the same time, the Consumers Power Company sought authorization to build two nuclear power plants in Midland, Michigan. Although hearings were instituted in both cases to hear concerns about the environmental effects of nuclear use, the AEC ruled that environmental groups did not present evidence about alternatives to nuclear power and thus granted an operating license to Vermont Yankee and a construction permit to Consumers Power. The Natural Resources Defense Council, an environmental group, sued both companies. A court of appeals ruled that in the case of Vermont Yankee, the AEC was compelled by NEPA to study the effects of processing nuclear fuel and disposing of nuclear waste. In the case of Consumers Power, it rebuked the AEC for not considering alternative methods of energy creation, including energy conservation, before awarding a construction permit.

Both companies, appearing as the Vermont Yankee Nuclear Power Corporation, appealed the ruling. The U.S. Supreme Court ruled unanimously (Justices Harry Blackmun and Lewis Powell did not participate, making the vote 7–0) that the court of appeals had overstepped its authority in both cases. Writing for the Court, Justice William H. Rehnquist held that under the Administrative Procedures Act (APA), administrative agencies had "broad discretion" in formulating their methods and procedures, and that NEPA had no such provision for the AEC to follow. Wrote Rehnquist, "In short, nothing in the APA, NEPA, the circumstances of this case, the nature of the issues being considered, past agency practice, or the statutory mandate under which the commission operates permitted the court [of appeals] to review and overturn the rulemaking proceeding on the basis of the procedural devices employed ... by the commission so long as [it] employed at least the statutory minima." As to the facts behind the granting of a construction permit to Consumers Power, the Court held that the AEC did not have to consider alternative power sources. Rehnquist opined, "Time and resources are simply too limited to hold that an impact statement fails because the agency failed to ferret out every possible alternative, regardless of how uncommon or unknown that alternative may have been at the time the project was approved." In a closing argument, Rehnquist explained that in this case, the safety of nuclear power was irrelevant. "Nuclear power may some day be a cheap, safe source of power or it may not," he rationalized. "But Congress has made a choice to at least try nuclear energy, establishing a reasonable review process in which courts are to play only a limited role. The fundamental policy questions appropriately resolved in Congress and in the state legislatures *are not* subject to reexamination in the federal courts under the guise of judicial review of agency action."

Vilas, William Freeman (1840–1908)

Besides being secretary of the interior, William F. Vilas was a successful businessman, U.S. senator from Wisconsin, and postmaster general. He was born on 9 July 1840 in the small Vermont village of Chelsea, the son of Levi Baker Vilas, a prominent Vermont attorney and two-term state senator, and Esther Green (nee Smilie) Vilas. At the age of 11, William Vilas and his family moved to Madison, Wisconsin. Later, he attended the University of Wisconsin, graduating first in his class in 1858. He then entered the University of Albany (New York) Law School, which awarded him a law degree in 1860. Although admitted to the Wisconsin bar, he gave up the practice of law at the outbreak of the Civil War to enlist in Company A of the 23d Wisconsin Volunteer Regiment, eventually rising to the rank of lieutenant colonel. He saw action at the Battles of Vicksburg, Black River Bridge, Port Gibson, and Champion Hill.

At the end of the war, Vilas returned to Wisconsin and ran unsuccessfully for the state legislature. His standing with the law community (he was in practice with his brother Edward) led to an offer of a seat on the Wisconsin Supreme Court, but he refused it, becoming instead an attorney for the Chicago and Northwestern Railroad. In 1868, he was named a professor of law at the newly opened University of Wisconsin Law School, a post he held for 17 years. In 1875, he was appointed by the state Supreme Court first to revise the state statutes, which culminated in his *Revised Statutes of the State of Wisconsin* (1878), and then to collect and publish the decisions of the court, which became the 20-volume *Wisconsin Reports* (1875–1876).

Vilas was a leading member of the Democratic Party in his state and was made the permanent chairman of the Democratic National Convention in 1884, which nominated New York Gov. Grover Cleveland for president. After Cleveland's election, Vilas was appointed postmaster general, which was then a cabinet post. He served from 6 March 1885 to 16 January 1888. With the elevation of

William Freeman Vilas

Interior Secretary Lucius Quintus Cincinnatus Lamar to the Supreme Court, Cleveland chose Vilas to fill the Interior post. According to historian Eugene Trani, during Vilas's short tenure (16 January 1888 to 6 March 1889), "He tightened rules. His economy campaign reduced the budget, a move the press praised, and he suggested far-reaching reform of the Department. He asked that the duties of the Assistant Secretaries be specifically defined for more efficient service. A backlog of legal business that had accumulated over the years was cleared up in Vilas' fourteen-month Secretaryship. He attempted to stop fraudulent land acquisitions and applied the new Dawes Act [also known as the General Allotment Act of 1887], which sought to deal with the American Indians" by distributing tracts of land to them as individuals, which allowed them to sell the plots as they wished. With the end of the Cleveland administration in March 1889, Vilas left office.

In 1891, the Wisconsin legislature elected Vilas to the U.S. Senate to replace John C. Spooner. He was a leader in the Senate against the free silver movement and served as a member of a Senate committee that attempted to find ways to solve the economic depression of

1893. A strong supporter of Grover Cleveland during the latter's second administration, Vilas was blamed for its failure and was defeated by Spooner in 1897.

William Vilas returned to private business. In 1896, he left the Democratic Party and joined the short-lived Gold Democratic Party, which opposed free silver and presidential candidate William Jennings Bryan that year. After the election, he ventured in timber and accumulated a sizable fortune. On 27 August 1908, Vilas suffered a cerebral hemorrhage and died at his home in Madison. He was 68 years old.

See also Department of the Interior; General Allotment Act.

Walcott, Charles Doolittle (1850–1927)

Noted paleontologist and geologist Charles D. Walcott was the third director of the U.S. Geological Survey. The youngest of four children of Charles Doolittle Walcott and Mary (nee Lane) Walcott, he was born in the small village of New York Mills in Oneida County, New York, on 31 March 1850. Charles Walcott was left fatherless as a small child and attended school sparsely until he enrolled at the Utica (New York) Academy, where he studied geology but did not graduate. After 1868, he learned about geology and paleontology from books and from scientists he associated with.

After a short period working in a hardware store, Walcott turned full time to geology, accumulating a collection of fossils that came to the attention of Professor Louis Agassiz of Harvard University shortly before Agassiz's death. Walcott then came under the influence of noted paleontologist James Hall (1811–1898), who directed Walcott toward investigations into Silurian fossils at Waldron, Indiana, as well as other inquiries in New York, Ohio, and Canada. With Hall's endorsement, Walcott was attached to the U.S. Geological Survey (USGS) in 1879; he worked in the Grand Canyon region under the direction of Clarence King. In 1882, he collaborated with fellow King survey member Arnold Hague on an expansive report on the Eureka mining district of Nevada while also working on the geology of the New England region. In 1888, he was appointed paleontologist in charge of the invertebrate paleontology division of the USGS; three years later, he was put in charge of all paleontology for the survey. In 1893, he was elevated to chief geologist in charge of all survey paleontology and geology studies. Among the more than 220 books and articles resulting from his studies are "The Fauna of the Lower Cambrian or Olenellus Zone," contained in the *10th Annual Report of the United States Geological Survey* (1890), and his massive *Cambrian Brachiopoda* (1912).

In May 1894, John Wesley Powell resigned as director of the USGS, and Walcott was immediately named as his replacement. His appointment was universally hailed. Wrote Clarence King biographer Thurman Wilkins, "King understood the wisdom of his appointment. For one thing, it promised to reverse Powell's sustained retreat from economic geology and to reestablish the survey's earlier alliance with the mineral industries. 'I think ... Walcott is a man of remarkable executive ability,' King announced in public later that year, 'and the mining interests may rest assured that he appreciates the national importance of the Geological Survey and will use his best efforts to carry out the work to the advantage of all the mining communities in the country.'" In his years as director of the USGS from 1894 to 1907, Walcott reestablished the survey as an all-encompassing examination of the nation's natural resources in many different fields. Through his advocacy, Congress expanded the survey's role in 1908 to include forestry studies. His work in the areas of reclamation and irrigation, forestry, and mining aided in the enactment of the Newlands Irrigation Act in 1902 and the establishment of the Bureau of Mines in 1910. In 1907, Walcott resigned to become the secretary of the Smithsonian Institution. He also served as treasurer, vice president, and president of the National Academy of Sciences and as a commissioner on the 1903 United States–Canada Boundary Commission.

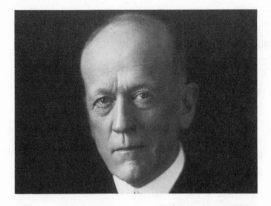

Charles D. Walcott, U.S. Geological Survey director, 1894–1907

He died suddenly in Washington, D.C., on 9 February 1927 at the age of 76. Historian Mary Rabbitt, in her history of the USGS, wrote, "the importance the Geological Survey achieved in its first 25 years, in fact its longevity, should be attributed not to the broad view of science taken by John Wesley Powell, but to the foresight of Clarence King in organizing the survey's research to aid in the industrial progress of the country while seeking ultimately the advancement of science, and the perspicacity, administrative skill, and seemingly limitless energy of Charles D. Walcott, who held that the Survey's field was geology and not all science, who directed its research toward the aid of not just the mineral industry, as envisioned by King, but of all industries and practical undertakings that would benefit from a knowledge of the Earth and its resources, and who insisted that basic and applied science cannot be separated."

See also King, Clarence Rivers; United States Geological Survey.

Walden
See Thoreau, Henry David.

Ward, Frank Lester (1841–1913)
Frank Ward was one of the philosopher-scientists whom Forest Service historian Terry West said "provided the theoretical framework of the conservation movement of the early 20th century," along with John Wesley Powell and William John McGee. Born on 18 June 1841 in Joliet, Illinois, he was the youngest son and one of ten children of Justus Ward, a mechanic, and Silence (nee Rolfe) Ward. Frank Ward attended schools in Joliet and in Buchanan County, Iowa, where he moved with his family at an early age. In 1858, he followed his older brother Cyrenus Osborne Ward (1831–1902), later a labor leader, to Pennsylvania, where Frank entered the Susquehanna Collegiate Institute at Towanda.

In 1862, Frank Ward enlisted in the Union army, serving until he was wounded at Chancellorsville; in 1864, he was discharged because of his wounds. Thereafter, he went to Washington, D.C., where he was hired by the Treasury Department as chief of the navigation and immigration division and later became librarian of the Bureau of Statistics. During this time, he attended Columbian (now George Washington) University, where he earned a bachelor's degree in 1869, a law degree in 1871, and a master's degree in 1872. In 1881, he left the Treasury Department to join the U.S. Geological Survey as an assistant geologist. Two years later he became a full geologist and, in 1892, a paleontologist. His major interest was in paleobotany, the study of fossils of prehistoric plants. Among his scientific monographs, which were considered major studies by conservationists, were *The Flora of Washington* (1881), *Sketch of Paleontological Botany* (1885), *Synopsis of the Flora of the Laramie Group* (1886), *Types of the Laramie Flora* (1887), *Geographical Distribution of Fossil Plants* (1889), and *Status of the Mesozoic Floras of the United States* (1905). His published books on sociology include *Dynamic Sociology* (1883), *The Psychic Factors of Civilization* (1893), and *Pure Sociology* (1903).

Frank Ward died in Washington, D.C., on 18 April 1913 at the age of 71. His work had formed a conceptual foundation for the conservation movement.

Warder, John Aston (1812–1883)

Physician and horticulturist John Aston Warder was the founder of the American Forestry Association and served as the organization's second president. Born 19 January 1812, in Philadelphia, Warder was the son of Jeremiah and Ann (nee Aston) Warder, both members of the Society of Friends (Quakers). At a young age, John Warder became interested in nature and horticulture because of his association with leading figures of the naturalist movement such as John James Audubon, Thomas Nuttall, and François André Michaux. Warder attended local schools in Philadelphia; in 1830, he moved with his family to a farm near Springfield, Ohio, but he returned to Philadelphia to complete his schooling at Jefferson Medical College, from which he earned a medical degree in 1836. A year later he opened a medical practice in Cincinnati, which lasted until 1855.

Warder became involved in the study of horticulture while on his family farm

John Aston Warder, founder of the American Forestry Association

in Ohio. In addition to publishing *A Practical Treatise of Laryngeal Phthisis ... and Diseases of the Voice* (1839), he edited the *Western Horticulture Review* from 1850 to 1853, founded and edited the *Horticulture Review and Botanical Magazine* in 1854, and wrote articles for the *American Journal of Horticulture*. A scientist of sorts in the area of apple growing, he also wrote about pomology, landscaping, and park beautification.

After 1870, Warder became involved in the fight to preserve the nation's forest reserves. In 1873, he was named a commissioner to the International Exposition in Vienna; the *Report of the Commissioners* (1876) contained a treatise on forests and forestry by him. In 1875, he called for those interested in forming an association dealing with forestry matters to meet in Chicago. (An American Forestry Council had been formed in New York in 1873 but had little influence.) On 10 September, this group met at the Grand Pacific Hotel in Chicago and created the American Forestry Association. According to the *Chicago Tribune* of 11 September 1875, Robert Douglas, a British horticulturist, was chosen the organization's first president. Warder announced that the group's purpose was "the fostering of all interests in forest planting and conservation on this continent." Douglas later resigned at that meeting, and the attendees unanimously chose Warder to succeed him. Warder remained president of the organization until he merged it with the American Forestry Congress, of which he was also the founder. Warder died a short time after, on 14 July 1883.

Warren, Francis Emroy (1844–1929)

In 1902, *Forestry and Irrigation* magazine said of Francis E. Warren, "He has been one of the most conspicuous and energetic advocates of irrigation legislation in the U.S. Senate. Coming from a state in which originate most of the large rivers whose waters are used in irrigation, it is fitting that he should originate many of

U.S. Senator Francis E. Warren

state militia while running a livestock farm. In search of his fortune, he headed west, first settling in Des Moines, Iowa, where he helped build the Rock Island Railroad; in 1868, he ended up in Cheyenne, Wyoming. Starting off as a laborer in a furniture store, he eventually purchased a half-share of the enterprise, which became Converse & Warren. In 1877, he acquired the rest of the business and renamed it F. E. Warren & Company. In 1883, he established the Warren Live Stock Company, which dealt in cattle and sheep, and merged it with the furniture business to form the F. E. Warren Mercantile Company. Wrote one source, "Associating himself with the growing cattle and sheep interests of that section, he became one of the great cattle men of the West, and sheep men of later day referred to him as the patriarch of their industry." In 1901, when the National Wool Growers' Association was established, Warren became its first president.

As one of Cheyenne's leading businessmen, Warren naturally became involved in the city's political scene—first working as a member of the board of trustees and later serving on the city council and as mayor. He was also a member of the territorial senate and served as territorial treasurer. In February 1885, he was named territorial governor by outgoing President Chester A. Arthur. Although he was removed by President Grover Cleveland in November 1886, President Benjamin Harrison reappointed him in March 1889, and he served until Wyoming achieved statehood in July 1890. The establishment of railroads across the state and the suppression of strikes over the issue of Chinese labor marked his gubernatorial tenure. With statehood achieved, he was elected the state's first governor but served only two weeks before being elected as Wyoming's second U.S. senator. He served in this post for the rest of his life. Working hand in hand with the state's other Republican senator, Joseph M. Carey (also a former mayor of Cheyenne

the ideas which have been embodied in national legislation relating to irrigation." A U.S. senator from Wyoming, he was born to Joseph Spencer Warren and Cynthia Estelle (nee Abbott) Warren on 20 June 1844 in Hinsdale, Massachusetts. Francis Warren grew up poor, and his education was interrupted when he had to work to supplement the family income. He eventually made enough money on his own to attend Hinsdale Academy for a brief period. When the Civil War erupted he volunteered for duty in the 49th Massachusetts Regiment; while under the command of Gen. Nathaniel Banks, Warren suffered a head wound and was awarded the Congressional Medal of Honor for "courage above and beyond the call of duty."

Discharged from the army because of his wound, Warren returned to Massachusetts, where he was a captain in the and later a governor of Wyoming), Warren was a leading advocate of a national irrigation policy. Although his name did not grace the Carey Act of 1894, he was instrumental in its drafting and passage. One historian in Wyoming even called him "the father of reclamation," although many other men played larger roles in the passage of the Carey Act and the subsequent Newlands Act. In Wyoming, Warren helped form the Wheatland Industrial Company, which through the Carey Act irrigated some 50,000 acres of land. Before World War I, Warren was chairman of the Senate Committee on Military Affairs, and after the war he was chairman of the committee on appropriations, making him one of the most powerful politicians in Washington for a number of years. He died in Washington, D.C., on 24 November 1929. In honor of his lifelong work, one of the highest peaks in the Wind River mountain range in Wyoming is named Mt. Warren.

See also Carey, Joseph Maull; Carey Irrigation Act.

Wastes, Hazardous
See Wastes, Toxic and Hazardous.

Wastes, Nuclear
The dimensions of the nuclear waste requiring disposal are too great to imagine. As of 1983, this problem extended to about 306,000 cubic meters of high-level waste (chiefly liquid) from the national nuclear defense program; 4,626 cubic meters of high-level radioactive spent fuel rods used in civilian nuclear power plants; 3,080,000 cubic meters of low-level wastes (materials contaminated by radioactivity) at both government and civilian plants; and 96,500,000 cubic meters of radioactive tailings or tail millings (the residue from the mining of uranium-rich ore) from uranium mines, mostly in the southwestern United States. High-level waste (HLW) is defined as "highly radioactive materials, containing fission-able products, traces of uranium and plutonium, and other transuranic [more radioactive than uranium and plutonium] elements, that results from chemical reprocessing of spent fuel." Low-level waste (LLW) is characterized as "all wastes not considered high-level or transuranic." LLW includes gloves of workers exposed to radioactive wastes, other material used during the operation of nuclear power plants or nuclear defense plants, and medical waste that has the radioactive imprint on it, including medical isotopes. As of 1989, the 99 nuclear reactors in the United States were generating about 30 tons of spent nuclear fuel each year. Scientists estimate that this waste will take up to *30 million years* to decay to the point of moderate harmlessness. Wrote Minard Hamilton, "Even after the [spent] fuel has cooled for five months, and a certain amount of radioactivity has decayed, there are 35,000 curies of strontium-90 and 48,000 curies of cesium-137 in one-half a metric ton of fuel. This is enough radioactivity to give a lethal dose in 10 seconds, if the material is unshielded."

For the first 50 years of the nuclear age, governments around the world solved their nuclear problems by storing the wastes in the earth or dumping them into the sea. The ramifications of these decisions are yet to be estimated. The near-meltdown at Three Mile Island in Pennsylvania awoke the public to the hazards of nuclear power plants in their backyards. The enactment of the Nuclear Waste Policy Act of 1982 allowed for the selection of a deep-earth repository for the disposal of HLW and LLW, both civilian and defense. In 1987, Congress nominated the Yucca Mountain site in Nevada as the first such repository, but as of 1994, there has been no disposal because of uncertainty over the area's geologic stability. Thus the debate over the disposal of these wastes rages on.

See also Nuclear Waste Policy Act; Three Mile Island; Uranium Tail Millings Radiation Control Act.

Wastes, Solid

"Solving the Garbage Glut" and "A Nation Buried in Plastic Garbage" are just two of the thousands of news stories dealing with perhaps the greatest problem facing state and city governments today: the rapid filling of landfills and how to cut down on the use of items that become solid waste. Used batteries, newspaper by the ton, lawn clippings and leaves—these are but a few of the main offenders. The Sunday edition of the *New York Times*, for instance, generates an estimated 8 million pounds of wastepaper.

The first legislation enacted by Congress to deal with the matter was the Solid Waste Disposal Act of 1965, which mandated the discovery of new methods to recycle materials disposed of in garbage dumps but did not specify which materials these might be. This law was supplemented by the Resource Conservation and Recovery Act (RCRA), which was enacted in 1976 to confront the nearly immeasurable amount of public solid waste that was being produced every year. In 1985, that number reached about 275 million metric tons—more than 1 metric ton for every man, woman, and child in the nation. In November 1984, Congress amended the act with the Hazardous and Solid Waste Amendments (HSWA), which dealt with the problem of underground storage tanks that contain oil, fuel, or other hazardous wastes that may be leaking into the environment. In 1988, Congress adjusted RCRA with the Subtitle J amendment, which covered the classification, tracing, and appropriate disposition of medical waste in government-approved incinerators and other acceptable disposal facilities.

Another side of solid waste is sewage. Because municipal sewage-treatment plants and their operations come under the jurisdiction of local governments, there is little chance a national policy would work. The problem caused by these by-products of human habitation is immense. Wrote author Wendy Petry, "Historically, Americans have not desired to pay any more for waste disposal than was

absolutely necessary. In taking what could be called a path of least resistance, 90 percent of our garbage is simply dumped in landfills." The citizens of New York City alone throw away 24,000 tons of garbage each day. As one source noted, garbage is "sent away" in three different ways: shipped to landfills, sent to special plants to have portions of the refuse recycled, or burnt in incinerators. The laws mentioned above have tried to address this situation, and industry has invented faster-deteriorating plastics to ameliorate the problem of rapidly filling landfill space. Whether these new methods will have any effect on this vexing problem remains to be seen.

See also Hazardous and Solid Waste Amendments; Recycling; Resource Conservation and Recovery Act; Solid Waste Disposal Act.

Wastes, Toxic and Hazardous

The "stew" of toxic and hazardous wastes produced in the United States includes some of the most potent chemicals in existence. U.S. industry alone generates some 200 million tons of waste every year, almost 1 ton for every man, woman, and child in the nation. According to one source, only 10 percent of this waste is handled in a safe manner, with thousands of tons just being dumped into the environment. Because of the seriousness of the situation at Love Canal near Niagara Falls, New York, Congress enacted the Comprehensive Environmental Response, Compensation, and Liability Act of 1980, also known as Superfund. This legislation, the only such law in existence, was supposed to fund the cleanups of sites such as Love Canal and Times Beach, Missouri, which was evacuated in the early 1980s because of concerns over dioxin. Unfortunately, Superfund has become bogged down in bureaucracy.

See also Resource Conservation and Recovery Act; Superfund.

Water Pollution Control Act (62 Stat. 1155)

Enacted on 30 June 1948, this act, also known as the Taft-Barkley Act, established a governmental program for the abatement or end of water pollution. It was the first such legislation in the area of federal water-quality policy. Thus, it became "the policy of Congress to recognize, preserve, and protect the primary responsibilities and rights of the States in controlling water pollution, to support and aid technical research to devise and perfect methods of treatment of industrial wastes which are not susceptible to known effective methods of treatment, and to provide Federal technical services to State and interstate agencies and to industries, and financial aid to State and interstate agencies and to municipalities, in the formulation and execution of their stream pollution abatement programs." The act was sponsored by Sen. Alben Barkley of Kentucky and Sen. Robert A. Taft of Ohio.

Water Pollution Control Act Amendments

On 9 June 1956, Congress amended the Water Pollution Control Act with the following language: "In the development of such comprehensive programs due regard shall be given to the improvements which are necessary to conserve such waters for public water supplies, propagation of fish and aquatic life and wildlife, recreational purposes, and agricultural, industrial, and other legitimate uses." On 23 February 1961, President John F. Kennedy sent a special message to Congress, asking for further amendments to the Water Pollution Control Act. On 20 July, Congress responded by passing new revisions that increased water research funds to $5 million a year, expanded grants to states from $3 million to $5 million a year, and approved $750 million in construction grants to the states through 30 June 1967. Governmental authority was also expanded: Water deemed "navigable," not just interstate water, was now under federal authority. This act failed to stem the tide of water pollution, however. In 1963, conservation writer Louis Clapper wrote,

"Live viruses, dumped into our coastal waters from sewers and dirty bilge tanks, have caused hepatitis in persons who ate clams harvested from sewage-laden waters on the Atlantic Coast, and from the polluted Gulf [of Mexico].... Nitrochlorbenzene, an extremely poisonous organic chemical, is detected in the Mississippi at New Orleans—and followed a thousand miles upstream past many city water intakes to an industrial waste discharge in St. Louis. No one can yet guess the damage this might do to man and wildlife.... This is only a sampling of what is happening in many parts of this country. The detailed story is much longer, equally appalling, and adds up to America's shame—the pollution of our waters."

See also Water Pollution Control Act.

Water Power Act (41 Stat. 1063)
See Federal Water Power Commission.

Water Protection and Flood Prevention Act (68 Stat. 666)

This act was passed on 4 August 1954 to use government resources to stop soil erosion and protect the nation's watersheds. The act authorized the secretary of agriculture to "cooperate with States and local agencies in the planning and carrying out of works of improvement for soil erosion." It stated that "erosion, floodwater, and sediment damages in the watersheds of the rivers and streams of the United States, causing loss of life and damage to property, constitute a menace to the national welfare; and that it is the sense of Congress that the Federal Government should cooperate with states and their political subdivisions, soil or water conservation districts, flood prevention or control districts, and other local public agencies for the purpose of preventing such damages and of furthering the conservation, development, utilization, and disposal of water and thereby of preserving and protecting the Nation's land and water resources."

Water Quality Act (Public Law 100-4)

This act supplemented the Federal Water Pollution Control Act Amendments of 1977. Enacted on 4 February 1987, it required new research on the effects of pollutants on water quality and established improvement projects for designated areas of the nation. Section 105 read, "In carrying out the provisions of section 104(a) of the Federal Water Pollution Control Act, the Administrator [of the Environmental Protection Agency] shall conduct research on the harmful effects on the health and welfare of persons caused by pollutants in water, in conjunction with the United States Fish and Wildlife Service, the National Oceanic and Atmospheric Administration, and other Federal, State, and interstate agencies carrying on such research. Such research shall include, and shall place special emphasis on, the effect that bioaccumulation of these pollutants in aquatic species has upon reducing the value of aquatic commercial and sport industries. Such research shall further study methods to reduce and remove these pollutants from the relevant aquatic species so as to restore and enhance these valuable resources." The act appropriated funds for the construction of "improvement projects" for the public waste-treatment works of such cities and areas as Avalon, California; Walker and Smithfield Townships, Pennsylvania; Taylor Mill, Kentucky; Nevada County, California; Wanaque, New Jersey; and Lena, Illinois.

Water Quality Improvement Act (79 Stat. 903)

This law, enacted on 2 October 1965, was intended to fix the worsening problem of water pollution in the United States—a problem addressed in the Water Pollution Control Acts of 1948 and 1956 and their accompanying amendments. Passed in response to two oil disasters—spills from the ships the *Torrey Canyon* and the *Ocean Eagle*—this act established the Federal Water Pollu-

tion Control Administration, whose duty was to "enhance the quality and value of our water resources and establish a national policy for the prevention, control, and abatement of water pollution." Further, the act provided for increased funds to the states for sewage-treatment plants. This legislation was later supplemented by the Clean Water Restoration Act of 1966.

See also Clean Water Restoration Act; Water Quality Act.

Water Resources Planning Act (42 U.S.C. 1962)

Congress passed this act of 22 July 1965 to set a national policy regarding water resources. "In order to meet the rapidly expanding demands for water throughout the Nation, it is hereby declared to be the policy of the Congress to encourage the conservation, development, and utilization of water and related land resources of the United States on a comprehensive and coordinated basis by the Federal Government, States, localities, and private enterprise with the cooperation of all affected Federal agencies, States, local governments, individuals, corporations, business enterprises, and others concerned," read the policy statement of the legislation. It established a Water Resources Council to "maintain a continuing study and prepare an assessment ... of the adequacy of supplies of water necessary to meet the water requirements of the United States."

See also National Water Commission.

Water Supply Act (43 U.S.C. 390(b))

Enacted on 3 July 1958, this legislation established national policy on the development of water supplies. Congress's formal statement said that "it is hereby declared to be the policy of the Congress to recognize the primary responsibilities of the States and local interests in developing water supplies for domestic, municipal, industrial, and other purposes and that the Federal Government should participate and cooperate with States and local interests in developing such water supplies in connection with the construction, maintenance, and operation of Federal navigation, flood control, irrigation, or multiple purpose projects." The legislation provided for the construction of reservoir projects that would be built by the Army Corps of Engineers or the Bureau of Reclamation.

Watershed Protection and Flood Prevention Act (16 U.S.C. 1002)

Enacted into law on 4 August 1954, this legislation set federal policy for cooperation with the states on watershed protection and flood-prevention programs. As Congress proclaimed in the declaration of policy, "Erosion, floodwater, and sediment damages in the watersheds of the rivers and streams of the United States causing loss of life and damage to property, constitute a menace to the national welfare; and it is the sense of the Congress that the Federal Government should cooperate with States and their political subdivisions, soil or water conservation districts, flood prevention or control districts, and other local public agencies for the purpose of preventing such damages, of furthering the conservation, development, utilization, and disposal of water, and the conservation and utilization of land and thereby of preserving, protecting, and improving the Nation's land and water resources and the quality of the environment." Appropriations would be made through the Soil Conservation Service and related agencies to establish watershed protection and flood-control projects. In the several years of this program, some 1,000 projects were authorized, with nearly 250 million acres affected at an immediate cost of $2.3 billion.

Waterways Commission

See Governors' Conference on the Environment; Newlands, Francis Griffith.

Watt, James Gaius (1938–)

Aside from Albert B. Fall and Harold L. Ickes, James G. Watt was perhaps the most controversial secretary of the interior. He was born on his family's ranch in the small town of Lusk, Wyoming, on 31 January 1938, the son of William Gaius Watt, a lawyer, and Lois (nee Williams) Watt. In 1952, the family moved to Wheatland, in Wyoming's Laramie River valley. In Wheatland, the seat of Platte County, William Watt opened a new law office and Lois Watt ran the Globe Hotel. After attending local schools, James Watt enrolled at the University of Wyoming's College of Commerce and Industry, where he was awarded a bachelor of science degree in 1960. One of his classmates was Alan Simpson, whose father, Milward, was a U.S. senator from Wyoming. He then entered the university's law school, where he edited the *Wyoming Law Journal* before graduating in 1962.

After graduation, Watt immediately went to work for Sen. Milward Simpson's campaign and later served as his legislative assistant and counsel until 1966. (Alan Simpson was later elected to the U.S. Senate and is still serving there as of this writing.) From 1966 to 1969, Watt was secretary to the Senate's Natural Resources Committee as well as to the Environmental Pollution Advisory Panel of the U.S. Chamber of Commerce. He was, as he later wrote, "responsible for coordinating the activities and interests of the National Chamber in the fields of mining, public lands, energy, water and environmental pollution." In 1969, he was named an advisor to incoming Secretary of the Interior Walter J. Hickel and subsequently served from 1969 to 1972 as deputy assistant secretary of the interior, with responsibility over several offices, including water power and development. In 1972, President Richard Nixon named Watt to head the Bureau of Outdoor Recreation, where, as his biographer Ron Arnold wrote, "Watt enjoyed a slightly larger role than a Deputy

Assistant Secretary, and much higher visibility." Most importantly, he helped complete the bureau's national policy report, *Outdoor Recreation—A Legacy for America.*

In 1975, Watt was named to the Federal Power Commission by President Gerald Ford and served until 1977. That year, he joined the Mountain States Legal Foundation, a conservative law firm in Denver, Colorado, that had been created by beer magnate Joseph Coors. The *New York Times* reported that the concern was a "public interest law center dedicated to bringing a balance to the courts in the defense of individual liberty and the private enterprise system." The foundation took on an array of lawsuits designed to counterbalance the power of environmental groups in the courts. As president and chief legal officer of the foundation, Watt was at the forefront of its activities.

On 22 December 1980, President-elect Ronald Reagan named Watt as secretary of the interior. Sen. Clifford P. Hansen of Wyoming had been Reagan's first choice, but the conservative senator owned cattle grazing leases, which would have been a conflict of interest in overseeing the Bureau of Land Management. Immediately, Watt was the target of environmental groups. William Turnage of the Wilderness Society called him "a joke" and defined Watt as the "caricature of an anticonservationist." Nevertheless, Watt was confirmed by the Senate, 83 to 12, and took office on 23 January 1981. He served as interior secretary until 9 October 1983. "He came to the department with a radical agenda for change—radical in the sense that it marked a sharp departure from federal land and resource policies of the recent past," reported the *New York Times* in 1983. "Declaring that the pendulum had swung too far toward conservation and away from the development of public resources needed for economic growth and national security, he moved swiftly to transfer some of these public resources to private industry." His plans to ease restrictions on offshore oil and gas leases, timber cutting and cattle grazing on public lands, and dam construction made him extremely controversial. "I want to change America," he proclaimed in an interview. "I believe we are battling for the form of government under which we and future generations will live." He was the target of increased scrutiny, and several remarks—including his outspoken opposition to the Beach Boys' playing on the Washington Mall and comments about the racial makeup of an Interior Department panel—eroded his effectiveness. In September 1983, the U.S. Senate, led by Republicans, voted to put a moratorium on the leasing of public coal mines to private interests. A month later, on 9 October, Watt resigned, saying that it was "a time for a new phase of management" at Interior. He returned to private life soon after.

See also Department of the Interior.

Weeks, John Wingate (1860–1926)

Congressman, Senator, and Secretary of War John Weeks of Massachusetts was the sponsor of several pieces of conservation legislation, including the Weeks Act of 1 March 1911, which allowed the federal government to purchase forestry reserves to protect them from fire, and the Weeks-McLean Act of 4 March 1913, which protected wild birds traveling across state lines from being killed. Weeks was born on 11 April 1860 on the family farm near Lancaster, New Hampshire, the son of William Dennis Weeks and Mary Helen (nee Fowler) Weeks. John Weeks taught school for a year at age 16 and then attended the Naval Academy at Annapolis, Maryland, which he graduated from in 1881. After leaving the navy in 1883, he spent several years in Florida doing surveying work.

In 1888, Weeks accepted a position with the Boston banking firm that later became Hornblower & Weeks. In the next decade, Weeks accumulated a small fortune. During the Spanish-American

War, he put his naval experience to use by protecting the Massachusetts coast from invasion. In 1900, he was elected an alderman-at-large for the city of Newton, Massachusetts; three years later he was elected mayor. This taste of politics led him to run for and win a seat in the U.S. House of Representatives in 1904. Serving until 1913, he was credited with the sponsorship and passage of the Weeks Act of 1911, which "authorized the Secretary of Agriculture to cooperate with the States to coordinate programs aimed at the prevention from fire of forested watersheds near navigable waters," and the Weeks-McLean Act of 1913. In 1913, the Massachusetts legislature elected Weeks to the U.S. Senate to fill the seat of Winthrop Murray Crane, who was retiring. Weeks served until 1918, when he was defeated by Democrat David I. Walsh. Weeks's campaign work and intense fund-raising for the Republican Party during the 1920 election led President Warren G. Harding to name the Massachusetts politican secretary of war, a post he held until 1925. Retiring because of ill health, he returned to his native Lancaster, New Hampshire, where he died a year later on 12 July 1926 at the age of 66.

See also Weeks Act; Weeks-McLean Act.

Weeks Act (36 Stat. 961–963)

According to one source, the Weeks Act, enacted by Congress on 1 March 1911, "not only authorized the purchase of new national forest land, but established a program for cooperation between the states and the Forest Service to protect watersheds from forest fires." Officially, it was called "an Act to enable any State to cooperate with any other State or States, or with the United States, for the protection of watersheds of navigable streams, and to appoint a commission for the acquisition of lands for the purpose of conserving the navigability of navigable rivers." It was sponsored in the House of Representatives by John Win-

gate Weeks of Massachusetts. The act set up the National Forest Reservation Commission, a governmental body that included the secretaries of war (now defense), interior, and agriculture and two members each from the U.S. House of Representatives and the Senate. The committee, which set land purchasing policy for the federal government, was abolished by the Federal Land Policy and Management Act of 1976.

See also Federal Land Policy and Management Act; National Forest Reservation Commission; Weeks, John Wingate.

Weeks-McLean Act (37 Stat. 828)

Enacted into law on 4 March 1913, the Weeks-McLean Act sought to give federal protection to wild birds migrating across state lines. Following the Supreme Court's 1896 decision in *Geer v. Connecticut*, which upheld the states' Tenth Amendment right to enact game and hunting laws, Congress passed the Lacey Act of 1900 to set limits on which birds could be hunted. To give that act more power and to provide more concrete federal protection for migratory birds that cross state lines, Rep. John Wingate Weeks of Massachusetts and Sen. George Payne McLean (1857–1932) of Connecticut sponsored this act. Its beginning reads, "All wild geese, wild swans, brant, wild ducks, snipe, plover, woodcock, rail, wild pigeons, and all other migratory game and insectivorous birds which in their northern and southern migrations pass through or do not remain permanently the entire year within the borders of any State or Territory, shall hereafter be deemed to be within the custody and protection of the Government of the United States, and shall not be destroyed or taken contrary to regulations hereinafter provided therefor." Because the Weeks-McLean Act seemed to regulate interstate commerce, it was widely ignored by states until they tried to mount a Supreme Court challenge. That challenge was short-circuited by the signing of the Migratory Bird Treaty Act between the United States and Great Britain in 1916.

See also Hornaday, William Temple; Lacey Game and Wild Birds Preservation and Disposition Act; Migratory Bird Treaty Act; Weeks, John Wingate.

Wentworth, Sir John (1737–1820)

Although known better as the last royal governor of New Hampshire, John Wentworth also served as the surveyor general of the king's woods in North America and was considered the first forestry expert on the continent. He was born in Portsmouth, New Hampshire, on 20 August 1737, the son of Mark Hunking Wentworth, a successful merchant, and Elizabeth (nee Rindge) Wentworth. Robert Monahan wrote that Mark Wentworth

Anno fecundo

Georgii II. Regis.

An Act for better Prefervation of His Majefty's Woods in *America*, and for the Encouragement of the Importation of Naval Stores from thence; and to encourage the Importation of Mafts, Yards, and Bowfprights, from that Part of *Great Britain* called *Scotland*.

WHEREAS by an Act paffed Preamble in the Eighth Year of His late Majefty's Reign [intituled, An Act giving further Encouragement for the Importation of Naval Stores, and for other Purpofes therein mentioned] it is enacted, That no Perfon or Perfons whatfoever, within any of His Majefty's Colonies of Nova Scotia, New Hampshire, the Maffachufets Bay, the Province of Main, Rhode Ifland, and Providence Plantation,

King George II ordered, in 1729, that Scotland and British colonies in North America protect timber suitable for masts, yards, and bowsprits for the Royal Navy. John Wentworth made a survey of American woodlands protected by the Naval Stores Acts in 1767.

"had made a fortune in the West Indies trade and by supplying His Majesty's navy with spar and masts from the virgin timberlands of North America." The family could be traced back to William Wentworth (1616–1697), a man of puritanical religious views from Alford, Lincolnshire, England, who immigrated to the colonies in 1636. John Wentworth's grandfather, also named John, served as lieutenant governor of New Hampshire, and his uncle, Benning Wentworth, served as governor. Elizabeth Rindge Wentworth also came from a prosperous and well-known family.

What early education John Wentworth received is unclear. He graduated from Harvard in 1755; among his classmates was future President John Adams. In 1758, he was awarded a master's degree from Harvard. Sometime before 1765, probably in 1763, young Wentworth was sent to England to represent his father in some business transactions there. Wrote one source, "he formed the acquaintance of influential and distinguished Englishmen, among them the Marquis of Rockingham, and he was appointed one of the agents for New Hampshire, serving with Barlow Trecothick, a London merchant." Another source added, "On the passage of the Stamp Act [1765], he [Wentworth] and Trecothick were instructed to use their influence for its repeal." Because of his contacts, on 11 August 1766, Wentworth was appointed in place of his uncle Benning as royal governor of New Hampshire and "Surveyor of the King's Woods" for all the lands in the colonies. On his way home to North America, the new governor stopped in Charleston, South Carolina, to inspect the forestry reserves there. Wentworth biographer Lawrence Shaw Mayo wrote, "No previous surveyor general of the King's woods had made [such] a thorough and intelligent investigation of the forests or a careful appraisal of their assets. ... He found no white pines—but yellow pines, tall and straight, carrying their proportionate size [a] sufficient length for 25 masts. There were also pitch pines, live oaks, and white oaks in immense quantities adjacent to all the Carolina rivers. The wood of the long-leafed pine he declared to be sound, but not clear of hard knots, whereas the pitch pines could be made into as good masts as those imported from Riga [in Russia]." After traveling north, on 13 June 1767, Wentworth took the oath of office as royal governor.

During his administration, Wentworth set up a system of "inventories" of royal timber, divided New Hampshire into five counties, and had the colony mapped. Unfortunately, he was the last royal governor of New Hampshire, as war broke out between England and the colonies during his tenure. In 1775, he was forced to flee to Boston and, finally, in 1778, to England. Later appointed surveyor of the king's woods in Halifax (now Canada) and lieutenant governor of Nova Scotia, he was knighted in 1795. Relieved of his post because of nepotism in 1808, Wentworth retired, dying in Halifax on 8 April 1820 at the age of 82. Wrote Monahan, "It was in this connection [as governor and surveyor] that Wentworth was a true pioneer, setting a working example of tact, perseverance and industry for the legion of foresters who were to follow him."

West, Roy Owen (1868–1958)

Roy O. West was secretary of the interior from 1928 to 1929 in the Calvin Coolidge administration. Born in Georgetown, Illinois, on 27 October 1868, West was the son of Pleasant West, an insurance salesman and lumber enterprise owner, and Helen Anna West. Roy West attended local schools before going to DePauw University in Greencastle, Indiana. In 1890, he earned both his bachelor of arts degree and law degree. He was admitted to the Illinois bar that year. Three years later, he also earned a master's degree from DePauw.

West opened a law practice with two men who would later be powerhouses in Illinois politics: Charles S. Deneen, a future governor and U.S. senator, and William Lorimer, a member of the U.S. House of Representatives. West's political career began in 1894 when he became assistant county attorney of Cook County, Illinois. Two years later, he was elected Chicago city attorney, serving until 1897. In 1898, he began the first of five terms as a member of the Cook County Board of Review. In 1904, West was named chairman of the Illinois Republican state committee, a post he held until 1914. In 1912, he and Deneen, in a power play, turned on their old friend Lorimer, who had been accused of election fraud, and helped oust him from the U.S. Senate. In 1924, West handled Deneen's senatorial campaign.

West's successful campaign work for Deneen brought him to the attention of the Coolidge administration. As a member of the Republican National Committee, West worked on budgetary matters in close connection with the president. On 24 July 1928, Secretary of the Interior Hubert Work resigned to become chairman of the Republican National Committee. Immediately, President Coolidge named Roy West as Work's successor. Although he served from July 1928 to 4 March 1929, he was officially confirmed by the Senate only five weeks before leaving office. A scandal over some of his law clients and the influence they might have over the secretary led the Senate to prolong his confirmation hearings. All along, West served in his post with aplomb.

Only seven months elapsed between West's appointment and resignation, and in that short time West "made no major changes," according to Interior Department historian Eugene Trani. Author James A. Rawley asserted, however, that his tenure was "marked by his quick grasp of the varied responsibilities of his office." West's only annual report, released in December, cataloged the nation's natural resources; one source called it "a remarkable inventory of the nation's wealth"; among historians of the department, it is considered, for the time devoted to it, a better than average document.

West's difficulty in getting confirmed (the Senate finally approved his appointment on 21 January 1929 by a vote of 53 to 27) made him a risky cabinet officer to hold over in the new Hoover administration. Instead, Hoover named physician Ray Lyman Wilbur to the Interior post. On 4 March 1929, West left office. He returned to Chicago, where he served as titular head of the Republican faction led by Deneen after the latter's death in 1940. West reappeared in the news when from 1941 to 1953 he served as a special assistant to the U.S. attorney general in the capacity of a judge hearing the appeals of conscientious objectors. He then retired to private life, dying on 29 November 1958 at the age of 90. Wrote the *Chicago Tribune*, "He was a foe of corruption, and believed in a minimum of federal interference with the nation's industry and economy."

See also Department of the Interior.

Western Governors' Conference (1913)

This important parley, one of a series held by western governors opposed to federal land policies, took place in Salt Lake City, Utah, from 5 to 7 June 1913. It called for a national land and conservation policy that favored the West. In essence, this meeting was the second "sagebrush rebellion," although it was limited in scope. The governors who attended were Taskar L. Oddie of Nevada, Joseph M. Carey of Wyoming, John M. Haines of Idaho, Elias Milton Ammons of Colorado, and William Spry of Utah. Other attendees were A. A. Jones, assistant secretary of the interior; Clay Tallman, commissioner of the General Land Office; and Lt. Gov. Stephen R. Fitzgerald of Colorado.

"Western Executives to Consider Uniform Demand for Federal Land Laws More Favorable to Their States," read

the *Salt Lake Herald-Republican* on the day the conference opened. Ammons opened the meeting by demanding that conservation laws be enacted to meet "local conditions." "For the purpose of this discussion, we may define conservation to mean [the] prevention of waste and monopoly, to which all good citizens should agree," he said in what the newspapers called a "forcible" manner. Ammons continued:

National conservation implies that this prevention of waste and monopoly and the control of our national resources shall be under federal jurisdiction. Having had some years of experience under this plan, it is proper to consider some of the objections to it as developed under its operation, being placed under the administration of bureaus operating from the national capital and ambitious to augment their power.... These bureaus have looked with disfavor upon every effort of the states or their citizens to secure title to lands or resources thereby taking such lands and resources from the jurisdiction of the bureaus and placing them within the power of the states. This condition has been intensified by time until these bureaus have become active missionaries against all forms of state control and the citizens of the states have developed a feeling of bitter opposition to what they term unwarranted interference in matters of local self-government.

Spry then took the podium and echoed Ammons's sentiments: "The movement for the conservation of the natural resources of the nation has been widely heralded and freely discussed [and] has raised a vital question of deepest importance to the people of the west—the most serious question, in my opinion, that has presented itself in the history of the west—the public land question. The foremost important point in connection with the national conservation policy is the handling of the public lands." The newspapers called this request for a new policy a "Fair Deal Asked for the West."

After the three-day conference ended, the leaders announced 14 resolutions and a final compact. It read:

We, the governors of public land states, in conference assembled, believing that upon the administration of the laws governing the disposal of the public lands, in a very large measure depends the future prosperity of our states, do hereby agree to the following statement of what we believe should be the policy of the national government in the administration of the public lands: (1) That the newer states, having been admitted in express terms on an equal footing in all respects whatever with the original states, no realization of that condition can be attained until the state jurisdiction shall extend to all their territory, the taxing power to all their lands, and their political power and influence be hereby secured. (2) That as rapidly as the states become prepared to take over the work of conservation, the federal government withdraw its bureaus from the field and turn the work over to the states. (3) The permanent withdrawal of any lands within our states from entry and sale, we believe to be contrary to the spirit and letter of the ordinance of 1787, the policy of which was followed for over a century, and we urge that such lands be returned to entry and opened to sale as speedily as possible. (4) Dilatory action on the part of executive departments of the government in passing title to purchasers of public lands is unfair to the states, as it permits purchasers to occupy the lands indefinitely without the states having the power to tax them. (5) We believe that the

best development of these states depends upon the disposal of the public lands as rapidly as the law can be complied with. (6) Bona fide homestead entry within forest reserve boundaries should be permitted in the same manner as on unreserved lands, subject only to protest where lands selected are heavily timbered with trees of commercial value or known to contain valuable mineral deposits. (7) That the government grant to the public land states 5 percent of the public land remaining in each, to be administered by the states as the school lands are now administered, for the purpose of building national public highways. (8) That liberal land grants be made for the purpose of establishing and maintaining forestry schools in the public land states. (9) That right-of-way for all lawful purposes be granted without unwarranted hindrance or delay. (10) That all mineral lands now withheld from entry or classified at prohibitive prices be reopened to entry at nominal prices, under strict provisions against monopolization. (11) That we express our appreciation of the splendid work done by the departments at Washington in co-operation with the several states in experimentation and instruction. This assistance has been most valuable in the education of our children and the development of our states, and we commend the same principle to the administration at Washington as being the most feasible plan for the advancement of true conservation. (12) We believe that the national government should provide for expert experimental work in the solution of the mining problems of the mineral states in the same manner that the agricultural department now assists the farmers in solving the agricultural problem. (13) We believe that the speedy settlement of these public lands constitutes the true and best interests of the republic. The wealth and strength of the country are its landowning population. (14) The best and most economical development of this western territory was accomplished under those methods in vogue when the states of the middle west were occupied and settled. In our opinion those methods have never been improved upon, and we advocate a return to those first principles of vested ownership with joint interest and with widely scattered individual responsibility.

The government ignored these demands, paving the way in the late 1970s for the third sagebrush rebellion.

See also Carey, Joseph Maull; Sagebrush Rebellion; Taylor, Edward Thomas.

Wetlands

The various agencies of the U.S. government that oversee the environment have different definitions of the term *wetland*. However, in his article "What Are Wetlands?" John Toliver wrote, "Although the EPA, Army Corps of Engineers, USDA Soil Conservation Service, and USDI Fish and Wildlife Service have formulated separate definitions for various laws, regulations and programs, they are conceptually the same and include three basic identifying characteristics: hydrophytic [water-bound] vegetation, hydric soils [those soils that are saturated, flooded, or ponded for part of the year], and wetland hydrology [lands saturated to the surface or inundated for one week or more during the growing season in an average rainfall year]." According to a 1990 U.S. Fish and Wildlife report, "a wide variety of wetlands exist, ranging from permafrost underlain wetlands in Alaska, to tropical rain forests in Hawaii, to riparian [habitats adjacent to rivers and streams] wetlands in the arid Southwest."

Some of the key federal legislation affecting wetlands includes the Wetlands Loan Act of 1961, the Federal Water Pollution Control Act Amendments of 1972 (known as the Clean Water Act), the Emergency Wetlands Resources Act of 1986, and the North American Wetlands Conservation Act of 1989. The first such law dealing with wetlands purchases, the Wetlands Loan Act of 1961, appropriated some $200 million over seven years for the acquisition of wetlands to add to national refuges. Section 404 of the Federal Water Pollution Control Act Amendments of 1972 "regulated the discharge of dredge or fill material into the nation's waters." This section was strengthened by a 1975 Supreme Court case, *Natural Resources Defense Council v. Callaway*, which expanded section 404's reach into wetlands. In the Emergency Wetlands Resources Act of 1986, Congress found that:

(1) wetlands play an integral role in maintaining the quality of life through material contributions to our national economy, food supply, water supply and quality, flood control, and fish, wildlife, and plant resources, and thus to the health, safety, recreation, and economic well-being of all of the citizens of the nation, and (2) that wetlands provide habitat essential for the breeding, spawning, nesting, migration, wintering and ultimate survival of a major portion of the migratory and resident fish and wildlife of the Nation; including migratory birds, endangered species, commercially and recreationally important finfish, shellfish and other aquatic organisms, and contain many unique species and communities of wild plants.

In 1989, Congress sought to slow the destruction of wetlands with the passage of the North American Wetlands Conservation Act. A new and growing policy to deal with the development of wetlands is mitigation banking, in which developers restore other damaged wetlands in exchange for being allowed to build on wetlands they purchased.

See also Emergency Wetlands Resources Act; Federal Water Pollution Control Act Amendments; Mitigation Banking Policy; North American Wetlands Conservation Act Wetlands Loan Act.

Wetlands Loan Act

This legislation was enacted by Congress in 1961 to acquire wetlands to add to the national refuge system. Appropriations of $105 million were made to the Migratory Bird Hunting Stamp Fund (later increased to $200 million) to purchase important wetlands over a seven-year period.

Wheeler, George Montegue (1842–1905)

Of the four great surveys of the American West completed for the federal gov-

ernment between 1867 and 1875, that of Lt. George Montegue Wheeler, the Geographical Surveys West of the 100th Meridian, remains the least known and examined. Wheeler was born on 9 October 1842 in Hopkinton, Massachusetts, the son of John and Miriam (nee Daniels) Wheeler. Before his twentieth birthday, he was accepted at the U.S. Military Academy at West Point. When he graduated four years later as a second lieutenant, he was ranked sixth in a class of 39. With the end of the Civil War the year before, the nation had tired of war and soldiers, and a downsizing of the military was under way. Wheeler, however, chose to stick with the military; shortly after graduating from West Point, he was assigned to the Army Corps of Engineers as second engineer on a survey being conducted in the area called Point Lobos around San Francisco.

The geological survey led by George Montegue Wheeler camped at Canyon de Chelly, Arizona Territory, in 1873. Canyon de Chelly became a national monument to preserve cliff dweller ruins in 1931.

During his work, which lasted until 1869, he was promoted to first lieutenant. He also traveled through the territories of Nevada and Utah, scouting for shorter routes to California.

In 1869, Wheeler advocated a survey of this area, and the U.S. government called on him to command a team that would analyze and map the area. This would become one of the four surveys of the West then being conducted by several of the great explorers and scientists: Clarence King, Ferdinand V. Hayden, and John Wesley Powell. Wheeler's survey would cover Nevada and Arizona and include topographical, geological, zoological, and ethnological matters. Wheeler was instructed that the "main object of this exploration will be to obtain correct topographical knowledge of the country traversed by your parties, and to prepare accurate maps of that section." Other matters were to be addressed as well. "It is at the same time intended that you ascertain as far as practicable everything relating to the physical features of the country, the numbers, habits and disposition of the Indians who may live in this section, the selection of such sites as may be of use for future military operations or occupation, and the facilities offered for making rail or common roads, to meet the wants of those who at some future period may occupy or traverse this part of our territory." The order further charged Wheeler to accumulate "all the information necessary before the settlement of the country, concerning the branches of mineralogy and mining, geology, paleontology, zoology, botany, archaeology, ethnology, philology, and ruins." Under the direction of Gen. Andrew Atkinson Humphreys (1810–1883), Wheeler "was authorized to employ ten assistants as topographers, geologists, and naturalists, in addition to packers, guides, laborers, and a photographer," according to writer James Horan. With this mandate, Wheeler assembled a small team of scientists to cover all facets of his mission, including geologists Grove Karl Gilbert and Archibald R. Marvine and photographer Timothy O'Sullivan. Later, artist John E. Weyss joined the expedition.

The party started out from Halleck Station, Nevada, on 3 May 1871 and reached Camp Lowell at Tucson, Arizona, on 5 December, covering 6,327 miles, with some 83,000 square miles of territory examined. During the next several years, Wheeler and his survey completed 13 other trips. In 1883, Wheeler wrote, "The field trips were often attended by the greatest hardship, deprivation, exposure and fatigue, in varying and often unhealthy climates from 31° N to 47° N and Altitudes from 200 ft. below sea level ... to nearly 15,000 ft. among the mountain peaks of the Sierra Madre." The account of all the explorations was told in Wheeler's *Preliminary Report of Explorations in Nevada and Arizona* (1872) and *Report upon United States Geographical Surveys West of the One Hundredth Meridian* (seven volumes, including a supplement and two atlases, 1875–1889).

Between 1867 and 1877, the four surveys reported their findings to Congress. Their explorations overlapped in many areas, however, and the National Academy of Sciences asked Congress to consolidate them under one government entity. With the passage of the Sundry Civil Appropriations Act of 3 March 1879, Congress created the U.S. Geological Survey and made Clarence King its first director. Wheeler, as well as Powell and Hayden, were left out in the cold. Although he spent the next several years consolidating his reports into the final treatise that he delivered to Congress in 1889, Wheeler's work was effectively finished. He was a member of the U.S. commission that went to the third International Geographical Congress and Exhibition in Venice, Italy, in 1881, which resulted in his 1885 work *Report upon the Third Geographical Congress and Exhibition at Venice, Italy, 1881*. Wheeler's health was broken by years of exploring the West. In 1888, he retired as a full major and spent his remaining years in New

York City, where he died on 3 May 1905 at the age of 62.

See also Gilbert, Grove Karl; Hayden, Ferdinand Vandeveer; King, Clarence Rivers; Powell, John Wesley; United States Geological Survey.

Wilbur, Ray Lyman (1875–1949)

Secretary of the Interior Ray L. Wilbur was a physician from agricultural roots. Born in Boonesboro, Iowa, on 13 April 1875, he was the fourth of six children of Dwight Locke Wilbur, a lawyer and part owner of a coal mine, and Edna Maria (nee Lyman) Wilbur, a teacher. Dwight Locke Wilbur served in the Civil War and was captured by the Confederates at Harpers Ferry, Virginia. Ray Lyman Wilbur's brother, Curtis Dwight Wilbur, served as secretary of the navy in the Coolidge administration and as chief justice of the California Supreme Court. The family moved first in 1883 to the settlement of Jamestown in the Dakota Territory, where Dwight Wilbur became an agent for the Northern Pacific Railroad, and then in 1887 to Riverside, California. Ray Wilbur attended schools in California before enrolling at Stanford University (then called Leland Stanford Junior University) to study medicine. He received his bachelor's degree in 1896, his master's in 1897, and his M.D. from San Francisco's Cooper Medical College (now part of Stanford) in 1899.

Wilbur practiced medicine briefly in San Francisco and served as an attending physician at Cooper College. In 1900, Stanford hired him as an assistant professor of physiology and internal medicine; at the same time, he began working on his Ph.D. In 1903, he went for a year of medical study in Frankfurt, Germany, and London, England. In 1904, he returned to the United States and resumed his medical practice, but he returned to Europe in 1909 for more training in Munich and Vienna. Cooper Medical College became part of Stanford University in 1908, and a year later, Wilbur was named professor of medicine; two years later, he was named dean of the school,

serving until 1916. In that year, he became president of Stanford.

As a result of his friendship with Herbert Hoover, then head of the Federal Food Administration, Wilbur became chief of the Food Administration's Conservation Division in 1917; he coined the famous motto, "Food Will Win the War." Following the end of World War I, he returned to his post as president of Stanford and served as president of the American Medical Association from 1923 to 1924. Wilbur was one of the five attending physicians when President Warren G. Harding died in San Francisco in August 1923.

A staunch Republican, Wilbur threw himself into the 1928 presidential campaign to help get his friend Hoover elected. Wilbur's work in California led the victorious Hoover to name the physician secretary of the interior. Wilbur served for the entire Hoover administration—5 March 1929 to 4 March 1933. Immediately, he set department policy by announcing that all new oil leases on government lands would be prohibited except where provided for by law. Since then, the department has maintained this stance. Wrote the *New York Times*, "Among the more controversial decisions made by Dr. Wilbur during his term ... was his allocation of the power of the Boulder Dam project. His opponents charged that he was allocating too much of the power to private utility enterprises.... He was particularly interested in the plight of the American Indian and ... he effected a complete reorganization of the Department's Bureau of Indian Affairs. He favored any moves that would encourage the Indian to become self-sufficient." A member of Hoover's "dark-horse" cabinet (most of the men were little known by the general public), Wilbur had a "conservation ethic" that was born of his agricultural upbringing. Writing once about "rugged individualism," he noted, "It is common talk that every individual is entitled to economic security. The only animals and

349

birds that I know that have economic security are those that have become domesticated—and the economic security they have is controlled by the barbed-wire fence, the butcher's knife, and the desires of others. They are milked, egged, skinned, and eaten up by their protectors."

With the end of the Hoover administration in 1933, Wilbur returned to Stanford, where he served until his retirement in 1943. He died six years later, on 26 June 1949.

See also Department of the Interior.

Wild and Scenic Rivers Act (82 Stat. 906)

Enacted by Congress on 2 October 1968, this legislation called for segments of a number of selected "wild" rivers to be preserved in their natural state. The act declared it to be "the policy of the United States that certain selected rivers of the nation which, with their immediate environments, possess outstandingly remarkable scenic, recreational, geologic, fish and wildlife, historic, cultural, or other similar values, shall be preserved in free-flowing condition, and that they and their immediate environments shall be protected for the benefit and enjoyment of present and future generations." The act listed the rivers to be protected, including the portion of the Rio Grande in New Mexico, the Salmon River in Idaho, and the Wolf River in Wisconsin. According to *Audubon* magazine, in 1993, the act covered 152 rivers—a total of 10,000 miles of the nation's 3.5 million miles of rivers and streams—with the possibility of 50 new rivers being added to the protected list.

Wilderness

"A wilderness, in contrast with those areas where man and his own works dominate the landscape, is hereby recognized as an area where the earth and its community of life are untrammeled by man, where man himself is a visitor who does not remain," reads Congress's definition of wilderness as contained in the Wilderness Act of 1964 (now 16 U.S.C. 1131(c)). "An area of wilderness is further defined to mean in this chapter an area of undeveloped Federal land retaining its primeval character and influence, without permanent improvements or human habitation, which is protected and managed so as to preserve its natural conditions and which (1) generally appears to have been affected by the forces of nature, with the imprint of man's work substantially unnoticeable; (2) has outstanding opportunities for solitude or a primitive and unconfined type of recreation; (3) has at least five thousand acres of land or is of sufficient size as to make practicable its preservation and use in an unimpaired condition; and (4) may also contain ecological, geological, or other features of scientific, educational, scenic, or historical value."

Wilderness Act (78 Stat. 890)

This major legislation was enacted on 3 September 1964. The act established the National Wilderness Preservation System, into which was deposited 9.1 million acres of federal land, with 51.9 million acres of other lands to follow. Further, it formulated a ten-year review program to oversee wildlife protection on these lands and mandated that any changes in the system be made through congressional action. This law was the product of several years of work, originating with a bill introduced in 1962 by Sen. Clinton Presba Anderson of New Mexico, a former secretary of agriculture. This bill, S. 174, spent several years in Congress being opposed for various reasons by major commercial interests, including the Chamber of Commerce and the American National Cattlemen's Association, as well as by the state legislatures of Idaho and Wyoming. S. 174 passed the Senate but failed in the

House; revised legislation, S. 4, was introduced in 1963 and passed both houses. The main lobbyist for the bill was conservationist Howard Clinton Zahniser, a writer and executive secretary of the Wilderness Society. Zahniser spent countless hours selling this plan to Congress, even going so far as to prepare an 11,000-word report to the House interior subcommittee. Zahniser died a week after the submission of his report, and it was his death that spurred the act's passage. The Wilderness Act is considered *the* landmark environmental legislation enacted in this century.

See also Anderson, Clinton Presba; Aspinall, Wayne Norviel; Zahniser, Howard Clinton.

Wilderness Society

This major environmental group was formed by wilderness advocates Robert Marshall, Benton MacKaye, Robert Sterling Yard, Olaus Murie, Harold Anderson, Ernest Oberholtzer, and Aldo Leopold in January 1935. In 1930, Marshall had written his landmark article "The Problem of the Wilderness," in which he explained the need for a national wilderness preservation program. Benton MacKaye later called the article the Magna Carta of the wilderness preservation movement. These men assembled in Washington, D.C., and, with Marshall's financial support, formed the Wilderness Society. Olaus Murie expressed the need for such a group when he wrote, "We took wilderness for granted in those days. Wilderness was there, for anyone who wished to experience it. I don't know just when the value of wilderness as a cultural influence was first expressed, but there are Biblical references to it. And of course in the earlier periods of [the] settlement of the American continent there were those far sighted ones who began to warn us about what was

Four of eight founders of the Wilderness Society: (left to right) Bernard Frank, Harvey Broome, Robert Marshall, and Benton MacKaye at Great Smoky National Park, North Carolina, 1936

happening to our human environment." One source noted that "the society soon involved itself in a number of disputes over the use of undeveloped and scenic lands, arguing that as much as possible of the wilderness should be preserved for its emotional, intellectual, and scientific values."

In September 1935, under the leadership and editorship of Robert Sterling Yard, the society began to publish its official journal, *The Living Wilderness*, now simply called *Wilderness*. In the magazine's first issue, Yard editorialized on the front cover: "The Wilderness Society is born of an emergency in conservation which admits of no delay. It consists of persons distressed by the exceedingly swift passing of wilderness in a country which recently abounded in the richest and noblest of wilderness forms, the primitive." Aldo Leopold asked, "Why the Wilderness Society?" and then answered, "This country has been swinging the hammer of development so long and so hard that it has forgotten the anvil of wilderness which gave value and significance to its labors. The momentum of our blows is so unprecedented that the remaining remnant of wilderness will be pounded into road dust long before we find out its values." Noted for its opposition to the construction of the Echo Park Dam in the 1950s and advocacy of the Wilderness Act in the 1960s, the society had more than 350,000 members nationwide in 1991. Its national headquarters are in Washington, D.C.

See also Anderson, Harold Cushman; MacKaye, Benton; Marshall, Robert; Murie, Olaus Johan; Oberholtzer, Ernest Carl; Yard, Robert Sterling.

Wilkes, Charles (1798–1877)

See United States Exploring Expedition.

Wilson, Alexander (1766–1813)

Of Scottish heritage, Alexander Wilson became an expert in American ornithology. Born in Seed Hills at Paisley, Renfrewshire, Scotland, on 6 July 1766, he was the son of Alexander and Mary (nee McNab) Wilson. Wilson's mother died when he was a child, and his father eventually remarried. Wilson's parents wanted him to have a life in the ministry, but he started out as a weaver apprenticed to his brother-in-law. He submitted anonymous articles on rural life to newspapers in Scotland. After three years as an apprentice, he set off to travel his homeland as an itinerant peddler. During this time, he collected stories on Scottish life that he published as the narrative poem *Watty and Meg* (1792), which has been attributed to another Scot, poet Robert Burns. Wilson returned to Paisley, but labor strife and the subsequent publication of his series of biting articles decrying capitalist excesses against local workers almost led to his hanging. So Wilson and his nephew William Duncan walked to the harbor at Port Patrick, took a ferry to Belfast, and left for the New World. They landed in Newcastle, Delaware, in July 1794.

Wilson settled in Philadelphia, where he worked as a copperplate printer and then turned to weaving. He eventually taught in small schools in New Jersey and Pennsylvania. While serving as a village schoolmaster in Kingsessing, Pennsylvania, he became acquainted with famed botanist and naturalist William Bartram, who allowed the Scot access to his library. Interested in nature in general and birds in particular, Wilson used Bartram's extensive collections while accumulating his own material on the birds of the northeastern United States. During his time off, he made short, leisurely trips around the region, making sure that the illustrations in his developing work were "chiefly colored by candle-light." Bored with teaching, he took a job as assistant editor with the publishing firm of Bradford and Inskeep of Philadelphia, which was issuing a new edition of Abraham Rhee's *New Cyclopaedia*. He was able to convince Bradford and Inskeep to publish his work on birds. *The Foresters* (1805) was based on his explorations of the territory from Philadelphia to the Niagara Falls.

From 1808 to 1813, Wilson published eight massive volumes of his *American Ornithology; or, the Natural History of the Birds of the United States*. This landmark work was the first such examination of American birds ever published. In 1808, while in Kentucky, Wilson witnessed a flock of passenger pigeons that took several hours to pass and later calculated that there had been 2 million birds in the one flock alone. Only 106 years later, the last passenger pigeon died in captivity. While working on the second volume in 1810, Wilson met John James Audubon in Louisville, Kentucky. Audubon wrote of Wilson, "How well I remember him as he walked up to me! His long, rather hooked nose, the keenness of his eyes, and his prominent cheek-bones, stamped his countenance with a peculiar character." The struggle to complete volume seven undermined Wilson's fragile health. Relying on the help of his friend and later biographer George Ord, Wilson struggled to render the illustrations for the eighth volume. When he saw a bird across a canal, he dived in and was able to grab it for sketching, but he caught pneumonia in the process and died on 23 August 1813. Ord completed the eighth volume and wrote a ninth. Three years after Wilson's death, *Poems; Chiefly in the Scottish Dialect, by Alexander Wilson, Author of American Ornithology, with an Account of His Life and Writings* (1816) appeared in London.

Although he is virtually unknown today, Wilson had more impact on the early study of American ornithology than anyone except Audubon. It took Audubon some 30 years to collect the material for his work, but Wilson did his studies in ten. Ornithologist Elliott Coues, who was later a member of the Hayden survey, said of the Scot's pioneering studies, "Perhaps no other work on ornithology of equal extent is equally free from error; and its truthfulness is illuminated by a spark of the 'fire divine.' ... Science would lose little, but, on the contrary, would gain much if every scrap of pre-Wilsonian writings about United States birds could be annihilated."

See also Ord, George.

Winters et al. v. United States (207 U.S. 564 [1908])

Decided by the Supreme Court on 6 January 1908, *Winters* involved the issue of whether water could be diverted to irrigate Indian reservations. In 1900, plaintiffs Henry Winters, John W. Acker, Agnes Downs, and others entered the area of the Milk River in Montana near the Fort Belknap Indian reservation and attempted to build a dam to divert the navigable waters of the river to irrigate Indian lands on the reservation. The U.S. government sued to stop construction of the dam on the grounds that although the act of 15 April 1874 granted the Indians of Montana their reservations apart from the state of Montana (the Fort Belknap reservation had been established in 1888), the reservations were for "the bare right of the use and occupation thereof at the will of and sufferance of the government of the United States." Winters sued in a Montana district court, and the court upheld the government's action enjoining construction of the dam. The Ninth Circuit Court of Appeals affirmed the lower court, and Winters sued to the U.S. Supreme Court. The Court ruled 8–1 (Justice David Josiah Brewer dissenting) that the Indians had no rights to land or waters outside of their reservations. Justice Joseph McKenna wrote the Court's opinion, in which he stated that "the power of the government to reserve the waters and exempt them from appropriation under the state laws is not denied, and could not be."

Wisconsin Public Intervenor et al. v. Ralph Mortier et al. (115 L Ed 2d 532, 111 S Ct 2476 [1991])

In this case, the Supreme Court held that the Federal Insecticide, Fungicide, and

Rodenticide Act of 1947 (FIFRA) did not preempt the regulation of pesticides by local governments. Under a law instituted by a town in Wisconsin, persons wishing to apply pesticides "to public lands, to private lands subject to public use, or ... aerial application of any pesticide to private lands" had to apply for a permit. Under the law, the town could grant the permit, refuse the permit, or grant the permit with certain conditions that had to be met. Ralph Mortier applied for a permit to aerially spray his private land. The town granted Mortier a permit to use the pesticide but refused to let him aerially spray and restricted the acreage he could spray. Mortier, in conjunction with pesticide manufacturers, sued the town under the notion that the ordinance was overruled by FIFRA. The trial court admitted the Wisconsin Public Intervenor, an assistant state attorney general commissioned to protect "environmental public rights," as the defendant. The court then struck down the town ordinance as an infringement of FIFRA's prohibition against local regulation of pesticides. The Supreme Court of Wisconsin upheld the judgment. On certiorari, the U.S. Supreme Court struck down the lower courts' decision on 21 June 1991. By an 8–1 vote (Justice Antonin Scalia concurred in part and dissented in part), the Court held in an opinion written by Justice Byron White that FIFRA did not preempt local regulation of pesticides and, if it was intended to, this congressional intent was not spelled out sufficiently to reach this conclusion. Scalia argued that although he agreed that FIFRA's language allowed for local regulation of pesticides, it was quite proper for a lower court to find that it did not, claiming that the congressional committees working on the bill that became FIFRA intended to override local regulation even if the final bill did not explicitly say this.

Wise-Use Movement

This antienvironmentalist movement preaches a gospel of the "wise use" of the nation's natural resources and a policy of laissez-faire by the federal government. In its declaration of beliefs, known as "The Wise-Use Agenda," it advocates the development of oil reserves in the Arctic National Wildlife Refuge in Alaska, the allowance of timber harvesting in the Tongass National Forest in Alaska, the upholding and codification of the 1872 General Mining Law, and the passage of a rural community stability act to permit the U.S. Forest Service to sell small timber reserves in ranger districts as well as a national industrial policy act to estimate the economic effects of congressional environmental legislation, the establishment of a national rangeland grazing system, and the formation of a national recreation trails trust fund for multiple-use trail development. The movement is in fact a loosely knit collection of pro-industry, pro-development, pro-recreation, anti–government regulation and antienvironmental groups nationwide. Among the leaders of the movement are Ron Arnold, a founding member and head of the Center for the Defense of Free Enterprise in Bellevue, Washington, and Charles Cushman of the National Inholders Association.

See also Sagebrush Rebellion.

Withdrawal Act (36 Stat. 847)

This act of 25 June 1910 "withdrew" certain government lands from use by anyone and reserved them for purposes the government chose. President William Howard Taft exempted some mineral lands from this law, and in 1912, an amendment allowed for the full opening of all public lands that had been withdrawn for the purpose of mining for mineral deposits only.

Wolcott, Edward Oliver (1848–1905)

A U.S. senator from Colorado, Edward Wolcott was a leader in the insurgent, or anticonservation, movement of the late nineteenth century. He was born in Longmeadow, Massachusetts, on 26 March 1848, the third son and one of eleven children of Samuel Wolcott, a Congregational minister, and Harriet Amanda (nee Pope) Wolcott. Wolcott was descended from Henry Wolcott, who immigrated to the colonies in 1636 and settled in Windsor, Connecticut, and was related to Roger Wolcott (1679–1767), a colonial governor of Connecticut (1750–1754); Oliver Wolcott (1726–1797), a signer of the Declaration of Independence; and the latter's son Oliver Wolcott (1760–1833), secretary of the treasury under George Washington. The family moved to Chicago when Edward was 11 and to Cleveland three years later. Edward attended Cleveland public schools until he was 16, when he volunteered as a private in the 115th Ohio Regiment. He served for only a few months before the Civil War ended, and it is unclear whether he saw any action. After the war, he enrolled at Yale University but left after only one year. He later attended Harvard Law School from 1870 to 1871; he graduated in the latter year but for some unknown reason was not awarded a law degree until 1875.

Following in the footsteps of his older brother Henry, who had gone to Colorado, Wolcott went west in 1871 and began to teach. Settling in the village of Georgetown, Colorado, he opened a law practice and found success as a result of his representation of the Denver and Rio Grande Railroad. After officially receiving his law degree from Harvard, Wolcott was elected district attorney of Georgetown. In a letter home in 1876, he explained his new job: "It is my duty to appear before the Grand Jury and the District Court of six counties (ten terms each year) and to prosecute on behalf of the people, all persons charged with felonies and misdemeanors. The state pays me a salary of $800 per annum, in addition to which I receive for each indictment (if it sticks), $10; for each trial of misdemeanors, $15; for each felony (except as below), $25; and for each capital charge, $50."

A little more than two years into his term, Wolcott was elected to the Colorado state senate, where he served from 1879 to 1882. In 1879, he moved to Denver after being named chief counsel for the Denver and Rio Grande Railroad. After leaving the state senate, this was his main occupation. A leader of the conservative faction of the state Republican Party, he dueled with Sen. Henry Moore Teller for control of the party apparatus. In 1889, Wolcott was elected to the U.S. Senate, succeeding Republican Thomas M. Bowen. His two terms (1889–1901) were marked by an avowed passion to fight against all conservation measures drafted by the government. During the battle over the McRae forestry bill in 1893, he was a leader in the Senate against its passage. In an 1895 speech on the Senate floor, he struck out bitterly against the agents who patrolled the western forests: "They tumble all over each other in the western states, broken-down politicians from the Eastern states, hunting for a man who has a mine which looks like it might produce ... who may cut a little timber from his mine, that they might compel him by blackmail into giving them a few hundred dollars to save himself from an indictment of the federal courts. These men are not fit to stay at home so they are unloaded on the western states. Talk about timber thieves! These people are worse than any timber thieves that Eastern men can imagine.... We do not want them in Colorado."

In 1901, Thomas Patterson, a fusion candidate, was elected to Wolcott's Senate seat in a bitter contest. To reward him for his years of loyalty to the party, President William McKinley named him U.S. representative to the International Bimetallic Commission, which looked into ways of creating a world monetary standard. Wolcott was in Monte Carlo on 1 March 1905 when he came down with influenza and died in his hotel room. He was three weeks shy of his fifty-seventh birthday. He was cremated, and his remains were returned to the

United States and interred at Woodlawn Cemetery in New York.

See also Bell, John Calhoun; McRae, Thomas Chipman; Teller, Henry Moore.

Woodhouse, Samuel Washington (1821–1904)

A member of the Sitgreaves Expedition of 1849–1850, Dr. Samuel Woodhouse, one of the lesser-known naturalists of the nineteenth century, nonetheless contributed important research to the scientific community. He was born on 27 June 1821 in Philadelphia, the son of Commodore Samuel Woodhouse of the U.S. Navy and H. Matilda (nee Roberts) Woodhouse. According to John S. Tomer and Michael J. Brodhead, Woodhouse's biographers and the editors of his journals, Samuel Woodhouse received his education at private schools in Philadelphia and West Haven, Connecticut.

Samuel Woodhouse apparently grew to love natural history and nature. In about 1840, he studied at the Academy of Natural Sciences in Philadelphia, where he became an intimate of such naturalists as Samuel George Morton, John Kirk Townsend, and Thomas Nuttall. Five years later, Woodhouse was chosen as a member of the academy. Sometime earlier, he had settled on a farm in Chester County, Pennsylvania, where he acquired plant and ornithological specimens. Illness forced him to leave the farm, and he entered into the study of medicine. After being awarded a Ph.D. from the University of Pennsylvania in 1847, he was recommended by Samuel Morton to join the Creek Boundary Expedition of 1849–50, directed by Capt. Lorenzo Sitgreaves of the U.S. Corps of Topographical Engineers. According to Kenneth P. Czech, "The Corps was ordered to survey and mark the northern and western boundaries of the Creek Indian lands in Indiana Territory [now Oklahoma], to comply with the requirements of the Creek Treaty of 1845." Woodhouse traveled with the expedition through the Creek territory, where he used the opportunity to explore virgin country and collect newly discovered plants and animals. He taught other members of the party how to skin and preserve bird specimens for the taxidermist. Later, all these samples were deposited in the Academy of Natural Sciences. When the expedition resumed in 1850 under the command of Lt. Israel Carle Woodruff, Woodhouse again joined the group. At its conclusion, he had collected more than 1,400 plant specimens, not to mention his countless ornithological discoveries.

In 1851, Woodhouse was selected by Capt. Sitgreaves to join him in a journey down the Zuni and Colorado Rivers. The Zuni Expedition, as it is called, went, according to one source, "from San Antonio, Texas, to San Francisco, and passed through territory largely unexplored by naturalists." The account of the exploration can be found in Sitgreaves's Report of an Expedition Down the Zuni and Colorado Rivers, Accompanied by Maps, Sketches, Views and Illustrations (Senate Executive Document no. 59, 32d Congress, 2d Session, 1853). "I have the honor to submit the accompanying map of the route explored by me from the pueblo of Zuni, New Mexico, to Camp Yuma, on the Colorado of the West, under instructions from you," Sitgreaves wrote to Secretary of War Charles M. Conrad. "The party was organized at Santa Fe, and consisted of Lieutenant J. G. Parke, Topographical Engineers; S. W. Woodhouse, M.D., physician and naturalist; Mr. R. H. Kern, draughtsman; Mr. Antoine Leroux, guide; five Americans and ten Mexicans as packers and arrieros." At Camp 35 on 9 November 1851, Sitgreaves reported that "while preparing for our departure before daylight, Dr. Woodhouse, who was warming himself by the fire, received an arrow through the leg, fortunately without doing him much injury."

In his summary that was attached to Sitgreaves's report, Woodhouse wrote, "On my arrival in Texas, and during my stay in San Antonio, I suffered much from intermittent fever, which was the cause of the loss to me of much time that might have been profitably spent in the pursuit of my favorite studies, for that country offers a great field for the naturalist." Later, he reported, "During our detention at the pueblo of Zuni I was unfortunately bitten by a rattlesnake (*Crotalus le Contei*) . . . this was a sad accident for me . . . I did not recover the use of my left hand for months afterwards, and this accounts for the small collection of birds, quadrupeds, and reptiles procured by me west of this place." During the sojourn, Woodhouse, Sitgreaves, and Kern marked their names on the northeast face of El Morro, a famous landmark in New Mexico. Among Woodhouse's discoveries was *Cyanocitta woodhouseii*, named Woodhouse's jay in his honor by scientist Spencer Fullerton Baird (it is now called *Aphelocomacoerulescens woodhouseii*, a subspecies of the California jay); *Bufo woodhousii*, called Woodhouse's toad; and a species of plant found in New Mexico that scientist Asa Gray named *Achyropappus woodhouseii*.

Woodhouse's last naturalist expedition was an 1853 journey to Central America under the command of Ephraim G. Squier, which traversed parts of Honduras. Although Woodhouse himself never published any papers on this mission, his work became part of Squier's *Notes on Central America* (1855). Woodhouse returned to Philadelphia and spent the last 50 years of his life as a physician. He was elected to the American Ornithologists' Union shortly before his death. His works were "discovered" in the years preceding the twentieth century and his name came to life again as one of the greatest living naturalists of his time. On 23 November 1904, he died at his home in Philadelphia at the age of 83.

Woodruff-McNary Act (45 Stat. 468)
This law, enacted on 30 April 1928 as an amendment to the 1911 Weeks Act, ex-

panded governmental authority over the purchase of wood- and timberlands and sanctioned the government to create new national forests. Further, the act provided $8 million in federal monies to purchase lands to be used in national forests, with such funds to be spent by 1931. Some senators objected that one portion of the act was unconstitutional—the spending of governmental funds for the purchase of lands where taxes had not been paid and the lands had been seized. McNary therefore set a limit for such purchases to 1 million acres per state. The legislation was cosponsored in the House by Roy Orchard Woodruff of Michigan (1876–1953) and in the Senate by Charles L. McNary of Oregon.

See also McNary, Charles Linza; Weeks Act.

Work, Hubert (1860–1942)
Noted more as a physician than as secretary of the interior, Hubert Work also served as postmaster general in the Harding administration. Born on his family's farm in Marion Center, Pennsylvania, on 3 July 1860, he was one of seven children and the only son of Moses Thompson Work, a farmer, and Tabitha Logan (nee Van Horn) Work. Sources report that he was descended from a long line of Pennsylvania pioneers. He attended local schools and the Indiana (Pennsylvania) state normal school before enrolling at the University of Michigan (1882–1884) and then the University of Pennsylvania to study medicine. In 1885, he was awarded his M.D. degree from the latter school. The western United States beckoned to Work, and he complied by traveling to Colorado, where he opened a medical practice in the town of Greeley. He moved on to Fort Morgan and, later, Pueblo, where in 1896 he helped found the Woodcroft Memorial Hospital for mental and nervous diseases; he operated the clinic until 1917. Work was elected president of the Colorado State Medical Society in 1896, president of the American Medico-Psychological

Society in 1911, and president of the American Medical Association in 1921.

Work's Republican background took root in Colorado when he chaired the Republican state convention in 1908. Five years later, he was elected chairman of the state Republican Party, serving until 1919. From 1912 to 1919, he was a member of the Republican National Committee. When the United States entered World War I, Work left his practice and enlisted with the rank of major in the Army Medical Corps, which was chosen by Gen. William Gorgas to oversee the medical examinations of draftees. At the war's end, Work retired as a full colonel in the Officers' Reserve Corps. Hubert Work's labors earned him national recognition, and he was called upon by Republican National Chairman Will Hays (who later headed the infamous "Hays Office" that set movie standards in the 1920s) to rally support among farmers and doctors for Republican presidential candidate Warren G. Harding in the 1920 election. With Harding's election, Hays was named postmaster general, and he selected Work as first assistant postmaster general. When Hays left the Post Office in March 1922, Work was named his replacement. Noted one source on his work there, "he made a complete investigation for the reorganization of the mail service, inspected buildings and devised plans for the general improvements of the service. His plan for government ownership [instead of the leasing] of post office buildings received the approval and support of the Treasury Department."

On 4 March 1923, in the midst of the burgeoning Teapot Dome scandal, Secretary of the Interior Albert B. Fall resigned. The next day, President Harding named Work as Fall's replacement. Work would serve as interior secretary from 5 March 1923 to 24 July 1928, through the presidencies of Harding and Calvin Coolidge. His medical knowledge was used in August 1923 when he was one of five physicians to attend to President Harding before his death in San Fran-

cisco. Work's efforts in cleaning up the mess left by Fall and his cronies have been universally praised by historians; efficiency and an honest work ethic were stressed under his leadership. His open-door policy on Indian affairs led to meetings with Indian leaders to formulate proper Indian policies, and his calls for the enactment of a national grazing law led to the passage of the Taylor Act several years after he left office. Overall, Work succeeded in bringing back respect and trust to the Interior Department, although the dismissal of Arthur P. Davis as head of the Reclamation Service and his replacement by David William Davis (no relation) was criticized until Davis was succeeded by water expert Elwood Mead in 1924.

On 24 July 1928, Work resigned as interior secretary to become chairman of the Republican National Committee to advise the presidential campaign of Herbert Hoover. This was Work's final political duty; he retired to Colorado and settled in Denver. He died there of a coronary thrombosis on 14 December 1942 at the age of 82.

See also Department of the Interior; Fall, Albert Bacon.

Wrather, William Embry (1883–1963)
Geologist William E. Wrather headed the U.S. Geological Survey from 1943 to 1956. Born on his maternal grandparents' farm on 20 January 1883, he was the son of Richard Anselm Wrather, a farmer, and Glovy Washington (nee Munford) Wrather. After attending public schools in Chicago, Wrather attended the University of Chicago with the intention of studying the law; instead, he studied geology and was awarded a Ph.B. degree in 1908. One of his classmates was Harold L. Ickes, a future secretary of the interior.

In the summer of 1907, while studying geology, Wrather joined the U.S. Geological Survey and went west, where he was one of the party that mapped the

Granite-Bimetallic Mine at Phillipsburg, Montana. In 1908, he planned to go with the survey to Tonopah and Goldfield in Nevada, but appropriations were slashed. Instead, he followed a fellow geologist to Oklahoma and became an oil-field geologist for the J. M. Guffey Petroleum Company, which evolved into Gulf Oil. Wrote Harold Burstyn, "For Gulf, Wrather leased a number of areas in Texas, on the basis of his geological mapping, that later proved productive. He also initiated the analysis of the deep well at Spur that proved important for the development of oil production in the Permian Basin." In 1916, Wrather helped form the American Association of Petroleum Geologists; soon after, he left Gulf and opened his own office as a consulting geologist for other oil firms. His partnership with the wildcatting firm of Benedum and Trees of Pittsburgh led to the finding of the Desdemona oil field in 1918 and the Nigger Creek field in 1926. Said the *New York Times*, "his work, sometimes called geologically directed wildcatting, led to the discovery of many important oil wells."

Wrather's years of work led to his appointment as associate chief of the metals and minerals division of the Board of Economic Warfare, a New Deal agency, in 1942. (The board later became the Foreign Economic Administration.) The board's main duty, according to Wrather, was "the procurement of metals and minerals from anywhere outside of the United States from foreign sources." Remarked Wrather on his new job, "It was very interesting. My work had largely been confined to oil earlier, and I relished the opportunity to learn more about the world of mineral resources." After just a few months on the job, his former school friend Harold L. Ickes, now secretary of the interior, approached Wrather to succeed Walter Curran Mendenhall as director of the U.S. Geological Survey. Wrather's tenure (1943–1956) marked the first time that the director was chosen from outside the agency. Wrather discovered that only 47 percent of the United States had been mapped topographically, so he began a large survey to map the remaining areas. To accelerate the work, he used a process called photogrammetric mapping (photographic mapping from the air). "As a result," reported one source, "during his tenure as Director, large strategic areas of the United States were comprehensively plotted at a speed of from three to five times faster than was possible by earlier techniques." Further, Wrather unified the separate surveys that analyzed groundwater and surface-water sources into one office. In 1954, he received the John Fritz Medal, awarded by engineering societies, for his body of work. Suffering from ill health, he retired from the Geological Survey in 1956. He died seven years later in Washington, D.C., on 28 November 1963 at the age of 80.

See also United States Geological Survey.

Yard, Robert Sterling (1861–1945)

Robert Sterling Yard, Robert Underwood Johnson's successor as editor of the influential magazine *Century*, was a staunch conservationist and one of the founders of the Wilderness Society. Born in New York on 1 February 1861, he graduated from Princeton University in 1883 and worked as a reporter for the *New York Sun*. In this capacity, he met and befriended Stephen T. Mather, the originator of the National Park Service, and became an ardent advocate for the preservation of the wilderness and the establishment of national parks.

In 1914, when Mather was called to Washington to handle the management of the national park system, he turned to his old friend Robert Yard, now the editor of *Century*, to act as the Interior Department's publicity chief. Using the pages of *Century*, Yard pounded out articles calling for the passage of a national parks act and the establishment of a national parks service to oversee park management. After passage of the National Park Service Act in 1916, Yard set out in several published works to advertise the park system to tourists. Among these books were *The National Parks Portfolio* (1916), *The Top of the Continent* (1917), *The Book of the National Parks* (1919), and *Our Federal Lands* (1928). Yard worked closely with Mather and his assistant, Horace Albright, in formulating park policy. To do this from a civilian point of view, Yard (at Mather's request) formed the National Parks Association (NPA) on 19 May 1919 at the Cosmos Club in Washington. Wrote historian Thomas Wikle, "The Washington-based organization was unique at the time, since most other conservation groups, including the Sierra Club, directed their attention to-wards areas in the western states." Added John Miles, "Yard was a pivotal figure in the National Parks Association. Though not a young man, the organized and resourceful Yard had remarkable energy and stamina. Under his leadership, the association became the principal citizen guardian of the national parks. He would remain a key figure in its history for nearly a quarter of a century."

Although a close friend of Mather's, Yard was what one biographer called a "purist." As such, he objected to the increasing commercialization of the parks and the use of predator control. Mather disliked this criticism and broke off from the NPA, leaving Yard no voice in government. Yard thus turned to other outlets for his advocacy. Because of his body of work for the national parks, he was a natural addition when several conservationists formed the Wilderness Society in Washington in January 1935. That September, he published the first issue of the society's journal, *The Living Wilderness*. Yard was a leading member of the society until his death on 17 May 1945.

See also Wilderness Society.

Yellowstone National Park Act (17 Stat. 32)

This law, enacted on 1 March 1872, was sponsored in the House of Representatives by Territorial Delegate William Horace Clagett (1838–1901) of Montana and in the Senate by Samuel Clarke Pomeroy (1816–1891) of Kansas, a member of the Public Lands Committee. Its purpose was to establish the Yellowstone National Park on "the tract of land in the Territories of Montana and Wyoming lying near the headwaters of the Yellowstone River." The movement to map Yellowstone began in earnest in 1871–1872, with the

William Henry Jackson, a member of the Hayden Survey, photographed Castle Geyser in the Yellowstone thermal area of Wyoming in 1872, the year Yellowstone became the first national park.

U.S. Geological and Geographical Survey of the Territories, better known as the Hayden survey. What followed was not just Hayden's enormous report to Congress but also the chronicling of the park by men such as biologist Clinton Hart Merriam and botanist John Merle Coulter. Over the next 20 years, many men in the conservation movement stepped forward to proclaim the park's treasures. Among these were George Bird Grinnell, editor of the sporting magazine *Forest and Stream*; Arnold Hague, a member of the U.S. Geological Survey; Carleton E. Watkins, who photographed the area in all its glory; and John Fletcher Lacey, a congressman from Iowa who sponsored the Yellowstone National Park Protection Act in 1894. One of the park's lesser-known advocates was William Hallett Phillips, a Washington, D.C., attorney; according to one source, he may have

been responsible for the passage in 1891 of the Timber Reserves Act. Phillips's untimely death in a boating accident in 1897 led to his obscurity.

See also Hayden, Ferdinand Vandeveer; Merriam, Clinton Hart.

Yellowstone National Park Protection Act (28 Stat. 73)

Known as the "Act to Protect the Birds and Other Animals in Yellowstone National Park" and the "National Park Protective Act," this legislation was enacted on 7 May 1894 and was sponsored by Rep. John Fletcher Lacey of Iowa. It ended all hunting in Yellowstone. Passage of the act, explained author Richard A. Bartlett, "meant that the Yellowstone ... was virtually secure and inviolate after 1894."

See also Lacey, John Fletcher.

Yosemite National Park Act
(26 Stat. 650)

Enacted on 1 October 1890, this legislation was the first move toward the formation of the massive Yosemite National Park in California. The Yosemite Valley had been ceded to the state of California, along with the Mariposa Big Tree Grove, in an act introduced in the Senate on 20 February 1864 by John Conness (1821–1909) of California and sponsored by Solomon Foot (1802–1866) of Vermont, a member of the Senate Public Lands Committee. William Henry Jackson's incredible photographs gave the nation a view of the area, providing impetus to the movement to declare it a preserve. In March 1890, Rep. William Vandever (1817–1893) of California introduced the Yosemite Park Act, which would create a sanctuary of about 288 acres but exclude the Tenaya and Tuolumne River watershed. The bill received the support of conservationist John Muir and landscape architect Frederick Law Olmsted, who had spent his own money to map the valley. In 1906, under a separate act (26 Stat. 650), Congress received back from California the Yosemite Valley and the Mariposa Big Tree Grove and added these lands to the park. Muir eventually lost a battle with the government over the Hetch-Hetchy Valley, which was dammed to reroute drinking water to San Francisco. Today, Yosemite National Park is one of the nation's most popular national parks.

See also Hetch-Hetchy Valley Controversy; Muir, John; Olmsted, Frederick Law.

Zahniser, Howard Clinton (1906–1964)

Howard Zahniser was, in the words of the *Washington Post*, "a leading advocate of the federal legislation to establish a national policy and program for the preservation of wilderness in the United States." Born on 25 February 1906 in Franklin, Pennsylvania, he was the son of Rev. A. H. M. Zahniser and Bertha Belle (nee Newton) Zahniser. Zahniser attended local schools and then went to Greenville (Illinois) College, where he was the editor of the school paper, *The Papyrus*. He returned to Pennsylvania briefly and worked at the *Pittsburgh Press*, but he soon returned to Greenville, where he worked on the town paper, the *Greenville Advocate*. Zahniser graduated from Greenville College in 1928 and taught there for a year before joining the Commerce Department in early 1930. From 1931 to 1942, he was a writer and editor at the Bureau of Biological Survey (one of the two agencies that later became part of the Fish and Wildlife Service). In 1942, he was named principal writer for the Bureau of Plant Industry, Soils, and Agricultural Engineering in the Department of Agriculture's Research Administration. He served in this post for three years.

In 1945, Zahniser joined the Wilderness Society as its executive secretary and editor of its magazine, *The Living Wilderness*, a post he held until his death. Zahniser, or "Zahnie," as he was known to his friends, became a "wilderness apostle." He was a founding member in 1946 of the Natural Resources Defense Council, an advocacy group. From 1951 to 1954, he was a member of the advisory committee on conservation to the secretary of the interior.

Starting in 1957, Zahniser began an intense campaign for the passage of a wilderness act to protect the wilderness areas of the United States. He attended congressional hearings and personally lobbied senators and congressmen. In the last week of April 1964, as support for the passage of a wilderness act grew, he submitted to the House interior subcommittee an 11,000-word report on the proposed bill. Later that year, on 3 September, Congress enacted the Wilderness Act. Zahniser, however, had not lived to see its passage. Just a week after submitting his report, on 5 May 1964, he died in his sleep in his Washington, D.C., home. He was 58 years old. His legacy was completed when President Lyndon Johnson signed the Wilderness Act into law.

See also Wilderness Act.

Zon, Raphael (1874–1956)

Russian immigrant Raphael Zon played an important part in the formation of the U.S. Forest Service experiment station program. Born in Simbirsk, Russia, on 1 December 1874, he was the son of Gabriel Zon, a musical instrument repairer, and Eugenia Zon. Raphael Zon attended the city's classical high school, where one of his classmates was Vladimir Ulyanov, who later called himself Lenin. This relationship led to Zon's socialistic philosophy. Wrote biographer Norman Schmaltz, "While a student in medical and natural sciences at the [Imperial] University of Kazan he was arrested for political activity but escaped before the sentence could be carried out." Zon escaped Russia and attended the University of Brussels in Belgium and the University of London in England before immigrating to the United States in 1898. Shortly thereafter, he enrolled at Cornell University and studied forestry under Bernhard Fernow and Filibert Roth. In

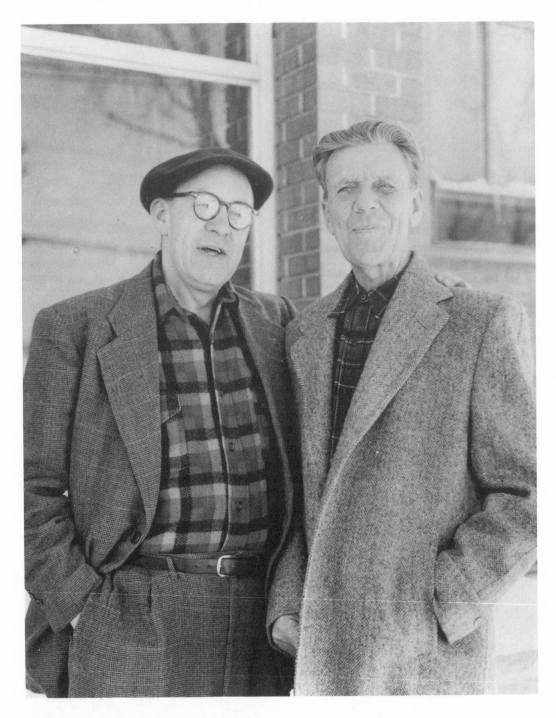

Howard Zahniser, left, with Olaus Murie in 1955

1901, Zon was awarded an F.E. (forest engineer) degree.

Shortly after his graduation from Cornell, Zon joined the U.S. Bureau of Forestry on the recommendation of Fernow. Starting as a student assistant, he was promoted to forest assistant in 1905 and chief of the bureau's Office of Silvics in 1907. In the latter position, he was responsible for the establishment of a national system of forest experiment stations. Wrote Gifford Pinchot:

> On May 6, 1908, a group of research men, under the leadership of Raphael Zon, laid before me a plan for establishing Forest Experiment Stations on the National Forests. The purpose, as set forth in their memorandum, was to carry on "experiments and studies leading to a full and exact knowledge of American silviculture, to the most economic utilization of the products of the forest, and to a fuller appreciation of the indirect benefits of the forest. Each station should be allowed an area sufficient for the proper handling of short-period experiments, for experiments requiring a number of years, and for the maintenance of model forests typical of the silviculture region. . . . These areas will furnish the most valuable, instructive, and convincing object lessons for the public in general, for professional foresters, lumbermen, and owners of forest land, and especially for the technical and administrative officers of the national forests. They should be made the meeting grounds for supervisors, rangers, and guards, where demonstrations may be given for the education of these men, and an active interest stimulated in the technical side of the forest work—an interest which could not be engendered by any amount of literary or oratorical effort."
>
> I had seen forest experiment stations abroad and I knew their value. The plan, therefore, was approved at once.

The first station was established at Fort Valley in the Coconino National Forest near Flagstaff, Arizona, in 1908.

Zon served in the Silvics Office until 1914, when he joined the bureau's Office of Forest Investigations. The work he did in this office and in his previous jobs resulted in the establishment of the Forest Products Laboratory in 1910 and the passage of the McSweeney-McNary Act of 1928, which put a congressional imprimatur on forest investigations and research. Further, he contributed to the Capper Report of 1920, which called for a national timber policy. Reported another source, "At the end of 1920, Zon was assigned to special work in forest economics. During the next couple of years, he produced an exhaustive treatise on *The Forest Resources of the World* (1923), in collaboration with William N. Sparhawk, which gained worldwide recognition." When the Forest Service created the Lake States Forest Experiment Station on the campus of the University of Minnesota at St. Paul in 1923, Zon was made its first director. As director, he took an interest in forests and flood control, resulting in his work *Forests and Water in the Light of Scientific Investigation* (1927). He also played a significant role in the establishment of the Great Plains Shelterbelt Program, also known as the Prairie States Forestry Program. Shelterbelts, the areas of the Midwest used mostly for farming, had been the focus of federal attention since Joseph S. Wilson, commissioner of the General Land Office, narrated in his 1866 annual report that trees should be planted throughout the Great Plains region to protect the soil from erosion. Under the Clarke-McNary Reforestation Act, Zon set out in 1934 to establish a national policy for tree planting. It was under way when he retired from the Forest Service in 1944.

In his final years, Zon worked with the United Nations, writing articles on forestry relief for underdeveloped countries. He died on 27 October 1956 at the age of 81. One source called him "a man of wide and diversified interests throughout his career."

See also Capper Report; Clarke-McNary Reforestation Act; McSweeney-McNary Forest Research Act.

Chronology

1626 Settlers of the Plymouth Colony pass an ordinance forbidding the cutting of timber on public lands without the colony's consent.

1729 The first English volume (the second was published in 1747) of Mark Catesby's *Natural History of Carolina, Florida, and the Bahama Islands* is published.

1799 Under the Federal Timber Purchases Act, the U.S. government appropriates $200,000 to purchase timber reserves for the use of the navy. The first tracts are on Blackbeard's and Grover's Islands off the coast of Georgia.

1803 Dr. Benjamin Barton's *Elements of Botany*, the first work on American flora, appears.

1807 (3 March) An act of Congress prohibits any settlement on the public lands not authorized by law.

1808–
1813 Scottish scientist Alexander Wilson's *American Ornithology; or the Natural History of the Birds of the United States* is published in eight volumes.

1812 (25 April) Congress establishes the General Land Office as part of the Treasury Department.

1817–
1828 Thomas Say's work *American Entomology; or Descriptions of the Insects of North America* is published in three volumes.

1822 Congress enacts "an Act for the preservation of timber of the United States in Florida" to protect government timber supplies in Florida.

1827 (3 March) Congress enacts the Timber Reservation Act, which authorizes the president to conserve live oaks on federal lands and restrict such lands from sale.

1827–
1838 John James Audubon's four-volume *The Birds of America* appears.

1830–
1834 Thomas Say's landmark work *American Conchology* is published in six volumes.

1832 George Catlin, noted painter of western landscapes and American Indians, writes in the *New York Daily Advertiser* about his ideas for the formation of an area to preserve the Missouri River country, the first such expressed plan for a national park.

1834 (28 June) Five thousand dollars is appropriated by the government "to be applied to geological and mineralogical survey and researches," hence, the first geological survey.

1838–
1842 The U.S. exploring expedition under the command of Lt. Charles Wilkes explores the coasts of South, Central, and North America and the Pacific Ocean.

1841 (4 September) The Preemption Act allows settlers to purchase surveyed public land set for auction at $1.25 an acre.

1844 The New York Association for the Protection of Game, the first known wildlife conservation group, is founded.

1845 Writer and nonconformist Henry David Thoreau begins his experiment in independence at Walden Pond. He writes of his studies in *Walden, or Life in the Woods*.

1849 (2 March) The first Swamp Land Act is enacted into law to allow Louisiana to construct levees and drains to reclaim swampland for use.

(3 March) Congress creates the Department of the Interior as a cabinet agency with responsibility over federal and public lands, the Bureau of Indian Affairs, and the General Land Office. Thomas Ewing of Ohio is named as the first interior secretary.

1850 In *United States v. Briggs*, the Supreme Court finds it illegal to cut any timber from public lands.

(28 September) The second of the Swamp Land Acts is passed to authorize the secretary of the interior to survey states with swampland and allow the states to drain and use them.

1857 Thomas Mayo Brewer's work on ornithology, *North American Oology*, appears.

1858 Naturalist James Graham Cooper's "On the Distribution of Forests and Trees of North America, With Notes on Its Physical Geography," a monograph in the Smithsonian Institution's annual report for 1858, discusses forestry reserves in the western United States.

1860 (12 March) The third and final of the Swamp Land Acts extends the rights passed under the first two acts to the states of Minnesota and Oregon, which had entered the Union in 1858 and 1859, respectively.

1862 (20 May) The Homestead Act effectively opens the West to settlement and development by allowing settlers to own 160 acres of land after living on it for five years.

1863 The National Academy of Sciences is established by such scientists as Spencer F. Baird.

1864 Politician and writer George Perkins Marsh writes of his concerns for the environment and how people were affecting it in *Man and Nature*, later expanded into *Physical Geography as Modified by Human Nature*. This work effectively launches the conservation movement.

(30 June) Congress approves a grant to the state of California of the Yosemite Valley and the area surrounding the Mariposa Tree Grove.

1866 (26 July) The Lode Mining Law of 1866 is enacted to protect the rights of lode and placer miners, primarily in the western United States.

1869 Part One of conchologist Thomas Bland's report "Land and Freshwater Shells of North America" is published.

1870 As part of the Tenth Census, the federal government surveys the nation's forestry reserves and resources.

1872 Nebraska becomes the first state to celebrate Arbor Day.

(1 March) Congress approves an act to segregate a tract of land near the headwaters of the Yellowstone River for what would become Yellowstone National Park.

(1 March) Congress enacts the Yellowstone National Park Act.

(10 May) Congress passes the General Mining Law of 1872 to protect the rights of miners not covered in the Lode Mining Law of 1866.

1873 (3 March) The Timber Culture Act, providing settlers with 160 acres of land if one-fourth of the acreage is planted with trees, is enacted.

1875 The American Forestry Association, under the leadership of Dr. John A. Warder, is founded in Chicago.

1876 (15 August) In an appropriations bill, Congress sets aside $2,000 to hire a forestry expert to report to Congress on forest conditions in the United States. U.S. Commissioner of Agriculture Frederick Watts appoints Franklin B. Hough to the post.

1877 (3 March) Congress enacts the Desert Land Act, which allowed settlers to purchase 640 acres of land that was basically unusable after working to irrigate it for three years.

1877–
1883 Hough submits three detailed analyses of American forestry to Congress.

1878 Ornithologist William Brewster authors the first book in the multivolume series "Descriptions of the First Plumage in Various North American Birds" (1878–1887).

John Wesley Powell's landmark work, *Report on the Lands of the Arid Region of the United States*, appears.

(3 June) Congress passes the Free Timber Act, which authorized settlers in the western states to cut down trees on public lands with mineral reserves. Congress also passes on the same date the Timber and Stone Act, which allowed the sale to individuals of 160 acres of uncultivable land with no timber or mineral interests in the western states of Oregon, Washington, California, and Nevada.

1879 (3 March) In a congressional appropriations act, monies are set aside for the creation of the United States Geological Survey and the Public Land Commission.

1879–
1880 The Public Land Commission of 1879 meets and a year later issues its pioneering report on the nation's public lands.

1880 John and Anna Comstock's *Report of the Entomologist* appears.

1881 The forestry survey in the Division of Agriculture is established as the Division of Forestry.

1884 Paleontologist Edward Drinker Cope's landmark work, *The Vertebrata of the Tertiary Formations of the West, Book I*, which is also referred to as "Cope's Bible" or "Cope's Primer," appears.

1885 (3 March) Congress appropriates monies to establish a Division of Economic Ornithology section in the Division of Agriculture. Naturalist Clinton Hart Merriam becomes the division chief.

1886 William Brewster writes *Bird Migration* (1886).

(30 June) The Division of Forestry is established as a permanent part of the Division of Agriculture. Bernhard E. Fernow is named chief. The economic ornithology section in the same agency is renamed the Division of Economic Ornithology and Mammalogy.

1887 (8 February) The Dawes Act, also known as the General Allotment Act, is passed by Congress; it breaks up Indian lands from tribal control and "allots" them to individual Indians in various parcels.

(2 March) The Hatch Act assists states in establishing agricultural experimentation stations.

1888 The Boone and Crockett Club, a sportsmen's conservation organization, is founded by Theodore Roosevelt and writer George Bird Grinnell.

1888–1894 Director of the U.S. Geological Survey John Wesley Powell conducts the so-called Powell Irrigation Surveys of the arid lands of the western United States.

1889 (2 March) The Department of Agriculture is given cabinet status.

1890 (3 February) In *Buford v. Houtz*, the Supreme Court upholds the right to graze unimpeded on public lands, calling it part of an "implied license."

(1 October) Congress enacts the Yosemite National Park Act, designed to eventually establish the Yosemite Valley of California as a national park.

1891 (3 March) The Forest Reserve Act, which repealed the Timber Culture Act of 1873, allows the president to set aside public lands marked by timber or forest.

1892 (29 February) Secretary of State James G. Blaine and British Foreign Secretary Julian Pauncefote sign the Convention for the Protection of Fur Seals in Alaska in an attempt to save the remaining seals there.

(4 June) Twenty-seven men found the Sierra Club with John Muir as its first president.

(24 December 1892) President Benjamin Harrison signs Proclamation no. 39, establishing the Afognak Forest and Fish Culture Reserve, the first unofficial wildlife refuge.

1894 (7 May) An Act to Protect the Birds and Other Animals in Yellowstone National Park, also known as the Yellowstone National Park Protection Act, is enacted by Congress.

(18 August) The Carey Act, which gives the president the power to allow the states to sell upwards of one million acres of public land for settlement and irrigation, is passed by Congress.

1896 (25 April) The Division of Economic Ornithology and Mammalogy is renamed the Division of Biological Survey.

1897 Prior to the end of his term, President Grover Cleveland creates new forest reserves totaling some 20 million acres. Anger rises among western politicians over this action.

(4 June) The Forest Management Act, or Pettigrew Act, is enacted by Congress to "unlock" the forestry reserves already created and to establish a system of management so that timber from the reserves could be harvested.

1898 Gifford Pinchot, a forestry expert, is named head of the Division of Forestry.

(1 July) Congress appropriates $75,000 to safeguard and manage the nation's forest reserves, the first time such an appropriation has been made.

1899 (16 February) Congress enacts the Dead and Down Timber Act, which gives the president the authority to dispose of dead or downed timber on Indian reservations.

(3 March) The Rivers and Harbors Act and the Refuse Act prohibit the dumping of wastes in the nation's harbors, canals, and waterways. The act is passed in response to the 1899

case of *United States v. Rio Grande Dam and Irrigation Company*.

(2 June) The National Irrigation Association is organized.

1900 (25 May) The Lacey Game and Wild Birds Preservation and Protection Act precludes the interstate shipping of wild animals and birds.

1901 (15 February) The Right-of-Way Act authorizes the secretary of the interior to "permit the use of rights of way through the public lands, forest and other reservations of the United States" for public utilities and other construction projects to aid the public welfare.

(2 March) The Division of Forestry in the Department of Agriculture is renamed the Bureau of Forestry.

1901–
1909 Theodore Roosevelt, perhaps the most conservation-minded president ever, sets aside 148 million acres of public land as national forests. Of these, 16 million acres are set aside to be used for farmland opened to homesteaders.

1902 William Brewster authors *Birds of the Cape Region of Lower California*.

(7 June) The Alaska Game Act is passed by Congress to protect certain game animals in Alaska.

(17 June) The Newlands Act, also known as the Reclamation Act, authorizes the secretary of the interior to create irrigation areas for agriculture. It also creates the Bureau of Reclamation as a part of the Interior Department, with reclamation expert Frederick Haynes Newell as the agency's first chief engineer.

1903 (14 March) President Roosevelt signs an executive order creating the Pelican Island National Refuge in Florida.

(22 October) President Roosevelt appoints William A. Richards,

Frederick Haynes Newell, and Gifford Pinchot to the Public Lands Commission.

1904 (7 March) The Public Lands Commission releases its final report.

1905 The Association of Audubon Societies is founded by William E. Dutcher.

(2–6 January) The American Forest Congress is held in Washington, D.C.

(1 February) The administration of forest reserves is transferred to the Bureau of Forestry.

(3 March) The Division of Biological Survey is renamed the Bureau of Biology Survey.

(1 July) The Bureau of Forestry is renamed the United States Forest Service.

1906 (8 June) Congress enacts the Antiquities Act, also known as the American Antiquities Act of the Lacey Antiquities Act, which establishes a program of preserving historic sites in America.

1908 (11 May) Congress amends the 1902 Alaska Game Act with the Act for the Protection of Game in Alaska, which lays out protection areas for animals, how their protection would be administered, and what hunting guidelines if any would be set forth.

(13–15 May) The Governors' Conference on the Environment is held at the White House.

(8 June) President Theodore Roosevelt appoints a 57-man group, led by Gifford Pinchot, to the National Conservation Commission.

(August) Roosevelt appoints Cornell scientist and agricultural expert Liberty Hyde Bailey to the seven-man Country Life Commission.

1910 (3 May) The United States Supreme Court finds in *United States v. Grimaud* and *United States v. Inda* that only Congress may set penalties for criminal behavior on grazing lands and may not delegate that power to a cabinet officer or other government employee.

(16 May) The Bureau of Mines is established within the Department of the Interior to oversee mines and mining issues.

1911 (1 March) The Weeks Act is enacted to preserve watersheds and navigable streams.

(25 September) The American Game Protective and Propagation Association is founded.

1913 (4 March) The Weeks-McLean Act, or Migratory Bird Act, is enacted to provide national protection for migratory birds.

1916 (25 August) Congress creates the National Park Service under the leadership of Stephen Tyng Mather and Horace Marden Albright.

1919 Under the leadership of Robert Sterling Yard and Robert Tyng Mather, the National Parks Association is founded.

1920 The Capper Report, officially called "Timber Depletion, Lumber Prices, Lumber Exports, and Concentration of Timber Ownership," delivers to Congress a detailed analysis of the nation's forestry reserves.

(25 February) Under the Mineral Leasing Act, the Congress promotes the mining of several named minerals located on the public domain.

(19 April) In *Cameron et al. v. United States*, the Supreme Court holds that the president has the power to create national parks and monuments, and any land claims on those lands no longer had any validity.

(10 June) The Federal Water Power Commission is established under the Water Power Act.

1922 The Izaak Walton League, a group of fisherman and hunters concerned about the environment, is founded.

(24 November) The Colorado River Compact is signed by representatives from Arizona, California, Colorado, Nevada, New Mexico, Utah, and Wyoming to apportion equally among these western states the waters of the Colorado River.

1924 William Brewster's posthumous and unfinished *Birds of the Lake Umbagog Region, Maine* appears.

(7 June) Congress enacts the Clarke-McNary Act, which allotted funds to state forestry officials for "the purpose of stimulating the acquisition, development, and proper administration and management of state forests," as well as the purchase of new forest lands that were opened up for sale. The same day, the Oil Pollution Act, enacted to control the pollution with oil of water, is passed.

1927 *Parmelian Prints of the High Sierra*, a collection of photographs by Ansel Adams, establishes him as the premier western nature photographer.

1928 (30 April) The enactment of the Woodruff-McNary Act of 1928 expands federal influence over the purchase of wood- and timberlands and permits the government to create new national forests. Further, the act appropriates $8 million in federal funds for the purchase of lands to be used for the management of national forests.

(22 May) The McSweeney-McNary Forest Research Act of 1928 authorizes the federal government to augment forestry research programs through the experiment station system and calls for the funding of a national timber survey.

1929 (18 February) The Migratory Bird Conservation Act, establishing the Migratory Bird Conservation Commission, is enacted.

1930 (10 July) The Shipstead-Nolan Act, which withdraws from entry (that is, public access) all public lands north of Township 60 in Minnesota, is passed by Congress.

1933 The Copeland Report, an analysis of the condition of the nation's forests, is issued by the U.S. Forest Service.

(31 March) Under the Unemployment Relief Act, Congress creates the Civilian Conservation Corps.

(12 May) Congress enacts the Agricultural Adjustment Act as a means of establishing a federal agency (the Agricultural Adjustment Administration [AAA]) to handle crop management for farmers.

(18 May) The Tennessee Valley Authority is established with the passage of the Tennessee Valley Authority Act.

1934 (10 March) The Fish and Wildlife Coordination Act authorizes the secretaries of agriculture and commerce to cooperate with states on establishing programs to coordinate national policies on fish and wildlife conservation.

(16 March) Under the Migratory Bird Hunting Stamp Act, hunters would have to purchase "duck stamps" to hunt, thereby funding the management of wildlife refuges and allowing for the purchase of additional lands.

(18 June) The Taylor Grazing Act establishes national policy on grazing in the western states.

1935 (January 21) Environmental activists Harold C. Anderson, Harvey Broome, Bernard Frank, Benton MacKaye, Robert Marshall, Olaus and Margaret Murie, Sigurd Olson, Ernest Oberholtzer, Theodor Swem, Robert Sterling Yard, and Howard Zahniser found the Wilderness Society.

(27 April) The Soil Conservation Act establishes national policy on the conservation and protection of soil and creates the Soil Conservation Service as part of the Agriculture Department.

(21 August) Congress enacts the Historic Sites and Buildings Act, an amendment to the Antiquities Act of 1906.

1936 (16 January) The Supreme Court strikes down the Agricultural Adjustment Act in *United States v. Butler* and *Rickert Rice Mills, Inc. v. Fontenot.*

(17 September) President Franklin Delano Roosevelt appoints an eight-man group headed by Morris L. Cooke, administrator of the Rural Electrification Administration, to the Great Plains Committee to report on how the resources of the Great Plains could be more efficiently utilized.

1937 (18 May) With the passage of the Norris-Doxey Act, or the Cooperative Farm Forestry Act, Congress endorses the use of funds for the furtherance of farm forestry programs.

(2 September) Congress enacts the Pittman-Robertson Act, or the Federal Aid in Wildlife Restoration Act, which levies an excise tax on firearms and ammunition to be used to acquire wildlife refuges and for wildlife management.

1938 Writer Louis Bromfield purchases a farm near his boyhood home in Ohio that later becomes Malabar Farm, a model for soil erosion studies, conservation, and innovative farming techniques.

(16 February) Congress enacts the second Agricultural Adjustment Act, which helps to supplement the Soil Conservation Act of 1935, with changes that would pass constitutional muster. This act fortifies the national policy "of conserving national resources, preventing the wasteful use of soil fertility, and of preserving,

maintaining, and rebuilding the farm and ranch land resources in the national interest, [and] to accomplish these [goals] through the encouragement of soil-building and soil-conserving crops and practices."

1939 (9 May) Under Reorganization Plan No. 3, Congress consolidates the Bureau of Fisheries and the Bureau of Biological Survey into the Fish and Wildlife Service.

1940 (8 June) Congress enacts the Bald and Golden Eagle Protection Act of 1940 to prevent the poaching and selling of rare bald and golden eagles.

1944 (29 March) The secretary of agriculture is mandated to establish sustained-yield forestry units with the enactment of the Sustained-Yield Forest Management Act.

1945 (28 September) With the signing of Executive Order 9634, President Harry S Truman establishes so-called Fishery Conservation Zones.

1946 (16 May) The Bureau of Land Management (BLM) is created by merging the General Land Office and the United States Grazing Service under Reorganization Plan No. 3.

1947 Marjory Stoneman Douglas's milestone book, *The Everglades: River of Grass*, describes the plight of the Everglades National Park in Florida.

(25 June) Congress passes the Federal Insecticide, Fungicide, and Pesticide Act to warn consumers about pesticides in foods and to prevent labeling fraud.

1948 (30 June) The Watwer Pollution Control Act, or Taft-Barkley Act, establishes a program of stream pollution suppression.

1950 (9 August) With the enactment of the Fish Restoration and Management Act, or Dingell-Johnson Act, Congress authorizes the raising of revenue from the sale of fishing equipment to be used for fish management programs.

1953 (22 May) The Submerged Lands Act of 1953 sets the congressional definition of submerged lands and grants states the rights and authority over navigation, flood control, and the production of power (federal conservation laws, such as those protecting fish and wildlife, would still have to be observed).

(7 August) The Outer Continental Shelf Lands Act of 1953 deals with congressional policy on lands just outside the land border of the nation, commonly known as the outer continental shelf.

1954 (15 March) In *Alabama v. Texas*, the Supreme Court upholds the Submerged Lands Act of 1953 as constitutional.

(4 August) The Watershed Protection and Flood Prevention Act authorizes the secretary of agriculture to collaborate with state and local governments to establish programs to prevent erosion and other damage to watersheds as well as to promote their careful management and use.

1956 (11 April) With the passage of the Colorado River Storage Project Act, Congress establishes a series of storehouses for waters from the Colorado River for public use.

1957 Robert Porter Allen's *On the Trail of Vanishing Birds* documents the preservation of the endangered whooping crane and the roseate spoonbill.

1958–1962 The Outdoor Recreation Resources Review Commission (ORRRC), composed of U.S. senators and congressmen, attempts to constitute governmental policy on outdoor recreation.

1959 (22 September) Congress passes the Marine Game Fish Research Act, which orders the secretary of the interior to "undertake continuing research on the biology fluctuations, status, and statistics of the migratory marine species of game fish of the United States and contiguous waters." The research mandated studies on fish migrations and requested data on all known environmental influences, both natural and artificial (including pollution).

1960 (12 June) Congress enacts the Multiple Use–Sustained Yield Act to add the recreational use of national forests to the list of uses of forests, which had been previously limited to such acts as timber collection and protection, watershed protection, grazing, and wildlife and fish protection.

1962 Author Rachel Carson's landmark work *Silent Spring* calls attention to the dangers of the use of pesticides and starts the modern environmental movement.

1963 (3 June) In *Arizona v. California*, the Supreme Court allows a "special master" to decide what portion several western states may receive from the Colorado River.

(17 December) The landmark Clean Air Act is enacted by Congress to set national policy as well as uniform standards for air pollution control.

1964 (3 September) Congress enacts the Wilderness Act, which sets up the National Wilderness Preservation System.

(19 September) The Classification and Multiple Use Act of 1964 instructs the Bureau of Land Management (BLM) to manage those public lands under its control under a policy of sustained yield and multiple use.

1965 (2 October) Congress enacts the Water Quality Improvement Act to address the problem of water pollution.

(20 October) The Solid Waste Disposal Act of 1965 is passed to set national policy on recycling.

(22 October) The beautification of the national highway system is stressed with the passage of the Highway Beautification Act.

(30 October) The passage of the Anadromous Fish and Great Lakes Act protects certain species of fish.

1966 Naturalist Edwin Way Teale is awarded the Pulitzer Prize for general nonfiction for *Wandering through Winter*, the last of a four-volume series on the seasons.

(4 May) With the signing of Executive Order 11278, President Lyndon B. Johnson establishes the President's Council on Recreation and Natural Beauty, headed by Vice President Hubert H. Humphrey.

(15 October) The first such legislation to protect endangered species, the Endangered Species Preservation Act of 1966 instructed the secretary of the interior to establish a program of preserving, restoring, and replenishing selected species of fish and wildlife that were deemed by the secretary to be endangered or threatened.

(3 November) Under the Clean Water Restoration Act, Congress appropriated funds for states and cities to construct waste-water treatment facilities.

1967 (21 November) Congress enacts the Air Quality Act (AQA) of 1967, amending the Clean Air Act of 1965. The AQA affixes enforcement mandates to past air pollution legislation.

1968 (3 August) The Estuary Protection Act of 1968 mandates federal protection for estuaries, which are tributary waterways teeming with wildlife.

(2 October) With the enactment of the Wild and Scenic Rivers Act, a

system of preservation for a number of selected rivers nationwide is established. The passage of the National Trails System Act accomplishes the same aim for chosen trails.

1969   (29 May) President Richard Nixon signs Executive Order 11472, creating the short-lived Cabinet Committee on the Environment, which tries to use the power of cabinet officers to promote the preservation of the environment.

(5 December) The Endangered Species Conservation Act, also known as the Endangered Species Act, establishes a national policy of identifying threatened and endangered species and adding them to the Endangered Species List.

1970   (1 January) The National Environmental Policy Act of 1969 takes effect. It mandates the creation of the Council on Environmental Quality (CEQ) and calls for environmental impact statements (EISs) to evaluate the potential impact of legislation and government policy on the environment.

(3 April) The Environmental Quality Act of 1970 (EQA) is passed by Congress. It calls for the execution, by all pertinent federal agencies involved in environmental and public works programs, of all environmental laws previously passed by Congress, as well as providing bureaucratic staff to the Council on Environmental Quality.

(9 April) President Richard Nixon signs Executive Order 11523, which establishes the National Industrial Pollution Control Council.

(22 April) Earth Day is celebrated for the first time.

(23 December) Under Executive Order 11574, President Richard M. Nixon lays out a permit schedule to manage the nation's refuse.

(31 December) Under the Mining and Minerals Policy Act of 1970,

Congress establishes a national mining and minerals policy.

1972   (8 February) President Richard Nixon signs Executive Order 11644, which deals with the use of off-road vehicles on public lands.

(18 October) With the passage of the Federal Water Pollution Control Act Amendments of 1972 (also known as the Clean Water Act Amendment), Congress attempts to limit effluent discharges and set water quality standards.

(21 October) The Marine Mammal Protection Act of 1972 establishes national policy on managing the declining species of marine mammals.

(23 October) Congress establishes policy toward marine wildlife sanctuaries with the enactment of the Marine Protection, Research, and Sanctuaries Act.

(27 October) Congress enacts the Coastal Zone Management Act of 1972 to formulate policy on the management of coastal zones.

1973   The Convention of International Trade in Endangered Species of Wild Flora and Fauna (CITES) is signed by 80 nations, including the United States, to end the selling, trafficking, and exploitation of endangered species of animals and plants.

1974   (17 August) Congress enacts the Forest and Rangeland Renewable Resources Planning Act of 1974 to facilitate long-range management for the utilization of natural resources in the National Park System.

(16 December) Under the Safe Drinking Water Act of 1974, Congress mandates regulations for clean water in public drinking systems and discusses which contaminants could be banned from these systems.

1975   (18 July) President Gerald Ford signs Executive Order 11870, which mandates precautions on procedures for controlling animal damage on federal lands.

1976 Congress passes the Resource Conservation and Recovery Act (RCRA) as an amendment to the Solid Waste Disposal Act of 1965 to deal with the growing amount of municipal and industrial waste.

(13 April) To deal with the problem of dwindling fish stocks in the waters around the United States, Congress enacts the Magnuson Fishery and Conservation and Management Act.

(11 October) With the enactment of the Toxic Substances Control Act, Congress establishes a national policy on toxic substances in the environment.

(21 October) The Federal Land Policy and Management Act, or Bureau of Land Management Organic Act, restates the federal policy (established in the Classification and Multiple Use Act of 1964) of sanctioning the multiple-use management of public lands under the jurisdiction of the Bureau of Land Management.

1977 (3 August) Congress establishes national policy on the control of surface mining and its impact on the environment with the enactment of the Surface Mining Control and Reclamation Act. The law also creates the Office of Surface Mining Reclamation and Enforcement (OSM), an agency in the Interior Department.

(18 November) The Soil and Water Resources Conservation Act of 1977 requests an appraisal of the nation's soil and water resources.

1978 (18 September) An amendment to the Fisherman's Protective Act of 1967, known as the Pelly Amendment, is enacted to protect endangered or threatened species of fish or wildlife.

1979 (28 March) A near meltdown at the Three Mile Island nuclear power

facility in Pennsylvania causes a national reassessment of the virtues of nuclear power.

1980 President Jimmy Carter awards Horace Albright the Medal of Freedom.

(2 December) Congress passes the Alaska National Interest Lands Conservation Act, also known as the Alaska Lands Act, or the D-2 Act, to expand the acreage in that state's wildlife refuge and park systems by some 97 million acres.

1982 The Earth Island Institute is founded in California by David R. Brower.

(18 October) The Coastal Barrier Resources Act of 1982 establishes a national policy for the protection and management of coastal barriers.

(8 November) With the passage of the Uranium Tail Millings Radiation Control Act of 1978, Congress for the first time addresses the problem of the disposal of uranium tailings, or tail millings, which are the residues of uranium brought to the surface for use.

1983 (30 March) In *Arizona v. California*, the Supreme Court holds that American Indian tribes may not relitigate their share of water diverted for irrigation from the Colorado River.

1986 (19 June) The passage of the Safe Drinking Water Act Amendment revises upward the quality of water systems authorized under the Safe Drinking Water Act of 1974.

(17 October) Congress enacts the Emergency Planning and Community Right-to-Know Act of 1986 to legislate the contingency handling of chemicals used across the nation.

(10 November) The Emergency Wetlands Resources Act of 1986 is passed to protect the nation's dwindling wetlands.

1989 (24 March) The *Exxon Valdez*, a 987-foot oil tanker, hits a reef in Alaska's Prince William Sound and spills 11 million gallons of oil in possibly the worst oil-spill to date.

1990 The Clean Air Act Amendments of 1990 revise the Clean Air Act of 1963 to deal with ethanol substitutes for diesel fuel.

(5 November) With the enactment of the Pollution Prevention Act of 1990, Congress seeks to establish a national policy on pollution emissions.

(29 November) Congress passes the Coastal Wetlands Planning, Protection and Restoration Act, which authorizes the secretary of the interior to establish projects in the area of "protection, restoration, or enhancement of aquatic and associated ecosystems."

1992 The Supreme Court strikes down higher disposal fees for waste generated outside a state than within it as a violation of the commerce clause of the Constitution in *Chemical Waste Management, Inc. v. Hunt, et al.*

1993 President Bill Clinton awards Marjory Stoneman Douglas the Medal of Freedom.

1994 (4 April) In *Oregon Waste Systems, Inc. v. Department of Environmental Quality of the State of Oregon et al.*, the Supreme Court reaffirms its 1992 *Hunt* decision.

(24 June) In *Dolan v. City of Tigard*, the Supreme Court holds that environmental protection was not a proper cause for taking private property without just compensation.

(11 August) A federal jury awards 10,000 Alaskan fishermen $286 million in compensation for the *Exxon Valdez* oil spill.

Bibliography

Agnew, Dwight L. "The Government Land Surveyor as Pioneer." *Mississippi Valley Historical Review* 28, 3 (December 1941): 369–382.

Albright, Horace M., as told to Robert Cahn. *The Birth of the National Park Service: The Founding Years, 1913–1933.* Salt Lake City: Howe, 1985.

Alexander, Thomas G. "The Powell Irrigation Survey and the People of the West." *Journal of the West* 7, 1 (January 1968): 48–54.

Alford, Terry L. "The West as a Desert in American Thought prior to Long's 1819–1820 Expedition." *Journal of the West* 8, 4 (October 1969): 515–525.

Allaby, Michael. *Dictionary of the Environment.* New York: New York University Press, 1989.

Allard, Dean C., Jr. "Spencer Fullerton Baird and the U.S. Fish Commission: A Study of the History of American Science." Ph.D. dissertation, George Washington University, 1967.

Allen, Arthur Augustus. "The Curlew's Secret." *National Geographic* 94, 6 (December 1948): 751–770.

Allen, Robert Porter. "Whooping Cranes Fight for Survival." *National Geographic* 96, 5 (November 1959): 650–669.

Allen, Thomas B. *Guardian of the Wild: The Story of the National Wildlife Federation, 1936–1986.* Bloomington: Indiana University Press, 1987.

"The American Forest Congress." *Forestry and Irrigation* 11, 1 (January 1905): 1–10.

American Men and Women of Science, 1992–1993. New Providence, NJ: Bowker, 1992.

American Wildlife Institute. *Proceedings of the North American Wildlife Conference.* Washington, DC: American Wildlife Institute, 1936.

Anderson, Martin J. "John Wesley Powell's Exploration of the Colorado River: Fact, Fiction or Fantasy?" *Journal of Arizona History,* 24, 4 (Winter 1983): 363–380.

Are We Cleaning Up? 10 Superfund Case Studies: Special Report. Office of Technology Assessment Report OTA-ITE-362. Washington, DC, June 1988.

Arnold, Ron. *At the Eye of the Storm: James Watt and the Environmentalists.* Chicago: Regnery Gateway, 1982.

———. *Ecology Wars: Environmentalism As If People Mattered.* Bellevue, WA: Free Enterprise, 1993.

Arnold, Ron, and Alan Gottlieb. *Trashing the Economy: How Runaway Environmentalism Is Wrecking America.* Bellevue, WA: Free Enterprise, 1993.

Ashworth, William. *The Encyclopedia of Environmental Studies.* New York: Facts on File, 1991.

Baker, Richard A. "The Conservation Congress of Anderson and Aspinall, 1963–64." *Journal of Forest History* 29, 3 (July 1986): 104–119.

Baker, Robert D., et al. *The National Forests of the Northern Region: Living Legacy.* U.S. Forest Service publication FS-500. Washington, DC: U.S. Department of Agriculture, November 1993.

Baker, Robert D.; Robert S. Maxwell; Victor H. Treat; and Henry C. Dethloff. *Timeless Heritage: A History of the Forest Service in the Southwest.* U.S. Forest Service Publication FS-409. Washington, DC: U.S. Department of Agriculture, 1988.

Barlow, J. W. *Reconnaissance of the Yellowstone River.* Senate Executive Document 66, 42d Congress, 2d Session, 1872.

Barney, Gerald O. *The Global 2000 Report to the President: A Report Prepared by the Council on Environmental Quality and the Department of State.* Washington, DC: Government Printing Office, 1980.

Bartlett, Richard A. *Yellowstone: A Wilderness Besieged.* Tucson: University of Arizona Press, 1985.

———. *Great Surveys of the American West.* Norman: University of Oklahoma Press, 1962.

Bartram, William. *Travels through North and South Carolina, Georgia, East and West Florida, the Cherokee Country, the Extensive Territory of the Muscogulges, or the Creek Confederacy, and the Country of the Choctaws, Containing an Account of the Soil and Natural Productions of Those Regions, Together with Observations on the Manners of the Indians.* New York: Dover, 1955. Reprinted.

Bates, James Leonard. "Fulfilling American Democracy: The Conservation Movement, 1907 to 1921." *Mississippi Valley Historical Review* 44, 1 (June 1957): 29–57.

Bedini, Silvio A. *Thinkers and Tinkers: Early American Men of Science.* New York: Scribner's, 1975.

Begley, Sharon. "On the Trail of Acid Rain." *National Wildlife* 25, 1 (February-March 1987): 6–12.

Beidleman, Richard G. "Some Biographical Sidelights on Thomas Nuttall, 1786–1859." *Proceedings of the American Philosophical Society* 104, 1 (February 1960): 86–100.

Bennett, Steven J. *Ecopreneuring: The Complete Guide to Small Business Opportunities from the Environmental Revolution.* New York: Wiley, 1991.

"Benton MacKaye: A Tribute." *Living Wilderness* 39, 132 (January-March 1976): 6–34.

Billington, Ray Allen. *Frederick Jackson Turner: Historian, Scholar, Teacher.* New York: Oxford University Press, 1973.

Blair, W. Reid. "William Temple Hornaday: A Tribute." *Bulletin of the New York Zoological Society* 40, 2 (March-April 1937):46–49.

Blake, Israel George. *The Holmans of Veraestau.* Oxford, OH: Mississippi Valley, 1943.

Bogue, Margaret Beattie. "To Save the Fish: Canada, the United States, the Great Lakes, and the Joint Commission of 1892." *Journal of American History* 79, 4 (March 1993): 1429–1454.

Bosmajian, Haig. "The Imprint of the Cascade Country on William O. Douglas." *Journal of the West* 32, 3 (July 1983): 80–86.

Boucher, Norman. "The Legacy of *Silent Spring.*" *Boston Globe Magazine*, 15 March 1987, 17–22.

Brooks, Paul. *Speaking for Nature: How Literary Naturalists from Henry Thoreau to Rachel Carson Have Shaped America.* Boston: Houghton Mifflin, 1980.

———. "The Written Word: The Courage of Rachel Carson." *Audubon* 89, 1 (January 1987): 12–15.

Brown, George Rothwell, ed. *Reminiscences of William M. Stewart of Nevada.* New York: Neale, 1908.

Browning, James A.; John C. Hendee; and Joe W. Roggenbuck. *103 Wilderness Laws:*

Milestones and Management Direction in Wilderness Legislation, 1964–1987. Station Bulletin 31. Moscow: University of Idaho, Idaho Forest, Wildlife and Range Experiment Station, October 1988.

Buenker, John D., ed. *Historical Dictionary of the Progressive Era, 1890–1920*. Westport, CT: Greenwood, 1988.

Burkholder, Harry C., ed. *High-Level Nuclear Waste Disposal*. Columbus, OH: Battelle, 1986.

Cart, Theodore Whaley. "The Federal Fisheries Service, 1871–1940." Master's thesis, University of North Carolina at Chapel Hill, 1968.

———. "The Lacey Act: America's First Nationwide Wildlife Statute." *Forest History* 17, 3 (October 1973): 4–13.

———. "The Struggle for Wildlife Protection in the United States, 1870–1900: Attitudes and Events Leading to the Lacey Act." Ph.D. dissertation, University of North Carolina at Chapel Hill, 1971.

"A Centennial Celebration: 1892–1992." *Sierra* 77, 3 (May-June 1992): 52–73.

Chapman, Frank M. "Origin of West India Bird Life." *National Geographic* 9, 5 (March 1898): 243–246.

Chemicals in Your Community: A Guide to the Emergency Planning and Community Right-to-Know Act. Environmental Protection Agency Report OS-120. Washington, DC, September 1988.

Chittenden, Hiram M. *The Yellowstone National Park, Historical and Descriptive, 1915*. Cincinnati: Stewart and Kidd, 1917.

CIS Index to Presidential Executive Orders & Proclamations. Part 1: April 30, 1789 to March 4, 1921 (George Washington to Woodrow Wilson). Washington, DC: Congressional Information Service, 1987.

Clapper, Louis S. "Pollution." *National Wildlife* 1, 6 (October-November 1963): 9–11.

Clark, John G. *The Frontier Challenge: Responses to the Trans-Mississippi West*. Lawrence: University of Kansas Press, 1971.

Clean Air through Transportation: Challenges in Meeting National Air Quality Standards. A Joint Report from the United States Department of Transportation and Environmental Protection Agency Pursuant to Section 108(f)(3) of the Clean Air Act. Washington, DC, 1993.

Clepper, Henry. "A Salute to Samuel T. Dana at 90." *Journal of Forestry* 71, 4 (April 1973): 200–202.

———. "Who Was Robert Douglas?" *Journal of Forest History* 19, 1 (January 1975): 22–23.

Clepper, Henry, ed. *Leaders of American Conservation*. New York: Ronald, 1971.

Coan, Eugene. "James Graham Cooper: Pioneer Naturalist and Forest Conservationist." *Journal of Forest History* 27, 3 (July 1983): 126–129.

Cohn, Roger, and Ted Williams. "Bruce Babbitt: Interior Views." *Audubon* 95, 3 (May-June 1993): 78–84.

Cole, Wayne S. "Senator Key Pittman and American Neutrality Policies, 1933–1940." *Mississippi Valley Historical Review* 46, 4 (March 1960): 644–662.

Coletta, Paolo E. *The Presidency of William Howard Taft*. Lawrence: University Press of Kansas, 1971.

Compendium of History and Biography of North Dakota, Containing a History of North Dakota, Embracing an Account of Early Explorations, Early Settlement, Indian Occupancy, Indian History and Traditions, Territorial and State Organization, a Review of the Political History, and a Concise History of the Growth and Development of the State. Chicago: Ogle, 1900.

Congress A to Z: Congressional Quarterly's Ready Reference Encyclopedia. Washington, DC: Congressional Quarterly, 1988.

Conniff, Richard. "RAP (Rapid Assessment Program): On the Fast Track in Ecuador's National Forests." *Smithsonian* 22, 3 (June 1991): 36–49.

Conover, Milton. *The General Land Office: Its History, Activities and Organization*. Baltimore: Johns Hopkins University Press, 1923.

Conrad, Barnaby, III. "C. M. Russell and the Buckskin Paradise of the West." *Horizon* 22, 5 (May 1979): 42–49.

Cooper, James Graham. *The Forests and Trees in North America, as Connected with Climate and Agriculture.* Report of the U.S. Patent Office, in *Agricultural Report for 1860.* House Executive Document 48, 36th Congress, 2d Session, 1861.

———. *On the Distribution of Forests and Trees of North America, with Notes on Its Physical Geography.* In *Annual Report of the Board of Regents of the Smithsonian Institution for 1858.* Senate Miscellaneous Document 49, 35th Congress, 2d Session, 1859.

———. *On the Forests and Trees of Florida and the Mexican Boundary.* In *Annual Report to the Board of Regents of the Smithsonian Institution for 1860.* Senate Miscellaneous Document 21, 36th Congress, 2d Session, 1861.

Cox, Thomas R. "The Stewardship of Private Forests: The Evolution of a Concept in the United States, 1864–1950." *Journal of Forest History* 25, 4 (October 1981): 188–196.

Crèvecoeur, Hector St. John de. *Letters from an American Farmer, and Sketches of Eighteenth-Century America.* New York: New American Library of World Literature, 1963. Reprinted.

Cross, Whitney R. "W J McGee and the Idea of Conservation." *Historian* 15, 2 (Spring 1953): 148–162.

Current, Richard N.; John A. Garraty; and Julius Weinberg, eds. *Words That Made American History since the Civil War.* Boston: Little, Brown, 1962.

Dahl, Thomas E. *Wetlands Losses in the United States, 1780s to 1980s.* U.S. Fish and Wildlife Service, National Wetlands Inventory. Washington, DC: U.S. Department of the Interior, 1990.

Dana, Samuel Trask. "(The American Forestry Association:) The First Eighty Years." *American Forests* 62, 4 (April 1956): 13–19, 42–48.

Darling, Arthur B., ed. *The Public Papers of Francis G. Newlands.* 2 vols. Boston: Houghton Mifflin, 1932.

Darrah, William Culp. *Powell of the Colorado.* Princeton, NJ: Princeton University Press, 1951.

Davis, Richard, ed. *Encyclopedia of American Forest and Conservation History.* 2 vols. New York: Macmillan, 1983.

Dawdy, Doris Ostrander. *Congress in Its Wisdom: The Bureau of Reclamation and the Public Interest.* Boulder, CO: Westview, 1989.

Dawson, Thomas Fulton. *Life and Character of Edward Oliver Wolcott.* New York: Knickerbocker, 1911.

Decker, Leslie E. "The Railroads and the Land Office: Administrative Policy and the Land Patent Controversy, 1864–1896." *Mississippi Valley Historical Review* 46, 4 (March 1960): 679–699.

Deetz, James. *In Small Things Forgotten: The Archeology of Early American Life.* Garden City, NY: Anchor Books, 1977.

Dixon, Albert. "The Conservation of Wilderness: A Study in Politics." Ph.D. dissertation, University of California, Berkeley, 1968.

Doenecke, Justus D. *The Presidencies of James A. Garfield and Chester A. Arthur.* Lawrence: Regents Press of Kansas, 1981.

Dolph, James Andrew. "The American Bison Society: Preserver of the American Buffalo and Pioneer in Wildlife Conservation." Master's thesis, University of Denver, 1965.

———. "Bringing Wildlife to Millions: William Temple Hornaday: The Early Years." Ph.D. dissertation, University of Massachusetts, 1975.

Donaldson, Thomas. *The Public Domain: Its History with Statistics.* New York: Johnson Reprint, 1970. Reprinted.

Dunlap, Thomas R. *DDT: Scientists, Citizens, and Public Policy.* Princeton, NJ: Princeton University Press, 1981.

Dutton, Clarence E. *General Description of the Volcanic Phenomena Found in That Portion of Central America Traversed by the Nicaragua Canal.* Senate Document 357, 57th Congress, 1st Session, 1902.

"Early Days in Forest School and Forest Service [Ralph S. Hosmer oral history]." *Forest History* 16, 3 (October 1972): 6–11.

Edwards, Walter Meayers. "Lake Powell: Waterway to Desert Wonders." *National Geographic* 132, 1 (July 1967): 44–75.

"The Eleventh National Irrigation Congress." *Forestry and Irrigation* 9, 10 (October 1903): 478–489.

Elliott, Clark A. *Biographical Dictionary of American Science: The Seventeenth through the Nineteenth Centuries.* Westport, CT: Greenwood, 1979.

Ellis, Elmer. *Henry Moore Teller: Defender of the West.* Caldwell, ID: Caxton, 1941.

Emmons, David M. "American Myth: Desert to Eden. Theories of Increased Rainfall and the Timber Culture Act of 1873." *Forest History* 15, 3 (October 1971): 6–14.

Emmons, Samuel Franklin. *Geology and Mining Industry of Leadville, Colorado, with Atlas.* United States Geological Survey Monograph 12. Washington, DC, 1886.

Emmons, Samuel Franklin; Whitman Cross; and George Homans Eldridge. *Geology of the Denver Basin in Colorado.* United States Geological Survey Monograph 27. Washington, DC, 1896.

Emory, Major William H. *Report on the United States and Mexican Boundary Survey, Made under the Secretary of the Interior.* Senate Executive Document 135, 34th Congress, 1st Session, 1857.

Environmental Protection Agency. *Access EPA.* Washington, DC: Environmental Protection Agency, 1992.

Environmental Statutes: 1993 Edition. Rockville, MD: Government Institutes, 1993.

Erikson, Kai. "Out of Sight, Out of Our Minds." *New York Times Magazine,* 6 March 1994, 34–39.

Errington, Paul L. "In Appreciation of Aldo Leopold." *Journal of Wildlife Management* 12, 4 (October 1948): 341–350.

Everhart, William C. *The National Park Service.* Boulder, CO: Westview, 1989.

Facing America's Trash: What's Next for Municipal Solid Waste? Office of Technology Assessment Report OTA-O-424. Washington, DC: Government Printing Office, October 1989.

Faust, Patricia L., ed. *Historical Times Illustrated Encyclopedia of the Civil War.* New York: Harper & Row, 1986.

Ferguson, Shirley. "Major John Wesley Powell's Exploration of the Grand Canyon." *Journal of Arizona History* 2, 2 (Summer 1961): 34–38.

Fernow, Bernhard E. *Report upon the Forestry Investigations, U.S. Department of Agriculture, 1877–1898.* House Document 181, 55th Congress, 3d Session, 1899.

Fitzpatrick, Virginia L. "Frederick Law Olmsted and the Louisville Park System." *Filson Club History Quarterly* 59, 1 (1985): 54–65.

Flader, Susan. "Thinking like a Mountain: A Biographical Study of Aldo Leopold." *Forest History* 17, 1 (April 1973): 14–28.

———. *Thinking like a Mountain: Aldo Leopold and the Evolution of an Ecological Attitude toward Deer, Wolves, and Forests.* Columbia: University of Missouri Press, 1974.

Fleming, Donald. "Roots of the New Conservation Movement." *Perspectives in American History* 6 (1972): 7–94.

Fleming, Thomas. "The Bartrams: They Blazed a Trail in Paradise." *National Wildlife* 16, 6 (October-November 1978): 24–30.

Foreman, Dave. *Confessions of an Eco-Warrior.* New York: Harmony Books, 1991.

Forest Research: Hearing before the Committee on Agriculture and Forestry of the United States Senate. Washington, DC: Government Printing Office, 1928.

Forest Service Fish Habitat and Aquatic Ecosystem Research. U.S. Forest Service pamphlet FS-556. Washington, DC: U.S. Department of Agriculture, 1994.

Forness, Norman Olaf. "The Origins and Early History of the United States Department of the Interior." Ph.D.

dissertation, Pennsylvania State University, 1964.

Fox, Stephen. *The American Conservation Movement: John Muir and His Legacy.* Madison: University of Wisconsin Press, 1981.

Fradkin, Philip L. *A River No More: The Colorado River and the West.* Tucson: University of Arizona Press, 1981.

Friedenberg, Daniel M. *Life, Liberty, and the Pursuit of Land: The Plunder of Early America.* Buffalo, NY: Prometheus Books, 1992.

From Sea to Shining Sea: A Report on the American Environment: Our Natural Heritage. Report by the President's Council on Recreation and Natural Beauty. Washington, DC: Government Printing Office, 1968.

Gabrielson, Ira N. *Wildlife Refuges.* New York: Macmillan, 1943.

Gales, Robert L. *The Gay Nineties in America: A Cultural Dictionary of the 1890s.* Westport, CT: Greenwood, 1992.

Gallagher, Annette. "Citizen of the Nation: John Fletcher Lacey, Conservationist." *Annals of Iowa* 46, 1 (Summer 1981): 9–24.

Gannett, Henry. "The Conservation League of America." *National Geographic* 19, 10 (October 1908): 737–739.

Ganoe, John T. "The Beginnings of Irrigation in the United States." *Mississippi Valley Historical Review* 25, 1 (June 1938): 59–78.

Gara, Larry. *The Presidency of Franklin Pierce.* Lawrence: University Press of Kansas, 1991.

Gates, Paul W. *History of Public Land Law Development.* Report of the Public Land Law Review Commission. Washington, DC: Government Printing Office, 1968.

Gibbons, Boyd. "Aldo Leopold: A Durable Scale of Values." *National Geographic* 160, 5 (November 1981): 682–708.

Gibson, Arrell Morgan. "The Centennial Legacy of the General Allotment Act." *Chronicles of Oklahoma* 65, 3 (Fall 1987): 228–251.

Gibson, Paris. "The Repeal of Our Objectionable Land Laws." *Forestry and Irrigation* 9, 10 (October 1903): 484–489.

Gilbert, Grove Karl. *Lake Bonneville.* United States Geological Survey Monograph 1. Washington, DC, 1880.

Gillespie, Charles Coulston, ed. *Dictionary of Scientific Biography.* 16 vols. New York: Charles Scribner's Sons, 1970–1980.

Gilmore, William E. *Life of Edward Tiffin, First Governor of Ohio.* Chillicothe, OH: Horney & Son, 1897.

Glass, Mary Ellen. "The Newlands Reclamation Project: Years of Innocence, 1903–1907." *Journal of the West* 7, 1 (January 1968): 55–63.

Glover, James M. *A Wilderness Original: The Life of Bob Marshall.* Seattle: Mountaineers Press, 1986.

Glover, James M., and Regina B. Glover. "Robert Marshall: Portrait of a Liberal Forester." *Journal of Forest History* 30, 3 (July 1986): 112–119.

Goethem, Larry Van. "The National Wildlife Federation's Conservation Hall of Fame Honors Ernest Swift." *National Wildlife* 17, 3 (April-May 1979): 24–25.

Goetzmann, William H. *Army Exploration in the American West, 1803–1863.* New Haven, CT: Yale University Press, 1959.

Gottlieb, Alan M., ed. *The Wise Use Agenda: The Citizen's Policy Guide to Environmental Resource Issues: A Task Force Report to the Bush Administration by the Wise Use Movement.* Bellevue, WA: Free Enterprise, 1989.

Gould, Alan B. " 'Trouble Portfolio' to Constructive Conservation: Secretary of the Interior Walter L. Fisher, 1911–1913." *Forest History* 16, 4 (January 1973): 4–12.

Gould, Lewis L. *The Presidency of William McKinley.* Lawrence: University Press of Kansas, 1980.

Graham, Frank, Jr. "What Matters Most: The Many Worlds of Archie and Marjorie Carr." *Audubon* 84, 2 (March 1982): 90–105.

———. "Mardy Murie and Her Sunrise of Promise." *Audubon* 82, 3 (May 1980): 106–127.

Grantham, Dewey W., Jr. *Hoke Smith and the Politics of the New South.* Baton Rouge: Louisiana State University Press, 1958.

Graustein, Jeannette E. *Thomas Nuttall, Naturalist: Explorations in America, 1808–1841.* Cambridge, MA: Harvard University Press, 1967.

Graves, Gregory Randall. "Anti-Conservation and Federal Forestry in the Progressive Era." Ph.D. dissertation, University of California at Santa Barbara, 1987.

Graves, Henry S. *Report of the Forester for 1910.* Washington, DC: Government Printing Office, 1910.

———. *Report of the Forester for 1911.* Washington, DC: Government Printing Office, 1911.

Graves, Henry Solon. "Dr. Sargent's Contribution to Forestry in America." *American Forestry* 27, 335 (November 1921): 684–687.

———. "The Public Domain." *Nation* 131 (6 August 1930): 147–149.

Green, Joseph G. "Joseph Wood Krutch, Critic of the Drama." Ph.D. dissertation, Indiana University, 1964.

Green, Samuel B. *Principles of American Forestry.* New York: John Wiley & Sons, 1910.

Griggs, Robert Fiske. "Competition and Succession on a Rocky Mountain Fellfield." *Ecology* 37, 1 (January 1956): 1–17.

Grosvenor, Gilbert H. "John Wesley Powell." *National Geographic* 13, 11 (November 1902): 393–394.

Guise, John D. W. "Turner's Forgotten Frontier: The Old West." *Historian* 51, 4 (August 1990): 602–612.

Haapoja, Margaret. "Conservation Easements: Are They for You?" *American Forests* 100, 1 & 2 (January-February 1994): 29–31, 38.

Hadley, Edith Jane. "John Muir's Views of Nature and Their Consequences." Ph.D. dissertation, University of Wisconsin, 1956.

Hage, Wayne. *Storm over Rangelands: Private Rights in Federal Lands.* Bellevue, WA: Free Enterprise, 1989.

Hague, Arnold. *Geology of the Eureka District, Nevada, with an Atlas.* United States Geological Survey Monograph 20. Washington, DC, 1892.

Hales, Peter B. *William Henry Jackson and the Transformation of the American Landscape.* Philadelphia: Temple University Press, 1988.

Hansbrough, Henry Clay. "A National Irrigation Policy." *Forestry and Irrigation* 8, 3 (March 1902): 102–104.

Hansen, Paul. "The Loss of a Conservation Giant." *Outdoor America* 47, 2 (March-April 1982): 24–25, 33.

Harwell, Albert Brantley, Jr. "Writing the Wilderness: A Study of Henry Thoreau, John Muir, and Mary Austin." Ph.D. dissertation, University of Tennessee, 1992.

Harwood, Michael. "Mr. Audubon's Last Hurrah." *Audubon* 87, 6 (November 1985): 80–116.

Haskell, Daniel C., ed. *The United States Exploring Expedition, 1838–1842, and Its Publications, 1844–1874.* New York: Greenwood, 1968.

Hassrick, Royal B. *The George Catlin Book of American Indians.* New York: Watson-Guptill Publications, 1977.

Hayden, Frederick V. *First Annual Report of the United States Geological Survey of the Territories, Embracing Nebraska.* In *Report of the Secretary of the Interior,* House Executive Document 1, 40th Congress, 2d Session, 1867.

———. *Second Annual Report of the United States Geological Survey of the Territories, Embracing Wyoming.* In *Report of the Secretary of the Interior,* House Executive Document 1, 40th Congress, 3d Session, 1868.

Hays, Samuel P. *Beauty, Health and Permanence: Environmental Politics in the United States, 1955–1985.* Cambridge, MA: Cambridge University Press, 1989.

———. *Conservation and the Gospel of Efficiency: The Progressive Conservation Movement, 1890–1920.* Cambridge, MA: Harvard University Press, 1959.

Heinze, Andrew R. "The Morality of Reservation: Western Lands in the Cleveland Period, 1885–1897." *Journal of the West* 31, 3 (July 1982): 81–89.

Hennessy, W. B., comp. *History of North Dakota, Embracing a Relation of the History of the State from the Earliest Times Down to the Present Day, Including the Biographies of the Builders of the Commonwealth.* Bismarck, ND: Bismarck Tribune, 1910.

Hession, Jack M. "The Legislative History of the Wilderness Act." Master's thesis, San Diego State University, 1967.

Historical Correspondence on Establishment of Forest Products Laboratory at Madison, Wisconsin. Madison, WI: Forest Products Laboratory, 1960.

The History of Engineering in the Forest Service. U.S. Forest Service publication EM-7100-13. Washington, DC: U.S. Department of Agriculture, October 1990.

"Hon. Redfield Proctor, United States Senator from Vermont, and Vice-President of the American Forestry Association for Vermont." *Forestry and Irrigation* 8, 3 (March 1902): 100–101.

Hoover, Roy O. "Public Law 273 Comes to Shelton: Implementing the Sustained-Yield Forest Management Act of 1944." *Journal of Forest History* 22, 2 (April 1978): 86–101.

Horan, James D. *Timothy O'Sullivan: America's Forgotten Photographer.* Garden City, NY: Doubleday, 1966.

Hornaday, William Temple. "Eighty Fascinating Years: An Autobiography." Unpublished manuscript, William Temple Hornaday Papers, Library of Congress.

———. "My Fifty-Four Years with Animal Life." *Mentor* 17, 4 (May 1929): 3–10.

Horton, Tom. "The Endangered Species Act: Too Tough, Too Weak, or Too Late?" *Audubon* 94, 2 (March-April 1992): 68–74.

Hoy, Mark. "The Most Famous Farm in America: Louis Bromfield's Malabar Farm." *Audubon* 91, 6 (November 1989): 64–67.

Hudanick, Andrew, Jr. "George Hebard Maxwell: Reclamation's Militant Evangelist." *Journal of the West* 14, 3 (July 1975): 108–121.

Hughes, Claire D. "The Bald Eagle: A Story in Recovery." *SKY Magazine* 22, 1 (January 1993): 78–85.

Hummel, Don. *Stealing the National Parks: The Destruction of Concessions and Public Access.* Bellevue, WA: Free Enterprise, 1987.

Hunter, Robert. *Warriors of the Rainbow: A Chronicle of the Greenpeace Movement.* New York: Holt, Rinehart and Winston, 1979.

Huth, Hans. "Sequoias in Germany." *Journal of Forest History* 20, 3 (July 1976): 143–148.

Hutson, James H. *To Make All Laws: The Congress of the United States, 1789–1989.* Washington, DC: Library of Congress, 1989.

Ickes, Harold L. "Farewell, Secretary Krug." *New Republic* 121, 22 (23 November 1949): 17.

"The Irrigation Bill." *Forestry and Irrigation* 8, 6 (June 1902): 231–234.

Ise, John. *Our National Park Policy: A Critical History.* Baltimore: Johns Hopkins University Press, 1961.

Jackson, Clarence S. *Picture Maker of the Old West: William H. Jackson.* New York: Charles Scribner's Sons, 1947.

Jacobsen, Edna L. "Franklin B. Hough, A Pioneer in American Scientific Forestry in America." *New York History* 15 (1934): 311–325.

Johnson, Herb. "The Sigurd Olson Story." *BWCA* [Boundary Waters Canoe Area] *Wilderness News,* Autumn 1980, 5–6.

Johnson, Robert Underwood. *Remembered Yesterdays.* Boston: Little, Brown, 1923.

"The Joint Conservation Conference." *Conservation* 15, 1 (January 1909): 3–47; 15, 2 (February 1909): 92–97.

Jones, Holway R. *John Muir and the Sierra Club: The Battle for Yosemite.* San Francisco: Sierra Club, 1965.

Kawashima, Yasuhide, and Ruth Tone. "Environmental Policy in Early America: A Survey of Colonial Statutes." *Journal of Forest History* 27, 4 (October 1983): 168–179.

Keeva, Steve. "After the Spill: New Issues in Environmental Law." *American Bar Association Journal* 77, 2 (February 1991): 66–69.

Kehr, Kurt. "Walden Three: Ecological Changes in the Landscape of Henry David Thoreau." *Journal of Forest History* 27, 1 (January 1983): 28–33.

Kerwyn, Jerome G. "Federal Water Power Legislation." Ph.D. dissertation, Columbia University, 1906.

King, Clarence. Letter from the Director of the United States Geological Survey transmitting, in response to a letter of the Hon. Henry G. Davis, answers to certain inquiries propounded in the said letter. Senate Miscellaneous Document 48, 46th Congress, 2d Session, 1880.

King, Judson. *The Legislative History of Muscle Shoals*. Knoxville, TN: Tennessee Valley Authority, 1936.

Kinney, Jay P. "The Administration of Indian Forests." *Journal of Forestry* 28, 8 (December 1930): 1041–1052.

———. *A Continent Lost—A Civilization Won: Indian Land Tenure in America*. Baltimore: Johns Hopkins University Press, 1937.

Koppes, Clayton R. "Oscar L. Chapman: A Liberal at the Interior Department, 1933–1953." Ph.D. dissertation, University of Kansas, 1974.

Kusler, Jon A., and Mary E. Kentula, eds. *Wetland Creation and Restoration: The Status of the Science*. Washington, DC: Island, 1990.

Lamm, Richard D., and Duane A. Smith. *Pioneers and Politicians: 10 Colorado Governors in Profile*. Boulder, CO: Pruett, 1984.

Lamm, Richard D., and Michael McCarthy. *The Angry West: A Vulnerable Land and Its Future*. Boston: Houghton Mifflin, 1982.

Laycock, George. "Promise of Tortuguero." *Audubon* 84, 2 (March 1982): 28–31.

Le Unes, Barbara. "The Conservation Philosophy of Stewart L. Udall, 1961–1968." Ph.D. dissertation, Texas A & M University, 1977.

Lee, Lawrence B. "William Ellsworth Smythe and the Irrigation Movement: A Reconsideration." *Pacific Historical Review* 41, 3 (August 1972): 289–311.

Lewis, Elmer A. *Laws Relating to Forestry Game Conservation, Flood Control and Related Subjects*. Washington, DC: Government Printing Office, 1938.

Lewis, Jack. "The Birth of EPA." *EPA Journal* 12, 9 (November 1985): 6–11.

———. "John Muir: Environmental Pioneer." *EPA Journal* 12, 9 (November 1985): 40–44.

Lichtenstein, Nelson, ed. *Political Profiles: The Kennedy Years*. New York: Facts on File, 1976.

Lindt, David L. *Ding: The Life of Jay Norwood Darling*. Ames: Iowa State University Press, 1979.

Lockmann, Ronald F. "Forests and Watershed in the Environmental Philosophy of Theodore P. Lukens." *Journal of Forestry* 23, 2 (April 1979): 82–91.

Lounsberry, Colonel Clement A. *North Dakota: History and People*. Chicago: S. J. Clarke, 1917.

Lowenthal, David. *George Perkins Marsh: Versatile Vermonter*. New York: Columbia University Press, 1958.

McCarthy, G. Michael. "Colorado's Populist Party and the Progressive Movement: (1) Who Were the Populists? and (2) Selective Progressivism and the Conservation Movement." *Journal of the West* 15, 1 (January 1976): 62–75.

———. "The Forest Reserve Controversy: Colorado under Cleveland and McKinley." *Journal of Forest History* 20, 2 (April 1976): 80–90.

———. *Hour of Trial: The Conservation Conflict in Colorado and the West, 1891–1907*. Norman: University of Oklahoma Press, 1977.

MacCleery, Douglas W. *American Forests: A History of Resiliency and Recovery.* United States Forest Service Pamphlet FS-540. Washington, DC: U.S. Department of Agriculture, 1992.

McClurg, Gilbert, ed. *The Official Proceedings of the Eleventh National Irrigation Congress Held at Ogden, Utah, September 15–18, 1903.* Ogden: Proceedings, 1904.

MacColl, E. K. "John Franklin Shafroth, Reform Governor of Colorado, 1909–1913." *Colorado Magazine* 29, 1 (January 1952): 37–51.

McCracken, Harold. *George Catlin and the Old Frontier.* New York: Bonanza Books, 1989. Reprinted.

McDowell, Bart. "C. M. Russell: Cowboy Artist." *National Geographic* 169, 1 (January 1986): 60–95.

McFeely, William S. *Grant: A Biography.* New York: W. W. Norton, 1982.

McGee, William John. "Professor O. C. Marsh." *National Geographic* 10, 6 (June 1899): 181–182.

McGuane, Thomas. "The Spell of Wild Rivers." *Audubon* 95, 6 (November-December 1993): 60–75.

McHenry, Robert, with Charles Van Doren, eds. *A Documentary History of Conservation in America.* New York: Praeger, 1972.

Mackintosh, Barry. "Harold L. Ickes and the National Park Service." *Journal of Forest History* 29, 2 (April 1985): 78–84.

———. *The National Parks: Shaping the System.* Washington, DC: National Park Service, U.S. Department of the Interior, 1985.

McLaughlin, Charles Capen. "Olmsted's Odyssey." *Wilson Quarterly* 6, 3 (Summer 1982): 78–87.

McPhee, John. "Profiles: Encounters with the Archdruid." *New Yorker* 47, 5 (20 March 1971): 42–91.

Madson, John. "Grandfather Country." *Audubon* 84, 3 (May 1982): 40–57.

"Major John Wesley Powell." *Forestry and Irrigation* 8, 2 (February 1902): 59.

Mandell, Daniel R. "Compelling a Public Timberlands Policy: United States v. Briggs, 1850." *Journal of Forest History* 26, 3 (July 1982): 140–147.

Margolis, John D. *Joseph Wood Krutch: A Writer's Life.* Knoxville: University of Tennessee Press, 1980.

Marsh, George Perkins. *Irrigation: Its Evils, the Remedies, and Compensations.* Senate Miscellaneous Document 55, 43d Congress, 1st Session, 1874.

———. *Man and Nature; or, Physical Geography as Modified by Human Action.* New York: Charles Scribner's Sons, 1864.

Marsh, Othniel Charles. *Dinocereta: A Monograph of an Extinct Order of Gigantic Mammals.* United States Geological Survey Monograph 10. Washington, DC, 1886.

———. Letter Transmitting the Report on the Scientific Surveys of the Territories, Made by the National Academy of the Sciences. Senate Miscellaneous Document 9, 45th Congress, 3d Session, 1878.

Marshall, Robert. "The Problem of the Wilderness." *Scientific Monthly* 30, 2 (February 1930): 141–148.

———. "The Universe of the Wilderness Is Vanishing." *Living Wilderness* 35, 114 (Summer 1971): 8–14.

Mason, David T., and Elwood Maunder. "Memoirs of a Forester." *Journal of Forest History,* Part 1, 10, 4 (January 1967): 6–12; Part 2, 13, 1&2 (April-July 1969): 28–39.

Maxwell, George H. "Forestry and Irrigation Save the Forests and Store the Floods: Reserve the Public Lands for the Home-Builders." *Forestry and Irrigation* 8, 1 (January 1902): 13–17.

———. *Our National Defense: The Patriotism of Peace.* Washington, DC: Rural Settlements Association, 1915.

———. "Reclamation of the Arid Region." *Forestry and Irrigation* 8, 11 (November 1902): 444–447.

"Meet Secretary Seaton." *American Forests* 62, 7 (July 1956): 8, 63.

Melosi, Martin V., ed. *Garbage in the Cities: Reform and the Environment, 1880–1980.*

College Station: Texas A & M University Press, 1981.

Merritt, J. I. "Turning Point: John Muir in the Sierra, 1871." *American West* 41, 4 (July-August 1979): 4–15.

Miles, John. "Charting the Course: A 75th Anniversary Retrospective." *National Parks* 67, 1 (November-December 1993): 1–12, 38–41.

Miller, Joseph A. "Congress and the Origins of Conservation: Natural Resource Policies, 1865–1900." Ph.D. dissertation, University of Minnesota at Minneapolis, 1973.

Milner, James A. *Report on the Fisheries of the Great Lakes: The Result of Inquiries Prosecuted in 1871 and 1872*. In *Report of the U.S. Commission of Fish and Fisheries*. Senate Miscellaneous Document 74, 42d Congress, 3d Session, 1873.

Monahan, Robert S. "John Wentworth: Colonial Forester." *American Forests* 74, 10 (October 1968): 33–34, 78–79.

Monnett, John H., and Michael McCarthy. *Colorado Profiles: Men and Women Who Shaped the Centennial State*. Evergreen, CO: Cordillera, 1987.

Morgan, Arthur E. *The Making of the TVA*. Buffalo, NY: Prometheus Books, 1974.

Morris, Edmund. *The Rise of Theodore Roosevelt*. New York: Coward, McCann & Geoghegan, 1979.

Morris, Lindsey Gardner. "John Wesley Powell." Master's thesis, Illinois State University, 1947.

Muhn, James, and Hanson R. Stuart. *Opportunity and Challenge: The Story of BLM*. Washington, DC: Bureau of Land Management, 1988.

Muir, John. "The Hetch-Hetchy Valley: A National Question." *American Forestry* 16, 5 (May 1910): 263–269.

———. *The Mountains of California*. New York: Barnes & Noble, 1993. Reprinted.

Murray, Raymond L. *Understanding Radioactive Waste*. Columbus, OH: Battelle, 1989.

Myers, John L., ed. *The Arizona Governors, 1912–1990*. Phoenix, AZ: Heritage, 1989.

The National Cyclopaedia of American Biography. New York: James T. White, 1898–1984.

National Water Commission. *Water Policies for the Future: The Final Report to the President and the Congress of the United States by the National Water Commission*. Port Washington, NY: Water Information Center, 1973.

"The National Wildlife Federation's Conservation Hall of Fame Honors Anna Botsford Comstock." *National Wildlife* 26, 6 (October-November 1988): 50–51.

"The National Wildlife Federation's Conservation Hall of Fame Honors George Perkins Marsh." *National Wildlife* 18, 5 (August-September 1980): 28–29.

Neal, Steve. *McNary of Oregon: A Political Biography*. Portland, OR: Western Imprints, 1985.

Nelson, Gaylord. "Earth Day '70: What It Meant." *EPA Journal* 6, 4 (April 1980): 6–7, 38.

Neuberger, Richard L. "McNary of Fir Cone." *Life* 9, 7 (12 August 1940): 76–84.

Newell, Alan; Richmond L. Clow; and Richard N. Ellis. *A Forest in Trust: Three-Quarters of a Century of Indian Forestry, 1910–1986*. Washington, DC: Bureau of Indian Affairs, Division of Forestry, 1986.

Newell, Frederick H. "The National Forest Reserves." *National Geographic* 8, 6 (June 1897): 177–186.

Newhall, Beaumont, and Diana E. Edkins. *William H. Jackson*. Dobbs Ferry, NY: Morgan & Morgan, 1974.

Newlands, Francis G. "National Irrigation Works." *Forestry and Irrigation* 8, 2 (February 1902): 63–66.

Nichols, Roy Franklin. *Franklin Pierce, Young Hickory of the Granite Hills*. Philadelphia: University of Pennsylvania Press, 1931.

Nimlos, Thomas J. "The 1985 Food Security Act: A Conservation Boon." *Western Wildlands* 14 (Summer 1988): 2–5.

Nobbe, George. "Interview with Interior Secretary Babbitt." *Wildlife Conservation* 96, 6 (November-December 1993): 76–77, 80–81.

Noggle, Burl. "The Origins of the Teapot Dome Investigation." *Mississippi Valley Historical Review* 44, 3 (September 1957): 237–266.

The Nuclear Waste Primer: A Handbook for Citizens. New York: Nick Lyons Books, 1985.

Nuttall, Thomas. *A Journal of Travels into the Arkansa Territory during the Year 1819.* Norman: University of Oklahoma Press, 1980. Reprinted.

O'Callahan, Kate. "(The Wise Use Movement:) Whose Agenda for America?" *Audubon* 94, 5 (September–October 1992): 80–91.

Office of Technology Assessment. *Wetlands: Their Use and Regulation.* Assessment Report OTA-O-206. Washington, DC: Office of Technology Assessment, 1984.

The Official World Wildlife Fund Guide to Endangered Species of North America. 3 vols. Washington, DC: Beacham, 1990–1992.

Ogden, Kate Nearpass. "Yosemite Valley as Image and Symbol: Paintings and Photographs from 1855 to 1889." Ph.D. dissertation, Columbia University, 1992.

Olson, James C. "Arbor Day: A Pioneer Expression of Concern for the Environment." *Nebraska History* 53, 1 (Spring 1972): 1–13.

Olson, James S. *Historical Dictionary of the 1920s: From World War I to the New Deal, 1919–1933.* Westport, CT: Greenwood, 1988.

Olson, Sigurd. "Battle for a Wilderness." *Living Wilderness* 32, 104 (Winter 1968–1969): 4–13.

Olson, Sigurd F. *Listening Point.* New York: Knopf, 1958.

Otis, Delos Sackett. *The Dawes Act and the Allotment of Indian Lands.* Norman: University of Oklahoma Press, 1973.

Palmer, Tim. *Endangered Rivers and the Conservation Movement.* Berkeley: University of California Press, 1986.

Parenteau, Patrick A. "NEPA at Twenty: Great Disappointment or Whopping Success?" *Audubon* 92, 2 (March 1990): 104–107.

Parker, Theodore A., III, and Brent Bailey. *Rapid Assessment Program: A Biological Assessment of the Alto Madidi Region and Adjacent Areas of Northwest Bolivia, May 18–June 15, 1990.* Rapid Assessment Program Working Papers 1. Washington, DC: Conservation International, 1991.

Paulson, George. "The Congressional Career of Joseph M. Carey." Master's thesis, University of Wyoming at Laramie, 1962.

Pavich, Paul. "Joseph Wood Krutch: Western Nature Essayist." Seminar paper, Colorado State University, 1968.

Pease, Theodore Calvin, and James G. Randall. *The Diary of Orville Hickman Browning, 1850–1881. Collections of the Illinois State Historical Library*, vols. 20, 22, 1925–1933.

Peek, George N. "The McNary-Haugen Plan for Relief." *Current History* 38, 11 (November 1928): 273–278.

Penick, James L., Jr. *Progressive Politics and Conservation: The Ballinger-Pinchot Affair.* Chicago: University of Chicago Press, 1968.

Perry, James M. "The Complexity of John D. Dingell." *Audubon* 84, 4 (July 1982): 100–105.

Perry, James M., and Richard Frank. "This Fella from Arizona." *Audubon* 83, 6 (November 1981): 64–73.

Peters, Betsey Ross. "Joseph M. Carey and the Progressive Movement in Wyoming." Ph.D. dissertation, University of Wyoming at Laramie, 1971.

Peters, J. Girvin. *Forest Fire Protection under the Weeks Law in Cooperation with States.* Washington, DC: Government Printing Office, 1913.

Petry, Wendy Helen. "The Euclid Incinerator, 1955–1988: A History of Solid Waste Disposal, The City of Euclid, Ohio." Master's thesis, Case Western Reserve University, 1989.

Pettigrew, Richard F. *Triumphant Plutocracy: The Story of American Public Life from 1870 to 1920.* New York: Academy, 1922.

Pinchot, Gifford. *Breaking New Ground.* Washington, DC: Island, 1987. Reprinted.

———. *Report of the Forester for 1909.* Washington, DC: Government Printing Office, 1909.

Pinkett, Harold T. "Gifford Pinchot and the Early Conservation Movement in the United States." Ph.D. dissertation, American Univerity, 1953.

Pisani, Donald J. "Forests and Conservation, 1865–1890." *Journal of American History* 72, 2 (September 1985): 340–359.

———. "Forests and Reclamation, 1891–1911." *Forest and Conservation History* 37, 2 (April 1993): 68–79.

———. "The Origins of Reclamation in the Arid West: William Ralston's Canal and the Federal Irrigation Commission of 1873." *Journal of the West* 22, 2 (April 1983): 9–19.

———. "Reclamation and Social Engineering in the Progressive Era." *Agricultural History* 57, 1 (January 1993): 46–63.

"A Place at the Table: A Sierra Club Roundtable on Race, Justice, and the Environment." *Sierra* 78, 3 (May-June 1993): 51–58, 90–91.

Ponder, Stephen Edward. "New Management in the Progressive Era, 1898–1909: Gifford Pinchot, Theodore Roosevelt, and the Conservation Crusade." Ph.D. dissertation, University of Washington, 1985.

Powell, John Wesley. *Report on the Lands of the Arid Region of the United States, with a More Detailed Account of the Lands of Utah, with Maps, by J. W. Powell.* House Executive Document 73, 45th Congress, 2d Session, 1878.

Preble, Edward A. "William Temple Hornaday: An Appreciation." *Nature Magazine* 20, 5 (May 1937): 303–304.

"The President's Quetico-Superior Committee." *Living Wilderness* 34, 112 (Winter 1970-1971): 54–57.

Pringle, Henry F. *Theodore Roosevelt: A Biography.* New York: Harcourt, Brace & World, 1956.

Proceedings of the Fourth National Irrigation Congress Held at Albuquerque, New Mexico, September 16–19, 1895. Santa Fe: New Mexican, 1896.

Proceedings of the Twelfth Annual Session of the National Irrigation Congress, El Paso, Texas, November 1904. El Paso, TX: Clark & Courts, 1904.

Progressive Men of Western Colorado. Chicago: Bowen, 1905.

Pumpelly, Raphael. *Across America and Asia: Notes of a Five Year Journey around the World and of Residence in Arizona, Japan, and China.* New York: Leypoldt & Holt, 1870.

———. *My Reminiscences.* 2 vols. New York: Holt, 1918.

Pumpelly, Raphael; J. E. Wolff; and T. Nelson Dale. *Geology of the Green Mountains in Massachusetts.* United States Geological Survey Monograph 23. Washington, DC, 1894.

Rabbitt, John C., and Mary C. Rabbitt. "The U.S. Geological Survey: 75 Years of Service to the Nation, 1879–1954." *Science* 119, 3100 (28 May 1954): 741–758.

Rabbitt, Mary C. *Minerals, Lands, and Geology for the Common Defence and Welfare: A History of the United States Geological Survey.* 3 vols. Washington, DC: Government Printing Office, 1979–1986.

Rabbitt, Mary C.; McKee, Edwin D.; and Leopold, Luna B. *The Colorado River Region and John Wesley Powell: A Collection of Papers Honoring Powell on the 100th Anniversary of His Exploration of the Colorado River.* United States Geological Survey Professional Paper 669. Washington, DC, 1969.

Rae, John B.. "Commissioner Sparks and the Railroad Land Grants." *Mississippi Valley Historical Review* 25, 2 (September 1938): 211–230.

Raven, Peter H. "Defining Biodiversity." *Nature Conservancy* 44, 1 (January-February 1974): 16–21.

Reiger, John F. *American Sportsmen and the Origins of Conservation.* New York: Winchester, 1975.

———. "George Bird Grinnell and the Development of American Conservation, 1870–1901." Ph.D dissertation, Northwestern University, 1970.

Reisner, Marc. *Cadillac Desert: The American West and Its Disappearing Water.* New York: Viking, 1986.

Reisner, Marc, and Sarah Bates. *Overtapped Oasis: Reform or Revolution for Western Water.* Washington, DC: Island, 1990.

Renehan, Edward J., Jr. *John Burroughs: An American Naturalist.* Post Mills, VT: Chelsea Green, 1992.

Report of the Mississippi Valley Committee of the Public Works Administration. Washington, DC: Government Printing Office, 1934.

Report to the Governor and General Assembly on the Erection of the Joseph Trimble Rothrock Memorial Authorized by an Act of Assembly No. 51-A, July 11, 1923 [Pennsylvania].

Resource Conservation and Recovery Act Orientation Manual. Washington, DC: Environmental Protection Agency, 1986.

Richardson, Elmo. " 'The Compleat Forester': David T. Mason's Early Career and Character." *Journal of Forest History* 27, 3 (July 1983): 112–121.

———. "The Interior Secretary as Conservation Villain: The Notorious Case of Douglas 'Giveaway' McKay." *Pacific Historical Review* 41, 3 (August 1972): 333–345.

———. *The Politics of Conservation: Crusades and Controversies, 1897–1913.* Berkeley: University of California Press, 1962.

Richardson, Robert D., Jr. *Henry Thoreau: A Life of the Mind.* Berkeley: University of California Press, 1986.

Richmond, Charles W. "In Memoriam: Edgar Alexander Mearns." *Auk* 35, 1 (January 1918): 1–18.

Robbins, William G. "Federal Forest Protection: The Fernow-Pinchot Years." *Journal of Forest History* 28, 4 (October 1984): 164–173.

Roberts, Walter Keith. "The Political Career of Charles Linza McNary, 1924–1944." Ph.D. dissertation, University of North Carolina at Chapel Hill, 1953.

Robertson, Alexander F. *Alexander Hugh Holmes Stuart, 1807–1891: A Biography.* Richmond, VA: William Byrd, 1925.

Robinson, Michael C. *Water for the West: The Bureau of Reclamation, 1902–1977.* Chicago: Public Works Historical Society, 1979.

Rodgers, Andrew Denny, III. *Bernhard Eduard Fernow: A Story of North American Forestry.* Princeton, NJ: Princeton University Press, 1951.

Rohrer, Daniel M., et al. *The Environmental Crisis: A Basic Overview of the Problem of Pollution.* Skokie, IL: National Textbook, 1970.

Roller, David C., and Robert W. Twyman. *The Encyclopedia of Southern History.* Baton Rouge: Louisiana State University Press, 1979.

Roper, Laura Wood. *FLO: A Biography of Frederick Law Olmsted.* Baltimore: Johns Hopkins University Press, 1973.

Roth, Dennis. "The National Forests and the Campaign for Wilderness Legislation." *Journal of Forest History* 28, 3 (July 1984): 112–125.

Roth, Dennis M. *The Wilderness Movement and the National Forests: 1980–1984.* United States Forest Service pamphlet FS-410. Washington, DC: U.S. Department of Agriculture, 1988.

Roth, Filibert. "Administration of U.S. Forest Reserves." *Forestry and Irrigation* 8, 5 (May 1902): 191–193.

Russell, Francis. *The Shadow of Blooming Grove: Warren G. Harding in His Times.* New York: McGraw-Hill, 1968.

Russell, Milton; E. William Colglazier; and Bruce E. Tonn. "The U.S. Hazardous Waste Legacy." *Environment* 34, 6 (July-August 1990): 12–15, 34–39.

Ruston, Guy J. "Espionage: Commissioner Thompson's Bizarre War Was Nearly Laughable—Up to a Recently Revealed Point." *Military History* 10, 6 (February 1994): 8, 79–82.

Salmond, John A. "The Civilian Conservation Corps and the Negro." *Journal of American History* 52, 1 (June 1965): 75–88.

Sample, V. Alaric. *Land Stewardship in the Next Era of Conservation.* Milford, PA: Pinchot Institute for Conservation, 1991.

Sanborn, John B. "Some Political Aspects of Homestead Legislation." *American Historical Review* 6, 10 (October 1900): 19–37.

Sargent, Charles Sprague. *Report on the Forests of North America.* Washington, DC: Government Printing Office, 1884.

Schapsmeier, Edward L., and Frederick H. Schapsmeier. *Encyclopedia of American Agricultural History.* Westport, CT: Greenwood, 1975.

Schlup, Leonard C. "Henry C. Hansbrough and the Fight against the Tariff in 1894." *North Dakota History* 45, 1 (Winter 1978): 4–9.

———. "Political Maverick: Senator Hansbrough and Republican Party Politics, 1907–1912." *North Dakota History* 45, 4 (Fall 1978): 32–39.

———. "Quiet Imperialist: Henry C. Hansbrough and the Question of Expansion." *North Dakota History* 45, 2 (Spring 1978): 26–31.

Schmaltz, Norman J. "Forest Researcher: Raphael Zon." *Journal of Forest History* 24, 1 (January 1980): 25–39.

Schoenebaum, Eleanora W., ed. *Political Profiles: The Eisenhower Years.* New York: Facts on File, 1977.

———. *Political Profiles: The Truman Years.* New York: Facts on File, 1978.

Schofield, Edmund A., and Robert C. Baron, eds. *Thoreau's World and Ours: A Natural Legacy.* Golden, CO: North American, 1992.

Schuchert, Charles, and Clara M. LeVene. *O. C. Marsh, Pioneer in Paleontology.* New Haven, CT: Yale University Press, 1940.

Shepherd, Jack. "A New Environment at Interior: Cecil Andrus Is Trying to Turn Things Around." *New York Times Magazine*, May 8, 1977, 36, 38–46.

Sherow, James E. "Marketplace Agricultural Reform: T. C. Henry and the Irrigation Crusade in Colorado, 1883–1914." *Journal of the West* 31, 4 (October 1992): 51–58.

Shipley, Donald D. "A Study of the Conservation Philosophies and Contributions of Some Important American Conservation Leaders." Ph.D. dissertation, Cornell University, 1953.

"The Sierra Club Bulletin: 100 Years of Activism and Adventure." *Sierra* 78, 5 (September-October 1993): 54–86.

Sifakis, Stewart. *Who Was Who in the Civil War.* New York: Facts on File, 1988.

Simon, James F. *Independent Journey: The Life of William O. Douglas.* New York: Harper & Row, 1980.

Sitgreaves, Captain Lorenzo. *Report of an Expedition down the Zuni and Colorado Rivers.* Senate Executive Document 59, 32d Congress, 2d Session, 1853.

"Sketch of Thomas Nuttall." *Popular Science Monthly* 46, 5 (March 1895): 689–696.

Smith, Frank E., ed. *Conservation in the United States: A Documentary History.* 5 vols. New York: Chelsea House, 1971.

Smith, Henry Nash. "Clarence King, John Wesley Powell, and the Establishment of the United States Geological Survey." *Mississippi Valley Historical Review* 34, 1 (June 1947): 37–58.

Smith, Page. *America Enters the World: A People's History of the Progressive Era and World War I.* New York: McGraw-Hill, 1985.

Smith, Robert Earl. "Colorado's Progressive Senators and Representatives." *Colorado Magazine* 45, 1 (Winter 1968): 27–41.

Sobel, Dava. "Marjory Stoneman Douglas: Still Fighting the Good Fight for the Everglades." *Audubon* 93, 4 (July-August 1991): 31–39.

Sobel, Robert, and John Raimo, eds. *Biographical Dictionary of the Governors of the United States, 1789–1978.* Westport, CT: Meckler, 1978.

Socolofsky, Homer E., and Allan B. Spetter. *The Presidency of Benjamin Harrison.* Lawrence: University Press of Kansas, 1987.

Soffer, Allan J. "Differing Views on the Gospel of Efficiency: Conservation Controversies between Agriculture and Interior, 1898–1938." Ph.D. dissertation, Texas Tech University, 1974.

Speech of Hon. Duncan U. Fletcher of Florida in the Senate of the United States, Thursday, January 19, 1911: Investigation of the Department of the Interior and of the Bureau of Forestry. Copy in the Records of the United States Senate, RG 46, National Archives, Washington, DC.

Stampp, Kenneth M. *America in 1857: A Nation on the Brink.* New York: Oxford University Press, 1990.

Stapleton, Richard M. "On the Western Front: Dispatches from the War with the Wise Use Movement." *National Parks* 67, 1 & 2 (January-February 1993): 32–36.

Starr, Frederick, Jr. *American Forests: Their Destruction and Preservation.* In *Report of the Commissioner of Agriculture for the Year 1865.* Washington, DC: Government Printing Office, 1866.

Start, Edwin A. "The New Forest Products Laboratory." *American Forestry* 16, 7 (July 1910): 387–403.

Steen, Harold K. *The Beginning of the National Forest System.* United States Forest Service pamphlet FS-488. Washington, DC: U.S. Department of Agriculture, 1991.

Stegner, Wallace. *Beyond the Hundredth Meridian: John Wesley Powell and the Second Opening of the West.* New York: Houghton Mifflin, 1953.

———. "Clarence Edward Dutton." Ph.D. dissertation, University of Iowa, 1935.

———. "John Sumner and John Wesley Powell." *Colorado Magazine* 26, 1 (January 1949): 61–69.

Stephen, Sir Leslie, and Sir Sidney Lee, eds. *The Dictionary of National Biography.* New York: Oxford University Press, 1917–1973. Reprinted.

Stephens, Mark. *Three Mile Island: The Hour-by-Hour Account of What Really Happened.* New York: Random House, 1980.

Sterling, Everett W. "The Powell Irrigation Survey, 1888–1893." *Mississippi Valley Historical Review* 27, 3 (December 1940): 421–434.

Stine, Jeffrey K. "Regulating Wetlands in 1970s: U.S. Army Corps of Engineers and the Environmental Organizations." *Journal of Forest History* 27, 2 (April 1983): 60–75.

Strong, Douglas H. *Dreamers and Defenders: American Conservationists.* Lincoln: University of Nebraska Press, 1988.

———. "The Sierra Club: A History. Part 1: Origins and Outings." *Sierra* 62, 8 (October 1977): 10–14.

Stroud, Richard, ed. *National Leaders of American Conservation.* Washington, DC: Smithsonian Institution, 1985.

Stubbs, Walter. *Congressional Committees, 1789–1982.* Westport, CT: Greenwood, 1985.

Stuckey, Ronald L. "Biography of Thomas Nuttall: A Review with Bibliography." *Rhodora* 70, 783 (1968): 429–438.

Swain, Robert E. "Ray Lyman Wilbur: 1875–1949." Science 111, 2883 (31 March 1950): 324–327.

Swerdlow, Joel L. "Central Park." *National Geographic* 183, 5 (May 1993): 2–37.

The Taming of the Salt: A Collection of Biographies of Pioneers Who Contributed Significantly to Water Development in the Salt River Valley. Phoenix, AZ: Salt River Project Communications and Public Affairs Department, 1979.

Tarr, Joel A. *Retrospective Assessment of Wastewater Technology in the United States: 1800–1972.* Pittsburgh: Carnegie-Mellon University Press, 1977.

Taylor, W. P. "Notes on Mammals Collected Principally in Washington and California between the years 1853 and 1874 by Dr. Jas. Graham Cooper." *Proceedings of the California Academy of Sciences* 9, 2 (12 July 1919): 69–121.

"The Tenth National Irrigation Congress." *Forestry and Irrigation* 8, 10 (October 1902): 400–404.

Thoreau, Henry David. *Thoreau in the Mountains: Writings by Henry David Thoreau, Commentary by William Howarth.* New York: Farrar Straus Giroux, 1982.

Tikalsky, Frank D. "Historical Controversy, Science, and John Wesley Powell." *Journal of Arizona History* 23, 4 (Winter 1982): 407–422.

Toliver, John. "What Are Wetlands? A Historical Overview." *Journal of Forestry* 91, 5 (May 1993): 12–14.

"Topic of the Day: The Third Knock-Out for McNary-Haugenism." *Literary Digest* 90, 2 (10 July 1926): 5–7.

Trani, Eugene P. "The Secretaries of the Department of the Interior, 1849–1969." Manuscript in the holdings of the Department of the Interior, 1975.

Trefethen, James B. *An American Crusade for Wildlife.* Alexandria, VA: Boone and Crockett Club, 1975.

Trefousse, Hans L. *Andrew Johnson: A Biography.* New York: W. W. Norton, 1989.

———. *Historical Dictionary of Reconstruction.* Westport, CT: Greenwood, 1991.

Turbak, Gary. "The National Wildlife Federation's Conservation Hall of Fame Honors Bob Marshall." *National Wildlife* 24, 2 (February-March 1986): 18–19.

Turnage, Robert. "Ansel Adams: The Role of the Artist in the Environmental Movement." Senior thesis in Environmental Studies, University of California at Santa Cruz, 1979.

Turner, Tom. *The Sierra Club: 100 Years of Protecting Nature.* New York: Abrams, 1991.

"Twentieth Century History of Vermont— Vermont Literature." *Vermonter* 9, 3 (October 1903): 69–83.

Udall, David King. *Arizona Pioneer Mormon: David King Udall (1851–1938), His Story and His Family.* Tucson: Arizona Silhouettes, 1959.

United States. Congress. House. *The Agricultural Adjustment Act of 1937: Representative Jones' Report.* House Report 1645, 75th Congress, 2d Session, 1937.

———. *Annual Report of the Commissioner of the General Land Office for the Fiscal Year 1887.* In *Report of the Secretary of the Interior,* House Executive Document 1 (part 5), 50th Congress, 1st Session, 1887.

———. *Annual Report of the Secretary of the Interior for 1902.* House Document 5, 57th Congress, 2d Session, 1903.

———. *A Bill for the Preservation of Forests of the Public Domain Adjacent to Navigable Rivers, introduced by Rep. Greenbury L. Fort of Illinois.* H.R. 2075, 44th Congress, 1st Session, 1876.

———. *Conference Report for the Department of Interior and Related Agencies Appropriations, 1994.* House Report 103-299, 103d Congress, 1st Session, 1993.

———. *Department of the Interior and Related Agencies Appropriation Bill, 1994.* House Report 103–158, 103d Congress, 1st Session, 1993.

———. *Disposal of Public Lands.* House Report 778, 50th Congress, 1st Session, 1888.

———. *Geographical and Geological Surveys.* House Report 612, 43d Congress, 1st Session, 1874.

———. *Geographical and Geological Surveys.* House Executive Document 81, 45th Congress, 2d Session, 1878.

———. *Letter from the Acting President of the National Academy of Sciences Transmitting a Report on the Surveys of the Territories.* House Miscellaneous Document 8, 45th Congress, 3d Session, 1878.

———. *Letter from the Secretary of the Interior Submitting a Report on Land Withdrawals from Settlement, Location, Sale, or Entry, Under the Provisions of Act of Congress Approved June 25, 1910 (36 Stat. 847).* House Document 1465, 63d Congress, 3d Session, 1914.

———. *The McNary-Haugen Bill: Representative Gilbert Haugen's Report.* House Report 631, 68th Congress, 1st Session, 1924.

———. *Report of the Joint Commission Relative to the Preservation of the Fisheries in Waters Contiguous to Canada and the United States.* House Document 315, 54th Congress, 2d Session, 1897.

———. *Report of the Secretary of the Interior for 1850.* House Executive Document 1, 31st Congress, 2d Session, 1850.

———. *Report of the Secretary of the Interior for 1851.* House Executive Document 2, 32d Congress, 1st Session, 1852.

United States. Congress. Senate. *The Agricultural Surplus Control Bill: Senator McNary's Report.* Senate Report 500, 70th Congress, 1st Session, 1928.

———. *A Bill Authorizing Appropriations for the Construction and Maintenance of Improvements Necessary for Protection of the National Forests from Fire, and for Other Purposes.* Senate Bill 3594, 71st Congress, 2d Session, 1930.

———. *A Bill to Insure Adequate Supplies of Timber and Other Forest Products for the People of the United States.* Senate Bill 3556, 70th Congress, 1st Session, 1928.

———. *A Brief History of the Committee on Agriculture and Forestry, United States Senate, and Landmark Agricultural Legislation, 1825–1970.* Senate Document 91-107, 91st Congress, 2d Session, 1970.

———. *Data Regarding the Sequoia Gigantea.* Senate Document 156, 58th Congress, 2d Session, 1904.

———. *Department of the Interior and Related Agencies Appropriations Bill, 1994.* Senate Report 103–114, 103d Congress, 1st Session, 1993.

———. *Equalization Fee, Debenture, and Farm Allotment Plans.* Senate Report 732, 72d Congress, 1st Session, 1932.

———. *Forest Research.* Senate Report 742, 70th Congress, 1st Session, 1928.

———. *Fur Seal Fisheries: Hearings before the Committee on Conservation of Natural Resources on a Bill entitled 'an Act to Protect the Seal Fisheries of Alaska, and for other purposes,'* Senate Document 605, 61st Congress, 2d Session, 1910.

———. *History of the Committee on Energy and Natural Resources, United States Senate, as of the 100th Congress, 1816–1988.* Senate Document 100-46, 100th Congress, 2d Session, 1989.

———. *History of the Committee on Environment and Public Works, United States Senate.* Senate Document 100-45, 100th Congress, 2d Session, 1988.

———. *Letter from the Secretary of War Communicating the Report of Lieutenant Gustavus C. Doane upon the So-called Yellowstone Expedition of 1870.* Senate Executive Document 51, 41st Congress, 3d Session, 1871.

———. *The National Appalachian Forest Reserve.* Senate Document 13, 58th Congress, 1st Session, 1903.

———. *Nomination of Bruce E. Babbitt to be Secretary of the Interior: Hearings before the United States Senate Committee on Energy and Natural Resources.* Senate Hearing No. 103-3, 103d Congress, 1st Session, 1993.

———. *Preliminary Report of the United States National Waterway Commission.* Senate Document 301, 61st Congress, 2d Session, 1910.

———. *Proceedings of the National Conference on Outdoor Recreation.* Senate Document 151, 68th Congress, 1st Session, 1924.

———. *Reform of the Land Laws— Conservation of National Resources: Extracts from Recommendations of the President, the Secretary of the Interior, and the Commissioner of the General Land Office, etc.* Senate Document 283, 61st Congress, 2d Session, 1910.

———. *Report of the Joint Committee to Consider the Present Organizations of the Signal Service, Geological Survey, Coast and Geodetic Survey, and the Hydrographic Office of the Navy Department.* Senate Report 1235, 49th Congress, 1st Session, 1886.

———. *Report of the National Conservation Commission.* Senate Document 676, 60th Congress, 2d Session, 1909.

———. *Report of the Public Lands Commission.* Senate Document 189, 58th Congress, 3d Session, 1905.

———. *Report of the Secretary of the Interior for 1849.* Senate Executive Document 1, 31st Congress, 1st Session, 1849.

———. *Report of the Secretary of the Interior, Showing the Quantity of Land Sold, the Amount of Money Received Therefore and the Amount of Incidental Expenses Thereon, in the Years 1847, '48, and '49, with the Quantity Remaining Unsold on 31st December, 1849.* Senate Executive

Document 63, 31st Congress, 1st Session, 1849.

——. *Report of the Secretary of the Interior for 1852.* Senate Executive Document 1, 32nd Congress, 2d Session, 1852.

——. *Report of the Special Committee of the United States Senate on the Irrigation and Reclamation of Arid Lands.* Senate Report 928, 51st Congress, 1st Session, 1890.

——. *Report upon the Colorado River of the West, Explored in 1857 and 1858 by Lieutenant Joseph C. Ives, Corps of Topographical Engineers, Under the Direction of the Office of Explorations and Surveys, A. A. Humphreys, Captain, Topographical Engineers, in Charge.* Senate Executive Document 1, 36th Congress, 1st Session, 1861.

——. *Senate Report of the Committee Appointed by the National Academy of Sciences upon the Inauguration of a Forest Policy for the Forested Lands of the United States.* Senate Document 105, 55th Congress, 1st Session, 1897.

——. *Veto Message Relating to the Agriculture Surplus Control Act: A Message from the President of the United States.* Senate Document 141, 70th Congress, 1st Session, 1928.

——. *Veto Message Relating to the Surplus Control Act: A Message from the President of the United States.* Senate Document 214, 69th Congress, 2d Session, 1927.

Vogt, Bill. "The National Wildlife Federation's Conservation Hall of Fame Honors Frederick Law Olmsted." *National Wildlife* 20, 4 (June-July 1982): 40–41.

Wakelyn, Jon L. *Biographical Dictionary of the Confederacy.* Westport, CT: Greenwood 1977.

Walcott, Charles Doolittle. *Paleontology of the Eureka District.* United States Geological Survey Monograph 8. Washington, DC, 1884.

Walker, Francis Amasa, and Henry Gannett. *Progress of the Nation, 1790 to 1880: Introduction to Statistics of the Population of the United States at the Tenth Census.* House Miscellaneous Document 42, vol. 1, 47th Congress, 2d Session, 1883.

Wallace, Andrew, ed. *Pumpelly's Arizona: An Excerpt from "Across America and Asia" by Raphael Pumpelly, Comprising Those Chapters Which Concern the Southwest.* Tucson, AZ: Palo Verde, 1965.

"The Watercolors of John James Audubon." *Audubon* 95, 4 (July-August 1993): 52–59.

Watkins, T. H. "Pulling Turtles out of the Soup." *National Wildlife* 30, 3 (April-May 1982): 18–25.

——. *Righteous Pilgrim: The Life and Times of Harold L. Ickes, 1874–1952.* New York: Holt, 1990.

Watson, Jim. "Dips, Spurs, and Angles— Some Chapters in the Story of the Great Anachronism: The General Mining Law of 1872." *Wilderness* 55, 197 (Summer 1992): 10–13.

Welch, Richard E., Jr. *The Presidencies of Grover Cleveland.* Lawrence: University Press of Kansas, 1988.

West, Terry L. *Centennial Mini-Histories of the Forest Service.* U.S. Forest Service pamphlet FS-518. Washington, DC: U.S. Department of Agriculture, 1992.

——. *Forest Service History Bibliography, 1891–1991.* Washington, DC: History Section of the Public Affairs Office, U.S. Forest Service, U.S. Department of Agriculture, 1993.

——. *Guide to the Curation of Forest Service Administrative History Artifacts and Records.* Washington, DC: History Section of the Public Affairs Office, U.S. Forest Service, U.S. Department of Agriculture, 1988.

——. "USDA Forest Service Management of the National Grasslands." *Agricultural History* 64, 2 (Spring 1990): 86–98.

Wheeler, George Montegue. *Preliminary Report of Explorations in Nevada and Arizona.* Senate Executive Document 65, 42d Congress, 2d Session, 1872.

——. *Report upon the Third Geographical Congress and Exhibition at Venice, Italy, 1881.* House Executive Document 270, 48th Congress, 2d Session, 1885.

Whipple, Amiel Weeks. *Report of the Whipple Survey of the Southwest from the Mississippi* to the Pacific Coast. Senate Document 78, 33d Congress, 2d Session, 1857.

Whitnah, Donald R., ed. Government Agencies. Westport, CT: Greenwood, 1983.

Whitney, Gordon G., and William C. Davis. "From Primitive Woods to Cultivated Woodlots: Thoreau and the Forest History of Concord, Massachusetts." Journal of Forest History 30, 2 (April 1986): 70–81.

Wikle, Thomas A. "Proposals, Abolishments, and Changing Standards for U.S. National Parks." Historian 54, 1 (Autumn 1991): 49–61.

Wild, Peter. Pioneer Conservationists of Western America. Missoula, MT: Mountain, 1979.

Wilkins, Thurman. Clarence King: A Biography. Albuquerque: University of New Mexico Press, 1988.

Wille, Chris. "Race to Save a Green Giant." National Wildlife 29, 6 (October-November 1991): 24–28.

Williams, Michael. Americans and Their Forests: A Historical Bibliography. New York: Cambridge University Press, 1989.

Williams, Ted. "The Sabotage of Superfund." Audubon 95, 4 (July-August 1993): 30–37.

Wirt, George H. Joseph Trimble Rothrock: Father of Forestry in Pennsylvania. Lewistown, PA: Mifflin County Historical Society, 1956.

Wolf, Peter. Land in America: Its Value, Use, and Control. New York: Pantheon, 1981.

Wolfe, Linnie Marsh, ed. John of the Mountains: The Unpublished Journals of John Muir. Boston: Houghton Mifflin, 1938.

Worobec, Mary Devine. Toxic Substances Controls Primer: Federal Regulations of Chemicals in the Environment. Washington, DC: Bureau of National Affairs, 1986.

Worster, Donald. Rivers of Empire: Water, Aridity and the Growth of the American West. New York: Pantheon, 1985.

Wortman, J. L. "Othniel Charles Marsh." Science 9, 225 (21 April 1899): 560–565.

Wunder, John E., ed. Historians of the American Frontier. Westport, CT: Greenwood, 1988.

Wyman, Roger E. "Insurgency in Minnesota: The Defeat of James A. Tawney in 1910." Minnesota History 40 (Fall 1967): 317–329.

Young, Kevin. "The Wolf: A Western Odyssey." SKY Magazine 22, 2 (February 1993): 68–77.

Young, Terence. "San Francisco's Golden Gate Park and the Search for a Good Society, 1865–80." Forest and Conservation History 37, 1 (January 1983): 4–13.

Zahniser, Howard Clinton. "The Need for Wilderness Areas." Living Wilderness 21, 59 (Winter-Spring 1956-1957): 37–43.

Zaitzevsky, Cynthia. Frederick Law Olmsted and the Boston Park System. Cambridge, MA: Harvard University Press, 1982.

Zaslowsky, Dyan. These American Lands: Parks, Wilderness and Public Lands. New York: Holt, 1986.

Zinn, Jeffrey A. "Conservation in the 1990 Farm Bill: The Revolution Continues." Journal of Soil and Water Conservation 46, 1 (January-February 1991): 45–48.

MANUSCRIPT AND ARCHIVAL COLLECTIONS

Governor Bruce Edward Babbitt Papers, RG 1, SG 23 (Governor's Papers), Arizona Department of Library, Archives and Public Records, Phoenix

John Calhoun Bell Papers, Denver Public Library

Thomas Bragg Diary, Southern Historical Collection, University of North Carolina at Chapel Hill

Irving Brant Papers, Library of Congress

Joseph Maull Carey Papers, Wyoming State Archives, Cheyenne

Thomas Henry Carter Papers, Library of Congress

Zachariah Chandler Papers, Library of Congress

Cornell University Alumni Files (Jay P Kinney)

Jacob Dolson Cox Papers, Oberlin College Archives, Oberlin, Ohio

Columbus Delano Papers, Library of Congress

Thomas Ewing Papers, Library of Congress

Walter Lowrie Fisher Papers, Library of Congress

James Rudolph Garfield Papers, Library of Congress

James Harlan Papers, Library of Congress

Ethan Allan Hitchcock, papers of, RG 200, National Archives

William Steele Holman Papers, Indiana State Library, Indianapolis

Holman-O'Brien Papers, Indiana Historical Society, Indianapolis

William Temple Hornaday Papers, Library of Congress

Izaak Walton League Archives, Arlington, Virginia

Samuel Jordan Kirkwood Papers, State Historical Society of Iowa, Iowa City

Julius Albert Krug Papers, Library of Congress

Joseph Wood Krutch Papers, Library of Congress

Joseph Fletcher Lacey Papers, State Historical Society of Iowa, Des Moines

Irvine L. Lenroot Family Papers, Library of Congress

Thomas Muldrup Logan Papers, Southern Historical Collection, University of North Carolina at Chapel Hill

Warren G. Magnuson Papers, University of Washington at Seattle

George H. Maxwell Papers, Records of the Bureau of Reclamation, RG 115, National Archives

George H. Maxwell Papers, Manuscript Group 1, Arizona Department of Library, Archives and Public Records, Phoenix

Robert McClelland Letterbook, Library of Congress

Robert McClelland Papers, Library of Congress

Governor Myron Hawley McCord Papers, RG 1, SG 4 (Governor's Papers), Arizona Department of Library, Archives and Public Records, Phoenix, Arizona

William John McGee Papers, Library of Congress

Charles L. McNary Papers, Library of Congress

Charles L. McNary Papers, University of Oregon Library, Portland

Frederick H. Newell Papers, Library of Congress

Francis G. Newlands Papers, Nevada Historical Society, Reno

John Willock Noble Miscellaneous Papers, Library of Congress

George William Norris Papers, Library of Congress

North Dakota State Water Commission Official Papers (containing information on Henry Clay Hansbrough), State Historical Society of North Dakota, Bismarck

Theodore S. Palmer Papers, Library of Congress

Gifford Pinchot Papers, Library of Congress

Key Pittman Papers, Library of Congress

Carl Schurz Papers, Library of Congress

John Franklin Shafroth Papers, Denver Public Library

Caleb Blood Smith Papers, Library of Congress

William Morris Stewart Papers, Nevada Historical Society, Reno

Alexander H. H. Stuart Papers, University of Virginia at Charlottesville

Alexander H. H. Stuart Miscellaneous Papers, Library of Congress

William Howard Taft Papers, Library of Congress

Edward Thomas Taylor Papers, Colorado Historical Society, Denver

Henry Moore Teller Papers, Denver Public Library

Edward Tiffin Papers, Western Reserve Historical Society, Cleveland, Ohio

Western History Department Papers, Denver Public Library

Wilderness Society Papers, Wilderness Society, Washington, D.C.

Wilderness Society Papers, Denver Public Library

Edward Oliver Wolcott Papers, Colorado Historical Society, Denver

NATIONAL ARCHIVES

Records of the Bureau of Reclamation, RG 115

Records of the Department of the Interior (including cartographic records), RG 48

Records of the General Land Office, RG 49

Records of the Geological Survey, RG 57

   Records of the Ferdinand Hayden Survey

   Records of the Clarence King Survey

Records of the John Wesley Powell Survey

Records of the United States Forest Service, RG 95

Records of the United States House of Representatives, RG 233

Records of the United States Senate, RG 46

ORAL HISTORY COLLECTIONS

Horace Albright Memoirs, Regional Oral History Office, University of California at Berkeley

Wayne N. Aspinall Oral History Memoirs, Oral History of Colorado Project, 11 June 1974

Jay P Kinney Oral History Interview, Forest History Society (Durham, North Carolina), 1960

William Embry Wrather Oral History Interview, Oral History Research Office, Columbia University

Illustration Credits

167 Iowa State Historical Society Portrait 291, Des Moines

169 Dr. Harry M. Walsh Collection, Chesapeake Bay Maritime Museum, St. Michaels, Maryland

174 Photograph by Robert McCabe. University of Wisconsin–Madison Archives #X25 1097

176 Photograph by Ava Long; Greenpeace, Washington, D.C.

181 National Anthropological Archives 2817A, Smithsonian Institution, Washington, D.C.

193 Photograph by Mable Mansfield; courtesy The Wilderness Society, Washington, D.C.

198 Research Division, Department of Library, Archives and Public Records, State of Arizona, Phoenix

200 Western History Collection F47151, Denver Public Library

206 Muir-Hanna Trust, Holt-Atherton Department of Special Collections, University of the Pacific, Stockton, California

209 Photograph by James Marshall; courtesy The Wilderness Society, Washington, D.C.

212 Drawing by Henry Worrall. William Edward Webb, *Buffalo Land* (Cincinnati, 1872), p. 314; Western History Collection F2436, Denver Public Library

217 Courtesy National Park Service

221 Nevada Historical Society N-18, Reno

238 Photograph by Kim Awbrey; courtesy Conservation International, Washington, D.C.

245 Forest History Society, Durham, North Carolina

247 Nevada Historical Society N-90, Reno

249 Photograph by J. K. Hillers; Photographic Collection, U.S. Geological Survey, Denver

260 University of Washington Information Services P-04912-A, Special Collections, University of Washington Libraries, Seattle

262 Photograph by Gary Rehn; courtesy Eco-Cycle, Boulder, Colorado

267 Theodore Roosevelt Collection, Harvard College Library, Cambridge, Massachusetts

271 Courtesy Environmental Protection Agency

276 Portrait 105, Photographic Collection, U.S. Geological Survey, Denver

282 Sierra Club Archives, San Francisco

287 Courtesy U.S. Forest Service

298 Colorado Historical Society F35,968, Denver

301 Courtesy Tennessee Valley Authority, Norris, Tennessee

304 Daguerreotype by Benjamin D. Maxham; Concord Free Public Library, Concord, Massachusetts

306 Photograph by Robert Visser; Greenpeace, Washington, D.C.

309 Western History Collection F13267, Denver Public Library

328 University of Wisconsin–Madison Archives 18359-c

332 Portrait 98, Photographic Collection, U.S. Geological Survey, Denver

333 Courtesy American Forests, Washington, D.C.

334 Wyoming State Museum, Cheyenne

342 Forest History Society, Durham, North Carolina

347 Photograph by Timothy O'Sullivan; Photograph Collection, U.S. Geological Survey, Denver

351 Courtesy The Wilderness Society, Washington, D.C.

362 Photograph by William Henry Jackson; Photographic Collection, U.S. Geological Survey, Denver

366 Courtesy The Wilderness Society, Washington, D.C.

Index

Powell Irrigation Survey, 250, 320
Powell Rocky Mountain survey, 250, 292
Power resources. *See* Energy resources
Prairie Island Park, 80
Prairie States Forestry Program, 367
Predatory animal control, 98, 175, 209
Preemption Act (1841), 239, 251
Preservationism, 62
Pribilof Islands, 63–65, 111, 317
Price, Overton Westfeldt, 21, 109, 252–253
Price, Robert Martin, 282, 283
Price, W. W., 283
Price-Anderson Act, 83
Prince William Sound, 98
Private conservation organizations, 61, 62
Private irrigation, 153, 215. *See also* Truckee
    Irrigation Project
Private lands
    Cameron v. United States on, 39
    common law and, xiii
    conservation easements on, 60
    Copeland Report on, 69
    Dolan v. City of Tigard on, 80–81
    Nature Conservancy and, 219
    Newlands Act on, 199
Proclamation No. 39, 2–3. *See also* Executive
    Orders
Proctor, Redfield, 253–254
Progressive Party
    Ballinger-Pinchot Affair and, 20, 21
    J. M. Carey and, 40
    conservation movement and, 62
    J. R. Garfield and, 115
    H. L. Ickes and, 149
    I. L. Lenroot and, 173
    G. Pinchot and, 245
    T. Roosevelt and, 268
Project 2012, 33
Promontory Stock Ranch Company, 32
Prospect Park, Brooklyn, 230
Proteus (ship), 200
Public employees. *See* Civil service
Public health, 227, 305, 337, 338
Public lands
    in Alaska, 11, 101
    animal damage control on, 98
    Cameron v. United States on, 39
    constitutional property clause on, xi, 4
    A. B. Fall quoted on, 100
    Garfield Public Land Commission quoted
        on, 115
    General Land Office and, 116
    P. Gibson quoted on, 119
    grazing on. *See* Grazing policy
    Homestead Act and, 143–144

House committees quoted on, 147–148
    Land Acts and, 170–171
    multiple use of, 101, 208, 256
    off-road vehicles on, 97–98, 265
    L. E. Payson quoted on, 240
    Public Land Commission quoted on, 254
    Public Land Law Review Commission
        quoted on, 255–256
    reclassification of, 178
    recreational use of, 282
    right of way through, 264
    ski areas on, 265
    withdrawal of, 354
    See also Forest reserves; Land fraud; Land
        laws; Land speculation; Mineral
        lands; Western public lands;
        Wilderness areas
Public parks. *See* National Park System;
    Urban parks
Public Rangelands Improvement Act (1978),
    256
Public recreation, 92–93, 103, 107, 208, 217
Public water supply. *See* Water supply
Public works, 11, 92, 220, 225. *See also*
    Construction projects; Water utilization
    projects
Puget Sound, 66–67
Pulitzer Prize, 29, 73, 79, 299, 312
Pumpelly, Raphael, 256–257, 274
Pumpellyite, 257
Pure Food Act (1904), 114
Pursh, Frederick, 21

Quetico-Superior Council, 229, 232

Radical Republicans, 31, 71, 76, 325
Radioactive waste disposal, 226–227,
    324–325, 335
Rafinesque, Constantine Samuel, 259–260
Railroads, 132, 290, 334
    municipal, 106
    transcontinental, 66, 67, 293
Rainbow Warrior (ship), 125
Rainforests, 118–119, 237
Rainy Lake, 229
Rand Corporation, 294
Ranger Rick (periodical), 219
Ransdell-Humphreys Act (1917), 106, 202
RARE (Roadless Area Review and
    Evaluation), 264–265
Rathbun, Richard, 324
Ray, Dixy Lee, 260–261
Raynolds, William F., 134
Reagan, Ronald, 17, 52, 141, 261, 340
Reagan administration, 177

Index